MARINE ECOLOGY

A Comprehensive, Integrated Treatise on Life in Oceans and Coastal Waters

Volume I   ENVIRONMENTAL FACTORS

Volume II   PHYSIOLOGICAL MECHANISMS

Volume III   CULTIVATION

Volume IV   DYNAMICS

Volume V   OCEAN MANAGEMENT

MARINE ECOLOGY

A Comprehensive, Integrated Treatise on Life in Oceans
and Coastal Waters

Editor

OTTO KINNE

Biologische Anstalt Helgoland
Hamburg, West Germany

VOLUME I

Environmental Factors

Part 3

1972

WILEY-INTERSCIENCE

a division of John Wiley & Sons Ltd.
London . New York . Sydney . Toronto

Library of Congress Catalog No. 79–121779

ISBN 0 471 480037

Printed in Great Britain by
Robert MacLehose and Co Ltd
The University Press, Glasgow

INTRODUCTION
to the
TREATISE

No words can introduce a Treatise on Marine Ecology more adequately than those by JOHANN WOLFGANG VON GOETHE[1] (1749–1832). I am quoting here the German text, as well as a favourite English translation by JOHN ANSTER:

'Alles ist aus dem Wasser entsprungen!!
Alles wird durch das Wasser erhalten!
Ozean, gönn uns dein ewiges Walten.
Wenn du nicht Wolken sendetest,
Nicht reiche Bäche spendetest,
Hin und her nicht Flüsse wendetest,
Die Ströme nicht vollendetest,
Was wären Gebirge, was Ebnen und Welt?
Du bist's, der das frischeste Leben erhält.'

'In Water all hath had its primal source;
And Water still keeps all things in their course.
Ocean, still round us let thy billows proud
Roll in their strength—still send up mist and cloud.
If the rich rivers thou didst cease to spread—
If floods no more were from thy bounty fed—
And the thin brooklet died in its dry bed—
Where then were mountains—valleys? Where would be
The world itself? Oh! thou dost still, great Sea,
Sustain alone the fresh life of all things.'

To modern science, the field of ecology comprises studies of organisms in relation to their environment, abiotic and biotic. Marine ecology is a branch of ecology dealing with the vast multiplicity of organisms living in oceans and coastal waters. The present treatise attempts to cover all major aspects of marine ecology. It consists of several volumes which, for convenience, in some cases have been sub-divided into parts. At present 5 volumes are envisaged:

Volume I   —Environmental Factors
Volume II —Physiological Mechanisms
Volume III—Cultivation
Volume IV —Dynamics
Volume V —Ocean Management

Environmental Factors is introduced on the following pages (Foreword). Volumes II to V are presently being organized and are scheduled to be published in the next few years.

Physiological Mechanisms will deal with support mechanisms, such as photosynthesis, respiration, reproduction; timing mechanisms, e.g. biological clocks, rhythms; orientation mechanisms; regulatory mechanisms, e.g. volume-, ion-, turgor-, osmo- and thermo-regulation; mechanisms of adaptation; and communication mechanisms, such as sound production and perception, as well as visual and chemical communications.

Cultivation will be concerned with maintaining, raising, rearing and breeding marine and brackish-water organisms in laboratories, ponds, under-sea farms, restricted sea areas, etc., both for scientific and economic purposes; this volume will also include sections on technical aspects and diseases.

[1] Werke (Hamburger Ausg., 4th ed.) Vol. 3. Faust. Part 2, p. 255. Wegner, Hamburg, 1959.

Dynamics will focus on production, transformation and decomposition of organic matter in the marine environment; population dynamics; food-chain relations; nutritional requirements; as well as on flow and balance of energy and matter.

Ocean Management—a rather ambitious term for a young and virgin aspect of marine ecological research—will present a brief synopsis of important taxonomic groups, zonations, organismic assemblages; sea-water pollution (sources, biological consequences, avoidance, control, conventions); organic resources of the seas (distribution, use, control, conservation); and a general discussion concerning possible ways and means for management of important sea areas.

A comprehensive, integrated treatise on life in oceans and coastal waters cannot be written by a single author; it must draw from a multitude of talents and sources, and hence requires interdisciplinary and international co-operation. Neither a compendium nor an encyclopaedia, the treatise is intended to be an exhaustive systematic exposition summarizing and evaluating information obtained thus far on living systems in the seas and littoral areas. It has been conceived with the growing number of individuals in mind who are professionally concerned with life in the marine environment, especially investigators, engineers, teachers, students, administrators and businessmen. Although, for the benefit of the reader, integrated into a methodically arranged general concept, each contribution is intended to represent a detailed, authentic critical account in its own right; all contributors are free in choice of material and emphasis.

The first tentative outline of the treatise was circulated among several hundred marine ecologists in November, 1965. The warm response received from the international scientific community and the stimulating support from the publishers have encouraged me to proceed with my plans. Criticism, advice and assistance of numerous colleagues have greatly affected and improved the first proposal. I gratefully acknowledge all this support. It is not possible to list here the names of even the most active supporters; they will be mentioned in the forewords to the respective volumes.

A treatise such as this needs continued criticism and advice. Any comments—especially on outline, coverage and new points of view—will be most welcome.

O. KINNE

FOREWORD

to

VOLUME I: ENVIRONMENTAL FACTORS

'Environmental Factors' summarizes and evaluates all important information available to date on the responses of ocean and coastal-water living organisms to intensity variations of the major abiotic and biotic ecological factors. It is subdivided into 3 parts which contain the following chapters:

Part 1

Chapter 1 : Oceans and Coastal Waters
               as Life-supporting Environments
Chapter 2 : Light
Chapter 3 : Temperature

Part 2

Chapter 4 : Salinity
Chapter 5 : Water Movement
Chapter 6 : Turbidity

Part 3

Chapter 7 : Substratum
Chapter 8 : Pressure
Chapter 9 : Dissolved Gases
Chapter 10: Organic Substances
Chapter 11: Ionizing Radiation
Chapter 12: Factor Combinations

Chapter 1 considers oceans and coastal waters as life-supporting environments. It describes briefly the ocean basins, their principal water masses and circulation, the sea–land boundary, the properties of sea water and the chemical cycles in the seas.

Chapters 2 to 11 deal with responses to environmental factors. Of course, only factors about which enough information is available can be treated. Each chapter begins with a general introduction informing the reader about (1) general aspects of the environmental factor concerned, (2) methods of measuring its intensities, and (3) its intensity patterns in oceans and coastal waters. The chapter outline, suggested to all contributors, distinguishes between functional and structural responses. Functional responses are subdivided into tolerance, metabolism and activity, reproduction, and distribution; structural responses are dealt with under the subheadings size (body length, width, volume), external structures (shape, differentiation, etc. of external body parts) and internal structures (organs, tissues, cells or parts thereof). The monofactorial approach used in Chapters 2 to 11 has been chosen because of the insufficient amount of information at hand on multifactorial relationships, and because organisms—whether bacteria, plants or animals—frequently exhibit comparable responses to intensity variations of environmental entities such as light, temperature or salinity. A monofactorial (univariable) design facilitates comparison, evaluation and generalization of reactions to a given environmental factor by members of different taxa. It is realized, of course, that in natural habitats organisms respond to their total environment rather than to single factors (selected by man for methodological, conceptual or historical reasons). Factor interactions, known or expected to be of special importance, are therefore referred to briefly in each chapter.

Chapter 12 presents a special, detailed account on organismic responses to factor combinations. There can be no doubt: investigation of responses to intensity variations of environmental factors acting in concert must be given priority if man wants to understand ecological dynamics and to achieve forecasting and controlling capacities in regard to life in the marine environment. There is great need for (i) conducting large-scale research projects based on multivariable designs and including all life-history stages of important food-web representatives, (ii) developing appropriate analyzing and evaluating techniques (computation, mathematical models and concepts of abstraction, formalization and generalization). Chapter 12 represents a pioneer effort to stimulate progress in this modern branch of ecological research.

Our intention to provide the reader with a well-organized source of information which enables him to find and compare facts and problems of interest to him quickly and easily created several difficulties. The first difficulty was to achieve general agreement in regard to gross taxonomic subdivisions. The subdivisions 'bacteria, fungi and blue-green algae', 'plants', and 'animals' have been adopted after long discussions; they are the result of a compromise between the need to keep the number of taxa as small as possible and to choose groups of organisms which can be conveniently treated by single authors; whenever necessary these groups are subdivided further, e.g. 'animals' into 'invertebrates' and 'fishes'. The second difficulty concerned the treatment of 'nutrition'. In bacteria, nutrients and substratum (Chapter 7) are hardly separable; in plants, nutrients overlap to a certain degree with salinity (Chapter 4), in animals with organic substances (Chapter 10). While some aspects of nutrition have been considered under various headings, nutritional aspects will be treated in detail in Volumes III and IV. The third difficulty was created by differences in thematic emphasis and in the usage of certain scientific terms in the fields of marine microbiology, botany or zoology. An example is the connotation of the term 'growth', which means increase in individual numbers in microbiology, but increase in organic matter of individuals in botany and zoology. Such terminology problems were solved by providing definitions or explanations.

The policy of placing the conceptual grid of the chapter outlines on the body of knowledge available and reviewing the material found near each 'point of intersection' (rather than following, as usual, the meandering path along which information happens to have accumulated) made us aware that many important areas of marine ecological research have hardly been touched upon, while others have attracted unparalleled attention; such disproportions are reflected in the lengths of the respective contributions. The Chapter 'Water Movement: Bacteria, Fungi and Blue-green Algae' had to remain unwritten because of insufficient knowledge available.

Lack of information also created a serious gap in regard to biotic factors (e.g. behavioural and biochemical interactions between organisms of a given ecosystem) which may affect, or even govern, intra- and interspecific patterns of organismic co-existence. Little pertinent information is at hand on marine mammals and birds; their responses to environmental stress often depend on homeostatic mechanisms.

'Environmental Factors' concentrates on responses of intact organisms. However, if considered relevant, information obtained at the individual level is complemented by findings at the sub- or supra-individual levels. Functional and structural responses are primarily considered under the aspect of quantitative variability, i.e. in terms of changes in rates or intensities of performance. The physiological mechanisms involved will be dealt with in Volume II. General trends that have become apparent are documented by referring to one or a few well worked out examples rather than by presenting a long list of parallel findings. All literature cited appears in alphabetical order at the end of each chapter; it is hoped that such a procedure will help to strengthen interdisciplinary contacts between the fields of marine microbiology, botany and zoology and to facilitate a fast and convenient survey of important pertinent literature.

While an effort has been made to concentrate on marine and brackish-water organisms, in some instances information obtained on limnic forms has been included, especially in situations where knowledge on salt-water living organisms is scarce, or in which it appears safe to assume that both groups of aquatic organisms would exhibit comparable responses.

Much of our present knowledge on responses of marine and coastal-water living organisms to environmental stress has been obtained during casual observation or in insufficiently equipped and staffed laboratories. More complete studies require modern scientific dimensions: more space, better facilities and teams of scientists and technicians.

I am deeply indebted to all contributors for their patience, dedication and willingness to co-operate far beyond the usual demands; despite technical difficulties it was possible in most cases to adhere closely to the outlines proposed. The publishers have supported me wholeheartedly and considerably reduced the many problems by not imposing any space or time limits; I am grateful for this confidence and for excellent co-operation. It is a pleasure to acknowledge support, advice and criticism received by many colleagues, especially by D. F. ALDERDICE, J. R. BRETT, A. W. COLLIER, M. GILLBRIGHT, E. HAGMEIER, M. HOPPENHEIT, H. W. JANNASCH, R. I. SMITH, R. W. TAYLOR and B. P. USHAKOV. During the years of organizing and preparing Volume I, Mrs. J. M. CHRISTIAN, Miss V. J. CLARK and Miss F. W. CROUSE have served as reliable and highly capable editorial secretaries and assistants, Mr. J. MARSCHALL has given generously of his time and talent in altering or improving illustrations and Mr. W. MEISS was an indispensable and conscientious helper in all matters related to bibliographical problems. It is with a deep sense of gratitude that I acknowledge all this assistance.

O. KINNE

CONTENTS
of
VOLUME I, PART 3

CONTRIBUTORS

to

VOLUME I, PART 3

ALDERDICE, D. F.,   *Fisheries Research Board of Canada, Biological Station; Nanaimo, British Columbia, Canada.*

BACESCU, M. C.,   *Musée d'Histoire Naturelle 'Grigore Antipa', 1 Chaussée Kisselef; Bucharest, Roumania.*

CHIPMAN, W. A.,   *1512 Evans Street; Morehead City, North Carolina 28557, USA.*

FLÜGEL, H.,   *Institut für Meereskunde an der Universität Kiel, Niemannsweg 11; 23 Kiel, West Germany.*

FOGG, G. E.,   *Department of Botany, Westfield College, University of London, Kidderpore Avenue; London NW3 7ST, England.*

GERLACH, S.,   *Institut für Meeresforschung, Am Handelshafen 12; 285 Bremerhaven-G, West Germany.*

GUNKEL, W.,   *Biologische Anstalt Helgoland (Meeresstation); 2192 Helgoland, West Germany.*

HARTOG, C. DEN,   *Rijksherbarium, Schelpenkade 6; Leiden, The Netherlands.*

KALLE, K.,   *Lyra Weg 12; 3118 Bevensen, West Germany.*

KINNE, O.,   *Biologische Anstalt Helgoland (Zentrale), Palmaille 9; 2000 Hamburg 50, West Germany.*

MORITA, R. Y.,   *Department of Microbiology, Oregon State University; Corvallis, Oregon 97331, USA.*

RHEINHEIMER, G.,   *Institut für Meereskunde an der Universität Kiel, Niemannsweg 11; 23 Kiel, West Germany.*

VERNBERG, F. J.,   *Belle W. Baruch Coastal Research Institute, University of South Carolina; Columbia, South Carolina 29208, USA.*

VIDAVER, W.,   *Department of Biological Sciences, Simon Fraser University; Burnaby 2, British Columbia, Canada.*

WILBER, C. G.,   *Department of Zoology, Colorado State University; Fort Collins, Colorado 80521, USA.*

WOOD, E. J. F.,   *Institute of Marine Sciences, University of Miami, 1 Rickenbacker Causeway; Miami, Florida 33149, USA.*

ZoBELL, C. E.,   *Scripps Institution of Oceanography, University of California, San Diego, P.O. Box 109; La Jolla, California 92037, USA.*

ENVIRONMENTAL FACTORS

7. SUBSTRATUM

7.0 GENERAL INTRODUCTION

S. A. GERLACH

In contrast to all other environmental factors to which chapters are devoted in Volume I, substratum is difficult to define in terms of physico-chemical properties.

The term 'substratum', as used in this chapter, refers to structures to which the organisms in question maintain, temporarily or permanently, a close contact, e.g. the interface between water and air, floating particles, seaweeds, animals, rocks, sand, and mud. It does not include, however, aquatic or nutritive media.

(1) General Aspects of Substratum

(a) Air–Water Interface

It may not be common use to treat the air–water interface as a substratum. However, snails, such as *Janthina prolongata*, creep along on the subsurfaces of the ocean, and air-breathing insects of the genus *Halobates* skate on the water surface using it as substrate. The community of organisms living immediately below or on the surface of the sea is called 'pleuston' or 'neuston'; quite a number of animals belong to it (ANKEL, 1962; BIERI, 1966; ZAITSEV, 1968; Chapter 7.3).

In recent years, it has become obvious that the neuston comprises not only animals but also a number of micro-organisms. Special devices have been developed to collect a fraction of the upper millimetre of the sea in order to study neuston micro-organisms (HARVEY, 1966). The neuston is also characterized by special physico-chemical conditions: organic solutes tend to get adsorbed at the air–water interface, together with phosphate ions, and to polymerize into insoluble films or particles, in such a way contributing to the amount of organic particles suspended in sea water (SUTCLIFFE and co-authors, 1963; Chapter 10).

(b) Floating Substrates

The various types of particulate matter floating in the sea provide substrata for bacteria, fungi and blue-green algae. In fact, a considerable number of micro-organisms occuring in the free water do not really lead a pelagic life; they are attached to a particle of substrate; in some cases such particles are below 100 μm in diameter. Floating substrates of larger size are detached leaves of eel-grass and seaweeds, and the permanent beds of drifting tropical marine plants of the genus *Sargassum*. Wood and other floating materials of terrigenous origin may drift for years at the ocean's surface. Whereas plants floating off from their habitat normally carry population fragments of their originally benthonic surroundings with them, truly oceanic floating substrates are characterized by a flora and fauna of their own, which maintain intricate relations to the neuston community.

Pelagic invertebrates as well as fish, tortoises, sea snakes, and whales represent floating substrates, usually colonized by plants and animals. If such living substrates move actively, attached suspension feeders enjoy the advantage of living in a permanent current of sea water; since the current flows predominantly in one direction, they can adjust their feeding mechanisms accordingly. Anchored buoys and ship hulls are used as substrates by a host of marine organisms. This fact interferes with various human activities and hence requires constantly tremendous industrial efforts. Buoys and ships are populated in part by members of the floating community and in part by representatives of plants and animals inhabiting solid surfaces on the shore.

(c) Solid Substrates

Various kinds of rocks, concrete, bricks, iron, wood, rubber and other marine engineering materials are used as substrata by numerous marine organisms. Depending on where they live, i.e. on or in these solid substrates, the organisms may be divided into three groups:

(1) **Epilithion**; its representatives move on, or are attached to, the substrate's surface.
(2) **Endolithion**; its representatives bore into the substrate; or they use burrows made by other organisms, or natural crevices.
(3) **Mesolithion**; its representatives live in the smallest natural crevices comparable to the interstitial system of sand.

For some organisms associated with solid substrates, the chemical composition of the substrate may be important; some of the marine borers, for example, are confined to wood (*Teredo*) others to limestone (*Lithodomus*). Of special importance is the quality of the substrate surface; settling planktonic larvae are often guided by smoothness, colour, organic films or bacterial coverings of solid substrate surfaces.

Ice can serve as a solid substrate, at least for micro-algae. These may form a felt even on the subsurface of sea ice in the Arctic and the Antarctic Oceans (MEGURO and co-authors, 1966).

(d) Organisms Serving as Substrates

Pelagic and benthic plants, as well as animals, frequently make substrates for a diversity of species (see also Chapter 7.3).

Eel-grass, algae, hydroids, Octocorallia and whole coral reefs are often abundantly colonized by epiphytes and epizoans. However, most living corals defend themselves against such colonization, more or less successfully, by the secretion of mucus (GERLACH, 1960). The basal parts of coral stocks, devoid of living polyps, are far more thickly covered by other organisms than their upper parts.

All substrate-forming organisms mentioned above make up the 'phytal', which maintains a flora and fauna distinct from those of solid rocks and sediments.

(e) Sediments

Sediments, commonly divided into sand (psammon) and mud (pelos), form the most widely distributed substrata on the sea bottom. They are inhabited by an impressive variety of benthonic organisms (see also Chapter 7.3).

(1) The **epipsammon** and **epipelos** consist of organisms moving on the sediment surface, or attached to it (sometimes by rhizoid structures).

(2) The **endopsammon** and **endopelos** (infauna) consist of organisms living within the sediment, either in distinct burrows or tubes, or just between the particles of the sediment without maintaining distinct dwellings.

(3) The **mesopsammon** (meio-, micro-, interstitial fauna and flora) consists of rather small organisms. They live in the interstitial spaces between the sediment particles, swimming, creeping or gliding through the interstices, or attaching themselves to the particles. For some of these organisms, it is not the sediment but the single sand grains which form the substrate on which they live.

Ecologically, a sediment consists at least of the following components: (i) inorganic particles of quite different size, form and origin (minerals, remains of mollusc shells, sea-urchin spines, corals, calcareous algae, skeletons of radiolarians, and foraminiferans, etc.); (ii) organic particles of different size, composition and origin (remains of wood and other plant tissues, faecal pellets and detritus); (iii) interstices between particles, filled with water (in the littoral zone, with air and water); (iv) interstitial water. Besides the free water that fills the pores, adsorbed water has to be considered which adheres to grain surfaces. The extent to which the special physico-chemical conditions at the water–solid interfaces affect the adsorption of nutrients and other dissolved components of sea water is largely unknown. Research concerning this important aspect is wanting. Different sediments are characterized by different animal, plant or microbial populations. Organisms are able to discern different sediments according to grain size, organic components and colonization by bacteria and other micro-organisms (CRISP, 1965; GRAY, 1965).

(2) Measuring Meaningful Properties of Substrates: Methods

Properties of substrates other than sediments cannot be measured by any standardized method; their diversity requires individual treatment. Properties of sediments are measured by sedimentological and geological methods.

Properties of a sediment, most important for the organisms inhabiting it, are: (i) suitability to move upon and to burrow in it; (ii) chemistry of the interstitial water, e.g. the oxygen regime; (iii) size of interstices; (iv) amount of decomposable organic particles. The methods commonly used by marine ecologists to measure sediment properties are far from taking these characteristics into proper consideration.

Measuring grain size (granulometry) gives the percentage distribution of solid sediment particles in terms of size classes, as well as information on various physical properties of the sediment: texture, size of interstices, amount of interchange between interstitial water and superlying water; however, detailed measurements are often difficult to make. Even a small amount of silt, for example, may fill the interstices in a coarse sand to such an extent, that the meiofauna is severely affected or even excluded.

Technical details concerning granulometry are described in sedimentological textbooks (e.g. MÜLLER, 1964): a sediment sample is cleaned of coagulent components and salts, for example, by washing with distilled water, H_2O_2, NH_4OH and isopropyl alcohol. To measure the fraction of sand grains with diameters larger than 0·1 mm, the sediment is sifted through a set of standardized sieves. The silt fraction is determined by floatation methods (e.g. 'Sedimentationswaage' of 'Sartorius', Germany), the clay fraction by centrifugation or microscopic measurement.

The results of such measurements (Fig. 7-1) are usually presented in a histogram, a cumulative curve, or a grain frequency graph (WALGER, 1964).

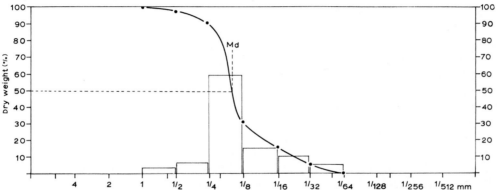

Fig. 7-1: Granulometric or grain size distribution presented as histogram and cumulative curve. Md: median. (Original.)

Unfortunately, mesh sizes of standard sieves, as well as classifications of grain size classes, vary in different countries (Fig. 7-2).

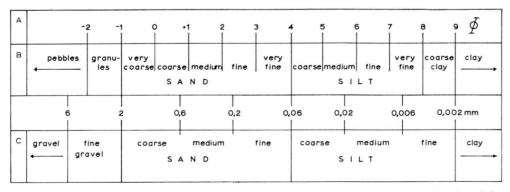

Fig. 7-2: Different systems of grain size classifications. A: Φ (phi) scale introduced by KRUMBEIN ($\Phi = -\log_2$ mm). B: WENTWORTH-scale, used in USA. C: Classification, according to DIN ('Deutsche Industrie-Norm') 4022, used in Germany. (Original.)

Measuring organic matter contained in sediments is, next to granulometry, of great importance. Organic components of sediments may serve as a food source; there exists, in fact, a great variety of animals which graze on sediment surfaces, licking the upgrowth from individual sand grains or swallowing entire portions of the sediment. Many heterotroph micro-organisms live on the organic contents of sediments. The determination of the quality and quantity of organic substances in sediments provides a clue as to the nutritional value of the substrates in question.

There are various methods for measuring the amount of organic matter in sediments:

(i) Determination of weight loss by combusting dry sediment.

(ii) Combustion of sediment and measurement of the amount of CO_2 formed (ELLINGBOE and WILSON, 1964). The amount of organic carbon in the sediment is considered proportional to the amount of organic material (1 mg organic carbon = 1·724 mg organic material).

(iii) Wet ignition with potassium dichromate and measurement of dichromate consumption by titration (ALLEN, 1964). This is the most practicable method, but the results are not very exact, due to the fact that organic material of different composition requires different equivalents of potassium dichromate for ignition.

(iv) Wet ignition with potassium dichromate and measurement of the amount of CO_2 formed (LEPORE, 1962; MÜLLER, 1964).

All these methods have drawbacks: (i) oxidation may not be restricted to organic material but also include other components of the sediment, i.e. FeO; (ii) the presence of NaCl may lead to higher values (OKUDA, 1964); (iii) the carbon of carbonates is also ignited to CO_2, and hence special precautions have to be taken in order to distinguish between organic and carbonate carbon. Consequently, all methods mentioned above produce questionable results if used in sediments with high carbonate contents, a very discouraging fact if one considers that quite a number of marine sediments (coral sand, globigerina ooze, etc.) consist, to a very high percentage, of calcareous matter.

However, these are not the only drawbacks. Organic material as a whole seems too large an entity to be meaningful as an ecological parameter. Organic carbon is contained in sugars, fats and proteins (nutrients of marine animals), in lignin, cellulose, humic substances (slowly decomposed by microbes), and in coal cast overboard by steamships. There is urgent need for developing more adequate methods for measuring organic matter in sediments. This is especially true in regard to that fraction of the organic material which can easily be used in metabolic processes (see also Chapter 10).

Another obstacle is that not only the dead organic material is determined but also the organic material contained in bacteria, fungi, algae and small animals living in the sediment tested. At present there are no sufficient means of separating these organisms from the sediment. An indirect method is to determine the weight of living organisms in parallel samples and to subtract this weight from the total weight of organic material, thus obtaining the value for the non-living particulate organic fraction. Simpler methods will have to be developed, and it seems justified to hope that they may be found: Adenosine triphosphate occurs only in living cells and is distributed in various organisms more or less proportional to their organic-matter content. Measurements of adenosine triphosphate, therefore, give a clue as to the quantity of living organic matter present in a sediment sample (HOLM-HANSEN and BOOTH, 1966; ERNST, 1970).

Other factors have been measured in regard to sediments. Some of them are, to our present knowledge, not directly important as ecological factors (e.g. the mineralogic composition of sediments); others, especially those concerning the interstitial water, are treated in the respective chapters of Volume 1.

(3) Distribution of Substrate Types in Oceans and Coastal Waters

The substratum 'air–water interface' is, of course, evenly distributed over the oceans. Nothing is known yet, however, about regional differences in microbial life associated with the sea surface. It seems safe to assume that microbial activity is higher in warmer areas. The neuston community (including the drifting beds of *Sargassum*) is confined to the tropical seas. The siphonophore genus *Velella* and probably other neuston animals have deep-sea living larvae. This peculiarity may help to compensate for adult drifting at the sea surface because larvae drift via deep counter-currents.

Floating substrates are also present in every ocean. Particulate matter, drift-wood and drifting seaweeds that come loose from the sea bottom are more abundant near the shores and especially in estuaries. Nevertheless, particulate matter is also present even in mid-ocean regions and down to abyssal depths. From 200 to 5000 m depth (North Atlantic and Caribbean Sea), particulate organic carbon was found to be rather uniformly distributed with 17 to 18 μg/l (MENZEL and GOERING, 1966).

Solid substrate surfaces are most common in the intertidal region and on the shelf, where wave action prevents deposition of sediments. Modern echo-sounding techniques revealed steep cliffs even in larger depths which probably are not covered by sediments, and deep-sea photography, as well as deep-sea submersibles, gave direct evidence of hard-bottom regions far below the shelf region. Furthermore, every piece of mollusc shell or other hard object lying on sand or mud bottoms forms a small area of solid substrate for epilithion-living organisms.

Living organisms which may serve as substrates are, of course, also distributed in all oceans and coastal waters. Their distributions, as far as they are affected by environmental factors, are discussed in various chapters of Volume I.

With regard to the sediments, sand and all transitional states between sand and mud are common in littoral and shelf regions. Muddy bottoms are found particularly in sheltered areas of the littoral zone as well as in deeper regions. Various kinds of mud substrata cover vast areas of the bottom of the seas. According to different origin and composition, sedimentologists distinguish between terrigenous and pelagic deposits (red clays; diatom, globigerina, radiolaria oozes; etc.). Since it is not yet known—apart from granulometry and organic content—to what extent the different origin and composition influences sediment living organisms, the distribution of the various kinds of mud will not be dealt with here.

7. SUBSTRATUM

7.1 BACTERIA, FUNGI AND BLUE-GREEN ALGAE

C. E. ZoBell

(1) Introduction

An appreciable proportion of bacteria, fungi and blue-green algae found in aquatic environments live on the surfaces of such submerged solids as sand, rocks, consolidated materials, both living and dead plants and animals or their remains, and a diversity of man-made structures. The latter include pilings, jetties, conduits, cordage, buoys, drogues, boat bottoms, submersibles, glass slides, metal plates, painted or coated materials, etc.

Organisms which attach to, or grow on, immersed objects have been described in the literature by several different terms, none of which is entirely apt for all surface-dwelling microbiota (ROLL, 1939; COOKE, 1956; SLÁDĚCKOVÁ, 1962). Sedentary, which means stationary or not free to move about, is not descriptive of all kinds of organisms which live on, or depend on, solid surfaces. Among the many exceptions are certain flagellated bacteria which attach to, detach, and re-attach (MEADOWS, 1968), and the Flexibacteria, filamentous Cyanophyta, and Labyrinthulae, which glide about on surfaces (SORIANO and LEWIN, 1965). Sessile is defined as permanently attached, not free to move about, or attached directly by the base and not raised upon a stalk or peduncle. Besides excluding gliding and transient forms found on substrata, the latter part of the definition excludes all of the surface-inhabiting stalked bacteria belonging either to the family Caulobacteraceae (BREED and co-authors, 1957) or to the order Hyphomicrobiales (STARR and SKERMAN, 1965) as well as numerous stalked algae and fungi. When used as an adjective, 'attachment' is an apt term to connote 'fondness for' or 'adherence to' solid surface (ZoBELL and ALLEN, 1933; ZoBELL, 1943).

The German word Aufwuchs, initially applied to biocoenoses of microbiota occurring on living substrata, has been anglicized by limnologists to aufwuchs, and is often applied to communities of micro-organisms that are firmly attached to the substratum but do not penetrate into it (ROLL, 1939; REID, 1961; SLÁDĚCKOVÁ, 1962; RUTTNER, 1963). More widely used in the literature is the term periphyton. Although literally meaning 'around a plant', this term is frequently applied to all kinds of microbiota found on any kind of solid or semisolid substrata. Certain workers distinguish between true periphyton, the attached immobile organisms, and pseudoperiphyton, the associated free-living, creeping, crawling, or grazing organisms on submerged substrata (SLÁDĚCKOVÁ, 1962).

Microbiota found growing on plant surfaces may be termed epiphytic (literally, upon plants), those on animal surfaces epizoic, those on rocks or stones epilithic, those on sand grains epipsammic, those on bottom sediments epipelic, etc. (ROUND, 1964, 1965). WOOD (1965, 1967) uses the term epontic (Greek epi, upon+ ontos, being or existing) to describe micro-organisms which live on solid surfaces. Bacteria adsorbed or otherwise attached to the surfaces of plankton or other solid substrata are designated 'exobactéries' by CAMPELLO and co-authors (1963), to distinguish them from 'endobactéries' occurring within the host or other substrata. Endophytic, endozoic, endolithic, etc., are terms which connote living within plants, animals, rocks, shells, etc., probably attached to internal solid surfaces. A perthophyte is a non-parasitic fungus, bacterium or other saprophyte that lives on dead or decaying tissue associated with a living organism.

The terms epibiont and epibiotic are applied to organisms which live on the surface of another organism, usually parasitically, but both terms are sometimes used by ecologists to describe any kind of organism living on a solid surface of either living or inanimate objects.

In physical consistency, the substratum may vary from that of hardest rock to that of the weakest gel. Substrata also differ greatly in shape and in surface texture or roughness. Probably no substratum is perfectly smooth. In various ways, the depth and distribution of the surface depressions influence the attachment of both inanimate and living materials. The substratum may be no larger than the smallest periphyte that it supports or it may be many m² or km² in area.

Biotic substrata in water may be either planktonic, nektonic, or benthonic. Abiotic substrata may be either stationary or mobile and they may be either benthonic or suspended in the water at any depth in the sea. The substratum may be fairly durable like certain rocks, silica sand, diatom tests, or glass slides, for example; or the substratum may be transistory like a chunk of ice, algal fronds, crustacean integuments, unpreserved wood, or other materials susceptible to deterioration. The electrostatic charge at the surface of the substratum may be either negative, positive, or neutral with respect to the periphyton. The redox potential, pH, and partial pressure of gases in the vicinity of the solid surface may be approximately the same as in the surrounding water, or these and other properties may be quite different, particularly in microniches on the solid surface.

(2) Functional Responses

(a) Tolerance

Since there are so many different kinds of substrata and since periphytic species differ so widely in their responses to substrata under diverse environmental conditions, ranges of tolerance to different substrata can be assessed only in general terms. Probably all marine bacteria, fungi, and blue-green algae are potentially periphytic. Many are facultative or part-time periphytes, being free living under certain environmental or life-history conditions and firmly attached to substrata under other conditions. Many species appear to be obligate periphytes which depend on a suitable substratum for normal growth. The absence of suitable substrata from a given region precludes the possibilities of periphytic forms thriving there.

The suitability of a substratum for a particular periphyte is usually determined by the chemical as well as the physical nature of the substratum. Many periphytic bacteria and most fungi depend on the substratum as a source of organic nutrients or as a concentration site for such nutrients. Chitinous substrata, for example, provide food for only chitinoclastic periphytes, unless other micro-organisms which provide assimilable chitin-decomposition products happen to be present

(CAMPBELL and WILLIAMS, 1951). The physical and physico-chemical nature of the substratum, its size, shape, smoothness, electrostatic properties, surface tension, toxicity, and other characteristics, play an important role in determining whether a given periphyte becomes attached or can flourish. Each organism evinces more or less exclusive or preferential requirements for special types of substrata.

Substrata treated with arsenicals, mercurials, copper salts, chlorinated phenols, or other toxic substances may repel or sometimes kill periphytic organisms. Ordinarily though, such toxic effects are not lasting, because to be effective the chemical toxicant must leach out of the substratum in injurious concentrations. Consequently the reservoir of toxicant is depleted sooner or later (WOODS HOLE OCEANOGRAPHIC INSTITUTION, 1952).

On natural substrata, the production of antibiotics or other antagonistic conditions created by certain organisms may be injurious to periphytic forms. Likewise, various kinds of predators such as protozoans, gastrotrichs, nematodes, and rotifers which graze on film-forming organisms may limit the abundance or kinds of periphytic microflora.

The tolerance of periphytic species to solid surfaces depends on the sensitivity of the particular species and a multiplicity of interrelated chemical, physico-chemical, physical, and biological conditions. The attachment and growth or survival of periphytes are also influenced by the characteristics of the substratum and the attachment structures and mechanisms of the periphytes.

Environmental aspects. The position or location of the substratum influences the attachment and growth of periphytes. Their well-being, like that of planktonic, nektonic, and benthonic organisms of all kinds, is influenced by temperature, light, free oxygen, hydrostatic pressure, salinity, pH, nutrients, water movements in relation to the substratum, and other factors. The tolerances and requirements of organisms in regard to these factors are treated in other chapters of this Volume. Some of these conditions (temperature and pressure, for example) may be essentially the same on the substratum as in the surrounding water. On the other hand, dissolved oxygen may, at times, be either more concentrated or less concentrated at or near the solid surface than in the surrounding water, owing to the physiological activities of diverse organisms. Likewise, the redox potential, the pH, or surface tension at a given site on the substratum is often measurably different from that in the surrounding water.

During the hours of sunlight, heavy growth of periphytic algae on substrata in the photosynthetic zone may liberate enough oxygen to increase its concentration substantially. Up to 50% more electric current was required to provide a protective potential on submerged steel plates during hours of daylight to keep them from being corroded by the oxygen generated by attached algae (ULANOVSKIY and GERASIMENKO, 1965). More commonly, dense populations of oxygen-consuming periphytes, including algae during hours of darkness, aerobic bacteria, fungi, protozoans, etc., may consume oxygen more rapidly than it is replaced by diffusion or water circulation, thereby creating anoxic conditions in localized areas on the substratum.

Light-dependent periphytes, which include a few species of photosynthetic bacteria and nearly all of the 300 species of blue-green algae known to inhabit the sea, thrive only in illuminated environments. Should such forms attach to mobile substrata in the photosynthetic zone and subsequently be carried into the aphotic zone, they would eventually perish.

Of much greater ecological importance than light as a limiting factor for periphytes on sinking substrata may be the large number of predators, the lack of oxygen, the presence of hydrogen sulphide, or other adverse conditions on the sea floor. The sinking of plant and animal remains or other particulate materials to the sea floor probably results in the death of large numbers of attached bacteria, fungi, and blue-green algae. Indeed, adsorption and sedimentation are believed to be major factors in destroying alien bacteria in polluted waters (WEISS, 1951; PAOLETTI, 1965) and in diminishing the indigenous microbial population (ZOBELL, 1946). PRESCOTT and co-authors (1946) list sedimentation of bacteria, as a result of their attachment to particles of higher specific gravity than water, as one of the principal factors influencing the diminution of bacteria in natural water. As stated by RENN (1937),

'Particulate substrates, necessary for the favorable development of large attached populations, tend to settle and carry large numbers of bacteria into the mud during sedimentation.'

According to HENRICI (1939), there is a marked tendency for bacteria to be adsorbed by or otherwise attached to solid particles in the water, and to be carried by these particles to the bottom, where they ultimately perish. Thus, while solid surfaces may promote the growth of certain periphytes, the sinking of substrata may be fatal for some of the attached organisms.

Among the many ways in which solid surfaces promote growth of the attached microflora is by providing a higher concentration of organic nutrients than that present in the surrounding water. This is the environmental condition on solid surfaces which is most unlike conditions in the surrounding water. Should the substratum be plant or animal, either living or dead, it will probably provide some of the periphytes with utilizable organic matter. Parasitic periphytes may invade living tissues, but most periphytic marine bacteria and fungi are harmless saprophytes. Virtually all kinds of organic tissues of dead plants and animals are susceptible to microbial attack. Refractory remains of plants and animals as well as other inanimate solids tend to adsorb organic matter from water (ZOBELL and ANDERSON, 1936; HEUKELEKIAN and HELLER, 1940; HARVEY, 1941; ZOBELL, 1943; NEWCOMBE, 1949, 1950; RUTTNER, 1963).

The movement of water with respect to the substratum is rated second in importance only to the kind and concentration of organic nutrients as an environmental factor affecting the attachment and fate of periphyton. The passage of water over the substratum is the principal mechanism whereby micro-organisms and nutrients come into contact with substrata. The proportion of those which adhere to the substratum will be influenced by various properties of the substratum, the micro-organisms, and the nutrients and particularly by the velocity of the water relative to the substratum. Up to certain limits, the greater the velocity the greater are the chances for particulate material coming into contact with the substratum, but above these limits the flow of water may prevent the attachment of particles or even dislodge them from solid surfaces. Water velocities of 25 to

50 cm/sec have been reported to suppress the attachment of certain kinds of organisms to smooth surfaces (Woods Hole Oceanographic Institution, 1952). Under otherwise comparable experimental conditions, certain pure cultures of bacteria examined by Zvyagintsev (1959) were dislodged from glass slides by a stream of gently flowing water (10 to 15 g/cm²), whereas other species were not washed off by a stronger current of water (30 to 50 g/cm²). Jetting at high pressure a stream of sterile water over glass slides was reported by Persoone (1965) to be more effective than swabbing for removing attached bacteria. From 70 to 90% of the bacteria which attached to glass slides submerged in raw sea water were removed by such treatment.

Characteristics of substrata. The adsorption or adhesion of gases, liquids, or solids to submerged substrata is influenced by (i) the texture or smoothness of the substratum, (ii) its cleanliness, (iii) chemical composition, (iv) ion-exchange, properties, (v) hardness or cohesiveness, (vi) its shape and area, (vii) surface tension, (viii) pH, (ix) electrostatic properties, and certain other characteristics or conditions. In general, the rougher or more uneven the solid surface, the more susceptible it is to the attachment of foreign molecules or larger objects, including micro-organisms.

Adam (1941) declares that discussions concerning thoroughly clean solid surfaces under natural conditions are purely academic, because it is so difficult to produce a clean surface experimentally, except by splitting a crystal. After brief exposure to the atmosphere, the cleanest substrata, even recently split crystals, adsorb gases and other molecules. When submerged in water, the possibilities of adsorption are greatly enhanced. So-called hydrophobic or non-wettable surfaces are only relatively so, inasmuch as a solid surface which permanently repels all kinds of molecules in sea water has yet to be demonstrated. The duration of submergence, the characteristics of the substratum, and the chemical, physical, and biological properties of the surrounding water will determine the kinds and quantities of film-forming materials that attach to submerged substrata.

In natural sea water having a surface tension of about 70 dynes/cm at 25° C, the wettability of glass and other chemically inert solids was found to be less than in sea water, whose surface tension was lowered to 35 or 40 dynes/cm by the addition of a variety of different surface tension depressants. The attachment and growth rate of bacteria on glass slides was found to be greatest in sea water having a surface tension between 60 and 65 dynes/cm. Although several different morphological types of bacteria reproduce rapidly in nutrient-enriched sea water at surface tensions lower than 45 dynes/cm, none attached to glass slides in enriched sea water at surface tensions lower than 45 dynes/cm.

In studies concerned primarily with mechanisms whereby commercial detergents are injurious to bacteria, Salton (1957) noted that most cationic and anionic detergents, in concentrations which appreciably reduce the surface tension of aqueous solutions, initiated the lysis of bacterial cells. Ostensibly, the surface-active agents disorganize the bacterial cell's permeability barrier, resulting in leakage of intracellular metabolites and in some instances the loss of enzymes. From their studies on *Micrococcus lysodeikticus*, Gilby and Few (1957) concluded that ionic detergents interact with a phosphatide component of the protoplast membrane, resulting in a disorganization of the membrane structure.

According to JAMES (1957a), the zeta potential (also called the electrokinetic or Helmholtz potential) of bacteria is influenced by (i) salt concentration, (ii) pH, and (iii) surface tension depressants or other surface-active substances. The zeta potential, as manifested by the electrophoretic mobility of bacteria, was found to be related to the lipid content of the cell wall. Ordinarily, cationic detergents tend to decrease the negative charge on bacteria whereas anionic detergents may have the opposite effect (JAMES, 1957b; Chapter 10.1; Volume V).

These observations have a bearing on the subject under discussion, because certain marine organisms produce surface tension depressants or other surface-active substances. Inasmuch as unlike charges tend to attract while like charges repel, the electrostatic properties of micro-organisms and substrata are of prime importance in affecting attachment. Glass slides as well as iron, steel, bronze, brass and copper immersed in sea water are electropositive at first whereas hard rubber, nickel, monel metal, and stainless steel are electronegative. Each substratum, though, has a different and fluctuating potential. Changes in potential occur as the substrata react with various chemicals in sea water and as they become coated with living organisms. Both cathodic and anodic potentials often develop in close proximity on the same test panel. Such electrostatic conditions affect the attachment of organisms as well as the corrosion of metal surfaces (WOODS HOLE OCEANOGRAPHIC INSTITUTION, 1952).

In general, the cells of most micro-organisms carry negative charges. Other things being equal, such cells are attracted to anodic surfaces. But some are electropositively charged and they may differ greatly in electrokinetic potential with age, stage of life cycle, cultural conditions, and species (MOYER, 1936a, b). The electrokinetic mobility of bacteria seems to be a function of the physical and chemical constitution of the bacterial cell surface (JAMES, 1957b).

According to NORDIN and co-authors (1967), the attachment of *Chlorella* to glass surfaces appears to be an electrostatic interaction between electrical double layers and various specific surface interactions resulting from surface heterogeneity and ion adsorption. Under most conditions, the algal cells and the glass surfaces have negative zeta potentials, and adhesion will not occur. However, if $FeCl_3$ is added to an algal-glass system immersed in 0·05 M NaCl solution, adhesion occurs because the algal cells and glass surfaces now possess very different zeta potentials. Adhesion was found to be strongest under those conditions which produced the greatest difference in zeta potential.

McLAREN and SKUJINS (1968) have reviewed the various mechanisms whereby the sorbtive properties of soil influence the growth and physiological activities of bacteria and fungi. Of paramount importance is the capacity of soil particles to adsorb a great variety of organic compounds, including proteases and other extracellular enzymes. McLAREN and SKUJINS also explain the effects of electrokinetic charges on interfaces between substrata and organic macromolecules.

DURHAM (1957) found a qualitative correlation between interfacial electrical conditions, especially the zeta potential, and the attachment of soil particles to solid surfaces, but factors other than electrical attraction influenced deposition. Among these factors were barrier effects of a solvated layer around the particle and barriers due to steric of spatial effects of adsorbed ions. Fine particulate soil was found by SHUTTLEWORTH and JONES (1957) to be much more difficult to remove from surfaces than larger particles. The attraction between particle and solid surface increased with decreasing size of particles, due primarily to sorptive electrostatic forces rather than the micro-occlusion or mechanical entrapment.

The adsorption of particles on substrata is often influenced by ion-exchange phenomena. The attraction of soil particles for bacteria was found by PEELE (1936) to be markedly influenced by base-exchange properties. Soils saturated with Fe and Al adsorbed more bacteria than soils saturated with Na, K, Li, or NH_4. Leaching soils with sodium, potassium, lithium, and ammonium chloride solutions removed many more bacteria than did water alone. Soils leached with ferric chloride or aluminium chloride solutions removed fewer bacteria than soils leached with water alone. Differences in the adsorptive properties are attributed by PEELE (1936) to differences in the electrostatic properties of the soil. By the use of ion-exchange resins, ZVYAGINTSEV (1962) has been able to demonstrate that the adsorption of bacteria on otherwise inert substrata is intimately associated with ion-exchange phenomena. The sorption between microbial cells and clay minerals was observed by SANTORO and STOTZKY (1968) to increase as the valency of the cations present on the clay and in the suspending electrolyte increased. Neither the size of the clay particles (considerably smaller than the bacteria) nor the size, motility or Gram reaction of the bacteria appeared to influence sorption.

Size or area of the substratum may influence the well-being of periphytic microflora in several different ways. Unless the substratum is appreciably larger than the microbial cell, there is no place for development and the attachment of progeny. Moreover, undersized substrata are not as effective in minimizing the diffusion of exo-enzymes or metabolites away from the attached cell(s) as are substrata which are larger than the attached cell(s). Cells attached to minute substrata are more susceptible to ingestion by protozoans and filter-feeders. On the other hand, a large substratum may invite colonization by grazing animals and competition for space by diverse sessile organisms. His own experiments as well as literature reviewed by ZoBELL (1943) indicate that solid surfaces larger than the bacteria are more beneficial to periphytic forms than more minute particulate materials.

Recent studies summarized by BREMNER (1954), ESTERMANN and McLAREN (1959), and STOTZKY and REM (1966) indicate that the chemical composition and physico-chemical properties of particles are more important than the size of the particles in affecting the ecology and physiological reaction rates of micro-organisms. The clay minerals, montmorillonite and kaolinite, were found by STOTZKY and REM (1966) to stimulate the respiration of a wide spectrum of bacterial species, including *Agrobacterium radiobacter, Arthrobacter globiformis, Escherichia coli, Mycobacterium phlei, Pseudomonas radiobacter* and *Pseudomonas striata.*

One of the many mechanisms whereby finely divided particles such as clay influence microbial ecology and certain enzymatic activities, notably reactions catalyzed by exo-enzymes, is the sorption of surface-active particles by the micro-organisms. Employing some micro-electrophoretic and microscopic techniques, MARSHALL (1969) was able to demonstrate that the surfaces of bacteria became progressively enveloped by a layer of adsorbed clay particles. For any particular microbial strain, almost identical amounts of illite and montmorillonite were adsorbed per cell. Such envelopes are believed to influence the uptake of nutrients and certain enzymatic reaction rates surrounding the microbial cell. The rate of sorption was influenced much more by the concentration of clay particles than by the size of the particles.

(b) Metabolism and Activity

The principal mechanisms whereby solid surfaces influence the metabolism and physiological activities of periphyton appear to be (i) concentration of organic nutrients on the substrata, (ii) orientation of large polar molecules on solid surfaces, (iii) minimization of the diffusion of exo-enzymes, hydrolyzates, and metabolites away from the periphyte on the substrata, (iv) neutralization of the zeta potential or other electrostatic conditions on the cell wall, thereby influencing permeability and leakage phenomena, and (v) creation of micro-environmental conditions (pH, redox potential, gas tension, surface tension, etc.) which may be quite unlike conditions in the surrounding water. Whether a given substratum promotes the metabolism of a given species or has detrimental effects on physiological activities will depend on the properties of the substratum, the characteristics of the attached organism(s), and the effects of other species which may be co-inhabitants of the substratum.

A sufficiently large substratum may accommodate an assemblage of diverse organisms forming a natural ecological unit or biocoenosis (BROCK, 1966). In this ecosystem, there may be symbiosis, mutualism, interdependence, and/or antagonism between bacteria, fungi, blue-green algae, diatoms, protozoans, and higher organisms. Indeed, the surface of a larger organism may be the substratum for the micro-organisms making up the ecosystem. Besides affecting micro-environmental conditions, micro-organisms in the ecosystem on the substratum may be competing for nutrients, or one species may be providing nutrients for another species. Many algae, for example, excrete organic matter (MOORE and TISCHER, 1964). From 0·2 to 25% of the carbon photo-assimilated by algae may be excreted as amino acids, peptides, proteins, various kinds of carbohydrates, lipids, or other organic compounds (HELLEBUST, 1965; NALEWAJKO, 1966). The kind and concentration of such ectocrines (LUCAS, 1961) and hydrolyzates produced by organisms in the ecosystem may have pronounced effects on the metabolism of organotrophs on the substratum. Similarly, the liberation of oxygen by photosynthetic algae may affect the respiration of nearby organisms.

If the substratum itself is organic, it may provide nutrients as well as a resting place for organotrophic periphytes. Its chemical composition must affect the metabolism of attached micro-organisms in many ways. Virtually all kinds of organic compounds, including such complex substances as lignoprotein, cellulose, chitin, keratin, algin, agar, and phospholipids are susceptible to microbial attack. Hydrolyzates, or simpler compounds resulting from the decomposition of complex molecules, may provide for the growth and metabolism of various kinds of organotrophic micro-organisms on the substratum (see also Chapter 10.1).

On nearly any kind of substratum, either organic or inorganic, there is usually a tendency for ectocrines and extracellular metabolites to be retained in the immediate vicinity of the microbiota on the substratum. Besides the physical attraction of the substratum for extracellular metabolites and exo-enzymes, the interstitial spaces at the tangent of the attached cells and the solid surface may serve as concentration centres for the exo-enzymes and hydrolyzates. This is believed to be of the utmost importance to the metabolism of saprophytic periphyton, particularly in extremely dilute nutrient solutions like natural sea water, which usually has an organic-carbon content of less than 1 mg per litre (WANGERSKY, 1965). Without solid surfaces for the concentration or retention of exo-enzymes and extracellular metabolites, very few micro-organisms are able to absorb enough nutrients to maintain their metabolism in such nutrient-poor medium. As stated by MITCHELL (1951),

> 'the effectiveness of growth enhancement by the surface phase in very dilute media evidently depends upon two primary factors: the tendency for essential nutrients to accumulate at the interface, and the tendency of the organisms to become adsorbed at the site of high nutrient concentration.'

To be effective, the proper enzymes must also be present where the concentration of nutrients occurs.

Chitinases, agarases, alginases, carageenases, cellulases, proteinases, certain nucleases and phospholipases are examples of microbial enzymes which occur extracellularly in media (POLLOCK, 1962). Whether the enzymes are truly extracellular, bound to the exterior of the cell wall, or liberated as a result of autolysis, unless adsorbed or otherwise retained on solid substrata, such exo-enzymes tend to become so dissipated in natural sea water that they are ineffective. According to HEUKELEKIAN and HELLER (1940), solid surfaces enable bacteria to develop in nutrient media otherwise too dilute for growth. Quoting these workers,

> 'Development takes place either as a bacterial slime or colonial growth attached to the surfaces. Once a biologically active slime is established on the surfaces, the rate of biological reaction is greatly accelerated.'

In this quotation, the term 'slime', as used by sanitary engineers, has reference to the aggregate of materials attached to solid surfaces and not merely the loose slime or specific polymers elaborated by bacteria (p. 1266).

The orientation of large molecules, especially enzymes, is influenced by solid surfaces. If the active centres of enzymes are to fall within an effective region, they must be adsorbed with the correct orientation with respect to the molecule(s) to be affected (DANIELLI, 1937). The angle of adsorption ranges from zero to 180°, but is usually less than 100°. According to ADAM (1941), most organic molecules in water form zero angles on glass, but certain larger molecules are adsorbed on glass with the long chains outward. For example, the contact angle of paraffin wax molecules to glass in water is 105°. The contact angles of various molecules are influenced by surface tension depressants or detergents. Contact angles are of importance in metabolism, because the susceptibility of large molecules to enzymatic action is a function of the orientation of the molecules with respect to the surface and to the enzymes.

Solid substrata sometimes promote the growth of micro-organisms by adsorbing nutrients which are poorly soluble in water like mineral and vegetable oils, for example, thereby substantially extending the oil–water interface or otherwise increasing the availability of large molecules to the action of enzymes. The addition of ignited sand, asbestos fibres, diatomaceous earth, or glass wool to mineral media containing liquid hydrocarbons as the only source of energy was found by ZoBELL and co-authors (1943) to result in 5 to 10-fold increases in bacterial growth. Dispersing solid lipids, paraffin wax, and asphalt in extremely thin layers on sand rendered these substances, which are usually refractory to microbial decomposition, readily susceptible to assimilation by certain bacteria. The majority of the bacteria grew firmly attached to the solid surfaces.

Besides requiring a resting place which provides a special combination of micro-environmental conditions, periphytic Flexibacteria, Labyrinthulae, and Cyanophyta are dependent upon solid surfaces for gliding.

(c) Reproduction

Obligate periphytes reproduce only when in contact with suitable solid substrata. Such substrata promote the reproduction of most facultative periphytes, particularly in dilute nutrient solutions such as natural sea water or oligotrophic lake waters. Organic solids may promote the reproduction of saprophytic fungi and bacteria by providing a source of essential nutrients. Purely inorganic, including chemically inert, solids also promote the growth of many aquatic micro-organisms.

It has been demonstrated by controlled laboratory experiments that such substances as ignited asbestos fibres (BREDEN and BUSWELL, 1933), silica sand, and diverse forms of glass (beads, slides, tubes, glass wool, etc.) promote microbial reproduction. Bacterial counts as follows were found by ZoBELL and ANDERSON (1936) in comparable samples of sea water incubated for 5 days at 16° C in the presence of glass to provide different areas of solid surface:

Ratio of volume of sea water in ml to area of glass in cm^2	Bacteria per ml found in water
1 : 0·24	$2·6–3·1 \times 10^5$
1 : 0·5	$4·9–5·5 \times 10^5$
1 : 1·7	$5·9–7·2 \times 10^5$
1 : 2·0	$8·5–9·7 \times 10^5$
1 : 4·1	$1·3–1·9 \times 10^6$
1 : 9·9	$2·4–3·1 \times 10^6$

Note that a 40-fold increase in solid surface area per unit volume of sea water resulted in a 10-fold increase in the number of bacteria which were found in the water. These data do not take into account the bacteria which grew on the glass so tenaciously that they resisted removal by vigorous shaking. Samples scraped from the surfaces of glass slides, direct microscopic observations, and oxygen-uptake tests indicated that from 40 to 200 times as many bacteria per unit volume of water developed firmly attached to solid surfaces as in the surrounding water.

Typically reproducing mainly on solid substrata are certain species of *Caulobacter* (JANNASCH, 1960; POINDEXTER, 1964), *Gallionella* (STARR and SKERMAN, 1965), *Leucothrix* (FUJITA and ZENITANI, 1967), *Thiothrix* (HAROLD and STANIER, 1955), and several other genera. The direct microscopic observation of glass slides which have been submerged in the sea reveal the presence of numerous, morphologically different, unidentified bacteria which do not reproduce in nutrient media commonly used for the cultivation of marine bacteria (ZoBELL, 1946; CvIIĆ, 1953; SKERMAN, 1956; KRISS, 1963). According to KRISS (1963), some of the unidentified bacteria found on submerged slides are rarely, if ever, seen on membrane filters through which water samples from the same place have been filtered, thereby suggesting that such bacteria reproduce only on solid surfaces. Reproduction of bacteria on glass slides submerged in the sea is affirmed by the appearance of pairs, short chains, and microcolonies which increase in size with duration of submergence.

After finding that sea water at all depths in the ocean contained suspended particles which were believed to provide for the accumulation of organic matter and the attachment of bacteria, and after finding that submerged glass slides provide favourable conditions for the reproduction of bacteria, KRISS and RUKINA (1952) employed this technique for estimating the rate of bacterial reproduction under the natural environmental conditions. Soon after the glass slides were submerged, bacteria were found attached in numbers proportional to their abundance in the water as demonstrated by other methods. Some of the attached bacteria commenced to reproduce, as indicated by the appearance of dividing cells and microcolonies. The generation time of some bacteria under the most favourable environmental conditions was found to be less than 30 mins. The generation times of bacteria on glass slides at shallow depths in temperate water (ca 20° C) of the western Pacific Ocean were found to range from 2·0 to 3·4 hrs during the first 8 hrs of submergence. During the first 25 hrs of submergence, the generation time ranged from 6·8 to 18·7 hrs. The slower rate of reproduction with increasing numbers of bacteria on the slides was attributed to increasing competition for space and organic matter adsorbed on the glass slides. Employing similar methods, KRISS and MARKIANOVICH (1954) estimated that the increase in bacterial biomass on submerged slides, due to reproduction, ranged from 13 to 80% per day in the Black and Caspian Seas. Water temperature, within the range of 11° to 25° C, was reported to have no effect on either the rate of settling or the growth of bacteria on submerged slides. To calculate the rate of production of bacterial biomass on submerged slides, they applied the modified formula of IERUSALIMSKII (1954):

$$\frac{P}{B} = \frac{4\sum \log m \cdot n}{Nt} \times 24$$

where P is the mean amount of biomass produced during 24 hrs, B is the biomass initially present, m is the mean number of cells in each microcolony, n is the number of microcolonies, N is the total number of single cells and microcolonies, and t is the time of submergence in hours. The daily coefficient P/B is thus the relative increase in bacterial biomass per day.

In Arctic Ocean water (0·6° to 2·4° C) in the region of the North Pole, KRISS and LAMBINA (1955) estimated the P/B to range from 0·12 to 0·72; that is, the daily gain in bacterial biomass was 12 to 72%. Although highly sporadic from layer to layer, the general tendency was for the rate of reproduction of bacteria on submerged glass slides to decrease with increasing depth of water.

The submerged glass-slide technique provides much useful information about the morphological types, abundance, and growth rates of bacteria in the sea, but

B

the method has some shortcomings. First, not all kinds of bacteria present in the water become attached to glass, either because they fail to come into contact with it or because they do not adhere. Second, sometimes it is difficult or impossible to distinguish certain bacteria from detritus or to enumerate all bacteria in large colonies or other aggregates. Third, sometimes submerged glass slides attract animals which subvert microbial growths in various ways. We find that, in 20-litre samples of sea water incubated in large glass carboys, bacteria reproduce much slower than those attached to glass slides submerged in such samples.

The reproduction of the vast majority of blue-green algae and nearly all species of fungi in marine habitats appears to be on or in solid substrata. The fungi obtain their food mainly from solid substrata.

(d) Distribution

Bacteria, fungi and blue-green algae are generally much more abundant in the littoral zone than in oceanic or pelagic waters. Suspended and benthonic solids, including higher plants and animals or their remains, are also much more abundant in the littoral zone than in the open ocean. Although the abundance of organo-trophic bacteria and fungi is mainly a reflection of available organic matter, the densest bacterial and fungal populations do occur on solid substrata, both organic and inorganic. Being photolithotrophic, the blue-green algae depend primarily on radiant energy rather than organic matter, but despite better illumination in the more transparent water in the open ocean, the densest populations and the majority of the Cyanophyta species are found in the littoral zone, mainly attached to solid substrata.

Some 300 species of blue-green algae have been reported to occur in saline situations, most of which are truly marine (DESIKACHARY, 1959). Only about 30 of these are free living or planktonic. A vast majority of the others are epilithic, epipelic, epiphytic, or epizoic. They occur singly, in filaments, and in mat-forming colonies on nearly all kinds of solid surfaces in shallow water, especially in the intertidal zone (ROUND, 1964). These mat- or carpet-forming periphytes are termed 'tapetic' by WOOD (1965), who discusses their wide-spread distribution in shallow waters. Most common in mats on rocks, coral reefs, and shallow bottoms are certain species of *Anacystis, Brachytrichia, Calothrix, Chamaesiphon, Cocco-chloris, Epilithia, Hydrocoleus, Merismopedia, Phormidium, Plectonema, Pleuro-capsa, Rivularia, Schizothrix,* and *Symploca.* Identified among the fouling cumula-tions on ships and buoys were species of *Calothrix, Chroococcus, Dermocarpa, Dichothrix, Hydrocoleus, Hyella, Lyngbya, Nodularia, Oscillatoria, Phormidium, Pleurocapsa, Scytonema, Spirulina, Symploca,* and *Trichodesmium* (WOODS HOLE OCEANOGRAPHIC INSTITUTION, 1952). Certain species of *Lyngbya, Chamaesiphon, Dermocarpa, Gloeocapsa,* and *Xenococcus* are epiphytic on other marine algae, sea-grasses, and marsh plants. The shells and carapaces of many molluscs and crustaceans often have dense growths of blue-green algae along with other peri-phytic forms. On many shallow-water bottoms and other substrata, the abundance of blue-green algae may be controlled only by grazing protozoans and other animals (see also Chapter 7.21).

Virtually all of the specimens of brown algae, including species of *Ascophyllum,*

Fucus, Laminaria, and *Pelvetia,* examined by KOHLMEYER (1968), were found to be laden with epiphytic or perthophytic fungi.

Although many kinds of fungal spores are ubiquitous, fungi are found growing in the sea only on solid substrata, and mainly in the littoral zone. Seaweeds drifting in the neritic zone sometimes provide for the growth of fungi. The brown algae *Sargassum* in the oceanic province harbours the ascomycete *Phyllachorella oceanica* (JOHNSON and SPARROW, 1961). These authors document the occurrence of Phycomycetes in planktonic diatoms. Of the 73 species of marine Phycomycetes listed by JOHNSON and SPARROW, 47 were found growing as epiphytes on algae and 5 on sea-grasses. The substrata for 10 of the Phycomycetes were tissues of dead animals.

Among the higher marine fungi (Ascomycetes, Deuteromycetes, and Basidiomycetes) numbering nearly 180 species, about 75% are associated with solid surfaces while about 25% are either parasites or perthophytes. Apparently there are no free-living species (KOHLMEYER and KOHLMEYER, 1964). These workers give a table of algae (31 genera), higher plants (16 genera), and wood or cellulosic materials which variously have been found to serve as substrata for 83 marine species of Ascomycetes and 29 species of Deuteromycetes or Fungi Imperfecti. Listed in the table are only two marine Basidiomycetes: the smut *Melanotaenium ruppiae,* found in rhizomes and leaf bases of the sea-grass *Ruppia,* and *Digitatispora marina,* found growing on submerged wood.

Brown, red, and green algae are the most common substrata for Fungi Imperfecti and Ascomycetes. Species of both classes also abound on driftwood, bark, sunken timbers, wood pilings, submerged cordage, and chitinous materials (KOHLMEYER, 1963; MEYERS and REYNOLDS, 1963). Some species thrive on organic-rich sediments. Dead as well as living roots and stems of sea-grasses (*Halophila, Thalassia, Phyllospadix, Zostera, Posidonia, Ruppia*), mangroves (*Avicennia, Laguncularia, Rhizophora*), and salt-marsh plants (*Limonium, Spartina, Juncus, Salicornia, Spergularia, Puccinella,* etc.) are inhabited by various species of Ascomycetes and Fungi Imperfecti. The spores of many species are provided with appendages or gelatinous sheaths which presumably facilitate their attachment to substrata.

A recommended method for obtaining cultures of marine fungi is to swab various submerged substrata such as seaweeds, wood, cordage, sea-shells, barnacles, fish, etc., and then plant the swab in appropriate nutrient medium. The occurrence of viable fungi and their ability to grow in marine habitats is usually demonstrated by 'trapping' them on presterilized pieces of wood, seaweed, crab shells, fish scales, pollen, and other vegetable or animal substrata. Many saprobic and facultatively parasitic fungi attach to and grow on such solid baits (HÖHNK, 1956; JOHNSON and co-authors, 1959; MEYERS and REYNOLDS, 1960; WILSON, 1960; MEYERS and co-authors, 1970).

Bacteria occur throughout the sea at all depths and at all latitudes. The cells of many species occur free in sea water, but the majority appear to be associated with solid substrata. The largest populations occur in sediments on the sea floor where, in general, the abundance of bacteria decreases with depth of the overlying water and also with increasing depth below the mud–water interface (ZOBELL, 1968). The high concentration of bacteria on the sea floor is attributable mainly to (i) the deposition of particulate materials, especially organic materials, bearing attached bacteria, and (ii) the presence of solid surfaces which promote the reproduction of bacteria. On shallow, sunlit bottoms, photosynthetic plants provide bacteria with a great variety of organic nutrients and solid substrata.

As pointed out by WAKSMAN and VARTIOVAARA (1938), a large proportion of the bacteria in bottom sediments are adsorbed or otherwise attached to solid particles. The microscopic examination of differentially stained sand grains from shallow marine and freshwater bottoms convinced MEADOWS and ANDERSON (1966) that bacteria, along with an assortment of organic materials, are nearly always present on submerged sand surfaces. The surface of sand grains in the littoral zone was observed by MEADOWS and ANDERSON (1968) to be a common habitat for bacteria, blue-green algae, and diatoms. Microcolonies growing tenaciously attached to such surfaces were found to contain from 2 to 100 or more bacterial cells. Various methods of observation, including the capillary-tube or 'peloscope' method of PERFIL'EV and GABE (1964), show that while a large number of bacteria may occur in interstitial water or in water overlying sediments, the largest populations are found and most bacterial growth takes place on solid surfaces.

The exterior surfaces of virtually all dead and many living seaweeds, holothurians, crustaceans, molluscs, fish, etc., are veritable bacterial gardens. In his review of the literature, SHEWAN (1961) documents the occurrence of from 10 to 10^8 viable bacteria per cm^2 of freshly caught fish surface. Saprobic bacteria appear to be associated with most kinds of animals ranging in complexity from protozoans to mammals. Relatively few species of epizoic bacteria have been shown to be responsible for skin infections in fish (AMLACHER, 1961; SINDERMANN, 1966), crustaceans (HESS, 1937), and other marine animals. Animal hosts not only help to augment the abundance of bacteria in a given habitat, but contribute greatly to the dissemination of bacteria.

From 10^4 to 10^9 bacteria have been detected per cm^2 on the fronds of the brown alga *Macrocystis pyrifera* growing off La Jolla, California, where the bacterial populations in the surrounding water ranged from 10^3 to 10^5 per ml. CHAN and McMANUS (1967) found from 10^5 to 10^7 bacteria per ml of freshly prepared homogenates of the brown alga *Ascophyllum nodosum* and the red alga *Polysiphonia lanosa*. These two algae were found to support more than 25 species of bacteria and a species of pink yeast belonging to the genus *Rhodotorula* (CHAN and McMANUS, 1969). Predominating among the bacteria were species of *Vibrio*, *Flavobacterium*, *Pseudomonas*, and *Achromobacter*. Ostensibly, the majority of bacteria were harmless periphytes or perthophytes since only healthy algae were examined in this study. Moribund or dead specimens often harbour 10 to 100 times more bacteria than healthy algae. BROCK (1966) says he has rarely seen an algal filament which was not colonized by bacteria, frequently at very high density.

Living diatoms, including *Nitzschia closterium*, and certain blue-green algae examined by SPENCER (1952) were found to be laden with tenaciously attached bacteria. Some also penetrated the gelatinous sheaths of blue-green algae. Most, but not all of the recently collected, unpreserved plankton diatoms examined in my laboratory have carried bacterial epiphytes. OPPENHEIMER and VANCE (1960) found bacteria attached to most dead diatoms but not to living diatoms, flagellates, or blue-green algae. Failing to find antibacterial ectocrines or other antibiotic substances led these workers to postulate bacteria-repelling electrokinetic potentials as a protective mechanism of the algae. Only a small percentage of the plankton diatoms tested by STEEMANN NIELSEN (1955) produced enough antibiotics to decrease substantially the activity of associated bacteria. Less than half of the 45 species of diatoms and 11 other phytoplankton species from the Mediterranean Sea produced antibiotics active against any of 17 species of bacteria which were tested (AUBERT and GAUTHIER, 1967).

Dating from the earliest quantitative investigations, bacteria were observed to be most abundant in sea water supporting the largest plankton populations. WAKSMAN and co-authors (1933) reported that bacteria occur only to a very limited extent free living in sea water, most of them occurring attached to larger plankton organisms. Not only is there a parallelism between the abundance of plankton and bacteria in the sea, but many investigators have demonstrated by various procedures that the careful removal of phytoplankton and zooplankton from samples of sea water also removes the majority of bacteria. Shaking, washing with various solutions including detergents, centrifuging, etc., remove many of the attached organisms, but no procedure frees all of the bacteria from plankton organisms. Between 80 and 99% of the bacteria associated with plankton collected off the French coast were found by CAMPELLO and co-authors (1963) to be 'exobactéries' which could be dislodged by washing 5 or 10 times with centrifuging. JONES (1958) removed 50,000 to 100,000 bacteria per g wet weight of zooplankton immediately after collection from pelagic sea water which contained only 50 to 100 bacteria per ml after removing the plankton. Sea-water samples collected along the Japanese coast were found by SEKI (1967) to contain between 10^3 and 10^5 bacteria per ml following passage through a No. XX-13 silk net (aperture size ca 100 μ) whereas the equivalent of from 10^6 and 10^8 bacteria per ml were associated with the retained plankton. RIGOMIER (1967) reported the occurrence of up to 10^4 times more bacteria attached to zooplankton than the number free-living in the water.

(3) Structural Responses

General

Either directly or indirectly the substrata supporting periphytes may affect their size, shape, flagellation, holdfasts, encapsulation or ensheathment, filament formation, aggregation, and colony characteristics. Inasmuch as so many natural substrata in the sea consist of highly complex and often biodegradable materials, it is extremely difficult to separate the direct effects of solid surfaces from many other factors which are involved. Among the most important of these extraneous factors are the physical consistency of the substratum and its chemical composition. Also obscuring the solid surface responses of periphytes under natural conditions are the activities of a great diversity of other kinds of organisms which may be symbiotic, competitive, antagonistic, predatory, or instrumental in affecting the pH, oxygen tension, redox potential, electrokinetic properties, surface tension, presence of organic materials, and other environmental conditions within the biocoenosis.

Bacteria

Most bacteria form various amounts and kinds of extracellular materials of a slimy viscous nature. These extracellular materials, composed mainly of poly-saccharides and gums, appear to be distinct from the bacterial cell wall in func-tion and chemical composition (SALTON, 1964). According to WILKINSON (1958), the extracellular polysaccharides and gums occur in two forms: (i) loose slime which is non-adherent to the bacterial cell and which may impart an increased viscosity or sticky consistency to media, and (ii) capsules or microcapsular materials which adhere to the cell wall, completely enveloping many cells. In addition to non-adherent slimes and adherent capsular materials, there are other surface components involved in the adherence of bacterial cells to one another (SALTON, 1964). All three categories of extracellular materials contribute to microbial attachment to solid surfaces.

The physical structure of capsular substances and the slime layer may be altered by cultural conditions. In both cultures, the slime layer decreases in density or concentration with increasing distance from the cells. On glass slides submerged in dilute nutrient sea-water medium, the slime layer tends to adhere to the glass, forming gently sloping islands surrounding the attached bacterial cells. These slime islands are commonly from 2 to 10 times the diameter of the cells. The slime layer is believed to help hold the bacteria on solid substrata. It seems to promote the attachment of certain other bacterial species, but the slime islands formed by some species are definitely repellent to others.

The capsular material, which ranges in thickness from $0 \cdot 2 \mu$ to 2 or 3μ, also has adhesive properties for certain substrata and it may hold some bacteria to-gether to form aggregates or microcolonies on substrata. The shape, structure, and size of such aggregates which develop on glass slides submerged in sea water are usually different from those which develop in ordinary nutrient media. Poly-saccharides excreted by bacteria have been shown by MARTIN and RICHARDS (1963) to be effective in binding soil particles into aggregates.

Gelatinous strains of *Chromobacterium violaceum*, examined by CORPE (1964), excreted copious quantities of polysaccharide which formed a fibrous, cross-linked matrix embedding the cells. The colonies became so tenaciously attached to the solid medium that they could be removed only if forcibly peeled off the agar. The polysaccharide not only bound the cells in the matrix but also penetrated the agar surface. Exopolysaccharide formation was influenced by the chemical composition of the medium in which the bacteria were growing.

CHRISTIE and FLOODGATE (1966) have described branching tree-like structures which developed on submerged surfaces, sometimes attaining a length of 1 mm. These 'microtrees' were composed of algae, bacteria, and detritus, including mineral crystals, amorphous organic material, fragments of diatom skeleton, etc. An appreciable part of the mineral matter consisted of calcite crystals which helped to cement detritus and living organisms to glass slides and other solid surfaces. An increase in pH in the immediate vicinity of attached bacteria and micro-algae was probably responsible for the precipitation of calcium carbonate and/or calcium phosphate, thereby promoting the attachment of more organisms and detritus to form the microtrees. Bacteria which liberate ammonia from proteins, or species which reduce nitrate and nitrite, could account for the increase in pH at the solid surfaces. The uptake of CO_2 by micro-organisms on such surfaces may also increase the pH of sea water sufficiently to promote the precipitation of calcium carbonate and calcium phosphate.

The attachment of most bacteria to substrata seems to be caused either by cements or sticky substances produced or excreted by the bacteria, or by extrinsic physical forces such as electrostatic attraction or the sticky nature of the substrata. Only a few groups of bacteria have special cytological structures for attachment. Stalks are formed by members of the family Caulobacteraceae having five small genera: *Caulobacter*, *Asticcacaulis*, *Gallionella*, *Siderophacus* and *Nevskia*.

In the *Caulobacter* group, the stalk arises from the pole of the cell, being a finely drawn out prolongation of one end of the rod-shaped or curved cell (POINDEXTER 1964). Adhesive material is secreted at the distal end of the stalk, which terminates in a small knob. Reproduction occurs by division of the stalked cell, giving rise to a non-stalked sibling which is motile by means of a single polar flagellum. At the base of the flagellum, secreted adhesive material provides for the attachment of the sibling to other micro-organisms, to inanimate substrata, or to each other's holdfasts to form rosettes consisting of from 2 to 20 or more cells. Dating from their discovery in freshwater lakes by HENRICI and JOHNSON (1935), Caulobacteria have been found in many other habitats, including sea water (JANNASCH, 1960; LEIFSON and co-authors, 1964; POINDEXTER, 1964).

Asticcacaulis differs from *Caulobacter* species primarily in the eccentric position of its single flagellum and stalk. The rod-shaped cells of *Asticcacaulis* secrete adhesive material at or near the pole of the cell at a site different from that where the stalk develops (POINDEXTER, 1964).

Stalks formed by *Gallionella* are not part of the cell, as in *Caulobacter*, but are slender, branched, twisted ribbons of excreted material, mainly ferric hydroxide (STARR and SKERMAN, 1965), which emerge from the side and attach to solid surfaces. Such ribbons may be more than 200 μ long. Although found chiefly in iron-bearing waters, species of *Gallionella* also occur in marine habitats (BUTKEVICH, 1928).

Siderophacus cells also excrete a stalk consisting mainly of ferric hydroxide, differing from *Gallionella* stalks by being unbranched and by not forming twisted bands (BEGER, 1944). They are restricted to iron-bearing waters, where they grow on solid substrata.

Nevskia cells excrete dichotomously branched, lobose stalks which are composed of gum. The long axis of the rod-shaped cells is set at right angles to the stalk. They grow in zoogleal masses in water, but have not been reported from marine habitats.

Hyphomicrobium, *Rhodomicrobium*, *Pasteuria*, and *Blastocaulis* are genera of filament-forming bacteria belonging to the order Hyphomicrobiales, which reproduce by budding or by longitudinal fission. The buds may be sessile or may be borne at the tip of a slender filament which arises from the pole of a mature cell or from a filament connecting two cells. Commonly the Hyphomicrobiales occur in aggregates consisting of groups of cells, attached to a solid substratum by stalks which appear to radiate from a common holdfast. Some have a motile stage. Representatives of this group have been found on solid substrata in mud and water, often attached to glass slides, algae, other organisms, or inanimate substrata (HENRICI and JOHNSON, 1935; DUCHOW and DOUGLAS, 1949).

Like the blue-green algal genus *Oscillatoria*, to which they appear to be closely related, most members of the order Beggiatoales have a strong predilection for solid surfaces in aquatic habitats. They lack flagella and other evident organs of locomotion, but when in contact with an appropriate substratum the trichomes of species of *Beggiatoa*, *Thioploca*, *Thiothrix*, *Vitreoscilla*, *Bactoscilla*, *Microscilla*, and *Leucothrix* are capable of gliding. The cells of *Achromatium*, which occur singly, show a rolling, jerky type of motility when in contact with a suitable substratum. A slimy sheath excreted by the Beggiatoales helps to hold them to the substratum and facilitates their gliding. Temporarily, the conidia of *Thiothrix* show creeping motility on solid substrata, to which the development of trichomes eventually grow attached at the base by means of gelatinous holdfasts. Similarly, the trichomes of *Leucothrix* are commonly attached basally to solid substrata by a gelatinous holdfast. The motile conidia of *Leucothrix* sometimes aggregate to form rosettes containing up to 50 cells, which become non-motile, develop holdfasts, and elongate to form trichomes (BREED and co-authors, 1957). Chemotaxis is believed to play a role in the aggregation of *Leucothrix* conidia (HAROLD and STANIER, 1955). *Leucothrix mucor* grows epiphytically on marine algae (FUJITA and ZENITANI, 1967).

Leucothrix mucor were always present in great numbers on substrates submerged in Ostend (Belgium) Harbour waters, according to PERSOONE (1968). These thigmotropic bacteria appeared to be an important source of food for the peritrichous ciliate *Zoothamnion commune*, which generally colonizes submerged surfaces as soon as the surfaces become coated with primary film-forming bacteria and algae.

Cytophaga is the only genus of the order Myxobacterales which is known to be represented in the sea. Members of this genus are flexible rods which glide on solid substrata. The moving cells may pave the substratum with a thin layer of slime on which they rest. They decompose cellulose and are commonly associated with marine vegetation and organic-rich bottom deposits.

Polysaccharides, lipoproteins, and possibly other chemical components of bacterial cell walls (SALTON, 1964) may account for the adhesiveness of certain attachment bacteria. Whether the nature of the substratum affects the structure of bacterial cell walls has not been determined. Recent observations on microstructures are highly suggestive of unexplored possibilities. For instance, fimbriae (DUGUID, 1959) have been found on the surface of several different species of bacteria. Fimbriae, also called pili (BRINTON, 1959), are filamentous appendages which are smaller, shorter, and more numerous than flagella. Having a diameter of only about 80 Å, they are visible only by electron microscopy. They are thought to be organs of attachment. All fimbriate strains examined appear to have adhesive properties for dead or living plant and animal cells, including red blood cells. Many fimbriate strains adhere to glass, cellulose, and other inanimate substrata.

JONES and co-authors (1969) have described a method for examining slime layers by electron microscope. Epoxy resin discs are placed in a holder and submerged in natural water for the desired period. Immediately after removal from the water, the attached slime is fixed, stained with ruthenium red glutaraldehyde-osmium tetroxide, dehydrated, and embedded in epoxy resin so that thin sections can be cut through the vertical plane of the slime mass. Such thin sections permit detailed examination of the attached layer, the surface–slime interface, and the spatial relationship between cells in the vertical slime structure. In their preliminary studies, JONES and co-authors observed no special attachment structures on various kinds of bacteria in aquatic slimes. The cells appeared to be attached to the surface by extracellular material alone. This material was observed in strands and netlike structures between cells. The ten electron micrographs, at magnifications of 19,400 to 105,000 × , are most impressive. The method should be applicable to the study of other kinds of micro-organisms besides bacteria which attach to solid substrata.

Fungi

Members of the genus *Labyrinthula* are gliding organisms whose somatic structure consists of a simple oval or spindle-shaped cell (ALEXOPOULOS, 1962). The cells secrete mucous filaments which unite to form a fine network on which the cells glide. This mucilage secures the cells and filaments to the substratum. About 10 marine species have been described. They occur mainly on marine vegetation as either saprobes or parasites. *Labyrinthula macrocystis* is believed to cause a destructive disease of the eel-grass *Zostera marina*. Other Labyrinthulae live on the surfaces or invade the tissues of dead or living algae. Parasitic cells penetrate the tissues and develop inside the host, forming a network of cells and connecting filaments. The form and structure of the substratum are of importance in determining the network pattern of living filaments, the formation of rhizopod-like extensions, and the position of cells of both parasitic and saprobic varieties. In a recent review of the genus *Labyrinthula*, POKORNY (1967) discusses the associations of 10 species with surfaces of marine vegetation consisting mainly of living algae.

The true fungi are represented in the sea by numerous species in three classes of Eumycetes or Eumycophyta: Phycomycetes, Ascomycetes, and Deuteromycetes (Fungi Imperfecti). The majority are usually filametous and multicellular. The fungal thallus or mycelium consists of filaments or hyphae which branch in all directions, spreading over the surface of or into the substratum. The substratum is nearly always organic matter upon which the fungus depends for nutrition. Aquatic plants in good health may support large populations of periphytic or perthophytic fungi. Some fungi invade only plants which have been damaged by trauma, boring animals, unfavourable temperature, or otherwise. The parts of the mycelium of saprobic fungi, like those of facultative or obligate fungal parasites which penetrate plant tissues or other substrata, are generally quite different in structure from the rest of the mycelium. The penetrating part, sometimes referred to as substrate mycelium, might grow between host cells or it might penetrate into the cells of the host. Many intercellular hyphae which penetrate the cells of certain plants obtain nourishment through a special hyphal outgrowth called a haustorium (ALEXOPOULOS, 1962). From these statements about the morphology and nourishment of fungi, it should be apparent that the structure of fungi may be substantially modified by the nature of the substratum. For further details concerning the structure and ecology of marine Phycomycetes, Ascomycetes, and

Fungi Imperfecti see JOHNSON and SPARROW (1961) and KOHLMEYER and KOHLMEYER (1964).

Blue-green algae

The chemical nature of the substratum, so important for saprobic periphytes, seems to have less influence on the form and structure of blue-green algae than on bacteria and fungi. The substratum serves mainly as a place for attachment of the algae. Usually the substratum provides no nutrients, the algae getting these almost exclusively from the surrounding sea water (FELDMANN 1951). The development and structure of blue-green algae are influenced by the position and location of the substratum with respect to sunlight, and by the physical nature of the substratum, particularly its smoothness or indentations, its hardness, and its penetrability by the algae. The basal filaments of Chamaesiphonaceae and members of certain other families penetrate the substratum, resulting in modifications in size and shape of the perforating parts. The structure of certain endolithic forms which inhabit calcareous substrata such as rocks, mollusc shells or corals may be influenced so much by the physical nature of the substrata on which algae are growing that it is often difficult to recognize species or even genera. In this category are several perforating forms which are mainly marine, e.g. *Brachynema, Epilithia, Krytuthrix, Lithonema, Plectonema, Pleurocapsa, Podocapsa, Schizothrix,* and *Solentia* (DESIKACHARY, 1959).

As implied by the term Myxophyceae (*myxo* = mucus or slime) sometimes applied to the Cyanophyta or Schizophyceae, many blue-green algae secrete slimy mucilage. This gelatinous mucilage differs greatly in amount, consistency, colour, and stratification, depending on the species and environmental conditions (FRITSCH, 1959).

The physical nature of the substratum affects the formation and especially the shape of algal sheaths, the arrangement of filaments, their branching, and the movements of blue-green algae. According to ROUND (1964), the filaments of almost all epipelic species, even apparently non-motile colonial forms, have some capacity of movement on solid surfaces. Indeed, periphytic forms are dependent on solid surfaces for their creeping or gliding movements.

The motile hormogonia of members of the family Rivulariaceae show marked apical-basal differentiation, being broadest at the base, and tapering towards the tip. In many species of *Calothrix, Dichothrix, Gloeotrichia,* and *Rivularia* occurring in marine situations, such filaments attach at the base to various kinds of substrata. The size and shape of the basal structure may be modified by the physical nature of the substratum.

Addendum

Attention is directed to a compilation of 15 articles on 'Adhesions in Biological Systems', edited by R. S. MANLY (Academic Press, New York, 1970), a 302-page book which was published after the galley proofs for this chapter were returned. Most of the articles have a bearing on the ecology of substrata and the mechanisms whereby organisms attach to solid surfaces. Particularly pertinent is the paper by W. A. CORPE entitled, 'Attachment of marine bacteria to solid surfaces', with special reference to cell structure and cements.

7. SUBSTRATUM

7.2 PLANTS

7.21 UNICELLULAR PLANTS

E. J. F. Wood

(1) Introduction

Because of the large surface–volume ratio of unicellular plants, their relation to the environment is more intimate than with multicellular plants. Many, possibly the majority of marine unicells, are motile and so can adjust themselves physically with respect to the substratum. Moreover, the marine unicellular plants include forms which are of the same size-order as the bacteria (1 to 5 μ) and exhibit many of the same phenomena with regard to physical parameters such as adsorption. For this reason, we have to consider organisms attached to floating substrates in the water, and therefore planktonic in their habitat, as well as the benthic habitats.

Floating substrates are primarily associated with planktonic habitats of unicellular plants. The types of particulate matter, usually found floating in the oceans, have been discussed by Wood (1964). These include organic particles, usually discoid or filamentous and composed of some refractory material such as chitin or cellulose (50 to 100 μ in size), and smaller inorganic particles, probably composed of ferric phosphate or other insoluble compounds (10 to 50 μ). Baylor and Sutcliffe (1963) have shown how organic aggregates may be formed; they believe that these are important for phytophagous organisms. In the oceans, it is possible that such aggregates also serve for the attachment of micro-algae either by adsorption or by attachment mechanisms. Some of of the particles described by Wood (1964) may have been formed in this way. Many unicellular plants may be only temporarily attached to particles, e.g. certain stages of the coccolithophores; some may adsorb and desorb under changing conditions or at different stages in their life history.

In estuaries, micro-algae attached to substrates—such as the leaves of *Zostera* and *Thalassia* species, and algal mats composed of species of *Ectocarpus*, *Enteromorpha*, *Cladophora*, *Ceramium*, *Polysiphonia* as well as other small filamentous algae—frequently detach in the early summer and form floating substrates of some magnitude, usually breaking up into fragments and distributing their load of epiphytes and adsorbates in various parts of an estuary, as dictated by wind patterns.

In the oceans, massive red tides of *Trichodesmium* species may occur; in their later stages, these may include other forms as a pseudoperiphyton, and large quantities of *Sargassum* species (e.g. in the Caribbean Sea) carry with them a whole plant association composed of benthic and epiphytic species of diatoms and other micro-algae.

Consolidated substrates include rocks (solid substrates), sands and subsurface sediments. They may be divided according to their situation into estuarine, shelf or abyssal substrates. Abyssal substrates are not usually considered as related to unicellular plants, but WOOD (1956) described diatoms which he believed were actively associated with sediments collected on the *'Galathea'* expedition from depths between 7000 and 10,000 m. More recently, it has been found that marine unicellular plants can usually be collected from great oceanic depths.

The sediments of the continental shelf are sometimes above and sometimes below the theoretical compensation point of unicellular plants. In either case it can usually be demonstrated that there exists an association of micro-algae.

Estuarine sediments often contain large numbers of unicellular plants belonging to a number of taxonomic groups. Associated with such sediments are the sea-grasses (Chapter 7.22) which form a habitat usually carrying an algal flora with a far larger biomass than the sea-grasses themselves (WOOD, 1959). Sea-grasses decay more or less rapidly, and the resulting material settles on the bottom as detritus, forming part of the sediments. In estuaries, origin and history of the substrate is also important, e.g. nature of the parent rock, associated river systems, tidal structure, rainfall through the watershed, and shape of the estuary (drowned valley, barrier island type, etc).

Sand and mud can be classified in detail by assessing the size ratio of the particles and the colour according to standards made up of ochre, carbon black and zinc oxide. Such standards are useful in characterizing the sediments as they may be correlated with unicellular plant associations; for example, a coarse, white sand allows considerable algal growth to a depth of several centimetres; purple and green bacteria are associated with dark sands and muds; and heavy, dark clays usually contain large organic fractions which affect the algal population and its composition.

(2) Functional Responses

In unicellular plants, quantitative aspects of performance in relation to the substratum are difficult to assess and have hardly been studied. Instead of following the general chapter outline, this subchapter will, therefore, emphasize the functional relations between substrate and unicellular plants on a more qualitative, descriptive basis.

(a) Relations to Floating Substrates

In the oceans, the particulate matter does not vary appreciably in character, except in the case of floating multicellular plants. We therefore find that the type of unicellular plants adsorbed or attached to the floating particles tends to be rather uniform, consisting of small flagellates, both coloured and colourless, and diatoms. Coccolithophores, small diatoms and dinoflagellates are found attached to planktonic crustaceans and larger diatoms (e.g. species of *Cocconeis*) are attached to whales.

In the Caribbean and Sargasso Seas, free-floating species of *Sargassum* form large tufts and rafts, and contribute a surprising amount to the planktonic flora of such areas. Attached forms such as the diatoms *Synedra frauenfeldii*, *Striatella unipunctata*, *Licmophora* species, *Climacosphenia moniligera*, and species of *Amphora* and *Amphiprora* may be abundant; they are usually associated with species of *Navicula* and *Mastogloia*, particularly *M. apiculata* and *M. angulata*, as well as with species of *Diploneis* and *Nitzschia*. In the Caribbean Sea, where there are strong currents, many of these forms become detached from their *Sargassum* substrate and occur in the phytoplankton. Large blooms of *Rhizosolenia* and *Trichodesmium* species and other unicells have associated with them—especially during the stationary and death phases of the growth curve—increasing quantities of small flagellates and ciliates; there is evidence that these forms represent a food source for animals such as the herring *Clupea harengus* and various *Calanus* species which do not graze on the bloom algae themselves.

In the estuarine environment, the floating masses of *Thalassia* or *Zostera* species and their epiphyton are affected by the water conditions. However, while floating, these weeds do not themselves appear to influence the nature of the water appreciably. On the other hand, algal mats consisting of species of *Enteromorpha*, *Cladophora*, *Ectocarpus*, etc., often rot rapidly while floating and hence cause major changes in their aquatic environment. The *Thalassia–Zostera* substrate tends to continue to be aerobic in character and to disintegrate slowly so that the epiphyton continues to remain attached, to function normally and to be transported around the estuary with tides and currents. In cooler waters, most epiphytes belong to the Diatomaceae and represent such forms as *Licmophora* species, *Climacosphenia moniligera*, *Synedra frauenfeldii*, *S. ulna*, *Striatella unipunctata*, and species of *Rhabdonema*, *Amphora*, *Amphiprora*, *Achnanthes*, etc. In warmer waters, and in fresh water, the diatoms are frequently replaced by Myxophyceae; it would seem that this replacement is related to the presence of calcium carbonate in so many tropical habitats.

With the mats composed of filamentous algae, a considerable micro-algal flora is attached to, or entrapped in, the mat formed by intertwining filaments grown primarily on sea-grasses. These mats, as they increase in quantity, form a carpet overlaying the sea-grass, and contain a number of animals as well as the attached and entrapped unicells. During the day, photosynthesis in this environment causes the pH to rise up to 9·4; at this value, the bicarbonate ion reaches a minimum and photosynthesis ceases. At night, however, the pH may drop to less than 7, and the redox potential to negative values. These fluctuations tend to kill some of the animals and to limit the growth of some plants; the subsequent decomposition, in turn, renders permanent the low redox potential, thus limiting the growth of unicellular plants. Finally, the macro-algal mats float off the sea-grass, break up and sink to the bottom, particularly in boat channels and natural deep holes. Here, further decomposition occurs and all but the most resistant plant and animal forms are killed. This condition appears to cause sporulation of a number of plant species; the spores sink to the bottom and await favourable conditions to germinate.

(b) Relations to Solid Substrates and Sediments

In solid substrates and sediments, derived primarily from inorganic sources, such as rocks, sands, clays and silts, the primary unicellular plant coverage (epipelic and epilithic organisms) is controlled by geological and physiographic characteristics and developed by microbial activity. Enlarging upon this, decomposition of rocks plus the entrapment of silt provides an ideal substrate for a sulphuretum in the sense of BAAS BECKING (1925) and BAAS BECKING and WOOD (1953), resulting in a thin felt of *Desulfovibrio* species below a layer of purple and green sulphur bacteria which, in turn, is usually overlain by a film, and later a felt, of micro-algae which can tolerate low (negative) redox potentials. Little work has been done with regard to the limits of redox potentials of algal species but BAAS BECKING and WOOD (1955) have recorded that species of the myxophyte *Lyngbya* can tolerate and grow at redox potentials as low as – 170 mV, and that diatoms can grow at – 70 mV and upwards. These findings fit well with the microscopic picture of such felts, in which the main component is a tangle of Myxophyceae accompanied by naviculoid diatoms. In tropical limestone areas such as coral reefs, the nigger-head corals owe their dark colour to a similar felt of Myxophyceae, mainly nitrogen-fixing species of *Nostoc*.

Coral reefs comprise a special category of limestone environment in which the algal component is represented by the zooxanthellae – spherical cells with a dinoflagellate stage in their life history. These plant cells comprise about 5 times as much biomass as the animal part of the coral (DI SALVO, 1965) and thus become the major plant component of the coral-reef association and the major source of primary productivity. Associated with these corals and other reef animals are flagellates and ciliates, which act as scavengers and, no doubt, are themselves consumed by many animals. This coral ecosystem is highly productive and characterized by an extremely high energy turnover, especially as the whole of it lies within the photic zone. Since the surrounding water is normally clear, light intensities are high. Temperatures and salinities frequently reach high values owing to insolation and shallowness of the lagoons. The local organisms must, therefore, be adapted to high intensities of light, temperature and salinity; one or more of these environmental factors may attain limiting values. These areas are highly oxygenated, and one would not expect nitrogen to be a limiting factor owing to the frequent presence of nitrogen-fixing blue-green algae.

An interesting phenomenon in the intertidal zone is the migration of naviculoid diatoms, especially species of *Hantzschia*, in sandy sediments. As the tide recedes, the flats appear bare, but gradually turn brown, due to the vertical movement of masses of this diatom. As the tide returns, the diatoms migrate downward again leaving the flats bare once more. This phenomenon has been studied by ALEEM (1950) and CALLAME and DEBYSER (1954); it occurs at Barnstable, Sapelo Island, and Fire Island (USA), as well as in certain areas of other countries, and would seem to be partly based on a biological rhythm (as it can be changed by altering the light-dark periods) and partly on an external stimulus, possibly a degree of desiccation. I have seen the phenomenon only when associated with areas of strong sulphate reduction (by species of *Desulfovibrio*) and oxidation by purple sulphur bacteria of the family Thiorhodaceae. It would seem that, with the recession of the tide, the surface becomes more aerobic and that migration is stimulated by changes in the redox-potential gradient.

In the subtidal zone of coarse sediments, a number of minute diatoms may frequently be found attached to quartz grains and form a considerable and important biomass. The diatoms seem to be confined to oxidized sediments and may be affected by the nature of the surface to which they become attached, the oxygen tension of the environment, and possibly by light (though light is probably unnecessary for a number of benthic species of diatoms and flagellates). LEWIN (1963) and others have shown that many diatoms and other unicellular plants are facultative heterotrophs and can live in the dark indefinitely. Members of the euglenid genus *Euglena* lose their chlorophyll under certain conditions and transform into colourless flagellates (*Astasia*); other flagellate species exhibit the same property. Even under conditions of minimal light intensities, many unicellular plants are able to obtain energy by assimilating organic carbon against an energy gradient. Such reactions require much less light energy than the photosynthesis of carbon dioxide. Other algae, including some diatoms as well as species of *Scenedesmus*, and possibly also of *Oscillatoria* (FRENKEL and co-authors, 1949), are capable of photoreduction, i.e. using hydrogen sulphide instead of water as hydrogen donor for a bacterial type photosynthesis. This again requires much less energy per unit of carbon dioxide fixed than photosynthesis of the plant type.

OPPENHEIMER and WOOD (1963) recorded unicellular plants in sediments of Texas bays (USA), to a depth of 11 cm, and WOOD and MAYNARD (1970) found 12 species of diatoms at a depth of 36 cm in cores from a mangrove swamp in salt regimes of the U.S. Florida Everglades. These plants are no doubt heterotrophic; their activity allows recolonization of the epipelic plant association. In tidal areas, much of the water movement is vertical, hence such algae can be transported upwards or downwards without invoking the cell's motility.

Because the sulphur cycle is primarily bacterial it will not be discussed here. However, as BAAS BECKING and WOOD (1955), and BAAS BECKING and co-authors (1957) point out, it is in the areas of active sulphur turnover that maximum biological activity occurs in an estuary. Such activity is associated with large amounts of organic matter on the surface of the sediment providing the low redox potential necessary for bacterial sulphate reduction. The organic matter is usually supplied via decomposition of benthic and epiphytic algae and of sea-grasses, and the decomposition is facilitated by bacteria and heterotrophic (saprotrophic and phagotrophic) micro-algae and Protozoa. Unicells are of great importance in the plant biomass in many estuaries, but quantitative information regarding the relative importance of non-living organic detritus (true detritus) to associated micro-algae and Protozoa is sadly lacking. It appears from some preliminary results (ZIEMAN, 1968) that decomposition rates of *Thalassia* species are increased markedly by the presence of epiphytic unicellular plants; whether this is directly due to their activity or to that of associated heterotrophic bacteria and fungi is not known. The activity of the sulphuretum may be estimated by the over-all potential of the system, i.e. changes in the redox potential throughout. The redox difference may be 1 volt, representing a large amount of energy when distributed over a whole *Thalassia* flat (ODUM, verbal communication).

A rather special substrate is the undersurface of ice in the Antarctic and Arctic Oceans. The undersurface carries a felt of diatoms and other micro-algae up to 1 cm thick. Such felt can be found even under ice layers of up to 2 m thickness. Micro-algae inhabiting these undersurfaces require a minimum light intensity of 2 footcandles (21·52 lux) (BUNT, 1965); their light optimum lies between 10 and 25 footcandles (100·76 to 269·00 lux). This special plant association seems to be capable of photo-assimilating organic compounds, a potential which may account for their ability to live and reproduce at low light intensities and temperatures (see also Chapter 2.2).

Solid substrates and sediments are relatively unimportant in the oceanic environment, but are of considerable importance in neritic areas, and—according to BAAS BECKING and WOOD (1955)—represent a dominant factor in the estuarine environment.

(c) Relations to Factor Combinations

In the open oceans, temperature, salinity, pH, degree of oxidation, etc. are relatively constant over large areas, and the rate of change is low. Thus planktonic associations tend to change little in quality though they may do so quantitatively. Little is known of the distribution of attached unicellular plants in the open oceans. Zooxanthellae have been found in Radiolaria, apparently autochthonous, at depths to 4800 m in the Rhodos Deep, obviously in an aphotic environment.

In neritic situations, currents and winds often stir up the bottom, and benthic species are found in the water column attached to or free from their substrate. The appearance of benthic diatom species of the genera *Pleurosigma* and *Diploneis* in the phytoplankton is an indication of such turbulence. EMERY and co-authors (1955) point out that the sediments of shallow seas such as the California Basin (USA) are more reducing than the water just above (see also BAAS BECKING and WOOD, 1955), and for this reason, most of the nitrogen of the sediments occurs in the form of ammonia, while that of the free water is largely nitrate. Such differences will exert influence on the types of micro-algae in and on the sediments according to their ability to use the different forms of nitrogen available. Phosphate is rendered available from bottom waters and sediments by chelation with hydroxyl groups attached to organic molecules of the organisms and their excretions, and as a result of sulphate reduction and hydrogen sulphide release. This latter reaction, described by BAAS BECKING and MACKAY (1956), results in the release of phosphoric acid from ferric phosphate by hydrogen sulphide, to form hydrotroilite.

In extreme cases, the pH of a closed estuarine system will drop to 5·8; free H_2S and colloidal hydrotroilite will appear in the water and oxygen will disappear. This causes a rapid change in the unicellular plant association, and, usually a violent bloom which may be multi- or unispecific. In estuaries, changes in salinity alter the attached and benthic flora and consequently the substrates for unicellular plants. Such modifications seem more important than temperature changes, to which the micro-algae become accustomed in such environments.

(3) Structural Responses

Practically nothing is known about structural responses of unicellular plants—such as modifications in cell architecture—to different types of substrata.

The structure of plant communities, i.e. their species composition, is, of course, affected by the type of substratum in many ways. This statement is documented by numerous examples presented in the preceding pages of this subchapter.

7. SUBSTRATUM

7.2 PLANTS

7.22 MULTICELLULAR PLANTS

C. den Hartog

(1) Introduction

Substratum is often considered a factor of minor importance in the ecology of marine multicellular plants (Lewis, 1964). It is generally accepted that these plants take up their nutrients directly from the surrounding medium, and need the substratum only as a holdfast. But many algae, if torn from their substrata, can survive in a loose-lying state in a sheltered habitat. There are only three categories of plants for which the substratum is a source of at least part of the nutrients required: marine phanerogams (sea-grasses), boring algae, and parasitic algae.

In general, marine plants are indifferent to the chemical composition of their substratum. This indifference may be illustrated by the following examples.

Calcareous algae (e.g. species of the genera *Corallina*, *Amphiroa*, *Lithothamnion* and *Melobesia* among the Rhodophyta, and *Halimeda* and *Cymopolia* of the Chlorophyta) are common on limestone as well as on granite. The calcium carbonate required for the calcification of their cell walls is obtained directly from the surrounding sea water. Along the Mediterranean coast between Nice and Cape Dramont, the same algal communities occur on limestone, on blue porphyry and on red porphyry (den Hartog, 1959). Consequently, one cannot distinguish between calciphilous and calciphobous species among marine plants. Molinier (1955) recorded *Rissoella verruculosa*, a palaeo-endemic rhodophyte of the western Mediterranean, as being calciphobous. He failed, however, to demonstrate that Ca ions are harmful to *Rissoella*. His data can also be interpreted in another way, namely, that the physical structure of limestone is usually unsuitable for the establishment of *Rissoella*; it is, in fact, able to grow quite well on lime-containing substrata.

There are a number of exceptions to the rule that marine plants are indifferent to the chemical composition of the substratum. The marine phanerogams grow only in a substratum that contains some organic matter. According to Feldmann (1937), the chlorophyte *Caulerpa prolifera* prefers organic sediment bottoms. The same may be true of the 'rooting' Chlorophyta of lagoons and intertidal flats along the tropical coasts. These Chlorophyta, mainly belonging to the genera *Caulerpa*, *Penicillus*, *Avrainvillea* and *Halimeda*, usually occur together with sea-grasses (e.g. Børgesen, 1911). The dependence of these algae on organic matter from the substratum has, however, still to be proved.

The boring, rock-penetrating Cyanophyta (e.g. *Entophysalis deusta*), Chlorophyta (e.g. *Gomontia polyrhiza*) and Rhodophyta (*Conchocelis* stages of *Bangia* and *Porphyra*) are confined almost exclusively to limestone and other solid substrata, such as the shells of barnacles and molluscs or the carapaces of crustaceans.

Further, the permanent absence of epiphytes on some algal species must be ascribed to poisonous substances secreted by these algae (e.g. *Dictyopteris membranacea*). The chemical composition of the host is probably often an important factor in the obligate association of some epiphytes and their hosts (e.g. *Lithosiphon laminariae* on *Alaria esculenta*).

The relative unimportance of the chemical composition of the substratum to marine plants is in marked contrast to the essential significance of the physical properties of the substratum for these plants. The degree of dispersion of the substratum is a major factor which determines whether a certain species can occur on it or not. Substrata are usually divided roughly as being either solid (rock) or finely dispersed (sediment bottoms), but of course many degrees exist between these extremes. The substrata can be arranged according to the degree of dispersion in the following series: solid rock, rock blocks, boulders, cobbles, pebbles, gravel, coarse sand, fine sand, mud (see also Chapter 7.0).

Another physical factor of decisive importance is the texture of the substratum. The attachment of algae to a rough rock surface is much easier than to a smooth rock surface (Fig. 7-3). On smooth surfaces plant growth is usually limited to fissures and small irregularities of the rock surface. In the intertidal belt, where the water-retaining capacity of the substratum is largely dependent on its surface texture, the influence of the latter is very obvious.

Fig. 7-3: The texture of the substratum is an important factor for the settlement of algae. The north-facing side of the mole of Roscoff (Finistère, France) is constructed of concrete and stone blocks with a rather smooth polished surface. The concrete is densely overgrown by *Fucus vesiculosus*, while the stone blocks are devoid of larger algae but covered with a green film of microscopical Chlorophyta (perhaps also the lichen *Verrucaria mucosa*). Roscoff, August 1959. (Original.)

The hardness of the substratum also plays a role. In contrast to the hard basaltic and granitic rocks, the soft chalk and sandstone rocks are easily eroded by the sea. Where the coast is more or less exposed to wave action, these soft rocks are less suitable for algal growth and are mostly rather bare. The rock-penetrating species are restricted to the soft limestone formations.

On sandy bottoms, the packing of the sand may be of importance. According to BLOIS and co-authors (1961), the establishment of a new bed of *Zostera marina* takes place by the settling of the seedlings on the giant ripple marks; from there, growth expands vegetatively by lateral branches of the rhizomes into the depressions between the giant ripples.

Other physical properties of the substratum do not seem to be significant, except under extreme environmental conditions. Thus the colour of the substratum may be important in deep water at the lower limit of occurrence of multi-cellular plants. The black mud, containing hardly any lime, at 100 to 150 m depth near Banyuls (France) is devoid of any algal growth. Near the Balearic Islands, the mud at the same depth is white and rich in lime, and carries an open growth of *Laminaria rodriguezii*. According to FELDMANN (1937), the occurrence of this alga may possibly be ascribed to the reflection of the light on the white bottom, but not to higher lime content. In the intertidal belt, the colour of the substratum is of importance with regard to heat absorption; this may be one of the factors which makes black basalt stones a less suitable environment for algae than, for example, the white Vilvordian limestone. At the higher levels of the eulittoral, the basalt bears hardly any algal growth or has no vegetation at all, while at the same levels the Vilvordian limestone is covered by a rather dense vegetation.

The water-retaining capacity is a factor, unimportant in the sublittoral, but gaining in importance in the eulittoral belt, where emersion and submersion alternate, with the increase in time of emersion. On the porose Vilvordian limestone, which retains much water, the algae rise to a considerably higher level than on the hard, compact, rough granite. In its turn, granite appears to be a more suitable substratum for algal growth than basalt.

It is not always easy to demonstrate the effect of the substratum on the marine plants. More often than not, the effects of the substratum are indissolubly connected with the effects of other major environmental factors. Muddy bottoms, for example, are usually confined to sheltered localities where the water is very calm (weak currents; limited, superficial wave action), while sandy bottoms occur at more exposed stations, where the water is always agitated (strong, turbulent currents; strong wave action and surf). *Zostera marina* plants from sandy bottoms have considerably narrower leaves than plants of the same species from muddy bottoms. In this case it cannot be decided whether the richer substratum or the sheltered environment causes the more luxuriant growth of the species on a muddy bottom.

(2) Functional Responses

(a) Tolerance

It is a well-established fact that the vegetation of the rocky shore is very different from that of muddy and sandy bottoms. This difference has been accepted as a convenient measure for the delimitation of the various ecological shore formations. The difference in vegetation was ascribed to the difference in substratum, but no attempt was made to further analyze this relation.

The composition of the benthic marine vegetation on a certain substratum is, in fact, determined primarily by three major factors, namely, the degree of substratum dispersion, its stability and topographical factors.

Degree of substrate dispersion. Whether a marine plant can grow on a particular substratum or not is determined by the degree of dispersion of that substratum. Most algae occur attached to the substratum by means of basal cells, discs or haptera (in the case of the crustaceous algae, whole plants are closely attached to the substratum) and either do not, or only very superficially, penetrate into the substratum. These 'haptophytes' (LUTHER, 1949; DEN HARTOG and SEGAL, 1964), therefore, are restricted to bed rock, blocks, medium-sized stones and gravel. Sand grains and mud particles are too small for successful colonization. On the other hand, the 'rhizophytes' (by this term is understood the marine spermatophytes and the 'rooting' algae) penetrate with roots or rhizoids into the substratum to a considerable depth (often more than 30 cm). These plants are usually restricted to sandy and muddy bottoms, but sometimes occur also on gravel. On gravelly bottoms, mixed vegetations of haptophytes and rhizophytes occur; the first category is attached to solid substrata, the latter is fixed between little stones and sand grains.

Stability of substratum. The stability of the substratum is the other major factor which determines whether or not a substratum will be tolerated by plants. In sheltered places the epilithic (saxicolous) algae not only colonize the coarser substrata, but some of them may establish themselves even on a sandy bottom. Species of *Enteromorpha*, for example, are able to germinate on sand grains.

Sandy bottoms are usually devoid of haptophytes because wave action, tidal currents and storms cause a high degree of instability of the bottom. Under very special circumstances, however, an *Enteromorpha* vegetation can develop on sand flats. NIENBURG (1927) described such an *Enteromorpha* vegetation from the upper part of the intertidal belt of the Königshafen on Sylt (Germany), an extremely sheltered locality. On the sand flat of Vlieland (Dutch Waddenzee) an *Enteromorpha* vegetation has developed in an area where current velocity is very low and mostly coupled with increased sedimentation and rather large salinity fluctuations (DEN HARTOG, unpublished).

In places exposed to surf or strong currents, the algae are unable to live on gravel or pebbles. Associations of fast-growing algae, consisting mainly of *Porphyra*, *Enteromorpha* and *Ulothrix* species, however, develop on cobbles and boulders, which are usually stationary, becoming mobile only when water agitation is intensified during gales or by increase of current velocity. These fast-growing species, which generally form ephemeral pioneer communities in disturbed places, are more or less permanent in the unstable boulder habitat (DEN HARTOG, 1967). The basal parts of rocks and boulders bordering on sand are mostly devoid of algae because of the continuous abrasion by sand.

The rhizophytes inhabit sandy and muddy bottoms, which are naturally unstable substrata. Most rhizophytes are restricted to sheltered places where the bottom is least disturbed. Some species of the genera *Posidonia* and *Phyllospadix*, for example, seem to prefer more exposed substrata, where they stabilize the bottom by their dense growth. In beds of *Posidonia oceanica*, however, severe damage can be caused by pebbles, e.g. during gales. Continuous movements of the pebbles (gale conditions) cause deep, round, bare depressions in the bottom of the grass bed (MOLINIER and PICARD, 1951, 1952).

Some substrata are easily subjected to erosion, e.g. chalk. When sheltered against wave action, chalk is suitable for algal growth (ANAND, 1937); but when exposed, the vegetation is either very poor or non-existent. The water around an exposed chalk cliff is often milky, thus light conditions are unfavourable for algal growth in deeper water, although there is no disturbance by wave action.

Topographical factors. The topographical position of the substratum comprises a complex of factors independent of the physical and chemical properties of the substratum. This factor complex is of paramount importance to many plants. The slope (inclination) and the aspect of the substratum determine whether or not it

Fig. 7-4: Big rocks serve as solid substratum in the intertidal belt of Ile de Beclem (Baie de Morlaix, Finistère, France). Plant growth is affected by topographical factors. On the rather horizontal, wave-washed surfaces, dense vegetations of Fucaceae occur, on top of the block in the form of *Pelvetia canaliculata* and *Fucus spiralis* belts, on the flat surface at the base of the block in the form of a mixed population of *Fucus vesiculosus* and *Ascophyllum nodosum*. The vertical surfaces of the blocks, which are exposed to wave shock, are covered by an association of the barnacle *Chthamalus stellatus* and the gastropod *Patella vulgata*. Dense, patchy growth of the lichen *Lichina pygmaea* is superposed on this. Ile de Beclem, September 1959. (Original.)

will receive direct sunlight. Further, the slope of the substratum plays an important part in relation to the tolerance of algae to wave action. On vertical substrata, the vegetation is beaten by waves and has to withstand wave shocks. On horizontal substrata, the vegetation is only 'washed'. These differences are clearly expressed in the species composition of the vegetation. In the intertidal belt along the French coast of the Channel, vertical rocks are usually covered by a dense growth of the barnacle *Chthamalus stellatus* and patches of the lichen *Lichina pygmaea*, while the horizontal rock surfaces are covered by a dense vegetation of Fucaceae (Fig. 7-4). All kinds of transitions exist in nature between these two extremes. Many species are probably restricted in their occurrence by the inclination of the substratum. Thus *Pelvetia canaliculata* does not occur on vertical rock surfaces (DONZE, personal communication).

It has already been pointed out that the chemical composition of the substratum is generally of minor importance to marine plants. Antifouling paints stop the establishment of algae only for a restricted period of time; as the poisonous substance is gradually washed out by the sea water, algal growth begins with filiform colonies of diatoms, succeeded by tubular, filiform, and membranous algae of the genera *Enteromorpha*, *Ulothrix*, *Ulva*, *Petalonia* and *Scytosiphon*.

(b) *Metabolism and Activity*

Metabolism

The effects of the substratum on the metabolism of the algae are insignificant; with the exception of parasites, lime-boring algae and, perhaps, 'rooting' Chlorophyta, they do not take up nutrients from their substratum. According to FRÉMY (1945), lime-boring endolithic algae are able to decompose the calcium carbonate of the substratum and to use the thus liberated carbon dioxide for photosynthesis. In marine spermatophytes, a richer bottom seems to result in a more vigorous growth, but exact analysis of this relation is still lacking.

Activity

It is very probable that marine plants exercise much more influence on their substratum than is generally supposed. The current opinion that the algal cover protects the shore against the violent action of surf and storm waves certainly needs reconsideration.

Algal growth on exposed rocks is usually rather scattered, and hence cannot offer much protection. Further, algae, in particular the larger ones, are often found torn from the rock by wave action together with a piece of the substratum. Thus, instead of protecting the substratum, these algae seem to function more as a point of impact for the destructive and erosive forces of the waves.

In sheltered places, where the substratum is completely covered by algae, the vegetation has a moderating effect on the erosion process. The vegetation forms, in fact, a buffer between the turbulent sea water and the substratum. It reduces the force of the waves before they reach the rock surface. Further, it diminishes the fluctuations of temperature of the substratum, and in the intertidal belt it prevents or delays the desiccation of the rock surface.

The erosion of rock in the intertidal belt can be ascribed mainly to the continuous alternation of emersion and submersion, fluctuations in temperature, and mechanical force of the water. When the rock is covered by a closed vegetation, e.g. a belt of Fucaceae, erosion is greatly reduced. When the vegetation is patchy, however, unequal desiccation of bare and algae-covered rock parts may cause heavy tensions in the substratum which, in the course of time, must be deleterious to the rock. In the Netherlands, the effect of patchy algal growth is demonstrated clearly on the dikes, the slopes of which are covered by a thick layer of bitumen. The tensions caused by patches of small algae such as *Blidingia minima*, *Ulothrix flacca*, *Porphyra purpurea* and *P. umbilicalis*, particularly in the upper part of the intertidal belt, have resulted everywhere in the formation of large and deep cracks in this soft substratum (DEN HARTOG, unpublished).

There are several algal species with sediment-fixing capacities. On flat or slightly sloping rocky surfaces, *Rhodochorton floridulum*, *Sphacelaria fusca*, *Cladostephus spongiosus*, *Halopteris scoparia*, *Jania rubens*, *Corallina officinalis*, etc., function as sand binders. These species are characterized by dense cushion-like growth. Spaces between the algal filaments are mostly filled with fine sand grains. Among the above-mentioned species, *Rhodochorton floridulum* is probably the only obligate sand-binding haptophyte (see also HOMMERIL and RIOULT, 1962); the other species also occur as epiphytes. These sand-fixing algae exist as an undergrowth in communities of larger algae, e.g. under *Ascophyllum nodosum* or *Fucus serratus* along the European west coast; they can also form quite independent communities without being sheltered by larger species, e.g. the association of *Corallina mediterranea* and the association of *Padina pavonia* and *Cladostephus verticillatus* along the Mediterranean coast of France (FELDMANN, 1937). The thickness of the fixed sand layer often exceeds 5 cm, but then the environment becomes less suitable for the algae. It is not known whether the algae die off and the sand is washed away at this stage so that the cycle has to begin anew, or whether a status quo is reached, whereby the algae survive, but further accumulation of sand does not take place. In the Mediterranean, these sediments, fixed by algae, are a very suitable substratum for the establishment of sea-grasses. According to MOLINIER and PICARD (1951, 1952), the sand-binding *Jania rubens* vegetations are often succeeded by the establishment of a bed of *Posidonia oceanica*. Sometimes *Posidonia* is preceded by the smaller, but less stenobiontic *Cymodocea nodosa*.

On sandy and muddy substrata, sediment-fixing plants play an important part in the stabilization and raising of the bottom. On sheltered sand flats, especially in the upper part of the intertidal belt, the bottom is often covered with a thin film of algae, in which diatoms and Cyanophyta (e.g. *Microcoleus chthonoplastes* and *Anabaena torulosa*) are dominant. Near the high-water mark, species of *Vaucheria* are sometimes found in this community. This association of small algae is able to grow through the thin layer of sediment deposited by each tide. The bottom-raising capacity of this association, however, must not be over-estimated, as it is easily destroyed during extreme weather conditions. A strong dry wind during the emersion period may strip the vegetation from the substratum and blow it away; the loosened papery sheaths of the algae are well known among the coastal inhabitants and have various local names. Increased wave action may also loosen the vegetation from its substratum. Disappearance of this association can be caused also by a form of self-destruction; air-bubble formation, as a result of excessive photosynthetic activities, combined with upward directed water movements (Chapter 5.0) suffice to detach the algal vegetation from the substratum.

On muddy bottoms, algal growth is very sparse, but near the high-water mark, extensive vegetations of *Vaucheria* species may occur, especially in somewhat brackish water. The attachment of *Vaucheria* in a muddy substratum is more secure than on sand, and consequently the vegetation is more resistant to the effect of unfavourable weather conditions. The sediment-fixing properties of the species show here to full advantage; the cushions grow out to humps, 20 to 30 cm in height, and not rarely into extensive carpets. In fact, *Vaucheria* is the most important sediment-fixing alga on the low salt marsh, and in this respect highly superior to the associated summer annual *Salicornia europaea*, so praised by phytocenologists. The *Vaucheria* humps form an excellent germination bed for other constructive salt-marsh plants, such as *Puccinellia maritima*. Apart from this initial stage, algae of the salt marshes are restricted to the undergrowth (*Rhizoclonium, Enteromorpha, Percursaria, Vaucheria, Oscillatoria, Lyngbya, Catenella, Bostrychia*, etc.), where they take part in the fixation of the sediments caught by the phanerogams.

The sea-grasses stabilize and modify their substratum to a large extent. The stabilizing effect of the sea-grasses has to be ascribed to their dense growth and the formation of a 'mat' of rhizomes in the upper layer of the bottom. The degree of protection offered by the sea-grass meadows is extremely high. THOMAS and co-authors (1961) reported that after hurricane Donna, which afflicted the southern part of Florida (USA), the beds of *Thalassia testudinum* showed hardly any damage. STODDART (1963) recorded that after hurricane Hattie, which swept the coast of British Honduras, the *Thalassia* beds appeared rather undisturbed, although on the nearby reefs 80% of the corals had been destroyed. A similar situation has been reported by GLYNN and co-authors (1964), who studied the effects of hurricane Edith on the biocoenoses of the south coast of Puerto Rico. They observed, however, that the beds of *Syringodium filiforme* had been seriously affected by this storm.

Sea-grasses can establish themselves on muddy and on sandy bottoms; some of them are even able to colonize the thin layer of sediment on solid substrata bound by algae. In the course of time, these various substrata tend to change into soft mud as a result of the continuous deposition of fine material between the plants (MARGALEF and RIVERO, 1958). The amount of organic matter in the bottom is increased by the addition of remains of the grass and its accompanying flora and fauna. The remains of epiphytic calcareous algae such as *Melobesia* and shells of molluscs increase the amount of calcium carbonate in the substratum (HUMM, 1964). The raising of the bottom, as caused by sea-grass vegetations, depends on the species and on the environmental circumstances. MOLINIER and PICARD (1952) reported the existence of continuous deposits of *Posidonia oceanica* several metres thick. In contrast to this, the vegetation of *Zostera noltii* in the intertidal belt of western Europe does not result in the bottom being raised, as the elevations formed by patches of this species become denuded again when they rise more than 5 cm above their surroundings (DEN HARTOG, 1970).

It has been reported several times that the calcified chlorophyte *Halimeda opuntia* var. *triloba* plays an important part in raising the bottom of some West

Indian lagoons. PHILLIPS (1959) found on the Marquesas Keys that, to a considerable depth, the bottom consists of dislodged segments of this alga. ALMODÓVAR (1962) mentioned a locality along the south coast of Puerto Rico where the substratum consists entirely of *Halimeda* fragments.

(c) Reproduction

Effects of the substratum on the reproduction of multicellular marine plants have not been recorded. It is, however, striking that, under natural circumstances, many loose-lying algae do not reproduce sexually, although their vegetative reproduction may be greatly increased. In enriched culture solutions, detached algae often form sexual organs.

(d) Distribution

Pattern of distribution

The effects of the substratum on the general distribution of marine plants are, in fact, very obvious. The sandy coast of the southern North Sea, for example, could not be colonized by algae until rocky substratum was provided in the form of harbour dams, dikes and breakwaters.

The influence of the physical structure of the substratum on the pattern of distribution within a certain area is often very evident under extreme conditions. In the upper part of the intertidal belt, this factor is often decisive for the height to which the various algal zones will rise with respect to a fixed hydrographical level. In the southern part of the Netherlands, it was noticed that, under similar environmental conditions, the various algal zones were situated at a higher level on Vilvordian limestone than on granite, although on both substrata the zonation was intact and the various belts contained the same species. It appeared that the lower border of the *Pelvetia canaliculata* belt on Vilvordian limestone was at a slightly higher level than the upper limit of the corresponding belt on granite. On granite, the *Pelvetia* and *Fucus spiralis* belts appeared to be distinctly narrower than on limestone. On the latter substratum, the upper limit of the *Ascophyllum nodosum* belt was higher than on granite. On basalt the upper limit of the *Ascophyllum* belt was considerably lower than on granite, and the belts of *Pelvetia canaliculata* and *Fucus spiralis* were usually absent (Fig. 7-5). On slopes facing north, these two belts may occur on basalt but then in a very impoverished form and at lower levels than they would have occupied on granite (DEN HARTOG, 1959; and later, unpublished observations). BOALCH (1957) has given another example. Along the south coast of Devon, England, he found *Ulva lactuca* as a dominant species on the reefs of the polychaete *Sabellaria alveolata*. This substratum appeared to be unsuitable for the development of fucoid algae.

Epiphytism and epizoism

Since marine algae are rather indifferent with respect to the chemical properties of their substratum, it is by no means surprising that epiphytism and epizoism are common features among this group. The relation between host plants and epiphytes depends on many factors. The length of life of the host plants must be

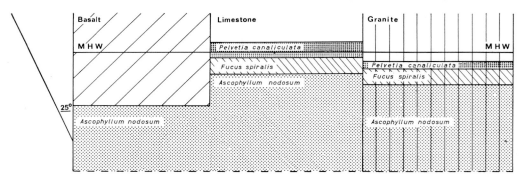

Fig. 7-5: The east-facing slope (25°) of the dike near Strijenham (Island of Tholen, The Netherlands) is constructed of basalt, limestone and granite. The basalt is bare and extends downward to about mean sea level (half tide line), where it is replaced by limestone covered by
Ascophyllum nodosum. On the granite, the algal vegetation extends higher, and belts of *Fucus
spiralis* and *Pelvetia canaliculata* are present. On the limestone, the algal belts occur at
significantly higher levels than on the granite boulders. Such vertical shifts in vegetation
levels are a common feature where substrata of different physical nature border on each
other. MHW: mean high-water mark. (Original.)

sufficient to enable the epiphytes to complete their life cycle. For this reason, epiphytes are usually absent or rare on annual and ephemeral algae. Epiphytes are
much more numerous on perennial plants, and their dense growth may even
'choke' their host. Another factor of vital importance is the size of the attachment
organs of the epiphytes, as this determines whether a certain epiphyte can grow
on a specified host or not. Species with small attachment organs can settle on a
great variety of algae. In contrast, species which develop a large attachment
organ, e.g. a disc or hapteron, are excluded from the small-sized filamentous and
dendritic algae. They may germinate on such hosts, but will die long before
reaching maturity. This is no doubt the cause of the fact that species of *Ectocarpus*,
Ulothrix and *Callithamnion* with their fragile filamentous structure are generally
devoid of epiphytes (diatoms excepted). In contrast, the larger algae offer suitable
substrata for epiphytes of various sizes, and may look like 'gardens'. The texture
of the surface of the algae is also of importance. *Laminaria digitata* and *L. hyperborea* have a very smooth thallus which is normally free of algae except where it
has been damaged by being repeatedly beaten against the rock. The first-mentioned
species also has a smooth stipe and is nearly always bare, but the stipe of *L.
hyperborea* has a very rough, almost tuberculate surface, which is always densely
set with Rhodophyta.

Other reasons for the lack of epiphytes on some plants can be the secretion of
poisonous substances by the potential host (e.g. *Dictyopteris membranacea*). Some
algae, e.g. *Enteromorpha*, *Ulva* and the heliophilous species of *Cladophora*, show a
high metabolic activity which causes considerable fluctuations of the pH in the
immediate surroundings of these algae. These unstable pH conditions prevent the
settlement of most epiphytes, as only very hardy species are able to tolerate these
conditions (LAMI, 1934a, b; FELDMANN, 1937). Further, some species may be
excluded as epiphytes by their growth rhythm, i.e. if their period of propagation
does not coincide with the period of vegetative growth of the potential host.

Taken together, these factors may explain to some extent why one species is abundantly covered with epiphytes and another devoid of them. They do not explain the existence of many taxa with an obligate epiphytic way of life, e.g. *Elachista*, *Myrionema*, *Lithosiphon*, *Erythrocladia*, *Acrochaetium* and *Melobesia*, nor do they give any indication as to the cause of the permanent associations of a certain epiphyte with a specified host (e.g. *Polysiphonia lanosa* on *Ascophyllum nodosum*). Although in a number of cases these epiphytes penetrate by means of rhizoids into the tissue of the host, this does not prove that they are parasites; it may be just a special mode of attachment.

On larger algae, epiphytes are sometimes differentiated into two or more communities, inhabiting different parts of the host. Such a differentiation is rarely a substratum effect (*Laminaria hyperborea*), but is mostly caused by the different light conditions to which the various parts of the host are exposed.

In general, the same factors which have been mentioned with regard to epiphytic algae apply to the epizoic algae. Epizoic algae occur on sedentary animals as well as on mobile ones, e.g. on tunicates, molluscs, echinoderms, various crustaceans and even on fishes. ALEEM (1952) reported the occurrence of *Erythrotrichia carnea* and some diatom species on the sea-horse *Hippocampus guttulatus*, and ROSEN-VINGE and LUND (1947) found *Giraudia sphacelarioides* on the ventral fins of the stickleback *Gasterosteus aculeatus*. Obligate epizoism seems to be rare. *Tellamia contorta* is restricted in western Europe to the periostracum of *Littorina obtusata*; in the Mediterranean, it occurs on *Pisania maculosa* (FELDMANN, 1937). *Rhodochorton membranaceum* is restricted to chitin-containing substrata, such as the coenosarcs of Hydrozoa and Bryozoa. The problems of endophytism and endozoism (zooxanthellae, zoochlorellae) will not be discussed here, as mainly unicellular algae are involved (Chapter 7.21; see also Chapter 7.3).

(3) Structural Responses

(a) Size

Although, in general, the size of the plants is not affected directly by the substratum, a certain relation seems to exist. When algae attached to substrata, such as sand grains, gravel or shells, grow too large, they and their substrata are lifted up by water movement (see Chapter 5). They drift away at the mercy of the currents and usually end their journey washed up on a beach. Such algae, which have become 'too big for their boots' can survive only under very special hydrological conditions, e.g. in the western part of the Dutch Waddenzee where *Gracilaria verrucosa*, *Chaetomorpha linum*, *Chorda filum* and various species of the genera *Ceramium*, *Polysiphonia*, *Cladophora* and *Enteromorpha* together form a drifting algal community (DEN HARTOG, 1958).

An epilithic alga with its size adapted harmoniously to a sand-grain substratum is *Cladophora pygmaea*, which occurs on undisturbed bottoms in the Kattegat and the southwestern part of the Baltic Sea (WAERN, 1940).

(b) External Structures

The effect of the substratum on the shape of plants, becomes evident when the plants are detached from their substrata. Usually, they no longer show a vertical axis (perpendicular to the substratum as a result of negative geotropy), but exhibit a globular growth with a definite centre (e.g. species of *Bryopsis*, *Cladophora*, *Sphacelaria*, *Ectocarpus*, *Chondrus*).

Fucus vesiculosus shows considerable structural variability both in the attached and detached states, depending on the habitat. In sufficiently wet places on salt marshes, it can persist in the undergrowth of phanerogams, as a form known as '*Fucus lutarius*'. This form arises from individuals of *F. vesiculosus* washed up on the salt marsh. A stranded plant develops a dense mass of proliferations along the margin of its thallus. These proliferations become free when the original part of the thallus decays; they grow out into narrow, strongly contorted thalli, which continue to grow at the apices, but which rot away at their base. New individuals are formed by 'dichotomic splitting' when the decay reaches a ramification of the thallus. In this way, very dense carpets may be formed (DEN HARTOG, 1959).

Fucus vesiculosus, stranded on a mussel bank and attached to it by the byssus of the mussels, undergoes a transformation to '*F. mytili*'. Under the influence of the environment, they become remarkably uniform in habit; the thallus becomes brittle and often contorted, the air bladders disappear, and the colour becomes lighter. The plants grow only at the top and rot away at the base; multiplication takes place by 'dichotomic splitting' just as in '*F. lutarius*'; the thalli, however, remain considerably wider than those of '*F. lutarius*' and do not proliferate (NIENBURG, 1925, 1927, 1932; DEN HARTOG, 1958, 1959). Although there are environmental factors other than the substratum which play a part in the transformation of detached *F. vesiculosus* plants, the influence of the substratum must be of major importance. On stones situated alongside a mussel-bed, perfectly normal *F. vesiculosus* plants have been found, while between the mussels in the immediate vicinity, '*F. mytili*' occurred (DEN HARTOG, unpublished).

Several Cyanophyta and Chrysophyta exhibit enormous morphological variation, which depends mainly on external influences, including the substratum. On limestone rocks along the Adriatic coast, ERCEGOVIĆ (1932a, b) discovered a zonation of several associations of Cyanophyta. He distinguished several new species and genera, which appear to be quite faithful to the various associations, but which nevertheless, for the greater part, seem to represent growth forms of the extremely plastic *Entophysalis deusta* (PRUD'HOMME VAN REINE and VAN DEN HOEK, 1966).

Among the Chrysophyta on the British chalk cliffs, ANAND (1937) recognized various very distinct genera with several species, each occurring in its special niche. From culture experiments by PARKE (1961), it appears that several of these taxa are probably nothing more than growth forms of other taxa. Thus the non-motile phase of *Syracosphaera carteri* may develop, depending on environmental circumstances, into a *Chrysotila* stage, an *Apistonema* stage, a *Gloeochrysis* (= *Ruttnera*) stage, a *Chrysonema* stage or a *Chrysosphaera* stage; some of these stages, obtained from laboratory cultures, appear to be identical with taxa described from natural habitats.

The means of attachment developed by the various multicellular plants show a high degree of diversity. Among the haptophytes, many species are simply fixed by a basal cell, which may be enlarged or otherwise modified. In some species, other cells also may develop rhizoids so that the plants become fixed at several places (e.g. *Acrosiphonia centralis, Rhizoclonium riparium*). The discs of *Fucus* and the haptera of *Laminaria* and *Cystoclonium* are much more evolved organs. The crustaceous algae are attached to the substratum by their entire thallus. Stolonization occurs among the haptophytes, but is a much more common feature among the 'rooting' algae (e.g. *Caulerpa*). The sea-grasses are anchored in the substratum by their rhizomes which develop roots at the nodes. Some sand-inhabiting algal species are plainly stuck into the substratum (e.g. *Halimeda opuntia*), but in other species a well-developed anchorage system occurs. In *Penicillus*, for example, the basal stalk emits, below the surface of the substratum, many rhizoids which penetrate deep into the bottom; this anchorage system simulates to some extent a tap-root of a phanerogam (Børgesen, 1911).

In two genera of sea-grasses, the diaspores have developed special structures which give them an effective grip on the substratum. The Australian genus *Amphibolis* is viviparous; when the young, about 8 cm high, plants become detached from the mother plant, they do not have a root system but, at their bases, there is a 'grappling apparatus' consisting of four pectinate lobes of a pericarpic nature, which seems very useful for anchoring the seedlings in the bottom (Tepper, 1882a, b; Ostenfeld, 1916). In *Phyllospadix*, a genus which inhabits wave-beaten rocky shores along the coasts of the northern Pacific, the fruits are crescent shaped. When detached, the exocarp rots away and the hard bristly endocarp of the two processes of the crescent is laid bare. The fruits become easily entangled between the dense growth of coralline algae by means of these stiff bristles. Germination takes place, and the young plants soon become firmly attached to the rocky substratum by their root system and begin to crowd out the original algal growth (Gibbs, 1902).

(c) *Internal Structures*

Practically nothing is known on possible effects of different substrata on internal structures such as cells and cellular components of multicellular plants.

7. SUBSTRATUM

7.3 ANIMALS

M. C. Bacescu

(1) Introduction

Substratum is one of the most important environmental factors in oceans and coastal waters: it largely determines distributions and associations of bottom-living (benthic) animals. As is the case in regard to other environmental factors, responses to the substratum cannot be viewed completely independently of the responses to other factors. Without good aeration (currents, waves), rich detritus or light, there would be no macrophytes nor phytobenthos—the main autotrophic link between free water and bottom in the food chain. Even where clear water allows photosynthesis also in greater depths, algae attached to solid substrates descend only to 200 m (e.g. in the Mediterranean and in tropical seas), except in areas where the bottom is lined with coral reefs (STRICKLAND, 1958). For mobile shore environments, grain size of sediments and depths to which light can penetrate into the substrate are of great importance (GOMOIU, 1967).

Rocky, sandy or muddy substrates greatly affect benthic animal life leading to coenoses with specific morphological and physiological adaptations, while other environmental components determine the density of populations and the variety of species present. Substrate components, living (bacteria, phytobenthos) or dead (organic remains), may constitute a rich food source and provide building material. Many animals feed on algae or on algal secretions, others are predators or feed on a mixed diet; however, the majority of benthic animals are detritus feeders. In detritus-feeding inhabitants of bottom areas strongly swept by waves or currents, mechanisms have evolved for optimum food-catching efficiency (prehensile extremities of crustaceans such as cirripedes and *Emerita*; strong aspiration organs (siphons) in the cumaceans and molluscs *Cardium, Venus*; etc.).

Variety and abundance of the bottom fauna greatly depend on the physical and chemical structure of the substratum. The existence of the mesopsammic fauna depends on size and arrangement of grains and interstitial spaces. Sedimentation of fine mud prevents interstitial spaces from becoming populated, while the occurrence of a thick layer of detritus supports development of the fauna.

In comparison to the pelagic environment, the benthic environment is considerably more differentiated, due to the great variety of substrata. In regard to animal life, a brief examination of the major types of substrates is necessary.

Air—water interface. The air—water interface is a rather special substrate: a thin surface film extending homogeneously over vast ocean areas. Animals may live both on its surface (hyperneuston) or on its subsurface (hyponeuston). Its biocoenosis has recently been studied by ZAITSEV (1961). Among the adaptations to life on the air—water interface substrate are floating cushions on the legs of some insects (*Halobates*, *Clunio*), extensive gliding surfaces and air sacks for floating (*Janthina*), saltatorial abilities (*Pontella mediterranea*, *Anomalocera*, etc.), as well as gas-filled (*Velella*) or oil-filled (eggs of many species) bladders (see also Chapter 7.0).

Floating Substrates. The most important floating substratum is formed by the immense agglomerations of Sargasso Sea algae (mainly three *Sargassum* species); it is characterized by faunal elements specific to it or derived from the benthos. The first category includes the molluscs *Spurilla sargassicola*, several *Lepas* species, Pycnogonida, Polychaeta, etc., and *Plumularia sargassia*, *Membranipora tehuelcha* and *Clytia* species making up 60 to 90% of the bryozoan and hydrozoan fauna. The second category includes the fishes *Antennarius marmoratus* and *Phyllopteryx eques*. Most of these animals affix themselves to the floating substrate just as closely as do their counterparts in the phytal or on littoral rocks; they have developed special tegument expansions (*Phyllopteryx*). Other animals creep (gastropods) or swim (*Neptunus*, shrimps) freely among the algae. Most are suspension feeders, consuming the floating seston (FRIEDRICH, 1965), and all exhibit remarkable specific adaptations in shape and colouration.

In the open sea, even floating feathers of oceanic birds (cormorans, albatross, *Oceanodroma*, *Larus modestus*, *Oceanites oceanus*, and others), negligible as they may appear, are covered with small cirripedes, amphipods, etc. Other typical floating substrates are buoys, wrecks, fishes (*Mola*), and whales; the body surface of the latter is often used as substratum by the barnacle *Coronula*.

Solid Substrates. They accommodate a variety of animals, affixed either by hooking (hydropolyps) or sucking devices (molluscs, fishes). *Patella* species and chitons have powerful sucking disks; the latter are able to attach themselves very closely even to the most slippery rocks. *Patella*, which 'grazes' the bacterial and algal epibiosis of the stones, cannot be removed unless pulled by a force exceeding 10 kg/cm² (WILBUR and YONGE, 1966). In *Trochus*, chitons and *Littorina*, adhering to the solid substrate is associated with hiding in rock crevices or under stones.

Even firmly attached forms move about in search of food, often eventually returning to their individual sites of attachment (e.g. *Patella*). In the littoral, many attached molluscs feed on suspended or deposit material; others, like *Petricola* and *Pholas*, bore into the solid substrate; perforation of solid substrates, such as rock, concrete, or wood, is usually preceded by acid secretion which softens the substrate's surface. Direct observation by diving helps to determine the relations between animals and their substrates (RIEDL, 1967). In fact, scientific diving represents the only precise method for the study of the rocky-substrate fauna. It has uncovered a special fauna and helped to analyze animal associations in submarine caves (Fig. 7-6). Typical of the rocky, wave-sprayed supralittoral, temporarily uncovered during low tide, are high numbers of attached animals belonging to a few species rigorously selected from eurybiont forms; an example is the 'pavement' zone of *Mytilus*, *Brachyodontes*, *Modiolus purpuratus*, etc. on European and American shores.

Living organisms as substrates. A special substrate, especially on littoral rocky bottoms, is provided by macrophytic algae, such as *Fucus*, *Ulva*, *Cystoseira*,

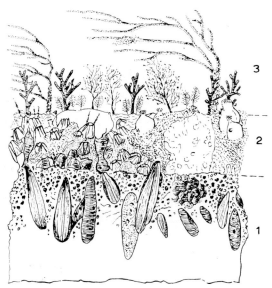

Fig. 7-6: Animals on solid substrates of marine
caves. Petricolous endo- and epifauna on the
walls of a submarine grotto. 1: rock with
species of *Lithodomus* and other borers;
2: *Balanus*, ascidians and sponges attached to
the rock surface; 3: algae, hydropolyps,
Aglaophenia and others. (After RIEDL, 1967;
modified.)

Phyllophora, Laminaria, Macrocystis, to mention only a few. Their thallus
is often crusty and covered by calcareous algae, Bryozoa, Serpulidae and tubes
of other worms, Balanidae, Spongia, or hydropolyps (*Aglaophenia, Cordylophora*).
Attached to the branches of the algae are Lophobranchia, *Palaemon* and Lucer-
nariidae. While such benthic animals are restricted to the substrates mentioned,
they themselves serve as substrate for other animals.

Next to the algae, the most important organisms serving as substrates are
corals and Spongia. Dozens of other lime-secreting animals (polychaetes, molluscs,
bryozoans) contribute to the building of coral reefs. The world of the coral reefs is
incredibly varied, displaying a wealth of forms and colours enchanting to the
diver's eye. Algae are present here in the form of monocells (zooxanthellae)
sheltered in the colourful mantles of *Tridacna*, in the tegument of the Opistho-
branchia and in the tissues of many coral reef animals (OTTER, 1937). In addition
to being permanent oxygen producers within animal cells exposed to light, they
represent also a direct intracellular food source. The madrepores shelter an
abundant association of filamentous algae with a biomass sometimes 10 times larger
than that of the coralobionts (ODUM and ODUM, 1955). Large Spongia provide
ideal biotopes, at all depths, for innumerable animal species. In a tubicole sponge
having a section surface of 700 cm², the author was able to count, in tropical waters
off Cuba, more than 400 different animals (especially Ophiurida, Amphipoda,
Isopoda, Alpheidae, Mysidacea) representing alone a biomass of 600 g/m².

c

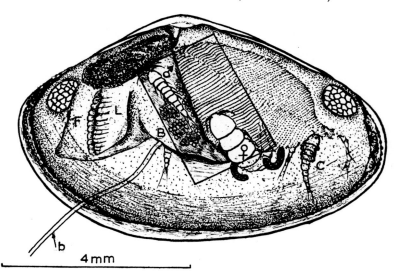

Fig. 7-7: The interior of the bivalve *Corbula mediterranea maeotica*
serves as substrate for the copepod *Leptinogaster histrio*. Left valve
and part of gills (square area) removed; F: foot; L: oral lobe; B:
byssogen pouch; b: byssus thread of the bivalve. Adult male (♂)
and female (♀) of *Leptinogaster histrio* attach themselves to their
living substrate; so do their copepodites (C). (After BACESCU and
co-authors, 1957; modified.)

The interior of many bivalves forms the substratum for numerous commensal
species. Thus 3 to 10 individuals of *Leptinogaster* sometimes occur in a single
Corbula (Fig. 7-7), 3 to 5 *L. pholadis* may live in a single *Pholas candida* and many
Mytilicola in *Mytilus* or *Modiolus*, etc.

A special algal substrate, piled upon the muddy bottoms of the northwest part
of the Black Sea, forms the so-called Zernov's Field of red algae (three *Phyllophora*
species). This field extends to depths of 30 to 50 m and covers an area of more than
10,000 km², surrounded by a circular current. It harbours faunal elements from the
nearby littoral fauna. Some species are common on the bottoms of shallower areas
(e.g. *Macropipus arcuatus, Synisoma capito, Gobius batrachocephalus*) and occur here
in great individual numbers; others appear to occur primarily on the algal beds
(e.g. species of *Pomatoschistus* and *Lepadogaster*, as well as *Ctenolabrus rupestris,
Paradactylopodia brevirostris, Apherusa*). All have in common the beautiful red
colour of their algal substrate (BACESCU and co-authors, 1971).

Another vast algal substrate, forming immense meadows of huge *Macrocystis* along
the isles of the South Atlantic, is characterized by many epibionts but a poor fauna.

The *Corophium acherusicum* and *C. bonelli* cushions forming characteristic
microbiocoenoses in the Black Sea do not exceed, as a rule, 1 cm in thickness
(Fig. 7-8). However, the relict Ponto-Caspian species of *Corophium curvispinum*
and *C. maeoticum* build true argile 'reefs', 4 to 10 cm thick. In the rocky depths of
the Danube River (relict marine biotopes at 20 to 70 m exposed to intense fluvial
currents), live at least 20 species, originally marine but now more or less adapted
to the freshwater environment, as epi- or endobionts (four amphipod species, in

Fig. 7-8: Tubiculous animals inhabit the mud-filled microcavities of the rocky substrate in the Black Sea. B: *Balanus*; C: *Corophium acherusicum*; D: colonial diatom; E: *Erichtonius*; F: *Fabricia sabella*; M: Miliollidae; P: *Polydora ciliata*; P': side view of its tube; T: *Tanais cavolin*. (After BACESCU and co-authors, 1963; redrawn.)

addition to *Corophium*; three polychaetes: *Manayunkia, Hypania, Hypaniola*; *Palaeodendrocoelum*; *Caspihalacarus*; *Jaera sarsi*; etc.; BACESCU, 1948).

Sediments. Mobile sandy sediments descend to 20 m (Black Sea), 100 to 200 m (other seas) and 500 m (Mauretanian coast). They consist of littoral and terrestrial components. Sand is largely made up of mineral (quartz) and organic (fragments of mollusc shells, corals, foraminiferans, ostracods, etc.) particles. Sandy beaches are well oxygenated and continuously remodelled by water movement; their fauna exhibits perhaps the greatest number of structural adaptations: body shapes are elongate, flattened towards the head or lance-shaped (*Branchiostoma, Onus, Crangon, Gastrosaccus, Eurydice*). Animals closely associated with sandy substrates are called 'psammobionts'. The dimensions of the spaces between the sand grains largely determine body size (usually 0·1 to 0·5 mm) and shape (mostly filiform) of the interstitial microfauna (Fig. 7-9; Fig. 7-11).

Fewer animals live on the surface of the sand (epipsammon) than inside it (endopsammon; mesopsammon). This quantitative relation is in contrast to that found in rocky or muddy substrates.

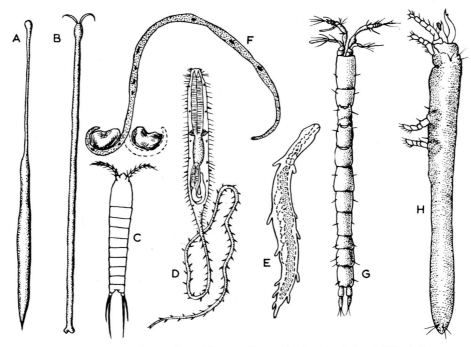

Fig. 7-9: Animals living in sandy sediments (interstitial animals) exhibit striking convergencies in body shape in spite of belonging to different taxonomical groups. A: *Trachelocerca* (Ciliata); B: *Protodrilus chaetifer* (Archiannelida); C: *Ameira parvula* (Harpacticoida); D: *Urodasys viviparus* (Gastrotricha); E: *Pseudovermis salamandrops* (Nudobranchia); F: *Marenda nematoides* (Foraminifera, in motion); G: *Microcerberus delamarei* (Isopoda); H: *Nematalycus nematoides* (Acarida). (C after BACESCU and co-authors, 1957; D after AX, 1966; the remainder after DELAMARE-DE-BOUTEVILLE, 1960.)

Endopsammic animals are provided with organs adapted to producing water currents: gills in the polychaete *Arenicola* and pleopods in the Thalassinidae (both live in U-shaped tubes in muddy-sandy substrates); tentacular fringes in some polychaetes, rotifers and in the Vorticellidae; cirriform extremities in cirripedes; ciliations in the Cladocera, Bivalvia, Brachiopoda and in *Branchiostoma* as well as in Spongia, Tunicata and in *Balanoglossus*. Sand-living molluscs maintain respiratory and nutritive water currents through long siphons sent to the substrate's surface and provided with efficient branchial filters. Blepharipoda (Fig. 7-10), *Hippa* and *Emerita*, as well as many forms inhabiting the supralittoral sands swept by waves (Gastrosaccinae, many Cirolanidae, Amphipoda, Cumacea), use their richly ciliated antennae to agitate water and search for food scattered on the sand surface with every retreating wave.

The mesopsammon represents an important evolution and differentiation centre for some independent systematic units (Ax, 1966). Some members of the microzoobenthos produce dozens of eggs per segment (*Microphthalmus aberrans*). The study of the benthic microfauna has hardly begun in several marine areas, and practically not at all in deeper areas. Its importance for the nutrition of abyssal

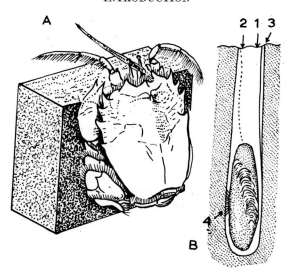

Fig. 7-10: Endopsammic animals. A: Psammobiont blepharipod from Peru; it remains sunk in the sand, like species of *Emerita*, at the limit of the surf area " brushing " with its long antennae towards its oral filter food particles left by each wave. B: Psammobiont lamellibranch *Lithophaga cumingiana*. 1: exhalant siphon; 2: inhalant siphon; 3: calcareous lining of its tube; 4: byssus threads. (After OTTER, 1937; redrawn.)

animals must be considerable, even more so than in the photic area. The littoral microphytobenthos (Chapter 7.2) has been largely neglected, although it constitutes an important link in the trophic chain and a major food source in the photic zone, especially for microbenthic animals. 80,000,000 living diatoms per m² were found on sandy bottoms in the Black Sea, many billion cells per m² in the silt of the stones (BODEANU, 1965). Many micro-endobionts (Fig. 7-9) are being studied more carefully since REMANE (1940), SWEDMARK (1956) and DELAMARE-DEBOUTEVILLE (1960) demonstrated the extraordinary adaptations which invertebrates have developed in this environment (Fig. 7-11).

The Protozoa constitute a major group in the psammon. Ciliata developed special adaptations for psammic life (Fig. 7-9). Some Foraminifera and Polychaeta incorporate sand grains in outer coverings, thereby modifying the physical properties of their substrate. They occur from the midlittoral to the ultra-abyssal (*Trochammina, Cyclostoma*) and are, as a rule, epipsammic and epipelic; only a few species are able to sink 2 to 3 mm into the substrate.

Sand-living, small gastropods and bivalves feed primarily on microphytobenthos (Chapter 7.21). The great mass of nematodes (hundreds of thousands per m²) represents an important food base for a variety of benthic animals. Other organisms which are of great importance in sandy substrates are bacteria, fungi and blue-green algae (Chapter 7.1). The spawns of *Hydrobia, Cyclonassa*, Blenniidae, and the numerous kinds of cocoons of gastropods, turbellarians, etc. also belong to the sandy microfauna. Bacteria, fungi and blue-green algae often use them as substrate. In sediments of great depths, the microfauna is expected to be rather rich. THIEL (1966) reported from 500 to 10,000 specimens per m²; these values are comparable to those found in the littoral zone.

A hard sediment, intermediate between muddy–sandy and rocky substrates, is formed by shell remains of molluscs or by microshell remains of ostracods, fora-miniferans, etc. On such a sediment, the fauna (mostly crustaceans) is less abundant; some gobiids and blennids lay their eggs on it, while actinians, *Calyptraea* and various spawns cling to older valve parts.

While rocks and sand are, as a rule, limited to the continental or precontinental shelf—with the exception of submarine mountain chains of which some exhibit the naked mother rock—the distribution of the mud fauna is unlimited; mud is found almost everywhere on the bottom of the oceans: mud is deposited between the stones on the shore, on rocks, in abyssal and ultra-abyssal depths; there are millimetre-thin mud layers (Indian Ocean) and muddy sediments thousands of metres thick (Atlantic Ocean). When mud occurs as a thin layer, it is burrowed by many minute tubicolous animals such as *Corophium, Fabricia, Tanais* (Fig. 7-8); when it occurs as a thick sediment layer, many animals build their tubes in it (e.g. polychaetes of the family Terebellidae or the genera *Spirographis* and *Sabella*; coelenterates (*Corymorpha*), and many actinians). Pure mud rarely occurs as a sediment. What is mud? It is a mixture of powders of various minerals of grain sizes below 0·03 mm and of very fine particles such as mica precoloids or colloidal argiles varying between 0·002 mm and 0·0002 mm, iron sulphur, coloids, etc., as well as humus substances such as peloid, which is linked to nearly all organic substances from the bottom (amino acids, chitinous remains, anaerobic

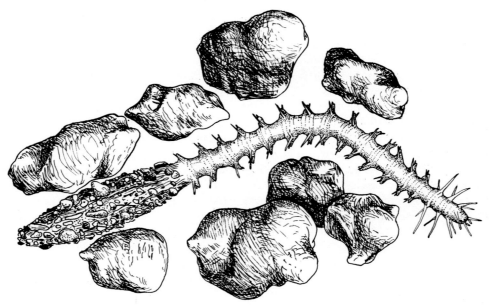

Fig. 7-11: Psammobiont polychaete *Microphthalmus sczelkowii* depositing cocoon, covered by various foreign bodies, among sand grains. (After Ax, 1966; redrawn.)

bacteria, etc.). In essence, mud is made of the finest sediment particles mixed with coloids.

One of the most obvious structural adaptations to life in muddy sediments is the worm-like body shape (holothurians, actinians, polychaetes); other adaptations are the ability to build tubes and the development of fastening filaments such as byssus threads in *Lithophaga* or *Modiolus*. Some mud living animals are blind (various malacostracans) and white in colouration (Apseudidae), others have very long legs that help to expand their supporting surface (Pycnogonida, some Isopoda and Decapoda).

(2) Functional Responses

(a) *Tolerance*

Ranges of tolerances to different substrata can be referred to only in rather general terms. The main reasons for the lack of knowledge concerning lethal limits of different substrata are easily listed: (i) the great diversity of animal responses to different substrates, (ii) the difficulties in measuring various physico-chemical aspects of substrates accurately, (iii) the absence of exact experimental data. Further, simultaneous responses to other environmental factors—such as temperature, water movement, light exposure and salinity—may greatly complicate the picture and hence make a detailed analysis very difficult, if not impossible.

Tolerances to different kinds of substrata are largely associated with exposure (in supra- and midlittoral animals; Chapters 4.2, 4.31) and water movement (Chapter 5.3). Water movement may act directly, as is the case in sessile animals exposed to wave actions; it may also act through substrate transport (sediment, silt).

Tolerances of supra- and midlittoral animals are largely determined by exposure to extreme intensities of temperature (Chapter 3), air moisture, salinity (Chapter 4) and light (Chapter 2). Both the absolute intensities and the fluctuation patterns of these factors are important (for factor combinations consult Chapter 12).

In regard to water movement (Chapter 5.3), supralittoral animals can often tolerate strong wave forces. However, if heavy storms cause water movement forces which rake the sand to depths of 5 to 10 m, or throw rocks against other rocks, whole populations may be torn off their solid or floating substrates, crushed to death and heaped up on the beaches.

Water movement in combination with low water temperatures may be lethal to many sand-living animals (KINNE, 1963, 1964; CRISP, 1964). During cold winters, death may be due to critical reductions in activities such as ciliary movement, burrowing, locomotion, etc. Thus filter mechanisms may get obstructed, adhesive disks lose their suction power, endobiotic animals lose their capacity to adjust their position in the sand, etc. If such conditions are combined with extensive substrate (silt, sand, mud) transport, the substrate itself may bring death to whole populations (see also COURTNEY and WEBB, 1964; CRISP, 1964; WOODHEAD, 1964a, b; ZIEGELMEIER, 1964; Chapters 3.3, 4.3).

In sandy shores, the preferred substrate requirements of endopsammic forms are adequate grain sizes and corresponding interstitial spaces allowing for locomotion and aeration. For example, in the Black Sea, *Mesodesma corneum*, *Otoplana*, archiannelids and *Bodotria arenosa* favour coarse, organogenic sand in the midlittoral zone, while *Corbula mediterranea, Pontogammarus maeoticus* and *Cumopsis goodsiri* favour fine, mineral sand. In summer, the latter occupy a zone 0·5 to 10 m deep; in winter, this zone extends from 8 to 15 m. Changes in grain size cause alterations in the species composition of the predominant animals of the coenosis.

Some animals, e.g. *Crassostrea virginica*, are capable of taking food even when the surrounding water contains as much as 400 mg of mud per litre (JØRGENSEN, 1966). However, sudden and massive sedimentation may cause severe damage. Thus, when in 1962 thousands of m³ of loess from the Roumanian shore were thrown into the sea, turbidity disturbed the waters in the rocky area off the Marine Research Station in Agigea for months, and the fauna and microflora suffered immediate and spectacular losses and changes in species composition (BACESCU, 1965). Rock coenoses were replaced by iliophilous ones; most members of the sessile rock fauna were killed on the spot (*Mytilus, Brachyodontes, Balanus,* sponges, bryozoans, actinians, colonial diatoms, etc.), as was the epibiosis. Of the vagile forms (crabs, shrimps), some escaped to the deep sea, others were killed. Crabs died because of silt hindering their respiration and the weight of epizoans settling on them. The vast, evacuated, ecological niche immediately became occupied by forms accustomed to turbidity and silt: sabellids, iliophilous cumaceans and Corophiidae (see also Chapter 6.3).

Examples of this kind, in connection with man's activities on shores (harbour constructions, building of beaches, dams, etc.) help to assess the tolerance to a given type of substrate. Thus, the dam of the Tide Power Works on the Rance River in France, while cancelling the effects of the tide, has brought about considerable changes in the over 15 km long estuary of that river. Water stagnation, a slight reduction in salinity and silting of sandy areas caused the death of several populations (Miliolidae, Bulimidae, porcellaneous or agglutinative forms of foraminiferans), and their replacement by more iliophilous and calcareous forms (Elphidiidae; later: Scrobiculariidae and *Enteromorpha*) (ROUVILLOIS, 1967).

Heavy silt sedimentation on the sandy bottom south of the Danube's mouth—associated with freshwater inflow during the long period of high waters of the river—has either killed or chased off many psammobionts, for example, *Corbula, Portunus* and even *Venus* and the Harpacticoida, from the microbenthos. When the freshwater current along the Roumanian shore persists for months, many of the more stenohaline forms, such as *Ostrea, Trochus, Eriphia, Clibanarius, Ophelia,* etc., either die or maintain a reduced population towards the south only. Many years were required for repopulation of the northern area.

The type of substratum largely determines the qualitative and quantitative composition of animal associations. This fact explains why even the most typical biocoenoses never show the same ratios of species predominance. For example, in one of the most characteristic biocoenoses of the Black Sea, that of *Modiolus phaseolinus*, a small hairy bivalve, the dominance coefficient of the common species is over 50% in most stations off the Roumanian sea-shore; off the Caucasian coast, the dominance coefficient is only 30 to 50%, while in the Bosphorus area it is below 40% (KISSELEVA, 1967). These differences occur in spite of the fact that the *M. phaseolinus* biocoenosis lives in deep water of 40 to 120 m, where environmental conditions are expected to be rather constant. The differences should, therefore, be viewed especially with regard to the composition of the substrate (percentages of calcium; sand in the silt; detritus; salinity; etc.). A similar case has been observed in the composition of psammobiont coenoses dominated in some seas by *Corbula* and in others by *Mytilus* or *Mya*; they depend especially on the characteristics and composition of the substratum.

The benthic stenobiont species, through their optimum and tolerance rates, are good indicators of the state of the bottom and of the degree of pollution of the sea. *Cystoseira barbata* and its epibiosis indicate oligotrophic waters; when waters become eutrophic, the qualitative and quantitative composition of coenoses changes (GAMULIN-BRIDA and co-authors, 1967). Even more sensitive indicators of the quality of the bottom are found among the microbenthos (BACESCU, 1965).

(b) Metabolism and Activity

Metabolism and activity of marine animals may be affected by different types of substrata in various ways. Substrates may serve as food source, or as dwelling-place; accordingly, they may cause quite different responses.

Let us consider some cases in which substrates modify metabolism or activity via nutritional aspects. The nutrition of benthic animals is undoubtedly one of the most interesting and important problems in oceanology, especially on ultra-abyssal bottoms. Where organic detritus is rich on the sea bottom, the micro- and macrofauna and predators (fish, squids, asteroids) and omnivorous animals (mysids and others) thrive at their expense. Some animals simply swallow the sediment (the young of the grey mullet *Mugil* which 'graze' the ripple-marked sand, various species of *Gobius*, and others) or take up large quantities of detritus (Echiuroidea, Polychaeta, Mysidacea, etc.; Fig. 7-12). These animals are 'psammi-

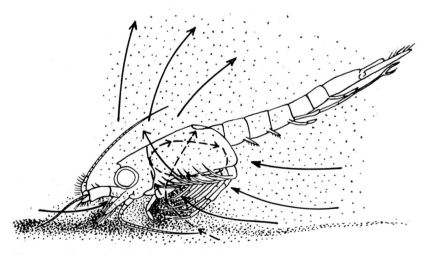

Fig. 7-12: Lateral view of *Hemimysis lamornae* swimming along the bottom and ploughing up food particles with the inner flagella of its antennules and antennal scales. Food is collected and held by the basket of the third to eighth endopodites and the mandibular palps. Solid arrows indicate water currents, broken arrows the swirl provoked by the beating of exopodites. (After CANNON and MANTON, 1927; redrawn.)

vorous' or 'iliophagous'. They ingest organic components, alive or dead, of the sediments. There is evidence that several marine animals are able to absorb, via mucous membranes, protein substances which are dissolved in the water. However, this ability has not yet been proved experimentally to occur in benthic forms, except for filter-feeders, some protozoans and corals (*Fungia*). Many Protozoa (Rhizopoda), Spongia, Turbellaria and others engulf the finest organic particulate matter and digest it in their vacuolar system. Intracellular nutrition occurs also in some

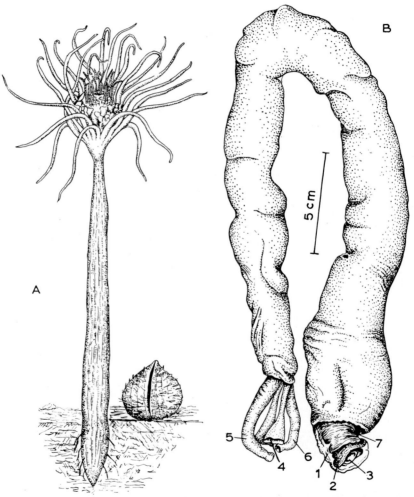

Fig. 7-13: Mud dwellers. A: The pivoting hydropolyp *Corymorpha nutans*, thrust into, and anchored in, mud with rhizoids; near it, a young *Modiolus phaseolinus*, the dominant species of the coenosis. (After BACESCU, 1952; redrawn.) B: *Kuphus polythalamia*, a lamellibranch mud borer with reduced, nearly smooth shells, surrounded posteriorly by a heavy muscular collar and very small adductor muscles. 1: cephalic hood; 2: shell; 3: foot; 4: pallet; 5: incurrent siphon; 6: excurrent siphon; 7: muscular collar. (After TURNER, 1966; redrawn.)

bivalves, for example, in those with zoochlorellae, either directly or through specialized amoebocytes.

Filter-feeders live and grow on microscopical floating substrates (suspended or sedimented particles). These particles, alive or dead, differ with latitude, distance from the shore and water depth. Primarily, they are of phytoplanktonic origin; but also of importance are organic detritus (derived from the decomposition of dead animals), colloidal substances, detrital chlorophyll concentrated in faecal pellets of the zooplankton and phytoplankton feeding fishes (e.g. *Engraulis*), protein molecules, various amino acids, carbohydrates, and organic substances absorbed on inert granules. The abyssal bottom is populated by microfungi, microzoobenthos, and especially bacteria. With increasing depth the amount of organic particles diminishes (WOLFF, 1960; BIRSHTEIN, 1963; see also VINOGRADOV and co-authors, 1970).

As a food source, suspended or sedimented substrate particles may affect growth rates and filtering activity. Laboratory experiments designed to assess in detail quantitative relations between metabolism and activity on the one hand, and type and abundance of particulate organic substrates on the other, are lacking.

As a dwelling place, substrates affect metabolism and activity of marine animals via their physico-chemical properties, particularly their resistance to animal activities such as boring, digging, locomotion, etc. The smoother, softer and muddier a substrate, the easier it is inhabited by tube-building forms. Many species occupying semifluid substrates anchor themselves with the help of rhizoids, for example, the hydropolyp *Corymorpha nutans* (Fig. 7-13), actinias such as *Phelia*, and members of more evolved groups like *Virgularia*, *Veretillum*; or they bore into the mud like the mollusc *Kuphus polythalamia* from the Solomon Islands (Fig. 7-13). Their mobility is restricted; however, the majority of soft-bottom animals are capable of swift movements in search of food or in order to secure convenient sites during gestation and the raising of offspring. Many dwellers of muddy substrates build solid or elastic tubes of mud (which they weave into a felt with threads produced by cnidoblasts, like *Cerianthus*); mud and mucus forming stiff tubes (Terrebellidae); mud with mollusc remains fixed on one edge resulting in more resistant flexible tubes (*Melinna*, Fig. 7-14); or of hardened, yet elastic secretions (some tubicolous amphipods like *Ampelisca* which inhabits muddy sands).

Less typical tubicoles exist in sandy substrates: e.g. *Pectinaria* with its hard tube thrust into the substratum, a few polychaetes (*Capitomastus*, *Leiochone*) with mobile tubes made of weakly agglutinated granules. The tubes of the latter species are slit in the part emerging from the substrate, but they can be closed hermetically.

The role of substrates as building material for benthic animals is obvious from the afore-mentioned examples. HEMPEL (1957) has studied the tubes of Spionidae from this viewpoint and found that substrate material used for building is not chosen at random but rather carefully selected. According to KISSELEVA (1967), the determining factors in the selection of building material are weight and quality of the substrate granules. For larvae of *Polydora ciliata*, the primary factor is the size of the substrate particles, not their mineralogical composition.

In solid substrates, only boring forms like Lythodidae, Veneridae, Pholadidae can drill holes. Some species drill the smoother substrates of the loess shores of the

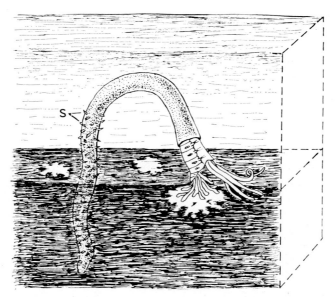

Fig. 7-14: The mud-living polychaete *Melinna palmata* builds resistant flexible tubes of mud and mollusc remains. S: shell fragments of *Abra alba* fastened obliquely on the elastic tube and anchoring it more firmly in the substrate. The specimen shown is engaged in taking sediment from the mud surface. White patches indicate areas from which the bioderm has already been removed. (Original.)

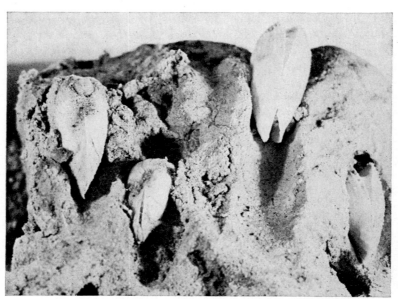

Fig. 7-15: The lamellibranch *Pholas candida* drills loess, a soft solid substratum, in the Black Sea near Constanta. (Original.)

Black Sea (*Pholas candida*; Fig. 7-15). On these shores, *Pholas candida* may attain densities of 4,000 specimens/m² and biomasses over 3 kg/m² (GOMOIU and MÜLLER, 1962). Other species bore argile substrates (*Pholas dactylus*) or even compact rocks such as the Sarmatian limestone (*Petricola lithophaga*, whose population density may exceed 300 specimens/m² off Agigea (BACESCU and co-authors, 1963).

Wooden substrates harbour a number of specialized borers such as the molluscs Xylophaga and Pholadidae. The mollusc *Teredo* (the shipworm) became entirely modified by its way of life in wooden substrates; it acquired the appearance of a worm and greatly reduced the size of its valves. Also several crustaceans are capable of boring into wooden substrates: species of *Limnoria*, *Sphaeroma terebrans*, *Chelura*, and others. All wood borers endanger man's maritime activities and annually cause considerable damage to harbour structures. Mud dwellers are frequently able to metabolize at rather low concentrations of ambient oxygen. Thus *Melinna*, a polychaete inhabiting poorly aerated circalittoral mud sediments in the Black Sea (Fig. 7-14), can exist under reduced oxygen levels down to anaerobic conditions and even H_2S formation. If conditions become critical, *Melinna* abandons its tube and actively moves away (DRAGOLI, 1960/61).

The substratum may also serve as refugium, protecting its permanent or temporary inhabitants from extreme temperatures, salinities or water movements. Thus *Pontogammarus maeoticus* and other amphipods from the midlittoral zone perceive storms approaching from the open sea hours ahead and escape into the sand. Some positively geotaxic archiannelids withdraw into their tubes at the slightest tectonic vibration. More detailed studies on such activities are highly welcome, especially of psammobionts which are particularly sensitive to vibrations and ultra-short waves.

(c) *Reproduction*

The type of substratum on or in which an animal lives may modify its pattern and rate of reproduction. Some specific adjustments of interstitial, sand-living turbellarians and polychaetes have been investigated by Ax and his students (consult for example Ax, 1966). *Hesionides arenaria*, a vagile polychaete very common in the sandy eulittoral of the Isle of Sylt, has a life span of 1 year; it breeds exclusively during summer in the sand near the zone of resurgence; its larvae grow in the upper midlittoral and the young ones winter in the respective substrate. Some mesopsammic polychaetes spawn huge cocoons, developed in the swollen posterior part of their body (Fig. 7-11). The reproductive pattern of the archiannelid *Nerilla* is of a primitive type: eggs are scattered and glued to sand grains. In some mesopsammic genera, eggs are retained on the posterior part of the mother individual. The care for their offspring shows great specialization. Some gastropods have developed particularly efficient means of ensuring their spawn: they glue each egg on smaller shells nearby or on some congeneric species (for example, *Cyclonassa* on *Nassa*); others wrap them into a sand cover (*Hydrobia*); still other species arrange their eggs into groups of gelatinous cocoons and attach them to the shells of other molluscs (*Buccinum*, *Rapana*) using these as substrate (Fig. 7-16; see also APELT, 1969).

The substratum plays an important part in the settling and metamorphosis of

Fig. 7-16: Egg cocoons of *Rapana thomasiana* attached
to a living substrate: the mussel *Mytilus galloprovincialis* of the Black Sea (Original.)

pelagic larvae of bottom animals. A widely discussed problem today in marine
biology is that of the factors which determine the settling of the larvae, their
distribution, and the completion of their benthic life cycles. According to THORSON
(1946), light plays a most significant role (see also Chapter 2) in archiannelids
(JÄGERSTEN, 1952), *Balanus, Corophium, Spirorbis, Ophelia,* gastropods, *Brachy-
notus,* and others. Indeed, the larvae of these forms seek actively until the end of
their metamorphosis for a suitable place to settle. The larvae of substrate-specific
(stenotopic) species such as *Brachyodontes lineatus* and *Polydora ciliata* show definite
substrate selectivity; in larvae of eurytopic species (*Mytilus edulis, Platynereis
dumerilli,* etc.) no such selectivity was observed.

Chemical and physical properties of the substrate represent the main factors
which accelerate larval sedimentation in certain biocoenoses. *Polydora ciliata*
select rocks according to their physical properties; the larvae of *Rissoa splendida,
Bitium reticulatum* and *Brachyodontes* are chemotactically attracted by the organic
substances excreted by living algae. The film of living bacteria and minute algae
covering the substrates everywhere determines the rapid sinking of the larvae of
many benthic animals, thus speeding up their metamorphosis. In the majority of
benthic invertebrates, swimming of the planktonic larvae over populations of adult
individuals stimulates their settlement (KISSELEVA, 1967).

The size of substrate granules is a determinant factor for meroplanktonic larvae,
especially for tubicolous polychaetes. They search for a convenient grain size in
order to build their tubes. Failure to find the appropriate type of substrate forces
planktonic larvae to prolong their free-swimming period and eventually to sediment
on inconvenient substrates while drifting with water currents; thus in the northwest sector of the Black Sea, myriads of *Balanus* larvae annually fall on the fine
sands and perish when they fail to contact *Corbula* shells which allow them to settle

and thus survive. In contrast, in larvae of the bottom-living flat fishes *Pleuronectes* and *Rhombus*, it is the completing of their pelagial metamorphosis which determines the time of their sinking to the bottom.

Some larvae need very strong light to undergo metamorphosis (*Mytilus edulis*, *Balanus improvisus*). They actively select a convenient substrate and are able to detect and settle on small areas of hard solid substrates even in the middle of otherwise granular sediments.

Areas with significant inflow of biogenic salts from nearby rivers or with currents causing high concentrations of microbenthos and detritivorous animals, represent suitable grounds ('nursery areas') for the young of many fishes and attract numerous other groups of marine animals such as shrimps, crabs and mysids for spawning. Moreover, such areas provide sediments for many meroplanktonic animals; they are centres of reproduction, nutrition and migration. Three examples may suffice. The mangrove area in Mozambique attracts many shrimps of commercial value (HUGHES, 1966). The Azov Sea and the predeltaic area of the Danube constitute nursery areas in the Black Sea. In spring, phytoplankton develops explosively and the bottom substrate shelters a rich and varied micro- and macrofauna—hundreds of thousands of *Microarthridium* and nematods per m²—and an abundant selection of relict Ponto-Caspian cumaceans; amphipods and mysids (*Mesopodopsis*) occur in great numbers attracting, in turn, masses of *Crangon*, *Macropipus* and *Upogebia* from greater depths, closely followed by sturgeons, plaice and others, and even by pelagic fishes such as *Sprattus*, Danube herring, etc. (BACESCU and co-authors, 1957). The third example is the La Jolla Gulf (California, USA) where the circulatory littoral system favours mass development and concentration of mysids, especially of *Metamysidopsis elongata* (CLUTTER, 1967).

(d) Distribution

The different kinds of substrata represent a major factor in regard to the distribution of marine animals. While most types of substrata affect primarily horizontal distribution patterns, floating substrates may also be of some importance in regard to vertical distribution, especially in epibionts.

The distributional patterns of the bottom fauna are infinitely more varied than those of the fauna of the oceanic pelagial. Among the most important entities of the bottom substrates which affect horizontal distributions are (see also Chapter 7-1): (i) the nature and size of granules, (ii) the kind, amount and form of organic matter associated with the substrate, (iii) the degree of mobility of the substrate, (iv) the degree of hardness of solid substrates, (v) the total area of a substrate of a given type, (vi) other environmental factors such as water movement, light, salinity, oxygen supply, pressure, etc.

Many biologists endeavoured to devise a classification of the benthic biocoenoses (e.g. EKMAN, 1935; ZENKEVITCH, 1956). One of the most successful classifications— not regional in scope and applicable to all seas—was discussed and improved at a Colloquium centred on the debating of benthic zonations (PÉRÈS, 1961). We shall use it here; it consists of 7 benthic floors: the **supralittoral** sprayed only by the waves; the **midlittoral** rhythmically covered and uncovered by tidal movements (=eulittoral of other authors); the **infralittoral** (=sublittoral) extending from the constantly covered zone to the limits of growth of marine phanerogams or photo-phileous algae; the **circalittoral** including the remaining littoral belt where algae (sciaphileous algae) survive (it corresponds largely to EKMAN's eulittoral; in the Black Sea it is bordered by the **periazoic**, BACESCU, 1963); the **bathyal** containing the organisms from the continental slope down to about 3000 m depth (= Archi-benthic); the **abyssal** reaching down to 6000 m depth; and the **hadal** (= ultra-abyssal) extending between 6000 and 11,045 m depth and characterized by gigantic hydrostatic pressures. The first 4 levels constitute the littoral or phytal system, the remaining 3 the profundal or aphytal system in which only bacteria and animals live.

Recent research has revealed that quantitative and qualitative variations of the benthos are much more pronounced. In general, the fauna decreases from the shore toward the abyss. From bottom productions of over 1 kg/m² (*Corbula* or *Modiolus* coenoses in the Black Sea or the Bering Sea) the annual production decreases to 50 g/m² at 200 m depth, 10 g/m² at 2000 m, 0·1 to 0·2 g/m² at 4000 m, and 0·25 to 0·93 g/m² at 10 km depth in the Kermadec Trench. There are seas with special conditions where the zoobenthos disappears completely at a certain depth. For example, in the Mediterranean Sea the benthos is quantitatively very poor below 200 m, especially in the eastern part. In many seas, the benthos becomes azoic below a certain critical depth, e.g. in the Black Sea below 200 m, in the Caspian Sea below 600 m, in the Red Sea below 1900 m.

With perfectioned dredges and the bottom sampler being used more often now, it becomes increasingly known that even the bottom substrates of abyssal and hadal areas harbour an unthought of diversity of marine animal life, similar to that found in or on littoral substrates. This diversity is thought to be due to the greater stability of the physical environment in abyssal and hadal regions (BELIAEV, 1966; HESSLER and SANDERS, 1967).

With the exception of bottom currents or of turbidity currents which sweep some trenches, conditions are rather uniform near the hadal substrate. Dominant environmental factors are: very high pressure, oceanic salinity (35°/₀₀), temperatures around − 1° to − 2° C, constant darkness, high concentrations of $CaCO_3$ and of dissolved organic matter (the principal food base of the Pogonophora and other forms inhabiting the substrate). Unfortunately, the chemical techniques available do not succeed in separating easily the labile organic components in the dissolved state, which play a considerable role in the nutrition of deep-sea substrate-living animals. The amounts of nitrogen and carbon present cannot be correlated to the abundance or the paucity of the benthos; this is true for littoral substrates and even more so as one traces the distribution of benthic animals to greater depths.

The water currents occurring in the long marine trenches help not only to provide oxygen for respiration and food for the deposit-feeders, they also contribute to the distribution of substrate-living forms from pole to pole. Thus, in the Chile-Peru Trench and in the California Trench, the recent finding (2° to 3° S of the Equator, by the 11th Cruise of RV 'Anton Bruun') of forms considered arctic (*Pseudomma*) or antarctic (*Hansenomysis, Petalophthalmus*) may be indicative of such poleward migrations (transportations) both to the north and to the south, rather than an example of a bipolar distribution as formerly assumed.

The bottom and turbidity currents fertilize the substrate and encourage exuberant growth of life wherever they occur. Even in the Black Sea, where animal and algal life ceases altogether beyond depths of 200 m, circular currents favour an exuberant development of the coenoses of *Mytilus* and *Phyllophora*. Pelagic larvae of substrate-living animals may be transported by water currents over great distances. Thus currents ascending from the Antarctic to the Equator (Humboldt Current) or from the Gulf of Mexico toward the polar regions (Gulf Stream) transport innumerable larval forms of benthic species, disseminating them over large areas.

The dominant groups in the hadal substrates are, by locality and trench of occurrence, Pogonophora, Crustacea, Echinodermata, Nematoda, Bivalvia, Echiuroidea and Polychaeta.

The epibiosis develops exuberantly only on solid substrates such as rocks or coraliferous facies. On sediments, epibionts may settle only on the restricted areas of solid substrates emerging here and there: blocks or rock, coke from ships, wrecks and cables, living or dead mollusc shells.

While the corals are zoogeographically confined to the warm areas of the world, substrates made up of living or dead animals may be found at any latitude. These organismic substrates range from minute bivalves living on the substrate, or half-way burrowed in it, to oysters and larger mussels, from the numerous kinds of crabs and tubicolous worms to the large bivalves like *Pinna* and sponges living in association with other different species. All these animal-surface substrates are often covered with a multitude of forms: sponges, colonies of coelenterates and bryozoans, sessile worms, gastropods, cirripedes, tunicates, spawns of all kinds, not to mention the endless list of protozoans and diatoms. Two examples may suffice: In the European seas, and especially in the Black Sea, the most characteristic animal, which also inhabits the extensive area of fine sand at the mouth of the River Danube, is *Corbula mediterranea* (Fig. 7-17). Although less than 1 cm in body size, the individuals of this species constitute the commonest substrate (thousands of specimens per m²) and provide the only suitable substrate for many micro-organisms and myriads of planktonic larvae restricted to settling on solid substrates, but drifted into sandy areas by water currents (BACESCU and co-authors, 1957). *Corbula mediterranea maeotica* are covered with epibionts, especially *Balanus*, the weight of which often doubles their own weight and may eventually kill them. Another example: In the surf areas, mussels are covered with barnacles and bryozoans (Fig. 7-18). In the deep sea, mussels live in large groups of dozens of individuals bound together by their byssus threads; their surfaces are covered by serpulid worms, actinias, hydropolyps, tunicates like *Ciona*, *Ascidiella*, and other animals. Colonial agglomerations other than coraliferous ones, in turn, may become substrates for other species. Thus, the amphipod *Corophium* or the polychaete *Mercierella* build tubes (the former of mud, the latter of limestone) fastened together and interlocked so as to form colonial aggregations on rocks or on ships; these may be more than 10 cm thick and harbour other species, such as crabs (especially *Pilumnus*), isopods (especially *Dynamene*), other amphipods, *Amphare-tidae*, and tunicates (in the Black Sea: *Ctenicella*). The Bryozoa, too, often form large colonies, especially on sheltered solid substrates; after death, they may shelter other colonial animals.

A ship's hull serves as substrate for all kinds of algae and planktonic larvae of bottom animals, which form textures (the so-called fouling), making the ship

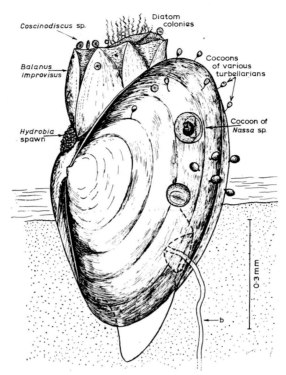

Fig. 7-17: The lamellibranch *Corbula mediterranea maeotica* lives in fine sand; it provides an opportunity for settlement to numerous solid-substrate dwellers, drifted into the fine sand area by water currents, and thus affects their distribution. The lamellibranch anchors itself with the aid of a byssus thread (b) produced in the byssus pouch (broken line). (After BACESCU and co-authors, 1957; modified.)

heavier and reducing its speed. The fight against such fouling is a heavy burden on maritime economy. Transport of marine animals by ship has undoubtedly affected their distribution and is likely to do so at an ever increasing rate.

Wood represents a substrate for animals adapted to boring into it. Wood borers are several isopods (e.g. *Limnoria*), amphipods (*Chelura*) and bivalves of the family Teredinidae, to mention only a few forms capable of damaging even the most resistant woods (Fig. 7-19). In brackish waters ($1 \cdot 8^0/_{00}$ to $17 \cdot 6^0/_{00}$ S), wooden substrates carry a less special fauna. The sessile and vagile macro- and microfauna of the wooden piles of the Kiel Canal (North Germany) was closely studied by SCHÜTZ and KINNE (1955). They found 72 microbenthic species—mainly nematods (30 species) and copepods (21 species)—and 21 macrobenthic species—mainly amphipods (10 species), isopods (4 species) and polychaetes (3 species). The distribution of the microfauna is a more sensitive indicator of saline zonations and of the degree of water pollution (e.g. KINNE and AURICH, 1968).

There exists an evolutionary trend from epibionts to commensals to parasites.

Fig. 7-18: Mussel *Mytilus galloprovincialis* from Constanta (Roumania), covered by two generations of *Balanus improvisus* and various bryozoans. (Original.)

Fig. 7-19: Wood-boring bivalve *Teredo navalis* from the Black Sea in a piece of wood which was cut open to illustrate the degree of tunnelling of the wooden substrate by this mollusc; c: fragments of the wood borer's calcareous wall (white) inside the tube drilled into the wood; m: intact individual with its worm-shaped body and its two reduced valves (v); t: intact calcareous wall. (Original.)

Many forms have changed from life on substrates to life inside substrates (endo-psammon, endolithion); in organisms serving as substrate, they have settled between appendages at the exterior, or inside the branchial chamber (e.g., *Carcino-nemertes*, several copepods, *Leptinogaster*, *Mytilicola*, Dajadidae), in the channels of sponges, or even inside internal organs (Sacculinidae, Trematoda, Acantho-cephala). Due to biochemical interreactions between the living substrate and its inhabitants, very specific and complex relations may exist.

(3) Structural Responses

Benthic animals respond to different substrata over long periods of time by structural adjustments in body size and shape as well as in anatomy. Such adjust-ments are largely genetic adaptations of basic structures and specialized morpho-physiological mechanisms.

On surf-beaten **rocky substrates**, most species have developed special structures for attachment or fixation. Fixation may be permanent or temporary; in the latter case, the animal can detach itself when environmental conditions become aggravating (intensive silting, extreme temperatures or salinities, etc.). In shell-bearing forms, one consequence of fixation to the substrate is that the shell tends to become helmet shaped; such structural adjustment is convergent in many different mollusc genera (*Patella*, *Trochita*, *Crepidula*, *Acmaea*, *Calyptraea*); the helicoidal line becomes reduced, the shell flattens, a sucking device is developed, etc. Other cases of structural adaptations are found in *Chthamalus* and *Balanus* (calcareous casks built on the substrate), in Bryozoa (crust-like colonies) and in Serpulidae (*Serpula* builds tubes cemented to the substrate). Some animals fasten themselves to solid substrates with adhesive plates in a similar way as do algae (hydroids, oysters), others attach themselves to the substrate with the aid of secreted byssus threads (e.g. *Brachyodontes*, *Mytilus*, *Pinna*); a special gland developed in these animals secreting a colagen substance which hardens to produce the threads.

Sponges and colonial ascidians such as *Botryllus* are fastened to solid substrates with broad contact surfaces; various foraminiferans have developed rhizoids. Actiniae cling to rocky substrates via an adhesive disk, and anchor themselves in muddy substrates by body parts thrust into it; still other animals cling to rock surfaces with flattened disk-like bodies, e.g. the copepod *Porcellidium*.

Another large group solved the problem of maintaining a hold on rocky sub-strates by boring, e.g. the sponge *Cliona*, the endolithic pholads, the molluscs *Lithodomus* and *Petricola*, the polychaete *Polydora*. All these endolithic forms show similar morphological adaptations: elongated bodies, rasping valves, acid secre-tions, defence tubes (*Polydora*). In addition to drilling into the substrate, *Litho-domus* binds itself to the bottom of the orifice by means of a byssus. Tube builders in solid or soft substrates often have special devices for closing their tubes (Poly-chaeta), or they defend their openings with the help of powerful chela and subchela (spongicolous Malacostraca, hermit crabs in gastropod shells). Many Inachidae and other forms have acquired shapes and colours simulating the substrate (camouflage).

The **sandy substrates** have caused numerous adaptations. One of the foremost

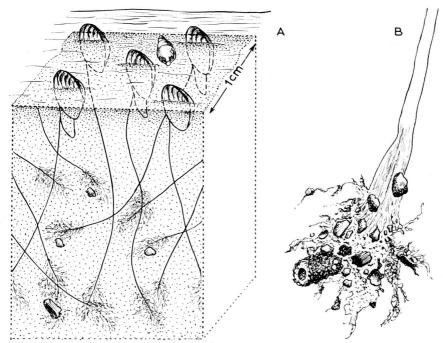

Fig. 7-20: Adjustments to life in moving sandy substrates. *Corbula mediterranea maeotica* from the Black Sea is capable of controlling its position in the sand by fast movements with its very mobile feet, and of firmly anchoring itself with the aid of long threads. A: Individuals anchored in the sand with their long byssus threads (population density: 25,000 specimens per m²); B: Terminal part of a byssus thread fastened to larger particles of the sandy substrate. (After BACESCU and co-authors, 1957; redrawn.)

concerns the shape of the body; the most frequent structural adjustment is small body size combined with an elongate, worm-shaped body (Fig. 7-9) revealing interesting convergencies; also in regard to locomotion, attachment devices and ciliation, lance and flattened shapes are typical of borers (*Eurydice, Gastrosaccus, Bathyporea, Canuella, Magellona, Arenonemertes,* etc.). A combined adaptation to moving sand substrates is shown by the bivalve *Corbula mediterranea*: in addition to its capacity to burrow fast into or out of the sand with the aid of its very mobile foot, it is also able to anchor itself firmly with the help of a long byssus thread (Fig. 7-20). The single thick byssus thread may extend 5 to 10 times the length of the valves. Many sand-living animals drill tubes (*Arenicola*) or burrow themselves into the sand (*Echinocardium cordatum, Emerita, Nassa, Upogebia,* polychaetes, cumaceans, *Dentalium, Branchiostoma,* etc.); other animals inhabiting sandy or muddy substrates are capable of boring (e.g. the carnivorous *Nereis,* Naticidae). The long siphon of sand-living bivalves represents an adaptation to life within the sediment. Still another kind of adaptation are the drilling blades (dactylus of the last pereiopod) of the Portunidae (*Macropipus*). In abyssal species (e.g. *Abra profundorum*) a definite correlation was observed between water depth and size of palpi, gills and digestive tube (ALLEN and SANDERS, 1966); in deep waters, *A. profundorum* has smaller gills but larger palpi.

Literature Cited (Chapter 7)

ADAM, N. K. (1941). *The Physics and Chemistry of Surfaces*, 3rd ed., Clarendon Press, Oxford.

ALEEM, A. A. (1950). The diatom community inhabiting the mud flats at Whitstable. *New Phytol.*, **49**, 176–188.

ALEEM, A. A. (1952). Présence d'une flore épiphyte sur *Hippocampus guttulatus* CUV. capturé à Banyuls. *Vie Milieu*, **3**, 210–211.

ALEXOPOULOS, C. J. (1962). *Introductory Mycology*, 2nd ed., Wiley, New York.

ALLEN, G. H. (1964). *An Oceanographic Study between the Points of Trinidad Head and the Eel River*. State Water Quality Control Board, Sacramento, California, Publ. No. **25**.

ALLEN, J. A. and SANDERS, H. L. (1966). Adaptations to abyssal life as shown by the bivalve *Abra profundorum* (SMITH). *Deep Sea Res.*, **13**, 1175–1184.

ALMODÓVAR, L. R. (1962). Notes on the algae of the coral reefs off La Parguera, Puerto Rico. *Q. Jl Fla Acad. Sci.*, **25**, 275–286.

AMLACHER, E. (1961). *Taschenbuch der Fischkrankheiten*, G. Fischer, Jena.

ANAND, P. L. (1937). An ecological study of the algae of the British chalk-cliffs. *J. Ecol.*, **25**, 153–188; 344–367.

ANKEL, W. E. (1962). Die blaue Flotte. *Natur Mus., Frankf.*, **92**, 351–366.

APELT, G. (1969). Fortpflanzungsbiologie, Entwicklungszyklen und vergleichende Frühentwicklung acoeler Turbellarien. *Mar. Biol.*, **4**, 267–325.

AUBERT, M. and GAUTHIER, M. (1967). Origine et nature des substances antibiotiques présentes dans le milieu marine. VIII. Etude systématique de l'action antibactérienne d'espèces phytoplanctoniques vis-à-vis de certain germes telluriques aérobies. *Revue int. Océanogr. méd.*, **6**, 63–71.

AX, P. (1966). Die Bedeutung der interstitiellen Sandfauna für allgemeine Probleme der Systematik, Ökologie und Biologie. *Veröff. Inst. Meeresforsch. Bremerh.* (Sonderbd), **2**, 15–65.

BAAS BECKING, L. G. M. (1925). Studies on the sulphur bacteria. *Ann. Bot.*, **69**, 613.

BAAS BECKING, L. G. M. and MACKAY, M. (1956). Biological processes in the estuarine environment. Pt 5. *Proc. K. ned. Akad. Wet.* (Sect. B), **59**, 190–213.

BAAS BECKING, L. G. M. and WOOD, E. J. F. (1953). Microbial ecology of the estuarine sulphuretum. *Int. Congr. Microbiol.*, Rome, **6**, 7; 22; 379–388.

BAAS BECKING, L. G. M. and WOOD, E. J. F. (1955). Biological processes in the estuarine environment. Pts 1–2. *Proc. K. ned. Akad. Wet.* (Sect. B), **58**, 168–181.

BAAS BECKING, L. G. M., WOOD, E. J. F. and KAPLAN, I. R. (1957). Biological processes in the estuarine environment. Pt 10. *Proc. K. ned. Akad. Wet.* (Sect. B), **60**, 88–102.

BACESCU, M. (1948). Quelques observations sur la faune benthonique du défilé roumain du Danube: son importance zoogeographique et pratique. *Annls scient. Univ. Jassy*, **31**, 240–252.

BACESCU, M. (1952). Hydropolype ou actinie—*Cerianthus* ou *Corymorpha*—est le coelentéré caractérisant les fonds à *Modiola* de la Mer Noire? *Comunle Acad. Rep. pop. rom.*, **2**, 233–237.

BACESCU, M. (1961). *Cerianthus* san *Corymorpha* este celenteratul caracteristie fundurilor au *Modiola* din U. Neagra? *Comunle Acad. Rep. pop. rom.*, **2**, 233–237.

BACESCU, M. (1963). Contribution à la biocénologie de la Mer Noire: l'etage pèriazoique. *Rapp. P.-v. Réun. Commn int. Explor. scient. Mer Méditerr.*, **17**, 107–122.

BACESCU, M. (1965). An instance of the effects of hydrotechnical works on the littoral sea life. *Studii Hidraul.*, **9**, 137–150.

BACESCU, M., DUMITRESCU, E., MARCUS, A., PALADIAN, G. and MAYER, R. (1963). Données quantitatives sur la faune pétricole de la Mer Noire à Agigea (secteur roumain), dans les conditions spéciales de l'année 1961. *Trav. Mus. Hist. nat. Gr. Antipa*, **4**, 131–155.

BACESCU, M., DUMITRESCU, H., MANEA, V., POR, F. and MAYER, R. (1957). Les sables à *Corbulomya* (*Aloidis*) *maeotica* MIL., base trophique de premier ordre pour les poissons de la Mer Noire. *Trav. Mus. Hist. nat. Gr. Antipa*, **1**, 305–374.

BACESCU, M., MÜLLER, G. and GOMIOU, M. (1971). Etude sur le benthos de la mer noire. L'analyse quantitative, qualitative et comparée de la faune pontique. *Ecologie marina*, **4**, 1–357.

BAYLOR, E. R. and SUTCLIFFE, W. H., JR. (1963). Dissolved organic matter in sea water as a source of particulate food. *Limnol. Oceanogr.*. **8**, 369–371.

BEGER, H. (1944). Beiträge zur Systematik und geographischen Verbreitung der Eisenbakterien. *Ber. dt. bot. Ges.* (Rundschreiben 1, 4), 62, (1944–49).

BELIAEV, G. M. (1966). *Donnaia fauna naibolshik glubin mirovogo Okeana*, Izd. Nauka Moscow.

BIERI, R. (1966). Feeding preferences and rates of the snail, *Ianthina prolongata*, the barnacle, *Lepas anserifera*, the nudibranchs, *Glaucus atlanticus* and *Fiona pinnata*, and the food web in the marine neuston. *Publs Seto mar. biol. Lab.*, **14**, 161–170.

BIRSHTEIN, I. A. (1963). Glubokovodnyie ravnonogie rakoobraznyie severozapadnoi chast' tikhovo okeana. Izd. Akad. Nauk USSR, Moscow.

BLOIS, J.-C., FRANCAZ, J.-M., GAUDICHON, M. & S. and BRIS, L. LE (1961). Observations sur les herbiers à Zostères de la région de Roscoff. *Cah. Biol. mar.*, **2**, 223–262.

BOALCH, G. T. (1957). Marine algal zonation and substratum in Beer Bay, South-east Devon. *J. mar. biol. Ass. U.K.*, **36**, 519–528.

BODEANU, N. (1965). Contributions à l'étude quantitative du microphytobenthos du littoral roumain de la Mer Noire. *Studii Cerc. Acad. RPR (Zool.)*, **16**, 553–563.

BØRGESEN, F. (1911). The algal vegetation of the lagoons in the Danish West Indies. *Biol. Arbejder t. E. Warming*, pp. 41–46.

BREDEN, C. R. and BUSWELL, A. M. (1933). The use of shredded asbestos in methane fermentations. *J. Bact.*, **26**, 379–383.

BREED, R. S., MURRAY, E. G. D. and SMITH, N. R. (1957). *Bergey's Manual of Determinative Bacteriology*, 7th ed., Williams and Wilkins, Baltimore.

BREMNER, J. M. (1954). A review of recent work on soil organic matter. *J. Soil. Sci.*, **5**, 214–232.

BRINTON, C. C. (1959). Non-flagellar appendages of bacteria. *Nature, Lond.*, **183**, 782–786.

BROCK, T. D. (1966). *Principles of Microbial Ecology*, Prentice-Hall, Englewood Cliffs, New Jersey.

BUNT, J. S. (1965). Measurement of photosynthesis and respiration in a marine diatom with the mass spectrometer and with carbon-14. *Nature, Lond.*, **207**, 373–375.

BUTKEVICH, V. S. (1928). Die Bildung der Eisenmangan-Ablagerungen am Meeresboden und die daran beteiligten Mikroorganismen. (Russ.; summary in German.) *Trudy morsk. nauch. Inst.*, **3**, 5–81.

CALLAME, B. and DEBYSER, J. (1954). Observation sur les mouvements des diatomées à la surface des sédiments marins de la zone intercotidiale. *Vie Milieu*, **5**, 243.

CAMPBELL, L. L. and WILLIAMS, O. B. (1951). A study of chitin decomposing micro-organisms of marine origin. *J. gen. Microbiol.*, **5**, 894–905.

CAMPELLO, F., BRISOU, J. and ROY, Y. DE LA (1963). Étude sur les relations existant entre le plancton et les bactéries dans les eaux portuaires de La Pallice. *C.r. Séanc. Soc. Biol.*, **157**, 618–623.

CANNON, H. G. and MANTON, S. M. (1927). On the feeding mechanism of a mysid crustacean, *Hemimysis lamornae*. *Trans. R. Soc. Edinb.*, **55**, 219–252.

CHAN, E. C. S. and MCMANUS, E. A. (1967). Development of a method for the total count of marine bacteria on algae. *Can. J. Microbiol.*, **13**, 295–301.

CHAN, E. C. S. and MCMANUS, E. A. (1969). Distribution, characterization, and nutrition of microorganisms from the algae *Polysiphonia lanosa* and *Ascophyllum nodosum*. *Can. J. Microbiol.*, **15**, 409–420.

CHRISTIE, A. O. and FLOODGATE, G. D. (1966). Formation of microtrees on surfaces submerged in the sea. *Nature, Lond.*, **212**, 308–310.

CLUTTER, R. J. (1967). Zonation of nearshore mysids. *Ecology*, **48**, 200–208.

COOKE, W. B. (1956). Colonization of artificial bare areas by microorganisms. *Bot. Rev.*, **22**, 613–638.

CORPE, W. A. (1964). Factors influencing growth and polysaccharide formation by strains of *Chromobacterium violaceum*. *J. Bact.*, **88**, 1433–1441.

COURTNEY, W. A. M. and WEBB, J. E. (1964). The effects of the cold winter of 1962/63 on the Helgoland population of *Branchiostoma lanceolatum* (PALLAS). *Helgoländer wiss. Meeresunters.*, **10**, 301–312.

CRISP, D. J. (1964). The effects of the winter of 1962/1963, on the British marine fauna. *Helgoländer wiss. Meeresunters.*, **10**, 313–327.

CRISP, D. J. (1965). Surface chemistry, a factor in the settlement of marine invertebrate larvae. *Botanica gothoburg.*, **3**, 51–65.

CVIIĆ, V. (1953). Attachment of bacteria to slides submerged in sea water. *Bilj. Inst. Oceanogr. Ribarst.*, **6**, 1–7.

DANIELLI, J. F. (1937). The relations between surface pH, ion concentrations and interfacial tension. *Proc. R. Soc. (B)*, **122**, 155–174.

DELAMARE-DEBOUTEVILLE, C. D. (1960). Biologie des eaux souterraines littorales et continentales. *Vie Milieu* (Suppl.), **9**, 1–740. (*Actual. scient. ind.* 1280.)

DESIKACHARY, T. V. (1959). *Cyanophyta.*, Indian Council of Agricultural Research, New Delhi.

DRAGOLI, A. L. (1960/61). Do biologii chernomorskoi polikheti *Melinna palmata* GRUBE. *Nauk. Zap. Odessa biol. St.*, **2**, 43–48; **3**, 71–83.

DUCHOW, E. and DOUGLAS, H. C. (1949). *Rhodomicrobium vannielii*, a new photoheterotrophic bacterium. *J. Bact.*, **58**, 409–416.

DUGUID, J. P. (1959). Fimbride and adhesive properties of *Klebsiella* strains. *J. gen. Microbiol.*, **21**, 271–286.

DURHAM, K. (1957). The influence of interfacial electrical conditions on the deposition of soil particles on to cotton from detergent solutions. *Proc. 2nd Int. Congr. Surf. Activ.*, **4**, 60–69.

EKMAN, S. (1935). *Tiergeographie des Meeres*, Akademische Verlagsgesellschaft, Leipzig.

ELLINGBOE, J. L. and WILSON, J. E. (1964). Direct method for the determination of organic carbon in sediments. *Analyt. Chem.*, **36**, 434–435.

EMERY, K. O., ORR, W. D. and RITTENBERG, S. C. (1955). Nutrient budgets in the ocean. In *Essays in the Natural Sciences in Honor of Capt. A. Hancock . . . 1955*. University of S. California Press, Los Angeles. pp. 299–309.

ERCEGOVIĆ, A. (1932a). Ekološke i sociološke Studije o litofitskim Cyanoficejama sa Jugoslavenske obale Jadrana. *Rad jugosl. Akad. Znan. Umjetn.*, **244**, 129–220.

ERCEGOVIĆ, A. (1932b). Etudes écologiques et sociologiques des Cyanophycées lithophytes de la cote Yougoslave de l'Adriatique. *Bull. int. Acad. yougosl. Sci.*, **26**, 33–56.

ERNST, W. (1970). ATP als Indikator für die Biomasse mariner Sedimente. *Oecologia (Berl.)*, **5**, 56–60.

ESTERMANN, E. F. and McLAREN, A. D. (1959). Stimulation of bacterial proteolysis by adsorbents. *J. Soil Sci.*, **10**, 64–78.

FELDMANN, J. (1937). Recherches sur la végétation marine de la Méditerranée. La côte des Albères. *Revue algol.*, **10**, 1–339.

FELDMANN, J. (1951). Ecology of marine algae. In G. M. Smith (Ed.), *Manual of Phycology*. Ronald Press, New York. pp. 313–334.

FRÉMY, P. (1945). Contribution à la physiologie des Thallophytes marins perforant et cariant des roches calcaires et les coquilles. *Annls Inst. océanogr., Monaco*, **22**, 107–144.

FRENKEL, A., GAFFRON, H. and BATTLEY, E. H. (1949). Photosynthesis and photoreduction by a species of blue-green algae. *Biol. Bull. mar. biol. Lab., Woods Hole*, **97**, 269.

FRIEDRICH, H. (1965). *Meeresbiologie. Eine Einführung in ihre Probleme und Ergebnisse.* Borntraeger, Berlin.

FRITSCH, F. E. (1959). *Structure and Reproduction of the Algae*, Vol. II, Cambridge University Press, Cambridge.

FUJITA, Y. and ZENITANI, B. (1967). On the filamentous bacterium *Leucothrix mucor* occurring as an epiphyte on thallus of *Porphyra*. I. On general microbial characteristics and environmental conditions of growth. (Japan.; Engl. title and abstract.) *Bull. Fac. Fish. Nagasaki Univ.*, **22**, 81–89.

GAMULIN-BRIDA, H., GIACCONE, G. and GOLUBIĆ, S. (1967). Contribution aux études des biocoenoses subtidales. *Helgoländer wiss. Meeresunters.*, **15**, 429–444.

GERLACH, S. A. (1960). Über das tropische Korallenriff als Lebensraum. *Zool. Anz.* (Suppl. Bd), **23**, 356–363. (Transl.: *Atoll Res. Bull.*, **80**.)

GIBBS, R. E. (1902). *Phyllospadix* as a beach-builder. *Am. Nat.*, **36**, 101–109.

GILBY, A. R. and FEW, A. V. (1957). Surface chemical studies on the protoplast membrane of *Micrococcus lysodeikticus*. *Proc. 2nd Int. Congr. Surf. Activ.*, **4**, 262–270.

GLYNN, P. W., ALMODÓVAR, L. R. and GONZÁLEZ, J. G. (1964). Effects of hurricane Edith on marine life in La Parguera, Puerto Rico. *Caribb. J. Sci.*, **42**, 335–345.

GOMOIU, M.-T. (1967). Some quantitative data on light penetration in sediments. *Helgoländer wiss. Meeresunters.*, **15**, 120–127.

GOMOIU, M.-T. and MÜLLER, G. I. (1962). Studies concerning the benthic association dominated by *Barnea candida* in the Black Sea. *Revue Biol. Buc.*, **7**, 255–271.

GRAY, J. S. (1965). Selection of sands by *Protodrilus symbioticus* GIARD. *Veröff. Inst. Meeresforsch. Bremerh.*, (Sonderbd) **2**, 105–116.

HAROLD, R. and STANIER, R. Y. (1955). The genera of *Leucothrix* and *Thiothrix*. *Bact. Rev.*, **19**, 49–58.

HARTOG, C. DEN (1958). De vegetatie van het Balgzand en de ooverterreinen van het Balgkanaal. *Wet. Meded. K. ned. natuurh. Veren.*, **27**, 1–28.

HARTOG, C. DEN (1959). The epilithic algal communities occurring along the coast of the Netherlands. *Wentia*, **1**, 1–241.

HARTOG, C. DEN (1967). Brackish water as an environment for algae. *Blumea*, **15**, 31–43.

HARTOG, C. DEN (1970). The sea-grasses of the world. *Verh. K. ned. Akad. Wet. afd. Natuurk.* (*Ser.* 2), **59**, 1–275.

HARTOG, C. DEN and SEGAL, S. (1964). A new classification of the water-plant communities. *Acta bot. neerl.*, **13**, 367–393.

HARVEY, G. W. (1966). Microlayer collection from the sea surface: a new method and initial results. *Limnol. Oceanogr.*, **11**, 608–613.

HARVEY, H. W. (1941). On changes taking place in sea water during storage. *J. mar. biol. Ass. U.K.*, **25**, 225–233.

HELLEBUST, J. A. (1965). Excretion of some organic compounds by marine phytoplankton. *Limnol. Oceanogr.*, **10**, 192–206.

HEMPEL, C. (1957). Über den Röhrenbau und die Nahrungsaufnahme einiger Spioniden (Polychaeta sedentaria) der deutschen Küsten. *Helgoländer wiss. Meeresunters.*, **6**, 100–135.

HENRICI, A. T. (1939). The distribution of bacteria in lakes. In *Problems of Lake Biology. Publs Am. Ass. Advmt Sci.*, **10**, 39–64.

HENRICI, A. T. and JOHNSON, D. E. (1935). Studies of freshwater bacteria. II. Stalked bacteria, a new order of Schizomycetes. *J. Bact.*, **30**, 61–93.

HESS, E. (1937). A shell disease of lobsters (*Homarus americanus*) caused by chitinovorous bacteria. *J. biol. Bd Can.*, **3**, 358–362.

HESSLER, R. and SANDERS, H. (1967). Faunal diversity in the deep sea. *Deep Sea Res.*, **14**, 65–78.

HEUKELEKIAN, H. and HELLER, A. (1940). Relation between food concentration and surface for bacterial growth. *J. Bact.*, **40**, 547–558.

HÖHNK, W. (1956). Mykologische Abwasserstudie. I. *Veröff. Inst. Meeresforsch. Bremerh.*, **4**, 67–110.

HOLM-HANSEN, O. and BOOTH, C. R. (1966). The measurement of adenosine triphosphate in the ocean and its ecological significance. *Limnol. Oceanogr.*, **11**, 510–519.

HOMMERIL, P. and RIOULT, M. (1962). Phénomènes d'érosion et de sedimentation marines entre Sainte-Honorine-des-Pertes et Port-en-Bessin (Calvados). Rôle de *Rhodothamnielle floridula* dans la retenue des sediments fins. *Cah. océanogr.*, **14**, 25–45.

HUGHES, D. A. (1966). Investigations of the 'Nursery Areas' and habitat preferences of juvenile penaeid prawns in Mozambique. *J. appl. Ecol.*, **3**, 349–354.

HUMM, H. J. (1964). Epiphytes on the sea-grass *Thalassia testudinum* in Florida. *Bull. mar. Sci. Gulf. Caribb.*, **14**, 306–341.

IERUSALIMSKII, N. D. (1954). An estimate of the rate of growth of aquatic microorganisms on submerged slides. (Russ.) *Mikrobiologiya*, **23**, 561–570.

JÄGERSTEN, G. (1952). Studies on the morphology, larval development and biology of *Protodrilus*. *Zool. Bidr. Upps.*, **29**, 426–512.

JAMES, A. M. (1957a). The identification of surface components on the bacterial cell wall. *Proc. 2nd Int. Congr. Surf. Activ.*, **4**, 254–261.

JAMES, A. M. (1957b). The electrochemistry of the bacterial surface. *Prog. Biophys. biophys. Chem.*, **8**, 95–142.

JANNASCH, H. W. (1960). *Caulobacter* sp. in sea water. *Limnol. Oceanogr.*, **5**, 432–433.

JØRGENSEN, C. D. (1966). *Biology of Suspension Feeding*, Pergamon Press, Oxford.

JOHNSON, T. W., FERCHAU, H. A. and GOLD, H. S. (1959). Isolation, culture, growth and nutrition of some lignicolous marine fungi. *Phyton, B. Aires*, **12**, 65–80.

JOHNSON, T. W. and SPARROW, F. K. (1961). *Fungi in Oceans and Estuaries.* J. Cramer, Weinheim, Germany.

JONES, G. E. (1958). Attachment of marine bacteria to zooplankton. *Spec. scient. Rep. U.S. Fish Wildl. Serv. Fisheries*, **279**, 77–78.

JONES, H. C., ROTH, I. L. and SANDERS, W. M. (1969). Electron microscopic study of a slime layer. *J. Bact.*, **99**, 316–325.

KINNE, O. (1963). The effects of temperature and salinity on marine and brackish water animals. I. Temperature. *Oceanogr. mar. Biol. A. Rev.*, **1**, 301–340.

KINNE, O. (1964). The effects of temperature and salinity on marine and brackish water animals. II. Salinity and temperature–salinity combinations. *Oceanogr. mar. Biol. A. Rev.*, **2**, 281–339.

KINNE, O. and AURICH, H. (Eds) (1968). Biological and hydrographical problems of water pollution in the North-Sea and adjacent waters. International Symposium, Helgoland, September 19–22, 1967. *Helgoländer wiss. Meeresunters.*, **17**, 1–530.

KISSELEVA, G. A. (1967). Vlijanie substrata na osedanie i metamorfoz lichinokbentosnykh zhivotnykh. In *Donnyie Biochenozy i Biologia Bentosnykh Organizmov Chernovo Moria*, Kiev, pp. 71–84.

KOHLMEYER, J. (1963). Importance of fungi in the sea. In C. H. Oppenheimer (Ed.), *Symposium on Marine Microbiology.* C. C. Thomas, Springfield, Ill., pp. 301–314.

KOHLMEYER, J. (1968). Revisions and descriptions of algicolous marine fungi. *Phytopath. Z.*, **63**, 341–363.

KOHLMEYER, J. and KOHLMEYER, E. (1964). *Synoptic Plates of Higher Marine Fungi.* J. Cramer, Weinheim, Germany.

KRISS, A. E. (1963). *Marine Microbiology (Deep Sea).* (Transl. from Russ.) Oliver and Boyd, London.

KRISS, A. E. and LAMBINA, V. A. (1955). Rapid increase of microorganisms in the ocean in the region of the North Pole. (Russ.) *Usp. sovrem. Biol.*, **39**, 366–373.

KRISS, A. E. and MARKIANOVICH, E. M. (1954). Observations on the rate of multiplication of microorganisms in marine environment. (Russ.) *Mikrobiologiya*, **23**, 551–560.

KRISS, A. E. and RUKINA, E. A. (1952). Biomass of microorganisms and their rates of reproduction in oceanic depths. (Russ.) *Zh. obshch. Biol.*, **13**, 346–362.

LAMI, R. (1934a). Sur l'hétérogénité de quelques caractères physiques des cuvettes littorales. *C. r. hebd. Séanc. Acad. Sci., Paris*, **198**, 1520–1529.

LAMI, R. (1934b). Sur l'alcalinisation spécifique et la répartition des algues dans les cuvettes littorales. *C. r. hebd. Séanc. Acad. Sci., Paris*, **199**, 615–617.

LEIFSON, E. B. J., COSENZA, R. M. and CLEVERDON, R. C. (1964). Motile marine bacteria. I. Techniques, ecology, and general characteristics. *J. Bact.*, **87**, 642–666.

LEPORE, N. (1962). Determinazione del carbonio organico e dei carbonati in sedimenti marini. *Archo Oceanogr. Limnol.*, **12**, 175–295.

LEWIN, J. C. (1963). Heterotrophy in marine diatoms. In C. H. Oppenheimer (Ed.), *Symposium on Marine Microbiology.* C. C. Thomas, Springfield, Ill. pp. 229–235.

LEWIS, J. R. (1964). *The Ecology of Rocky Shores*, English Universities Press, London.

LUCAS, C. E. (1961). Interrelationships between aquatic organisms mediated by external metabolites. In M. Sears (Ed.), *Oceanography.* A.A.A.S., Washington, D.C. pp. 499–517. (*Publs. Am. Ass. Advmt Sci.*, **67**.)

LUTHER, H. (1949). Vorschlag zu einer ökologischen Grundeinteilung der Hydrophyten. *Acta bot. fenn.*, **44**, 1–15.

McLAREN, A. D. and SKUJINS, J. (1968). The physical environment of micro-organisms in soil. In T. R. G. Gray and D. Parkinson (Eds), *The Ecology of Soil.* University of Toronto Press, Toronto, pp. 3–24.

MARGALEF, R. and RIVERO, J. A. (1958). Succession and composition of the *Thalassia* community. Assoc. Is. mar. Lab., Second Meeting, pp. 19–21 (unpubl. lecture).

MARSHALL, K. C. (1969). Studies by microelectrophoretic and microscopic techniques on the sorption of illite and montmorillonite to rhizobia. *J. gen. Microbiol.*, **56**, 301–306.

MARTIN, J. P. and RICHARDS, S. J. (1963). Decomposition and binding action of a polysaccharide from *Chromobacterium violaceum*. *J. Bact.*, **85**, 1288–1294.

MEADOWS, P. S. (1968). The attachment of bacteria to surfaces, and a way of measuring the rate of beat of bacterial flagella. *J. gen. Microbiol.*, **50**, 7.

MEADOWS, P. S. and ANDERSON, J. G. (1966). Micro-organisms attached to marine and freshwater sand grains. *Nature, Lond.*, **212**, 1059–1060.

MEADOWS, P. S. and ANDERSON, J. G. (1968). Micro-organisms attached to marine sand grains. *J. mar. biol. Ass. U.K.*, **48**, 161–175.

MEGURO, H., ITO, K. and FUKUSHIMA, H. (1966). Diatoms and the ecological conditions of their growth in sea ice in the Arctic Ocean. *Science, N.Y.*, **152**, 1089–1090.

MENZEL, D. W. and GOERING, J. J. (1966). The distribution of organic detritus in the ocean. *Limnol. Oceanogr.*, **11**, 1–333.

MEYERS, S. P., HOPPER, B. E. and CEFALU, R. (1970). Ecological investigations of the marine nematode *Metoncholaimus scissus*. *Mar. Biol.*, **6**, 43–47.

MEYERS, S. P. and REYNOLDS, E. S. (1960). Occurrence of lignicolous fungi in northern Atlantic and Pacific marine localities. *Can. J. Bot.*, **38**, 217–226.

MEYERS, S. P. and REYNOLDS, E. S. (1963). Degradation of lignocellulose material by marine fungi. In C. H. Oppenheimer (Ed.), *Symposium on Marine Microbiology*. C. C. Thomas, Springfield, Ill. pp. 315–328.

MITCHELL, P. (1951). Physical factors affecting growth and death. In C. H. Werkman and P. W. Wilson (Eds), *Bacterial Physiology*. Academic Press, New York. pp. 126–177.

MOLINIER, R. (1955). Note sur la répartition du *Rissoella verruculosa* (BERT.) J. AG. sur les côtes de la Méditerranée occidentale. *Bull. Mus. Hist. nat. Marseille*, **15**, 1–4.

MOLINIER, R. and PICARD, J. (1951). Biologie des herbiers de Zostéracées des côtes françaises de la Méditerranée. *C. r. hebd. Séanc. Acad. Sci., Paris*, **233**, 1212–1214.

MOLINIER, R. and PICARD, J. (1952). Recherches sur les herbiers de phanérogames marines du littoral méditerranéen français. *Annls Inst. océanogr., Monaco*, **27**, 157–234.

MOORE, B. G. and TISCHER, R. G. (1964). Extracellular polysaccharides of algae: Effects on life-support systems. *Science, N.Y.*, **145**, 586–587.

MOYER, L. S. (1936a). A suggested standard method for the investigation of electrophoresis. *J. Bact.*, **31**, 531–546.

MOYER, L. S. (1936b). Changes in the electrokinetic potential of bacteria at various phases of the culture cycle. *J. Bact.*, **32**, 433–464.

MÜLLER, G. (1964). *Methoden der Sediment-Untersuchung*, Schweizerbart, Stuttgart.

NALEWAJKO, C. (1966). Photosynthesis and excretion in various planktonic algae. *Limnol. Oceanogr.*, **11**, 1–10.

NEWCOMBE, C. L. (1949). Attachment materials in relation to water productivity. *Trans. Am. microsc. Soc.*, **68**, 355–361.

NEWCOMBE, C. L. (1950). A quantitative study of attachment materials in Sodon Lake, Michigan. *Ecology*, **31**, 204–215.

NIENBURG, W. (1925). Eine eigenartige Lebensgemeinschaft zwischen *Fucus* und *Mytilus*. *Ber. dt. bot. Ges.*, **43**, 292–298.

NIENBURG, W. (1927). Zur Ökologie der Flora des Wattenmeeres. I. Der Königshafen bei List auf Sylt. *Wiss. Meeresunters. Abt. Kiel*, **20**, 145–197.

NIENBURG, W. (1932). *Fucus mytili* spec. nov. *Ber. dt. bot. Ges.*, **50**, 28–41.

NORDIN, J. S., TSUCHIYA, H. M. and FREDRICKSON, A. G. (1967). Interfacial phenomenon governing adhesion of *Chlorella* to glass-surfaces. *Biotechnol. Bioengin.*, **9**, 545–558.

ODUM, E. P. and ODUM, H. T. (1955). Trophic structure and productivity of a windward coral reef community on Eniewetok Atoll. *Ecol. Monogr.*, **25**, 1–3.

OKUDA, T. (1964). Some problems for the determination of organic carbon in marine sediments. *Bolm Inst. Oceanogr., Cuma*, **3**, 106–117.

OPPENHEIMER, C. H. and VANCE, M. H. (1960). Attachment of bacteria to the surfaces of living and dead microorganisms in marine sediments. *Z. allg. Mikrobiol.*, **1**, 47–52.

OPPENHEIMER, C. H and WOOD, E. J. F. (1963). Note on the effect of contamination on a marine slough and the vertical distribution of unicellular plants in the sediment. *Z. allg. Mikrobiol.*, **2**, 45–47.

OSTENFELD, C. H. (1916). The sea-grasses of West Australia. *Dansk bot. Ark.*, **2**(4), 5–44.

OTTER, G. W. (1937). Rock destroying organisms in relation to coral reefs. *Scient. Rep. Gt Barrier Reef Exped.*, **1**, 1–323.

PAOLETTI, A. (1965). Pollution facale du littoral du Cumes et considérations sur le pouvoir d'autoépuration du milieu marin. *Revue int. Océanogr. med.*, **1**, 44–55.

PARKE, M. (1961). Some remarks concerning the class Chrysophyceae. *Br. Phycol. Bull.*, **2**(2), 47–55.

PEELE, T. C. (1936). Adsorption of bacteria by soils. *Mem. Cornell Univ. agric. Exp. Stn*, **197**, 1–18.

PÉRÈS, J. M. (1961). *Océanographie Biologique et Biologie Marine.* I. La Vie Benthique. Presses Université de France.

PERFIL'EV, B. V. and GABE, D. R. (1964). Study by the method of microbial landscape of bacteria which accumulate manganese and iron in bottom sedments. (Russ.) In M. S. Gurevich (Ed.), *The Role of Microorganisms in the Formation of Iron-Manganese Lake Ores.* Izd. 'Nauka', Moscow. pp. 16–53.

PERSOONE, G. (1965). Contributions à l'étude des bactéries marines du littoral Belge. II. Comparaison de plusieurs méthodes pour détacher et obtenir en suspension les bactéries contaminant des surfaces. *Bull. Inst. r. Sci. nat. Belg.*, **41**(12), 1–12.

PERSOONE, G. (1968). Ecologie des infusoires dans les salissures de substrats immergés dans un port de mer. I. Le film primaire et le recouvrement primaire. *Protistologica*, **4**, 187–194.

PHILLIPS, R. C. (1959). Notes on the marine flora of the Marquesas Keys, Florida. *Q. Jl Fla Acad. Sci.*, **22**, 155–162.

POINDEXTER, J. S. (1964). Biological properties and classification of the *Caulobacter* group. *Bact. Rev.*, **28**, 231–295.

POKORNY, K. (1967). *Labyrinthula.* *J. Protozool.*, **14**, 697–708.

POLLOCK, M. R. (1962). Exoenzymes. In I. C. Gunsalus and R. Y. Stanier (Eds), *The Bacteria*, Vol. IV. Academic Press, New York. pp. 121–295.

PRESCOTT, S. C., WINSLOW, C.-E. A. and McCRADY, M. H. (1946). *Water Bacteriology with Special Reference to Sanitary Water Analysis*, 6th ed., Wiley, New York.

PRUD'HOMME VAN REINE, W. F. and HOEK, C. VAN DEN (1966). Cultural evidence for the morphological plasticity of *Entophysalis deusta* (MENEGHINI) DOUET and DAILY (Chroococcales, Cyanophyceae). *Blumea*, **14**, 277–283.

REID, G. K. (1961). *Ecology of Inland Waters and Estuaries*, Reinhold, New York.

REMANE, A. (1940). Einführung in die zoologische Ökologie der Nord- und Ostsee. *Tierwelt N.– u. Ostsee*, 1a, 1–238.

RENN, C. E. (1937). Conditions controlling the marine bacterial population and its activity in the sea. *J. Bact.*, **33**, 86–87.

RIEDL, R. (1967). Die Tauchmethode, ihre Aufgaben und Leistungen bei der Erforschung des Litorals: eine kritische Untersuchung. *Helgoländer wiss. Meeresunters.*, **15**, 294–352.

RIGOMIER, D. (1967). Premier bilan de la population bactérienne hétérotrophe du zooplancton. *C. r. Séanc. Soc. Biol.*, **161**, 679–683.

ROLL, H. (1939). Zur Terminologie des Periphytons. *Arch. Hydrobiol.*, **35**, 59–69.

ROSENVINGE, L. K. and LUND, S. (1947). *The Marine Algae of Denmark.* II. Phaeophyceae, Pt 3.

ROUND, F. E. (1964). The ecology of benthic algae. In D. F. Jackson (Ed.), *Algae and Man.* Academic Press, New York. pp. 138–184.

ROUND, F. E. (1965). The epipsammon; a relatively unknown freshwater algal association. *Br. phycol. Bull.*, **2**, 456–462.

ROUVILLOIS, A. (1967). Observations morphologiques, sédimentologiques et écologiques sur la plage de la Ville Ger dans l'estuaire de la Rance (Côtes-du-Nord). *Cah. océanogr.*, **19**, 375–389.

RUTTNER, F. (1963). *Fundamentals of Limnology*, 3rd ed. (Transl. by D. G. Frey and F. E. J. Fry.) University of Toronto Press, Toronto.

SALTON, M. R. J. (1957). The action of lytic agents on the surface structures of the bacterial cell. *Proc. 2nd Int. Congr. Surf. Activ.*, **4**, 245–253.

SALTON, M. R. J. (1964). *The Bacterial Cell Wall*, Elsevier, New York.

SALVO, L. H. DI (1965). Preliminary studies on some microecological aspects of Windward Atoll reefs. M.Sci. thesis, University of Arizona, Tucson.

SANTORO, I. and STOTZKY, G. (1968). Sorption between microorganisms and clay minerals as determined by the electrical sensing zone particle analyzer. *Can. J. Microbiol.*, **14**, 299–307.

SCHÜTZ, L. and KINNE, O. (1955). Über die Mikro- und Makrofauna der Holzpfähle des Nord-Ostsee-Kanals und der Kieler Förde. *Kieler Meeresforsch.*, **11**, 110–135.

SEKI, H. (1967). Ecological studies on the lipolytic activity of microorganisms in the sea of Aburatsubo Inlet. *Rec. oceanogr. Wks Japan*, **9**, 75–113.

SHEWAN, J. M. (1961). The microbiology of sea-water fish. In G. Borgstrom (Ed.), *Fish as Food*. Academic Press, New York. pp. 487–560.

SHUTTLEWORTH, T. H. and JONES, T. G. (1957). Particle size and detergency. *Proc. 2nd Int. Congr. Surf. Activ.*, **4**, 52–59.

SINDERMANN, C. J. (1966). Diseases of marine fishes. *Adv. mar. Biol.*, **4**, 1–89.

SKERMAN, T. M. (1956). The nature and development of primary films on surfaces submerged in the sea. *N.Z. Jl Sci. Technol. (B)*, **38**, 44–57.

SLÁDEČKOVÁ, A. (1962). Limnological investigation methods for the periphyton 'Aufwuchs' community. *Bot. Rev.*, **28**, 286–350.

SORIANO, S. and LEWIN, R. A. (1965). Gliding microbes: some taxonomic reconsiderations. *Antonie van Leeuwenhoek*, **31**, 66–80.

SPENCER, C, P. (1952). On the use of antibiotics for isolating bacteria free cultures of marine phytoplankton organisms. *J. mar. biol. Ass. U.K.*, **31**, 97–106.

STARR, M. P. and SKERMAN, V. B. D. (1965). Bacterial diversity: The natural history of selected morphologically unusual bacteria. *A. Rev. Microbiol.*, **19**, 407–454.

STEEMANN NIELSEN, E. (1955). The production of antibiotics by plankton algae and its effect upon bacterial activity in the sea. In *Papers in Marine Biology and Oceanography. Deep Sea Res.* (Suppl.), **3**, 281–286.

STODDART, D. R. (1963). Effects of hurricane Hattie on the British Honduras reefs and cays. October 30–31, 1961. *Atoll. Res. Bull.*, **95**, 1–142.

STOTZKY, G. and REM, L. T. (1966). Influence of clay minerals on micro-organisms. I. Montmorillonite and kaolinite on bacteria. *Can. J. Microbiol.*, **12**, 547–563.

STRICKLAND, J. D. H. (1958). Solar radiation penetrating the ocean. *J. Fish. Res. Bd Can.*, **15**, 453.

SUTCLIFFE, W. H., BAYLOR, E. R. and MENZEL, D. W. (1963). Sea surface chemistry and Langmuir circulation. *Deep Sea Res.*, **10**, 233–245.

SWEDMARK, B. (1956). Etude de la microfaune des sables marins de la région de Marseille. *Archs Zool. exp. gén.*, **93**, 70–95.

TEPPER, J. G. O. (1882a). Some observations on the propagation of *Cymodocea antarctica* (ENDL.). *Trans. Proc. R. Soc. S. Austr.*, **1880–81**, 1–4.

TEPPER, J. G. O. (1882b). Further observations on the propagation of *Cymodocea antarctica. Trans. Proc. R. Soc. S. Austr.*, **1880–81**, 47–49.

THIEL, H. (1966). Quantitative Untersuchungen über die Meiofauna von Tiefseeböden (*Meteor*-Exp.). *Veröff. Inst. Meeresforsch. Bremerh.* (Sonderbd), **2**, 131–147.

THOMAS, L. P., MOORE, D. R. and WORK, R. C. (1961). Effects of hurricane Donna on the turtle-grass beds of Biscayne Bay, Florida. *Bull. mar. Sci. Gulf Caribb.*, **11**, 191–197.

THORSON, G. (1946). Reproduction and larval development of Danish marine bottom invertebrates. *Meddr Kommn Danm. Fisk.-og Havunders.* (Ser. Plankton), **4**, 1–523.

TURNER, R. D. (1966). *A Survey and Illustrated Catalogue of the Teredinidae (Mollusca: Bivalvia)*, Vols I, II. Museum of Comparative Zoology, Cambridge, Mass.

ULANOVSKIY, I. B. and GERASIMENKO, A. D. (1965). The influence of algae on the corrosion of carbon steel in sea water and the effects of ultrasonic vibrations on the intensity of algal photosynthesis. In I. V. Starostin (Ed.), *Marine Fouling and Borers*, Vol. 70. Trudy Instituta Okeanologii. (Transl. 221, U.S. Naval Oceanographic Office, Washington, D.C.)

VINOGRADOV, M. E., GITELZON, I. I. and SOROKIN, YU.I. (1970). The vertical structure of a pelagic community in the tropical ocean. *Mar. Biol.*, **6**, 187–194.

WAERN, M. (1940). *Cladophora pygmaea* und *Leptonema lucifugum* an der schwedischen Westküste. *Acta phytogeogr. suec.* **13**, 1–6.

WAKSMAN, S. A., REUSZER, H. W., CAREY, C. L., HOTCHKISS, M. and RENN, C. E. (1933). Studies on the biology and chemistry of the Gulf of Maine. III. Bacteriological investigations of the sea water and marine bottoms. *Biol. Bull. mar. biol. Lab., Woods Hole*, **64**, 183–205.

WAKSMAN, S. A. and VARTIOVAARA, U. (1938). The adsorption of bacteria by marine bottom. *Biol. Bull. mar. biol. Lab., Woods Hole*, **74**, 56–63.

WALGER, E. (1964). Zur Darstellung von Korngrößenverteilungen. *Geol. Rdsch.*, **54**, 976–1002.

WANGERSKY, P. J. (1965). The organic chemistry of sea water. *Am. Scient.*, **53**, 358–374.

WEISS, C. M. (1951). Adsorption of *E. coli* on river and estuarine silts. *Sewage ind. Wastes*, **23**, 227–237.

WILBUR, K. M. and YONGE, C. M. (1966). *Physiology of Mollusca*, Vols I, II. Academic Press, London.

WILKINSON, J. F. (1958). The extracellular polysaccharides of bacteria. *Bact. Rev.*, **22**, 46–73.

WILSON, I. M. (1960). Marine fungi: a review of the present position. *Proc. Linn. Soc. Lond.*, **171**, 53–70.

WOLFF, T. (1960). The hadal community: an introduction. *Deep Sea Res.*, **6**, 95–124.

WOOD, E. J. F. (1956). Diatoms in the ocean deeps. *Pacif. Sci.*, **10**, 377–381.

WOOD, E. J. F. (1959). Some aspects of the ecology of Lake Macquarie, N.S.W. 6. *Aust. J. mar. Freshwat. Res.*, **10**, 322–340.

WOOD, E. J. F. (1964). Studies in microbial ecology of the Australasian region. *Nova Hedwigia*, **8**, 5–54; 453–568.

WOOD, E. J. F. (1965). *Marine Microbial Ecology*, Reinhold, New York.

WOOD, E. J. F. (1967). *Microbiology of Oceans and Estuaries*, Elsevier, New York.

WOODHEAD, P. M. J. (1964a). The death of North Sea fish during the winter of 1962/63, particularly with reference to the sole, *Solea vulgaris*. *Helgoländer wiss. Meeresunters.*, **10**, 283–300.

WOODHEAD, P. M. J. (1964b). Changes in the behaviour of the sole, *Solea vulgaris*, during cold winters, and the relation between the winter catch and sea temperatures. *Helgoländer wiss. Meeresunters.*, **10**, 328–342.

WOODS HOLE OCEANOGRAPHIC INSTITUTION (1952). *Marine Fouling and its Prevention*, United States Naval Institute, Annapolis.

ZAITSEV, P. (1961). Surface pelagic biocoenose of the Black Sea. *Zool. Zh.*, **40**, 818–825.

ZAITSEV, P. (1968). La neustonologie marine: objet, méthodes, réalisations principales et problèmes. *Pelagos (Alger)*, **8**, 1–48.

ZENKEVITCH, L. A. (1956). *Morja SSR ich fauna i flora*. Moscow.

ZIEGELMEIER, E. (1964). Einwirkungen des kalten Winters 1962/63 auf das Makrobenthos im Ostteil der Deutschen Bucht. *Helgoländer wiss. Meeresunters.*, **10**, 276–282.

ZIEMAN, J. C., JR. (1968). A study of the growth and decomposition of the sea grass *Thalassia testudinum*. M. Sci. thesis, University of Miami, Miami, Fla.

ZOBELL, C. E. (1943). The effect of solid surfaces upon bacterial activity. *J. Bact.*, **46**, 39–56.

ZOBELL, C. E. (1946). *Marine Microbiology*, Chronica Botanica, Waltham, Mass.

ZOBELL, C. E. (1968). Bacterial life in the deep sea. *Bull. Misaki mar. biol. Inst.*, **15**, 77–96.

ZOBELL, C. E. and ALLEN, E. C. (1933). Attachment of marine bacteria to submerged slides. *Proc. Soc. exp. Biol. Med.*, **30**, 1409–1411.

ZOBELL, C. E. and ANDERSON, D. Q. (1936). Observations on the multiplication of bacteria in different volumes of stored sea water and the influence of oxygen tension and solid surfaces. *Biol. Bull. mar. biol. Lab., Woods Hole*, **71**, 324–342.

ZOBELL, C. E., GRANT, C. W. and HAAS, H. F. (1943). Marine microorganisms which oxidize petroleum hydrocarbons. *Bull. Am. Ass. Petrol. Geol.*, **27**, 1175–1193.

ZVYAGINTSEV, D. G. (1959). Adsorption of microorganisms by glass surface. (Russ.) *Mikrobiologiya*, **28**, 112–115.

ZVYAGINTSEV, D. G. (1962). Some regularities of adsorption of microorganisms on ion exchange resins. (Russ.) *Mikrobiologiya*, **31**, 275–277.

8. PRESSURE

8.0 GENERAL INTRODUCTION

O. KINNE

(1) General Aspects of Pressure

(a) Physical Aspects

The static pressure p at a defined point in a liquid is given by the force F acting in a normal direction on a surface area A:

$$p = \frac{F}{A} \qquad (1)$$

In contrast to the vectors F and A, p is a scalar quantity. By measuring the force F acting on a piston with an effective area A, the pressure under the piston can be determined. If F is the weight of a water column with a specific weight γ, its value can be obtained by employing the equation

$$F = \gamma\, Ad \qquad (2)$$

where $A =$ horizontal area, and $d =$ vertical depth of the water column. If the value for F is introduced into (1), we obtain

$$p = \gamma d \qquad (3)$$

In a resting fluid, Equation (3) defines **hydrostatic pressure**. It acts in all directions with the same intensity. Unless otherwise specified, the term 'pressure' employed in Chapter 8 refers to hydrostatic pressure.

The following classes of pressure may be distinguished (TROSKOLANSKI, 1960):
  (i) absolute pressure, p_a, measured relative to perfect vacuum;
 (ii) positive pressure, p_{pos}, i.e. the excess of p_a over the barometric pressure p_b at a given time and place, whereby p_b is always measured relative to perfect vacuum ($p_{pos} = p_a - p_b$);
(iii) negative pressure, p_{neg}, i.e. the difference between the barometric pressure p_b and the absolute pressure p_a ($p_{neg} = p_b - p_a$).*

One and the same pressure can be expressed by p_a and p_{neg} (on condition that $p_a < p_b$). For example, the absolute pressure $p_a = 0.8$ at, at a barometric pressure $p_b = 1.0$ at, can also be expressed by the negative pressure $p_{neg} = 1.0 - 0.8 = 0.2$ at. Below the free surface of water of specific weight γ, the absolute pressure at a depth d is the sum of the barometric and hydrostatic pressures

$$p_a = p_b + \gamma d \qquad (4)$$

---

* 'Positive pressure' should not be confused with the notion 'a pressure higher than the atmospheric'. It must also be differentiated between 'negative pressure' and 'a pressure lower than atmospheric', since it follows from the inequality $p_a < p_b$ that $p_{neg} = p_b - p_a$ has a physical meaning only within the limits $p_a = 0$ to $p_a = p_b$; hence negative pressures can only assume values from $p_{neg} = p_b$ to $p_{neg} = 0$ (TROSKOLANSKI, 1960).

Table 8-1

Basic units of pressure (After TROSKOLANSKI, 1960; modified)

	System of units	Basic units	Symbol	Derived units	Symbol
Absolute or physical dimension system (LMT)	m–kg–sec	pascal or newton/m²	N/m²	—	—
	cm–g–sec	dyne/cm² or microbar	b	1 bar = 10^6 b = 10^6 dynes/cm² 1 millibar = 10^{-3} bar = 10^3 b	—
	m–ton–sec	pièze or sten/m²	pz	1 hectopièze = 100 pz = 1·01972 at or 1 bar	hpz
	yard–pound–second	pound per square foot	lb/sq. ft	1 pound per square inch = 144 lb/sq. ft	—
Engineering or gravitational dimension system (LFT)	m–kp–sec (metre–kilopond–second)	kilogram-force per square metre kilopond* per square metre	kg/m² kp/m²	1 metric or practical atmosphere = 1 kp/cm² = 10^4 kp/m²	at
	yard–poundal–second	poundal per square foot	pdl/sq. ft	1 pdl/sq. ft = $\dfrac{1}{32 \cdot 174}$ lb/sq. ft	—
	yard–pound force–second	pound force per square foot	lbf/sq. ft	1 lbf/sq. in = 144 lbf/sq. ft	—

* Kilopond is the term suggested in Poland for the kilogram-force.

Dividing Equation (4) by γ gives the absolute pressure head (TROSKOLANSKI, 1960)

$$\frac{p_a}{\gamma} = \frac{p_b}{\gamma} + d \tag{5}$$

Although the absolute pressure head has the dimension of length, it can express the degree of pressure in terms of the height of a manometric column, usually mercury or water.

Marine ecologists and oceanographers use a number of units to designate pressure. Basic units of pressure are listed in Table 8-1; units of pressure head, in Table 8-2. For conversion of pressure units, consult Table 8-3.

Table 8-2

Units of pressure head (After TROSKOLANSKI, 1960; modified)

Basic units	Symbol	Derived units
Millimetre of mercury column $(t = 0°C; g = 980·655 \text{ cm/sec}^2)$	mm Hg (Torr)	1 standard atmosphere (atm) $= 760$ Torr $= 29·9201$ in Hg
Inch of mercury column $(60°F; g = 32·174 \text{ ft/sec}^2)$	in Hg	—
Millimetre of water column $(4°C; g = 980·655 \text{ cm/sec}^2) \approx 1 \text{ kp/m}^2$	mm H$_2$O	—
Foot of water column $(60°F; g = 32·174 \text{ ft/sec}^2)$	ft H$_2$O	inch of water column

In marine ecology, the most frequently used unit of pressure is the atmosphere (atm). An atmosphere is defined as the pressure exerted per cm^2 by a 760 mm high column of mercury at 0°C (Hg density $= 13·5951$ g/cm^3), where the acceleration of gravity is 980·665 cm/sec^2. The acceleration of gravity depends primarily on the geographic latitude (rotation and ellipticity of the earth); for marine ecological purposes, it may be taken to be 980 cm/sec^2 in middle latitudes, 978 at the Equator and 983 at the Poles. In English-speaking countries, many biologists have used pounds per square inch (psi) as pressure unit. 1 atm is equivalent to 14·696 psi, 1·033227 kg/cm^2, 1·013250 bars $= 10·13250$ decibars, or 1,013,250 dynes/cm^2.

In chemical oceanography, the Torr—a unit related to the atm—is employed; it equals the pressure exerted per cm^2 by a column of mercury 1 mm high at 0°C and at an acceleration of gravity of 980·665 cm/sec^2. In physical oceanography, the main unit of sea pressure is 1 decibar (dbar) $= 0·1$ bar $= 10^5$ dyne/cm^2. The pressure exerted per cm^2 by a 1 m high column of sea water very nearly equals 1 decibar; consequently, the pressure in the sea increases by 1 decibar for approximately every metre of depth. Hence water depth in metres and pressure in decibars are expressed by nearly the same numerical value (pp. 1357/8). In meteorology, the unit of pressure is 1 mbar $= 10^3$ dyne/cm^2. In acoustics, sound pressure is expressed in μbar. The unit of the MKS (m–kg–sec) system is N/m^2 (Newton per m^2), the unit of the CGS (cm–g–sec) system, dyne/cm^2. 1 dyne/cm$^2 = 0·1$ N/m$^2 = 1$ μbar $= 10^{-6}$

D

Table 8-3   Conversion table for pressure

Pressure unit	Symbol	N/m²	atm	at	pdl/ft²	lbf/ft²
Absolute unit (m–kg–sec)	N/m²	1	9.86923×10^{-6}	1.0197×10^{-5}	6.7197×10^{-1}	2.0885×10^{-2}
Standard atmosphere	atm	1.01325×10^5	1	1.03323	6.8087×10^4	2.1162×10^3
Metric atmosphere	at	9.80665×10^4	9.67841×10^{-1}	1	6.5898×10^4	2.0482×10^2
Poundal per sq. ft	pdl/ft²	1.48816	1.4687×10^{-5}	1.5175×10^{-5}	1	3.1081×10^{-2}
Pound-force per sq. ft	lbf/ft²	4.78803×10	4.7234×10^{-4}	4.8824×10^{-4}	3.2174×10	1
psi (Pound-force per sq. in.)	psi lbf/in²	6.89476×10^3	6.8046×10^{-2}	7.0307×10^{-2}	4.6331×10^3	144
Ton-force per sq. in.	tonf/in²	1.54443×10^7	1.5242×10^2	1.5749×10^2	1.0378×10^7	3.2256×10^5
Millimeter of mercury column	mm Hg (Torr)	1.33322×10^2	1.31579×10^{-3}	1.3595×10^{-3}	8.9588×10	2.7845
Inch of mercury column	in Hg	3.38639×10^3	3.3421×10^{-2}	3.4531×10^{-2}	2.2756×10^3	7.0726×10
Millimeter of water column	mm H₂O	9.80665	9.6784×10^{-5}	1.0000×10^{-4}	6.5898	2.0482×10^{-1}
Foot of water column	ft H₂O	2.98907×10^3	2.9500×10^{-2}	3.0480×10^{-2}	2.0086×10^3	6.2428×10
Inch of water column	in H₂O	2.49089×10^2	2.4583×10^{-3}	2.540×10^{-3}	1.6738×10^2	5.2023

bar or 1 kp/cm² = 1 at (metric or technical atmosphere) = 0·981 bar; 1000 bar are also referred to as 1 kilobar (kbar).

One of the most important physical aspects of variations in pressure is volume change. Since CANTON (1762) discovered that liquids are compressible, a multitude of experiments have been conducted on volume changes of physical, chemical and biological systems (see also *Chemical Aspects* and *Biological Aspects*). The application of increased pressures reflects the remarkable physical properties of water (Figs 8-1, 8-2, 8-3, 8-4, 8-5; see also Chapter 1). For further information concerning high pressure properties of water consult the review by LAWSON and HUGHES (1963). Aspects of mechanical engineering, emphasizing stress analysis of pressure vessels and their components, have been dealt with by GILL (1970).

(b) Chemical Aspects

High pressures may displace chemical equilibria. SPRING and VAN'T HOFF (1887), for example, showed that the volume of the double salt copper-calcium acetate

$$Ca \cdot Cu(CH_3CO_2)_4 \cdot 8H_2O$$

units (Compiled from various sources)

psi (lbf/in²)	tonf/in²	mm Hg (Torr)	in Hg	mm H₂O	ft H₂O	in H₂O
$1\cdot4504 \times 10^{-4}$	$6\cdot4749 \times 10^{-5}$	$7\cdot5006 \times 10^{-3}$	$2\cdot9530 \times 10^{-2}$	$1\cdot0197 \times 10^{-1}$	$3\cdot3456 \times 10^{-4}$	$4\cdot0146 \times 10^{-3}$
$1\cdot4696 \times 10$	$6\cdot5607 \times 10^{-3}$	760	$2\cdot9921 \times 10$	$1\cdot03323 \times 10^{4}$	$3\cdot3899 \times 10$	$4\cdot0678 \times 10^{2}$
$1\cdot4223 \times 10$	$6\cdot3497 \times 10^{-3}$	$7\cdot3556 \times 10^{2}$	$2\cdot8959 \times 10$	$1\cdot0000 \times 10^{4}$	$3\cdot2808 \times 10$	$3\cdot9370 \times 10^{2}$
$2\cdot1582 \times 10^{-4}$	$9\cdot6348 \times 10^{-8}$	$1\cdot1162 \times 10^{-2}$	$4\cdot3941 \times 10^{-4}$	$1\cdot5175 \times 10^{-1}$	$4\cdot9787 \times 10^{-4}$	$5\cdot9744 \times 10^{-3}$
$6\cdot9444 \times 10^{-3}$	$3\cdot1002 \times 10^{-6}$	$3\cdot5913 \times 10^{-1}$	$1\cdot4139 \times 10^{-2}$	$4\cdot8824$	$1\cdot6018 \times 10^{-2}$	$1\cdot9222 \times 10^{-1}$
1	$4\cdot4643 \times 10^{-4}$	$5\cdot1715 \times 10$	$2\cdot0360$	$7\cdot0307 \times 10^{2}$	$2\cdot3067$	$2\cdot7680 \times 10$
2240	1	$1\cdot1584 \times 10^{5}$	$4\cdot5607 \times 10^{3}$	$1\cdot5749 \times 10^{6}$	$5\cdot1669 \times 10^{3}$	$6\cdot2003 \times 10^{4}$
$1\cdot9337 \times 10^{-2}$	$8\cdot6324 \times 10^{-6}$	1	$3\cdot9370 \times 10^{-2}$	$1\cdot3595 \times 10$	$4\cdot4603 \times 10^{-2}$	$5\cdot3524 \times 10^{-1}$
$4\cdot5115 \times 10^{-1}$	$2\cdot1926 \times 10^{-4}$	$25\cdot4$	1	$3\cdot4531 \times 10^{2}$	$1\cdot1329$	$13\cdot5951$
$1\cdot4223 \times 10^{-3}$	$6\cdot3497 \times 10^{-7}$	$7\cdot3556 \times 10^{-2}$	$2\cdot8959 \times 10^{-3}$	1	$3\cdot2808 \times 10^{-3}$	$3\cdot9370 \times 10^{-2}$
$4\cdot3353 \times 10^{-1}$	$1\cdot9354 \times 10^{-4}$	$2\cdot2420 \times 10$	$8\cdot8267 \times 10^{-1}$	$3\cdot048 \times 10^{2}$	1	12
$3\cdot6127 \times 10^{-2}$	$1\cdot6128 \times 10^{-5}$	$1\cdot8683$	$7\cdot3556 \times 10^{-12}$	$25\cdot4$	$8\cdot3333 \times 10^{-2}$	1

is larger than that of its components

$$Ca(CH_3CO_2)_2 \cdot H_2O + Cu(CH_3CO_2)_2 \cdot H_2O + 6H_2O$$

and predicted that the salt should, therefore, dissociate by an increase in pressure. In their experiments, they found that dissociation proceeds rapidly at about 6000 atm and 40°C. SPRING had noticed earlier that pressure increase favours the association of substances whose volumes are smaller than those of their components. PLANCK (1887) demonstrated the existence of an exact quantitative relationship between the degree of the volume change accompanying a chemical reaction and the effect of pressure on its equilibrium constant; his formula remains one of the most important in the field of high pressure chemistry; it has been amply verified by experiments (HAMANN, 1963a).

In his review of pressure effects on chemical equilibria, HAMANN (1963a) considers some general thermodynamic relations in concentrated and diluted solutions; he also discusses reactions between nearly non-polar molecules, reactions involving polar molecules, and reactions involving ions. Although direct information concerning pressure effects on equilibria between nearly non-polar molecules in the liquid phase is lacking, pronounced effects can only be expected for reactions in

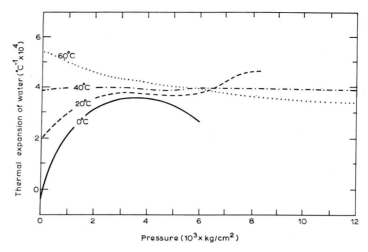

Fig. 8-1: Thermal expansion coefficient of pure water as a
function of pressure at different temperatures. The relationship is complicated and irregular. (After BRIDGMAN, 1949;
modified.)

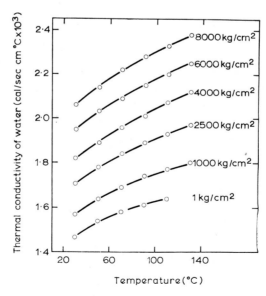

Fig. 8-2: Adjusted thermal conductivity of
pure water as a function of temperature
at different pressures. Data adjusted to
agree with most probable values at atmospheric pressure. (After LAWSON and co-
authors, 1959; modified.)

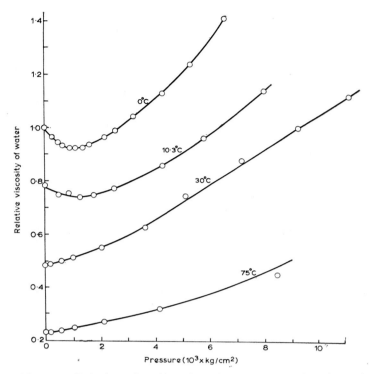

Fig. 8-3: Relative viscosity of pure water as a function of
pressure at different temperatures. In the lower part of the
pressure scale, increasing pressure causes a decrease in viscosity at low temperatures; according to LAWSON and HUGHES
(1963), this unique fact may be explained as a decrease in
structural viscosity in the framework of the two-fluid model.
(After BRIDGMAN, 1949; modified.)

which covalent bonds are formed or broken. At the molecular level, such effects
arise from the large differences between the 'van der Waals radii' of atoms and
their covalent radii. For further details consult HAMANN's (1963a) paper.

The rate of chemical reactions may be modified considerably by variations in
pressure. According to HAMANN (1963b), pressure effects on chemical kinetics can
be just as great and as varied as those caused by changes in temperature;

'in fact there are reasons to believe that very high pressures will cause the
spontaneous breaking of covalent bonds (HIRSCHFELDER et al., 1954) and in
this way bring about extremely fast reactions. The pressures needed to do
this are probably above a million atmospheres, and they have not yet been
reached under static conditions' (HAMANN, 1963b, p. 163).

According to MARSLAND (1970), it is possible to predict the effects of pressure
upon isolated physical or chemical reactions on the basis of thermodynamic
reasoning, when such reactions are classified, according to their energy–volume
relationships, into three types. Type 1 reactions liberate energy and are generally
accompanied by a volume decrement in the system. Accordingly, exergonic reac-

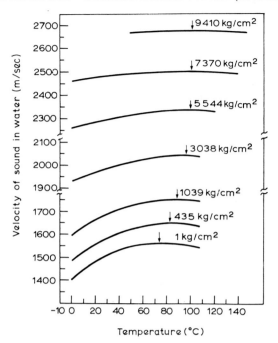

Fig. 8-4: Velocity of sound in pure water as a function of temperature at different pressures. The temperatures at which the maxima of sound velocity occur (arrows) are only slightly different from those for the minima in compressibility due to the difference between the isothermal and adiabatic moduli. (After SMITH and LAWSON, 1954; modified.)

tions tend to occur spontaneously, without an additional energy source. The steady states of such reactions are shifted forward by increasing pressure, but backward by increasing temperature. The magnitude of the pressure effect is determined by the extent of the volume change. Type 2 reactions absorb energy as they proceed and, generally, there is a volume increment in the system; hence type 2 reactions do not continue spontaneously but must receive appropriate energy, which permits them to do work against the ambient pressure. Consequently, the steady states tend to be shifted in the sol direction by increasing pressure and in the gel direction by increasing temperature. Type 3 reactions appear to be relatively unimportant in relation to effects of high pressure. They are virtually isogonic and little, if any, volume change occurs. Hence type 3 reactions are scarcely affected by variations in ambient pressure or temperature.

Of course, biological reactions are not isolated and, consequently, the manner in which they are affected by pressure and temperature is less predictable than is the case for the afore-mentioned physical or chemical reactions.

Variations in pressure modify a number of physical and chemical parameters of sea water, for example, density, friction, viscosity (e.g. HATSCHEK, 1958; STEELE and WEBB, 1963b; HORNE and JOHNSON, 1966, 1967), compressibility (e.g. STEELE

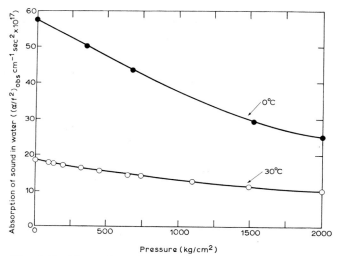

Fig. 8-5: Absorption of sound in pure water: absorption coefficient per cm divided by frequency squared as a function of pressure at 0° and 30°C. Sound absorption in water is anomalously high and depends on pressure, especially at low temperatures. (After LITOVITZ and CARNEVALE, 1955; modified.)

and WEBB, 1963a), heat expansion, specific heat, adiabatic temperature, thermal conductivity, acoustical properties (e.g. LAWSON and HUGHES, 1963), electrical conductivity (e.g. HAMON, 1956; GROH, 1963; HORNE and FRYSINGER, 1963; VOIGT, 1964), ionization (e.g. OWEN and BRINKLEY, 1941; HAMANN, 1963c; DISTÈCHE and DISTÈCHE, 1965), and pH (BUCH and GRIPENBERG, 1932; DISTÈCHE, 1959; PYTKOWICZ and CONNERS, 1964). See also HORNE's 'Marine Chemistry' (1969).

For further details concerning physical and chemical aspects of hydrostatic pressure, the reader is referred to NEWITT (1940), BRIDGMAN (1949), COMINGS (1956), HAMANN (1957, 1964), TROSKOLANSKI (1960), BUNDY and co-authors (1961), BRUNS (1962, 1968), WENTORF (1962), BRADLEY (1963), WESTPHAL (1963), FAIRBRIDGE (1966), GERTHSEN (1966), ARONSON (1967), NEUMANN (1968).

(c) Biological Aspects

All organisms which live below the sea's surface are subject to pressures in excess of 1 atm. Changes in pressure due to vertical migrations or to occupancy of different depths may significantly affect life processes in a number of ways (e.g. CATTELL, 1936; EBBECKE, 1944; JOHNSON and co-authors, 1954; ZENKEVITCH and BIRSHTEIN, 1956; BRUUN, 1957; MENZIES, 1965; KNIGHT-JONES and MORGAN, 1966; DIGBY, 1967; KUHN and STROTKOETTER, 1967; MORITA, 1967; SCHLIEPER, 1968a; ZIMMERMAN, 1970a; Chapters 8.1, 8.2, 8.3). There can be no doubt: sea pressure is of considerable importance as an ecological variable in oceans and coastal waters. This environmental factor requires much more attention in future studies than it has hitherto received.

Terminology

The study of pressure effects on organisms is known as **barobiology**. Pressures higher than standard atmospheric are called **hyperbaric**, those lower than standard atmospheric **hypobaric**. Organisms which inhabit wide ranges of pressures are **eurybaric** or, expressed in terms of wide ranges of depth, **eurybath**; those restricted to narrow pressure ranges are **stenobaric** (**stenobath**). Stenobaric forms limited to high pressures are **polystenobaric** (**polystenobath**), those restricted to low pressures **oligostenobaric** (**oligostenobath**).

In microbiology, organisms which grow well at pressures higher than 400 to 500 atm are referred to as **barophilic**, while organisms which grow poorly or not at all at pressures exceeding 300 to 400 atm are called **barophobic**; bacteria which tolerate a wide range of pressure have also been termed **barotolerant** or **baroduric** (Chapter 8.1).

General considerations

What are the biological properties which allow some organisms to inhabit wide depth ranges, whereas others— sometimes close relatives—are restricted to specific thin layers of water? Why can polystenobaric forms exist only in the deep sea under considerable pressures? What limits oligostenobaric organisms to near-surface waters? We do not know. Our present knowledge on ecologically meaningful aspects of pressure effects is quite limited. The few pioneering studies on organismic responses to variations in pressure intensities have not yet produced a basis sufficiently solid to allow detailed analyses.

'In view of the vast extent of the biosphere having a hydrostatic pressure higher than atmospheric, it seems paradoxical that so little is known about the effects of increased pressure on the structure, survival, growth, and biochemical activities of organisms' (ZoBELL, 1970, p. 85/86).

The inadequacy of the information available on ecological perspectives of pressure effects is due primarily to (i) historical aspects: the early marine ecologists have focussed their efforts largely on the description of marine life rather than on analytical studies, or have restricted their attention to a few 'classical' environmental factors, such as temperature and salinity; and (ii) technical or methodological problems: it is very difficult to collect, maintain and test deep-sea organisms under conditions which are ecologically meaningful.

However, progress is in sight. The importance of pressure as an environmental factor in the sea is recognized more and more among marine ecologists, and an increasing number of scientists are becoming interested in pressure studies. Furthermore, space sciences and 'man-in-the-sea' programmes of various countries are about to promote, and to give new impetus to pressure research. All these new developments are rapidly opening up new vistas beyond our present horizon.

The first experiments on biological aspects of high pressures were stimulated by the discovery of living organisms in great oceanic depths. They were conducted, in the last two decades of the nineteenth century, by CERTES (1884a) and REGNARD (1884a-d; see also Chapter 8.3). The pressures applied in these early studies did not exceed 600 to 1000 atm. Later investigations revealed that the pressure range between 1 and 600 atm is perhaps the most interesting one from an ecological point of view, since the responses elicited are often immediate, of great intensity, and—if not maintained for too long—largely reversible upon decompression (JOHNSON and EYRING, 1970).

It was soon discovered that increased pressures tend to retard or accelerate certain biological processes. The fact that, under otherwise identical conditions, a biological process can be 'influenced by a given pressure in quite an opposite manner at different temperatures', surprisingly, was not established until some decades later, in studies on muscle physiology and bacterial luminescence (JOHNSON and EYRING, 1970, p. 3). Today, we are keenly aware of the fact that there exists a fairly general antagonistic relationship between organismic responses to increased pressures and increased temperatures at all levels of organization (molecular, cellular, individual), and that pressure effects on living systems may also be modified by a variety of other concomitantly effective environmental factors, especially pH, salinity (ionic composition as well as total osmoconcentration), and dissolved gases.

A significant change in pressure automatically modifies a variety of properties of liquid systems which, in turn, may influence the responses of the organisms tested. It is not technically possible to eliminate completely such side-effects. Consequently, it is extremely difficult, if not impossible in many cases, to determine primary pressure effects. Secondary, tertiary, etc., pressure effects are caused by concomitant changes in pH, dissolved gases, density, viscosity, electrical conductivity, ionization of water and other substances, structural qualities of water, volume changes, basic chemical reactions, etc. For these reasons, analytical interpretations and assessments of the ecological significance of responses to changes in pressure require careful consideration of the whole set of circumstances to which the test organism is subjected.

Pressure–temperature relations have been dealt with in a number of papers, e.g. by CATTELL and EDWARDS (1930), BROWN and co-authors (1942, 1958), JOHNSON and co-authors (1942a, 1945, 1948a), TONGUR and KASATOCHKIN (1950, 1954), STREHLER and JOHNSON (1954), JOHNSON (1957a), MARSLAND (1957, 1970), SIE and co-authors (1958), KENNINGTON (1961), HAIGHT and MORITA (1962), ZOBELL and COBET (1962), MORITA and MATHEMEIER (1964), ALBRIGHT and MORITA (1968), SCHLIEPER (1968a), FLÜGEL and SCHLIEPER (1970), HOCHACHKA and co-authors (1970), JOHNSON and EYRING (1970), ZOBELL (1970), GILLEN (1971); Chapters 8.1, 8.2 and 8.3.

Pressure–pH relations have been studied by many authors. The following few examples must suffice here: JOHNSON and co-authors (1945, 1948b, 1954), OKADA (1954b), PONAT and THEEDE (1967), FLÜGEL and SCHLIEPER (1970), LANDAU (1970), MORITA and BECKER (1970), MURAKAMI and ZIMMERMAN (1970), ZOBELL (1970).

Pressure–salinity relations have received attention by FLÜGEL and SCHLIEPER (1970) and ZOBELL (1970). Unpublished work by ZOBELL and his students indicates that adverse effects of pressure may be accentuated by unfavourable salinities. ZOBELL (1970) has suggested, therefore, that increased pressures may interfere with the selective permeability of cell walls. Other studies concerned with permeability under elevated pressures (YAMATO, 1952a, b; OKADA, 1954a, c-e;

MIYATAKE, 1957; MURAKAMI, 1963; LANDAU, 1970; ZIMMERMAN, 1970b), at present, can neither reject nor support ZoBELL's suggestion unequivocally.

Under a given set of environmental circumstances, elevated pressure tends, in general, to elicit more pronounced responses in actively metabolizing systems than in less active or resting ones. Thus dividing cells are frequently more sensitive to pressure exposure than non-dividing cells, and organisms with high metabolic rates tend to be more affected by supranormal pressures than their comparable counterparts which exhibit lower metabolic rates. In complex organisms, there seems to be some indication that nervous tissue and gonads are more liable to be affected by elevated pressure than other, less specialized tissues. However, more critical studies are required before we can make such generalizations with a sufficient degree of certainty.

Protoplasmic gelations

A basic biological aspect of variations in pressure is the resulting effect on protoplasmic gelations. All protoplasmic gelations appear to involve processes of poly-

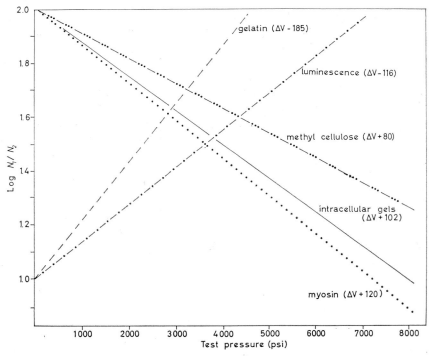

Fig. 8-6: Pressure effects on isolated gels. N_1: gel strength at test pressure; N_2: control values at atmospheric pressure. ΔV: cm³/mole change in volume which the examined substance undergoes in passing from sol to gel state. Gelatin: 4%, 22·0°C, pH 6·4; luminescence: reactivation of light intensity in *Photobacterium phosphoreum* at 34°C; methyl cellulose: 1·66% 'solution', 54°C; intracellular gels: averages obtained from plasmagel systems of *Amoeba* (two species), *Arbacia punctulata* eggs, and *Elodea canadensis* leaf cells; myosin: 2·1%, 36·7°C, pH 6·72, from rabbit muscle. (After MARSLAND and BROWN, 1942; modified.)

merization, whereby elongate macromolecular complexes are formed by the bonding together of a number of protein subunits, or monomers, present in the system prior to its gelation (MARSLAND, 1970). Apparently, bonding cannot occur until the shells of bound water, which protect the potential bonding sites of the separate monomeric units, are dispersed. Upon gelation, the dispersal of the protective water shells accounts, at least in part, for the resulting volume increase of the system. Due to the energy requirement of such dispersal, protoplasmic gelation is an endergonic process and proceeds more readily at increased temperatures. Although other types of bonding may be involved, polymer bonding is partly comprised of —S—S— bridges (MAZIA and ZIMMERMAN, 1958) and partly of hydrogen bonds (ANSEVIN and LAUFFER, 1959). The effects of pressure on isolated gels are illustrated in Fig. 8-6.

Gel structures of cells are of basic biological importance for intra- and extracellular kinetics (e.g. plasma streaming, amoeboid movement, contractility, pigment movement, locomotion). Gelations are weakened progressively by increasing pressure and strengthened by increasing temperature.

'The cell, apparently, expends metabolic energy in the formation of its gel structures and derives mechanical energy when the gels contract. Sustaining the source of mechanical energy involves a rebuilding of the gel structures, since in the course of contraction they revert toward the sol condition. Thus a continued cycle of gelation and solation is essential to the continued performance of mechanical work' (MARSLAND, 1970, p. 264).

Pressure effects at the molecular and enzyme levels

In recent years, much information has been gained from studies at the molecular and enzyme levels. While the ecologist is concerned primarily with responses at the individual or population level, studies on molecule and enzyme behaviour under sub- or supranormal pressures are of importance to him since they provide the basis for interpreting or predicting responses of integrated, complex systems.

With respect to micro-organisms, MORITA and BECKER (1970) have pointed out that studies both at the molecular and at the cellular (individual) level have certain drawbacks. The molecular approach, especially when working with purified preparations, does not take into consideration permeability, metabolic control mechanisms, protein–protein interactions, protein–nucleic acid interactions, protein–carbohydrate interactions, and other aspects of intra- or intercellular integration. The cellular approach, on the other hand, may suffer from interferences due to permeability factors and effects of added substrates or cofactors (substrates or cofactors may protect an enzyme by combining with it to form a more stable state); it hardly allows determination of the direct effect of pressure on the various cellular components.

Macromolecular biosynthesis. Recent advances in molecular biology have laid the groundwork for a more precise interpretation of the effect of pressure on macromolecular biosynthesis (LANDAU, 1970; Chapter 8.1). Augmentation of pressure tends to inhibit a reaction if molecular interactions require volume increase (i.e. increase of the volume occupied by a molecule), but stimulates a reaction requiring volume decrease; molecular interactions without volume changes tend to remain

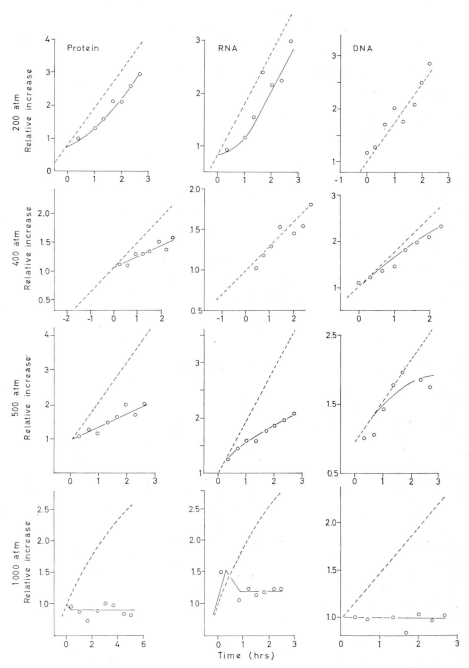

Fig. 8-7: Pressure effects on relative rates of synthesis of protein, RNA and DNA in the marine bacterium *Vibrio marinus* MP–4; 15°C. Broken lines: controls at 1 atm. (After ALBRIGHT and MORITA, 1968; modified.)

unaffected. If exposed to significant changes in pressure, temperature, pH or salinity, all cellular components undergo variations in volume. Among the environmental variables mentioned, pressure and temperature act as direct antagonists: while increased pressure reduces the molecular volume (negative ΔV), temperature raise increases it (positive ΔV). As a general rule, reactions with a negative ΔV are promoted by increased pressures, while those with a positive ΔV are inhibited.

In order to form the enzyme–substrate complex, the enzyme must unfold to accept the substrate. This unfolding involves a positive ΔV; hence the formation of the enzyme–substrate complex is usually the critical process which becomes inhibited under increased pressures (Chapter 8.1).

ALBRIGHT and MORITA (1968) studied, at 15°C, the effects of pressures of 200, 400, 500 and 1000 atm on synthesis rates of net protein, RNA and DNA in multiplying cells of the marine psychrophilic bacterium *Vibrio marinus* MP–4, isolated from water of 3·24°C collected at a depth of 1200 m off the coast of Oregon (USA). At 1000 atm, protein and DNA synthesis are completely inhibited, while RNA synthesis continues for a short time (Fig. 8-7). The data suggest that 400 to 500 atm reduce the rate of protein synthesis, which is based on the lowering of the rates of RNA and DNA synthesis. At 200 atm, protein and RNA synthesis decrease for ca 60 mins and then resume 1-atm rates, whereas DNA synthesis remains unaffected (see also Chapter 8.1).

ZIMMERMAN and ZIMMERMAN (1970) report on pressure effects on ribosomes, and on synthesis of protein, RNA and DNA in synchronized* populations of the protozoan *Tetrahymena pyriformis*. Pressure effects on ribosomes were assessed on the basis of changes in the ratio of the polysomal area to the total ribosomal area. Evaluation of the ratios (polysome/total ribosome) reveals that the cells are quite

Table 8-4

Changes in the ratio polysome/total ribosome area in heat-synchronized *Tetrahymena pyriformis* as a function of pressure. EH: last heat shock. Synchronized cells were pulsed at 45 mins after EH for 2 or 5 mins and homogenized at 60 mins after EH. The amount of polysomal material at 13 mins after decompression (60 mins after EH) was comparable to that of the non-pressurized atmospheric controls (After ZIMMERMAN and ZIMMERMAN, 1970; modified).

Pressure (psi)	duration (mins)	Time after EH (mins)	Polysome area / Total ribosome area
atmospheric controls	—	60	0·422
2,000	2	60	0·374
5,000	2	60	0·205
10,000	2	60	0·229
14,000	5	60	0·272

* Synchronization (SCHERBAUM and ZEUTHEN, 1954) consists of subjecting the cells to eight heat shocks (34°C) of 30-min duration, alternated with 30-min intervals (28°C); 70 to 75 mins after the last heat shock, 89 to 90% of the cells divide synchronously. Such heat synchronization permits biochemical analysis of large numbers of cells at comparable stages of development.

sensitive to short pulses of pressure. The ratios become reduced at a pressure of 5000 psi, but remain essentially unchanged at still higher pressures; at 2000 psi, the ratio is only slightly less than in the atmospheric controls (Table 8-4). Immediately after pressure treatment, the quantity of measurable polysomes becomes reduced. Short-term pressure pulses readily disrupt the newly formed polysomes; however, new polyribosomes can be formed following high pressure treatment (HERMOLIN and ZIMMERMAN, 1969). The ability of ribosomes to synthesize protein (incorporation of phenylalanin) remains similar in pressure-treated *T. pyriformis* cells and controls (LETTS, 1969). It is doubtful, therefore, that a decrease in rates of protein synthesis following exposure to high pressures can be attributed to ribosome inhibition (ZIMMERMAN and ZIMMERMAN, 1970).

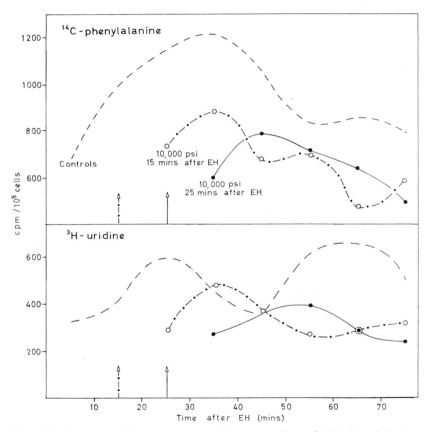

Fig. 8-8: Pressure effects on rates of incorporation of ^{14}C-phenylalanine and ^{3}H-uridine in synchronized *Tetrahymena pyriformis*. Controls at atmospheric pressures (broken lines) were pulsed for 10 mins in the radioisotope, and the radio-activity (cpm/10^5 cells) of the acid-insoluble fraction was plotted as a function of time after the last heat shock (EH). In the pressure series, the cells were subjected to 10,000 psi (for 2 mins) at 15 mins and at 25 mins after EH (arrows). After pressure treatment, the cells were incubated in ^{14}C-phenylalanine (0·25 μCi/ml) and ^{3}H-uridine (10 μCi/ml) for 10 mins and prepared for counting. (Based on data of LOWE, 1968; from ZIMMERMAN, 1969; modified.)

Pressure treatment reduces the incorporation rate and the cumulative incorporation of precursors of protein and RNA in synchronized *Tetrahymena pyriformis* (LOWE, 1968; ZIMMERMAN, 1969; Figs 8-8, 8-9); incorporation patterns depend on the age of the cells. Elevated pressures cause a reduction in newly synthesized RNA (Fig. 8-10); apparently, they retard RNA synthesis in both the low molecular weight (4S and 5S) and high molecular weight region, although there is no effect on pre-existing bulk RNA (YUYAMA and ZIMMERMAN, 1969). Rates of DNA synthesis of log phase *T. pyriformis* decrease with increasing pressure (Fig. 8-11). Percentage inhibition of ^{14}C-thymidine incorporation ranges from 10% at 2000 psi to 80% at 10,000 psi (MURAKAMI and ZIMMERMAN, unpublished; ZIMMERMAN and ZIMMERMAN, 1970).

The information at hand indicates that supranormal pressures do not affect primary protein structures. They may, however, significantly influence secondary (hydrogen bonds), tertiary and quaternary (non-covalent bonds) structures of proteins (Chapter 8).

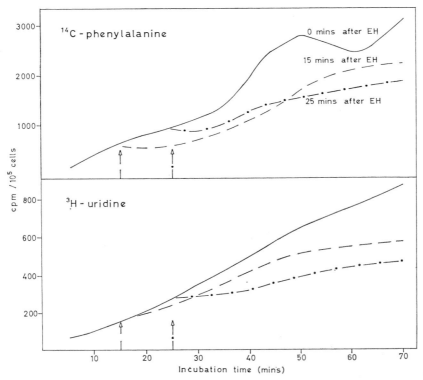

Fig. 8-9: Pressure effects on cumulative incorporation of ^{14}C-phenylalanine and ^{3}H-uridine in synchronized *Tetrahymena pyriformis*. Cells were incubated in ^{14}C-phenylalanine (0.15 μCi/ml) or in ^{3}H-uridine (3·3 μCi/ml) at 0 mins after the last heat shock (EH). At 15 or 25 mins after EH (arrows), the labelled cells were subjected to a pulse of 10,000 psi for 2 mins. Aliquots were removed at various times and the radio-activity (cpm/10^{5} cells) of the acid-insoluble fractions was plotted as a function of incubation time (time after EH). (Based on data of LOWE, 1968; from ZIMMERMAN and ZIMMERMAN, 1970; modified.)

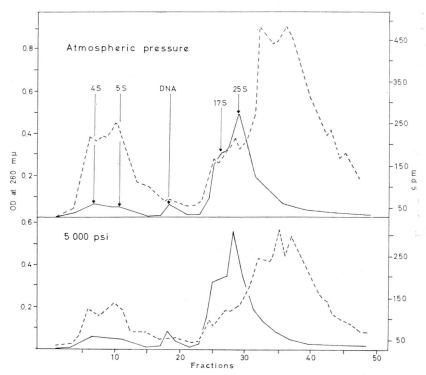

Fig. 8-10: Pressure effects on RNA synthesis in synchronized *Tetrahymena pyriformis*. MAK column chromatography of total cellular RNA from cells labelled for 10 mins with ³H-uridine (specific activity 20 Ci/mmole). 40 mins after the last heat shock (EH), the cells were placed into ³H-uridine (10 μCi/ml); one sample remained at atmospheric pressure (upper graph), the other was subjected to 5000 psi for 7 mins. 50 mins after EH, both samples were chilled in an ice bath and lysed in 1% sodium dodecyl sulphate. Nucleic acids were extracted with cold phenol and fractionated on the MAK column. Broken line: radio-activity in counts per min (cpm). (After YUYAMA and ZIMMERMAN, 1969; modified.)

Denaturation of protein systems and inactivation of enzymes. In Japan, C. and K. SUZUKI and their associates have conducted extensive studies on denaturation of protein systems and inactivation of enzymes under pressures exceeding 16,000 psi (MURAKAMI, 1970). Heat denaturation of haemoglobin is retarded at pressures below 29,400 psi and accelerated at higher pressures; the reaction of pressure denaturation follows first-order kinetics (SUZUKI and KITAMURA, 1960a, b). While it is generally accepted that protein denaturation results from changes in conformational states of molecular side chains (Chapter 3.0, p. 326),

'in all probability, different conformational changes will occur depending upon the methods used to denature proteins' (MURAKAMI, 1970, p. 136).

Thus the optical rotation of ovalbumin varies markedly, depending on whether the protein is denatured via pressure, heat or urea (SUZUKI and SUZUKI, 1962; see also SHULYNDIN, 1967). According to SUZUKI and co-authors (1963), maximum

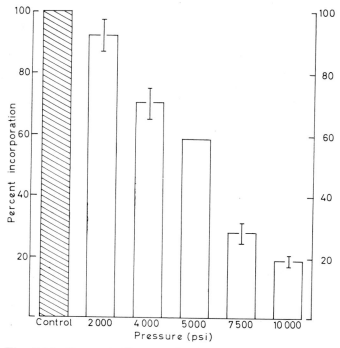

Fig. 8-11: Pressure effects on DNA synthesis in log phase cultures of *Tetrahymena pyriformis*. Cells were incubated in ^{14}C-thymidine (2 μCi/ml) for 10 mins under varying pressures. The radio-activity in the acid-insoluble fraction was determined as a percentage of the non-pressurized control cells. Vertical ranges at column tops: standard errors. (Based on unpublished data of MURAKAMI and ZIMMERMAN; from ZIMMERMAN and ZIMMERMAN, 1970; modified.)

resistance of ovalbumin to pressure denaturation (86,000 to 106,700 psi at 30°C for 5 mins) occurs at pH 9.0, whereby the rate of denaturation is proportional to the square root of the pH value. While urea and ethanol accelerate pressure denaturation of ovalbumin, sulphate and glucose retard or inhibit it. The degree of inactivation of trypsin and chymotrypsin increases with increasing pressures; above ca 114,000 psi for trypsin and 85,300 psi for chymotrypsin, no additional inactivation occurs following a single pressure impulse (MIYAGAWA and SUZUKI, 1963a, b). In these two enzymes, the thermodynamic quantities of the inactivation process are similar to the protein denaturation of ovalbumin and haemoglobin, except that the enthalpy* is positive for the enzymes, but negative for the proteins. RNase activity (beef pancreas)—determined in a specially designed spectrophotometer–pressure apparatus—increases under pressures of 7100, 14,200 and 21,300 psi (MURAKAMI, 1960). Magnesium-activated muscle ATPase and freshly prepared myosin ATPase are inhibited under a pressure of 14,300 psi (MURAKAMI,

* Enthalpy, h: the sum of internal energy and p/ρ; i.e., $h = e + p/\rho$ (dyne cm/g, ft-lb/slug). Internal energy, e: the energy per unit mass of a fluid due to the thermal motion of its molecules. Pressure p. Density, ρ: the mass per unit volume.

1958). Succinic dehydrogenase activity is accelerated at 4300 psi, but at pressures exceeding 11,400 psi the degree of activity becomes retarded in proportion to the magnitude of the pressure applied (TOKUMOTO, 1962).

In general, hydrolytic and proteolytic enzyme reactions are accelerated by subnormal pressures, but retarded or suppressed by supranormal pressures, while oxidative reactions are suppressed 'by all kinds of pressure' (SHULYNDIN, 1967). According to SHULYNDIN, inhibition by elevated pressures of the various enzyme reactions is due to enzyme denaturation, whereby the mechanism of pressure effects may depend on correlations between proteins and ambient water molecules.

Information on some of the non-covalent molecular interactions which stabilize native proteins (see also Chapter 3.0) may be provided by studying systems involving hydrophobic interactions. This has been suggested by MORITA and BECKER (1970), who have summarized the newest results, obtained in their laboratory, on pressure effects at the molecular and enzyme levels; the following paragraphs are quoted from their paper.

In 1965, KETTMAN and co-authors reported pressure effects on the aggregation reaction of poly-L-valyl-ribonuclease (PVRNase). This enzyme exhibits a turbidity at temperatures as low as 30°C, whereas RNase aggregates at higher temperatures. The low-temperature aggregation of PVRNase is attributed to increased apolar interactions due to polyvalyl chains attached to the enzyme (BECKER and SAWADA, 1963). The effect of pressure on PVRNase aggregation at 39°C is illustrated in Fig. 8-12. Although turbidity rate decreases at 150 and 300 atm, pressure release

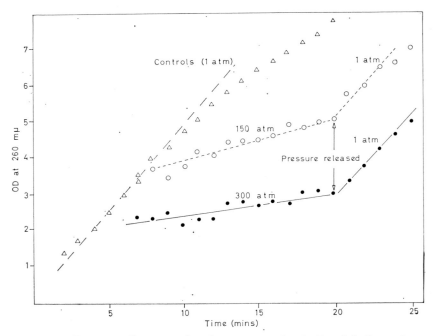

Fig. 8-12: Pressure effects on the time course of poly-L-valyl-ribonuclease (PVRNase) aggregation; 39°C. (After KETTMAN and co-authors, 1965; modified.)

immediately allows the system to aggregate at the rate exhibited at 1 atm. The fact that pressure increase reduces the aggregation rate indicates that the transition of the rate-limiting step exhibits a positive volume of activation; calculation of the ΔV for the system is 203 ml/mole. GILL and GLOGOVSKY (1965) studied the thermal transition of bovine pancreatic ribonuclease (RNase) under high pressure employing optical rotation methods. They noted a molar volume change of -30 ml/mole for this intramolecular conformation change.

An absorbance increase in the spectra below 280 nm on the thermal unfolding of RNase-A due to elevated pressure has been demonstrated by NISHIKAWA (unpublished). This increased absorbance indicates a spectral change rather than an increase in turbidity (TIMASHEFF, 1966). When native RNase-A was pressurized to 1000 atm at 30°C and pH 2, two distinct changes in the ultra-violet spectra were observed: the denaturation blue shift of the 278-nm band was enhanced but was reversed upon depressurization. A non-reversible general hyperchromicity occurred below 280 nm, the largest circular dichroic change taking place at the 240-nm band. At pH 7.2, no changes in the circular dichroic spectra could be observed between 1 and 1000 atm.

At 1000 atm and pH 2, and at normal pressure at pH 2, changes in ellipticities can be noted for RNase (Table 8-5). The reversibility of the 240-nm band of RNase at 1000 atm and pH 2, when neutralized, suggests that pressure induces a metastable conformational state in RNase expanded by low pH, but probably does not involve changes in most of the peptides located in the α-helix or β-structure regions; the latter is suggested by the ellipticity observed at 216 nm (MORITA and BECKER, 1970).

Table 8-5

Changes in optical parameters of RNase under different conditions of pressure and pH (After MORITA and BECKER, 1970; modified)

Pressure (atm)	pH	$[\theta]_{240}$	$[\theta]_{216}$	$[\theta]_{275}$	$\Delta\epsilon_{287}$
1000	7	$+73$	-9890	-228	0
normal	2	$+35$	-8630	-217	-246
1000	2	$+8$	-8570	-202	-594
1000[a]	2	$+74$	-8570	—	—

[a] Material neutralized, then observed again.

Studies on the thermal transition of ribonuclease-S and S-protein over a range of acid pH were conducted by MORITA and associates because of the importance to pressure studies of knowing the denaturation state of ribonuclease-S and S-protein with respect to temperature and pH. The data obtained were used to set up an equilibrium expression for the reversible thermal denaturation, in a manner employed by previous workers (e.g. BRANDTS, 1964; GILL and GLOGOVSKY, 1965). From the expression, the enthalpy of denaturation was determined as a function of pH for ribonuclease-S, and the effect of pressure on the equilibrium constant for the reversible denaturation was followed for ribonuclease-S and S-protein at given temperature and pH conditions. The change in the equilibrium constant with pressure was related to a change in the partial molar volume of the system using an expression first derived by PLANCK (1887). The difference in the effect of pressure on the denaturation state of ribonuclease-S and S-protein under similar conditions is assumed to reflect the dissociation of the S-peptide portion of ribonuclease-S, and the difference in partial molar volume changes of the two systems is interpreted as providing a direct indication of the magnitude and direction of volume change occurring when the S-peptide dissociates from ribonuclease-S (MORITA and BECKER, 1970).

From the change in molar absorbance at 287 nm and 286 nm, equilibrium constants as a function of pressure were computed for ribonuclease-S and S-protein. Using a plot of log K versus pressure at 25°C, a volume change of -80 ± 2 ml/mole was calculated for ribonuclease-S at pressures ranging from 300 to 900 atm, while S-protein was found to undergo an apparent volume change of -45 ± 2 ml/mole at comparable pressures (MORITA and BECKER, 1970).

LINDERSTRØM-LANG and JACOBSEN (1941) employed dilatometric techniques to monitor the hydrolysis of β-lactoglobulin by trypsin or chymotrypsin and suggested that the initial hydrolysis of β-lactoglobulin involves considerable rearrangement in the protein structure. LINDERSTRØM-LANG and JACOBSEN, as well as NOGUCHI and YANG (1962), GERBER and NOGUCHI (1967) and others, demonstrated that results obtained by the use of dilatometers tend to be in general agreement with the data produced by pressure studies. Volume change due to interaction of S-peptide and S-protein was found to be $+32 \pm 6$ ml/mole S-protein and appears to agree quite well with the value of -35 ± 2 ml/mole ribonuclease-S obtained from the pressure studies for the assumed dissociation of S-peptide and S-protein. MORITA and BECKER (1970) conclude that a volume increase of 17 ml/mole to 35 ml/mole accompanies the interaction of S-peptide with S-protein and suggest that a volume change of -27 to -45 ml/mole protein must accompany the reversible denaturation of S-protein.

'From the data obtained it is difficult to speculate on the individual contributions to the volume change from the various hydrophobic groups. In a gross manner, it is felt that a value of $+8$ ml/mole . . . for a volume change in a hydrophobic environment is in the proper range. For any case, the general agreement of both hydrostatic pressure experiments and the direct dilatometric measurements in this system provides a basis for studies on more complex systems where hydrophobic interactions are implicated' (MORITA and BECKER, 1970, p. 80/81).

While extreme high pressures affect the physical properties of proteins in bacteria, GILLEN (1971) found no correlation between the effects of elevated pressures on the activity of cell-free extracts of muscle lactic dehydrogenase from four bathypelagic fishes (*Photonectes margarita*, *Scopelogadus mizolepis*, *Pseudoscopelus* sp., unidentified species of the family Myctophidae) and three shallow-water fishes (*Cypselurus heterurus*, *Lagodon rhomboides*, *Lepomis macrochirus*).

Previous studies on pressure effects at the molecular and enzyme levels have been reviewed by ZoBELL (1970) and in Chapter 8.1. According to ZoBELL (1970), compression to 300 to 3000 atm at normal temperatures arrests or inhibits most bio-chemical reactions, the upper limits depending on the pressure tolerance of enzyme systems and other essential proteins. DNA synthesis becomes suppressed by elevated pressures (ZoBELL and COBET, 1964). Considering the various ways in which high pressure may influence the rates of synthesis and properties of DNA, HEDÉN (1964) proposed that, at certain temperatures, increased pressures pre-serve the double helix and that, in such a structure, the bases are partly protected from contact with the solvent; this would not be the case in a single strand and hence would lead to a higher rate of breakdown.

Conclusions. The information obtained at the molecular and enzyme levels opens up interesting and promising perspectives for future research. It is unfortunate, however, that the majority of the studies have been conducted under conditions which must be considered more or less inadequate from an ecological point of view:

(i) In almost all cases, preparations or organisms have been transferred from normal to stress pressure immediately. It is well known from comparable studies on other environmental factors, e.g. temperature (Chapter 3), that organismic responses may be different under fast or slow changes in factor intensity.

(ii) Most experiments deal with short-term pressure exposures (mins, hrs), without allowing the system tested to complete adaptive functional and struc-tural adjustments. If we want to understand how variations in pressure affect survival, metabolic performance, reproduction and morphology of marine organ-isms, we must learn more about the responses of fully stabilized systems. There is great need for studying pressure effects as a function of time.

(iii) The vast majority of organisms tested so far inhabit near-surface waters. We know practically nothing about the responses to elevated pressures of true deep-sea organisms.

(iv) Most studies have been conducted at temperatures between 20° and 50°C. As has been pointed out in Chapter 3.0 (p. 337), the average temperature of the water masses of all oceans is about 3.8°C; even at the Equator, the average tem-perature of the whole water column amounts to only about 4·9°C. At depths beyond 1000 m, temperatures below 5°C prevail, and on the deep-sea floor, below 2° or 3°C. There is a deplorable lack of pressure studies at such low temperatures.

(v) In the deep sea, temperatures are nearly constant. The fast temperature change applied in many of the experiments represents, therefore, an additional, entirely 'unnatural' stress.

(vi) Long-term pressure studies are likely to yield ecologically meaningful results only if quantitatively and qualitatively adequate energy (food) sources are available. Much of the present work, especially on micro-organisms, has been done under abnormal nutrient conditions.

In the deep sea, in addition to increased pressures, other environmental factors attain extreme or rather constant values. Light intensity diminishes practically to zero, food becomes scarce and often less variable, water movement is reduced and rather uniform, temperature decreases and remains quite constant. There is urgent need for conducting pressure studies on deep-sea organisms exposed to environmental factor combinations which attempt to simulate conditions actually met in their respective habitats.

Biological aspects of **reduced** pressures, i.e. below 1 atm, are of theoretical interest only to the marine ecologist, since pressures below 1 atm do not occur in the marine environment. There is considerably less information available on biological aspects of subatmospheric pressures than on the effects of pressures above 1 atm (e.g. ZoBELL, 1970; Chapter 8.2).

(2) Measuring Pressure: Methods

There is no standard incorporating* the unit of pressure. In the last analysis, all pressure measurements are, therefore, indirect ones, based on a certain relation between the value of pressure and the value of the measured quantity, e.g. the height of the mercury column in a manometer (TROSKOLANSKI, 1960).

This section considers briefly (i) the production of high pressures, (ii) some principles of measuring pressure, (iii) methods employed by biologists to conduct experiments under a variety of pressure conditions, and (iv) bathymetric methods for measuring depths in oceans and coastal waters.

Production of high pressures

High pressures may be produced in a number of ways. MUNRO (1963) distinguishes five practical methods of pressure generation: gravitational, thermodynamic, shock-wave, single-stage piston, and multi-stage piston methods.

Gravitational methods. According to MUNRO (1963), a column of liquid in a gravitational field provides the most simply generated, and—for the range 10^{-2} to $10^{+1.5}$ bars—the most accurately determinable pressures. While other liquids may be used, mercury is preferred on account of its high density. The simplicity of gravitational methods is based on the constancy of the density of the mercury column and of the acceleration of gravity (p. 1325), as well as on the absence of need to correct for friction. In view of the convenience and accuracy of producing low pressures, attempts have been made to extend this technique to the production of pressures in excess of 30 bars; however, the height of the liquid column required soon becomes prohibitive.

The most accurate and far-reaching application of gravitational methods used is a mercury column in a steel tube over 9 m high, connected at each end to free piston gauges which can be interchanged. Successive stages facilitate a pressure calibration scale extending up to 2500 ± 0.15 bars. This procedure provides an independent method of pressure generation which can serve to control the accuracy of free piston gauges. However, at high pressures, correction becomes necessary for density changes (the density of mercury changes by 0.4% over the first kilobar; a similar variation in density is caused by a change in temperature).

Thermodynamic methods. These are based on changes in pressure–volume relations upon variations in thermal energy. The technique of thermal expansion consists of sealing the test substance at temperatures below 0°C and subsequent heating to 100°C or more; such procedure subjects the sample to elevated pressures. If

---

* The incorporation of the length unit is the standard metre or the Imperial yard, that of the mass unit, the standard kilogram or the Imperial pound.

no expansion of the pressure vessel occurs, pressures of a few kilobars may be produced. In some ways, phase change methods represent more promising means for pressure generation than simple thermal expansion. Since the proportional changes in volume—or in pressure at constant volume—are larger, expansion of the vessel no longer subtracts significantly from the pressure produced (MUNRO, 1963). The considerable changes in temperature required largely disqualify thermodynamic methods from being used in biological research.

Shock-wave methods provide a useful approach to the problem of calibration at very high pressures; next to gravitational and piston methods, they rank as a third independent (primary) means of pressure calibration. Since shock waves cannot produce sustained constant pressure, they are of limited importance to ecological studies.

Single-stage piston methods. A piston is forced into a cylinder creating pressure in the remaining inner section. This simple technique requires a mechanical force and proper sealing between piston and cylinder to avoid loss of cylinder contents by leakage. MUNRO (1963) discusses various sealing arrangements and lists a number of pumps (hand pumps, motor pumps, fluid-powered pumps, etc.) designed for generating pressures; he also presents various examples of pressure-generating apparatus.

Multi-stage piston methods. In a single-stage apparatus, pressure production is limited by the strength of the materials employed. In order to overcome this limitation, multi-stage systems have been developed in which one pressure unit is, in sequence, completely surrounded by the next, with the maximum pressure generated in the innermost unit. Unfortunately, in multi-stage designs, the maximum pressure increases linearly to the number of stages, 'while the bulk of the apparatus increases exponentially' (MUNRO, 1963, p. 34), rapidly rendering the system impracticable. A convenient compromise between single and multi-stage designs is the use of part of the mechanical force on the pressure-producing apparatus to provide supporting pressure. Various examples of high-pressure generating apparatus have been presented and discussed in MUNRO's paper. Methods based on multi-stage techniques may possibly yield pressures of a million or more atmospheres. Although outside the realm of ecology, such tremendous pressures are likely to cause very interesting and unusual effects on physical or chemical systems.

Some principles of measuring pressure

Most principles of measuring pressure are based on the balancing of a pressure difference, employing a column of liquid whose weight is known in terms of its height, density and gravity acting upon it, or on the deformation of elastic material with known elasticity constants. There are three basic groups of instruments available for measuring pressure (TROSKOLANSKI, 1960): manometers, dead-weight pressure gauges, and elastic pressure gauges. Hydrostatic manometers (glass tube gauges) make use of the relationship $p = \gamma d$; see Equation (3) on p. 1323. Dead-weight pressure gauges are based on the principle of measuring the force acting on a piston with a given effective surface area (p. 1325). The essential element of elastic pressure gauges is a tube or diaphragm made of elastic material; the gauges measure the relationship between the degree of deformation of the elastic element and the pressure causing it. The deformation is proportional to the elasticity constants of the elastic material.

There exists today a large variety of manometers, barometers, dead-weight piston gauges, Bourdon tube gauges, as well as of vacuum gauges (for measurement of reduced pressures). At very high pressures, the controlled clearance piston gauge allows measurements with considerable accuracy; variable pressure is applied to its exterior cylinder surface in order to reduce errors due to elastic distortion. Reduced pressures down to a few mm Hg can be measured by simple mercury manometers. More sensitive mercury manometers employ two mercury surfaces which are kept precisely in contact with fine vertical pointers; the difference in pressure is then determined by the angle through which the manometer must be tilted in order to restore the level disturbed by the presence of gas on one side. Such gauges have been used to measure pressures from 1·5 to 0·001 mm. Pressures too low to be measured accurately by ordinary U-tube manometers were first successfully determined by the McLeod gauge, which, in the mid 20th century, was the standard instrument for measurement of gas pressures down to 10^{-6} mm Hg. Modern techniques have produced a considerable number of new types of manometers for measuring exceedingly low pressures in gases and liquids—e.g. viscosity gauges, Parani gauge, conductivity gauges, ionization gauges, Knudsen gauge—and of vacuum pumps, such as rotary oil-sealed pumps ('Kapselpumpen'), liquid mercury pumps, mercury diffusion and condensation pumps, oil vapour pumps, mercury vapour pumps and booster pumps.

We may distinguish primary and secondary instruments for measuring pressure (MUNRO, 1963). The primary means of producing pressures, i.e. by gravitational, shock-wave and piston methods, represent, at the same time, the primary instrumentarium for measuring pressure; they provide the basic scales of pressure measurement. All secondary instruments (based on phase change methods, Bourdon gauges, and resistance gauges measuring electrical resistance or conductivity) must be calibrated by employing primary instruments.

Pressures of a few bars up to 13 kbar are accurately transmitted by liquids and measured by a free piston gauge. Near 8 kbar, the accuracy is ± 0·1% according to BRIDGMAN (1912, in: MUNRO, 1963) and ± 0·7% according to JOHNSON and NEWHALL (1953, in: MUNRO, 1963). Up to 30 kbar, the production of known pressures by a free piston gauge acting on solid materials leads to the determination of the 'bismuth point' (e.g. MUNRO, 1963), for which the most satisfactory value is 25·410 ± 0·095 kbar (KENNEDY and LA MORI, 1961). The bismuth transition facilitates an extended calibration of the manganin resistance gauge by transmission of pressure through a liquid. Piston methods, in which allowance is made for friction and other loss of force, are implied in the apparatus described by BOYD and ENGLAND (1960) and by BALCHAN and DRICKAMER (1961) to fix the thallium point at 37·1 kbar, the barium point at 59·0 kbar and the upper bismuth point at 90 kbar, with an accuracy of ca ± 3% (MUNRO, 1963). At higher pressures, calibration depends on shock-wave measurements. A secondary scale of calibration points, twice removed from primary pressure production, is provided by the resistance discontinuities in iron, barium, lead, rubidium, and calcium, recorded by BALCHAN and DRICKAMER.

In moving water, three aspects of pressure must be distinguished: total pressure, static pressure, and pressure head ('Staudruck'). The **total pressure** is determined with a tube, the anterior part of which is bent 90°; its opening is placed vertical to the direction of the water current. The other end is connected to a manometer (Pitot tube). The **static pressure** is measured by employing a pressure probe (sonde). A tube with a rounded tip carries lateral holes in a sufficient distance from the tip. If the tube is directed towards the current, static pressure is created in its interior, since the medium can pass through without significantly changing its kinetic energy. The **pressure head** equals the difference between total pressure and static pressure. It is measured with the Prandtl tube—a combination of Pitot tube and pressure probe.

Reviews on, or concerned with, methods of producing and measuring elevated pressures have been written by MUNRO (1963), STEELE and WEBB (1963a), VODAR and SAUREL (1963), WYLLIE (1963) and MORITA (1967, 1970).

Methods employed by biologists to conduct experiments under a variety of pressure conditions

The methods used for conducting experiments on living systems exposed to a variety of pressures are based on the principles outlined above. In laboratory experiments on marine organisms, the intensity of pressure has been modified by varying the weight on a piston, by moving columns of mercury or water, and by employing compressed air or vacuum systems. In most cases, hydraulic pumps or vacuum pumps have been connected to pressure chambers equipped with a pressure-measuring device. Examples of methods employed in pressure experiments are listed in Table 8-6.

A useful compilation of equipment employed by biologists has been presented by MORITA (1970), with special emphasis on microbial cultures. MORITA describes pressure-pump assemblies, pressure cylinders, containers for use in pressure cylinders and the general procedure for pressurization of various materials.

A **pressure-pump assembly,** widely used in biological studies, represents a modification of a high-pressure (40,000 psi) hydraulic truck jack (Enerpac Model P228, Blackhawk Industrial Products, Butler, Wisconsin, USA). It employs a mixture of glycerol and water (1:1) as hydraulic fluid, in order to avoid undesirable effects of the hydraulic oil (normally used as a hydraulic fluid in this pump) on the test material inside the pressure cylinder. As far as pressure stability is concerned, the pump itself is the weakest link in the assembly, which employs super-pressure fittings and tubing. All items can be purchased from various USA manufacturers specializing in super-pressure equipment. Fig. 8-13 illustrates schematically the parts of the pump. Other pressure-pump assemblies can be prefabricated according to the investigator's requirements. Recently, also stainless steel pumps, that can use water as hydraulic fluid, have become available; however, their initial cost is much greater than the conversion of an hydraulic truck jack.

One of the most commonly employed **pressure cylinders** has been described by ZOBELL and OPPENHEIMER (1950). It represents a modification of the cylinder used by JOHNSON and LEWIN (1946a). Machined from stainless steel, the cylinder

Table 8-6

Examples of methods employed in experiments conducted under elevated pressures
(Compiled from the various sources indicated)

Apparatus	Remarks	Authors
Early pressure chamber	Simple hydraulic pump connected to pressure vessel	CERTES (1884a)
Early pressure chamber	—	REGNARD (1884a-d)
Pressure centrifuge vessel, for determining the gelational state of different cytoplasmic regions and for gel strength measurements	T-shaped, with pressure and control sections. Pressures up to 1000 atm	FONTAINE (1930) BROWN (1934b)
Pressure chamber, for microscopic observation of amoebas under high pressure. Total capacity 3 ml	With two windows	MARSLAND and BROWN (1936)
Improved pressure chamber, for microscopic studies. Total capacity 400 ml	7–10 mm thick tempered plate glass windows allow direct observation at magnifications up to 600 × and pressures up to 20,000 psi	MARSLAND (1938)
Pressure–temperature apparatus, for luminescence studies	—	BROWN and co-authors (1942)
Pressure chamber, for incubating bacteria, enzymes, etc.	—	JOHNSON and LEWIN (1946a)
Pressure 'bomb', for studying actomyosin and amphibian larvae. Inner diameter: 3·5 inches (8·9 cm)	Two herculite plate glass windows at opposite ends. Pressures up to 5000 psi	BORTS and co-authors (in: JOHNSON and FLAGLER, 1951)
Pressure centrifuge apparatus	Same design as BROWN's (1934b) except for added temperature control	MARSLAND (1950)
Pressure vessel, for exposing micro-organisms to high pressures; inside diameter: 1⅜ inches (3·9 cm)	Stainless steel cylinders fitted with an 'O' ring seal held in place by a steel cap. Hydraulic fluid: 50/50 glycerol and water. Pressures up to 1000 atm	ZOBELL and OPPENHEIMER (1950) ZOBELL (1954)
Glass micropipette connected to mercury manometer, for determining internal pressure in fish eggs	Measurement of pressures in eggs, following insemination or micropuncture	KAO and CHAMBERS (1954)
Stainless steel pressure vessel, for measuring electrical conductivity of sea water. Capacity 500 cm³	Glass conductivity cell with platinized platinum electrodes attached to thermometer protruding from lid of pressure vessel	HAMON (1956)
Pressure flask, for studying plants under a few atm	Mercury column creates pressures up to about 2 atm	FERLING (1957)

Method	Remarks	Author
Optical high pressure cylinder, for spectrophotometric measurements	Optical white sapphire windows and a neutral piston to prevent the hydraulic fluid from contaminating the enzyme reaction mixture	MORITA (1957a)
Indwelling cannula, for measurement of pressure in swim bladders of free-swimming fish	Cannula connected to hypodermic syringe and manometer	McCUTCHEON (1958)
Pressure–pH apparatus, for measuring pH under pressure	Electrodes are described which allow the determination of pH under pressure	DISTÈCHE (1959)
Spectrophotometer–pressure apparatus, for studies on enzyme activity	Pressures in excess of 21,000 psi	MURAKAMI (1960)
Pressurized teflon and sapphire cell, for optical absorption studies under high pressure	—	GILL and RUMMEL (1961)
Pressure apparatus with pulsating aeration, for growing cells under hyperbaric conditions	Small volumes of culture can be aerated continuously	HEDÉN and MALMBORG (1961)
Pressure cell with electrodes, for discharging electrical currents through aqueous systems under pressure	—	BRANDT and co-authors (1962)
Pressure chamber, for electrochemical measurements	Buffers are tested under pressure	DISTÈCHE (1962)
Pressure–fixation chamber, for histological fixation of test material under pressure	Two chambers are separated by a cover glass that can be broken by a steel ball, thus allowing fixation	LANDAU and THIBODEAU (1962)
Pressure cylinder with temperature control, for protein denaturation studies	Pressures above 1000 atm	SUZUKI and SUZUKI (1962)
High pressure equipment (Harwood Engineering, Inc., Walpole, Mass., USA) with a working capacity of 1 inch (2·54 cm) in diameter and 9 inches (22·9 cm) length	Compound cylinder of tapered shrink construction (pressure range 1 to 13,800 bars) immersed in a thermostatic bath	HORNE and FRYSINGER (1963)
Quick-pressurized small sample-freezer, for studies requiring rapid chilling and fast disassembly (biochemical analysis of pressurized material)	Hand-operated hydraulic pump. Pressures of 10,000 to 12,000 psi. Disadvantage: hydraulic fluid tends to pollute test material	LANDAU and PEABODY (1963)
Windowed pressure chamber, for permeability studies	Permits direct observation of pressure effects on plasmolysis and deplasmolysis	MURAKAMI (1963)
Pressure chamber with windows, for spectral measurements	Description of various types of windows and housing constructions	ROBERTSON (1963)
Optical rotation cell, for measurement of light transmission at different wavelengths	—	GILL and GLOGOVSKY (1964)
Tidal pressure apparatus, for creating periodic pressure changes	A mercury column is carried round on a long lever which moves slowly in a vertical plane about a central pivot	MORGAN and co-authors (1964)

Table 8-6—Continued

Apparatus	Remarks	Authors
Pressure cylinder, for incubating biological systems	Metal to metal seal	MORITA and MATHEMEIER (1964)
Pressure chamber, for studying $CaCO_3$ solubility	At 1000 atm, saturation of $CaCO_3$ is 2·7 times larger than at 1 atm	PYTKOWICZ and CONNERS (1964)
Pressure cell, for optical rotatory studies	Pressures up to 130 atm	RIFKIND and APPLEQUIST (1964)
Pressure viscometer, for viscosity measurements	Studies on variations of viscosity as a function of pressure	HORNE and JOHNSON (1966)
Portable pressure-tolerance meter, for experiments on organismic sensitivity to small pressure changes	Test organisms are placed in water in a glass bottle and air pumped into the top of the container	DIGBY (1967)
Low pressure aquarium, for behavioural studies on fishes	Fishes are trained to feed at certain water depths or pressures. Pressures deviate only little from atmospheric	KUHN and STROTKOETTER (1967)
Pressure–counting apparatus, for determining the number of protozoan cells developing under pressure	A sampling device is connected to a Model A Coulter Counter	MACDONALD (1967)
AMINCO pressure apparatus, for studying oxygen consumption	Permits experiments in running sea water at 100 to 400 atm. (Designed by American Instrument Company, Washington, D.C., USA)	NAROSKA (1968) SCHLIEPER (1968a)
Pressure–tolerance meter, for comparative studies on pressure resistance	Hydraulic hand pump. For experiments in stagnant sea water	SCHLIEPER (1968b)
Pressurized polarographic cell and optical pressure cuvette	—	VIDAVER (1969)
Observational pressure vessel, for studying behaviour of plankton (inner dimensions ca 41 × 5 × 4·5 cm)	Suitable for continuous observation at moderate pressures (up to 800 psi). 19 mm windows of armourplate glass	LINCOLN and GILCHRIST (1970) LINCOLN (1970)
Microscopic pressure chamber, for studying galvanometric responses of protozoans	With electrodes and windows. Permits observations at magnifications up to 600 × and pressures up to 20,000 psi	MURAKAMI and ZIMMERMAN (1970)
Pressure vessel, for photobiological studies	Controlled light intensity and quality, temperature, and sterility for up to several days. Pressures up to 1200 atm	VIDAVER (unpublished in: MORITA, 1970)

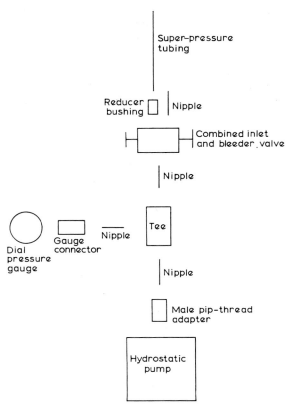

Fig. 8-13: Assembly parts of a pressure pump.
Diagram. Dial pressure gauge: Bourdon type.
Desired length of super-pressure tubing: about
1·5 to 2 m. (After MORITA, 1970; modified.)

can resist pressures up to 2000 atm, at room temperature, with a wide margin of safety (Bair Machine Co., Lincoln, Nebraska, USA). Schematic longitudinal sections of such pressure cylinders, as modified by SCHLIEPER and his associates, are presented in Fig. 8-14. A pressure vessel for studying migrations of planktonic animals (LINCOLN, 1970) is illustrated in Fig. 8-15. MACDONALD (1967) developed a pressure cylinder containing a modified sampling device connected to a Model A Coulter Counter. This apparatus enables the investigator to record changes in individual numbers of protozoan populations maintained under elevated pressures (Fig. 8-16); it could easily be modified to determine quantitative changes in bacteria populations under varying conditions of pressure.

In microbiology, containers for use in pressure cylinders are usually small (10 × 75 mm) Pyrex test tubes fitted with a white, non-toxic (neoprene) rubber stopper. For further details, consult MORITA (1970). In other biological studies, a variety of containers have been employed. The interested specialist should consult the references listed in Table 8-6 and HORNE (1969).

A pressure apparatus used by SCHLIEPER and his associates is illustrated in Fig. 8-17. NAROSKA (1968) employed an AMINCO pressure apparatus which

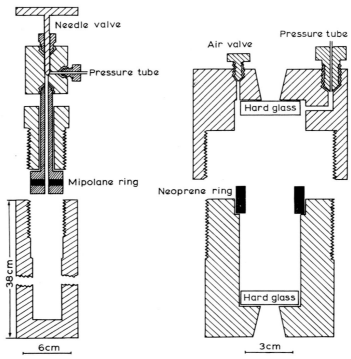

Fig. 8-14: Schematic longitudinal sections of simple pressure
cylinder (left) and windowed pressure cylinder (right).
(After SCHLIEPER, 1968a; modified.)

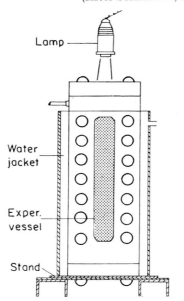

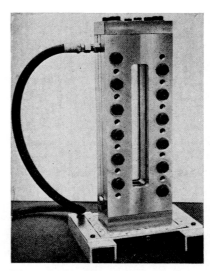

Fig. 8-15(a): Observational
pressure vessel for studying the migrational be-
haviour of planktonic
animals. Diagram of ap-
paratus. (After LINCOLN,
1970; modified.)

Fig. 8-15(b): Photograph of observational pressure vessel without
water jacket and lamp, but
with hydraulic connection tube.
(After LINCOLN and GILCHRIST,
1970.)

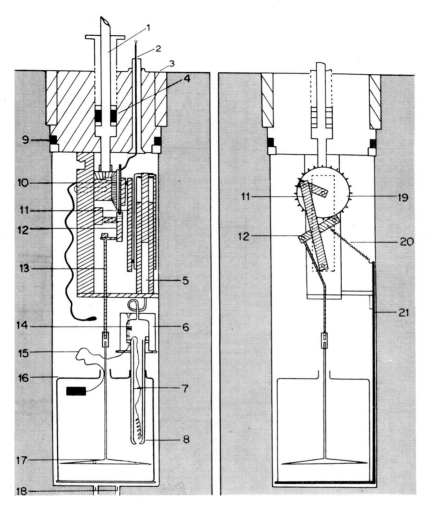

Fig. 8-16: Diagram of pressure vessel with inside counting device for determining cell numbers of protozoan populations maintained under elevated pressures. Left: sectional diagram; right: view at right angles to the former. 1: Drive shaft, 2: thermocouple wires, 3: retaining collar, 4: rotary seal, 5: piston connecting to the perspex cap (6) which fits over the orifice tube (8), 7: oil/water interface inside the orifice tube, 9: packing ring, 10: bevel gear, 11: pushrod, 12: lever which actuates the plunger-shaft (13), 14: a clamping screw which holds the electrode against the brass collar fitted to the rim of the perspex cap (6), 15: wires from electrodes, 16: glass vessels containing the cell suspension, 17: plunger, 18: inlet and outlet pipes connecting to the high pressure pump and drain cock respectively, 19: wheel bearing cams, 20: spring, 21: support for the platform which holds the growth vessel (16). (After MacDonald 1967; modified.)

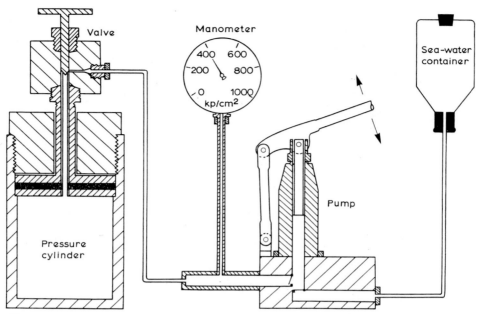

Fig. 8-17: Pressure apparatus with hydraulic hand pump (modified truck jack), pressure cylinder and manometer, for determining pressure tolerance of marine organisms. (After SCHLIEPER, 1968b; modified.)

Fig. 8-18: AMINCO pressure apparatus for studying oxygen consumption of marine organisms under sustained elevated pressures in steadily running sea water. a: piston pump, b: manometer, c: pressure cylinder, d: cooling coil, e: pressure control unit, f: sea water outlet control valve. (After NAROSKA, 1968.)

facilitates the study of long-term variations in rates of metabolism and activity of intact organisms under sustained elevated pressures in running sea water (Fig. 8-18).

Bathymetric methods for measuring depths in oceans and coastal waters

Bathymetric methods for measuring depths in the sea are largely based on 'sounding'. Sounding is accomplished either by direct measurement, e.g. with a lead line (direct sounding), or by indirect measurement via determining the time required for a pulse of sound to reflect from the sea bottom and to return to the site of the sound source (indirect sounding, echo-sounding). The methods of shallow-water bathymetry have changed little since the fundamental principles were laid down in the early nineteenth century, except that methods of navigation control and sounding have become more sophisticated (ADAMS, 1942; FAIR-BRIDGE, 1966). In deeper waters, precision depth recorders are employed. They record, with high resolution, acoustic echo time versus ship's time on continuous graphic profiles (echograms). Sound sources may be explosives, simple percussions, electric arcs, transducers (exploiting piezoelectric or magnetostrictive properties of certain materials, such as quartz, tourmaline, Rochelle salts, barium titanate or nickel) or pneumatic sources (sudden release of highly compressed air). A complete echo-sounding system comprises a sound transmitter and a sound receiver (hydrophone), electronics for timing and amplification, and an indicator or recorder. Modern echo-sounding equipment includes precision recorders, narrow-beam transducers, towed transducers, compact transceivers, inverted echo-sounding, and complete chart-making systems. For further details consult FAIR-BRIDGE (1966).

Thermometric pressure measurements employ two deep-sea reversing mercury thermometers, one of which is exposed to sea pressure at the depth to be determined, while the other is pressure-protected. The unprotected thermometer records a higher temperature (due to compression) than the protected one. In general, the difference amounts to ca 0·01 C°/dbar. Temperature difference and pressure coefficient allow the calculation of the actual sea depth fairly accurately.

Sea depths can be determined, furthermore, by employing the principle of mechanical deformation. In combination with electrical recording systems, such methods have been used for continuous depth recording and for determining the depth of instruments towed by, or lowered from, research vessels, and for measuring *in situ* pressure variations caused by tidal fluctuations; the latter can also be recorded by employing volume changes of enclosed air.

(3) Pressure in Oceans and Coastal Waters

Below the sea surface, the static sea pressure is given by the weight of a vertical water column of unit cross section between surface and the depth considered. The sea pressure varies as a function of water depth, temperature, salinity and acceleration of gravity (p. 1325).

The pressure of 1 dbar corresponds approximately to the pressure exerted by a water column of 1 m height. Hence the pressure in the sea increases by 1 dbar for approximately every m of depth. In other words, the numerical values of pressure

E

in dbar, of geometric depth in m and dynamic depth in dynamic m* are nearly identical. For example, in a 'standard ocean' with a temperature of 0°C, a salinity of 35‰ and an acceleration of gravity of 980 cm/sec², a pressure of 1000 dbar occurs at a geometric depth of 990 m and a dynamic depth of 970 dynamic m. In the 'real ocean', the distribution of temperature and salinity and the slight compressibility of water are among the main factors responsible for the slight differences between the increases in depth and pressure. Although the closeness of these numerical values is quite convenient, the small deviations are most important in describing the actual relative field of pressure in the sea (NEUMANN, 1968). In regard to the pressure in oceans and coastal waters, the atmospheric pressure is neglected and the pressure at the sea surface is entered as zero.

As has been pointed out on page 1325, the unit of pressure most frequently used in marine ecology is the atmosphere. Since 1 atm = 10·1325 dbars, it has become a

Table 8-7

Maximum depths in the major ocean trenches (Compiled from various sources; after WISEMAN and OVEY, 1955 and FISHER and HESS, 1963; modified)

Name of trench	Maximum depth (m)	Name of trench	Maximum depth (m)
Marianas (specifically Challenger Deep)	11,034±50 10,915±20 10,915 10,863±35 10,850±20	North Solomons (Bougainville)	9,103 8,940±20
		Yap (West Caroline)	8,527
		New Britain	8,320 8,245±20
Tonga	10,882±50 10,800±100		
		South Solomons	8,310±20
Kuril-Kamchatka	10,542±100 9,750±100[a]	South Sandwich	8,264
Philippine (vicinity of Cape Johnson Deep)	10,497±100 10,265±45 10,030±10	Peru-Chile	8,055±10
		Palau	8,054 8,050±10
Kermadec	10,047		
Idzu-Bonin (includes 'Ramapo-Deep' of the Japan Trench) (vicinity of 'Ramapo Deep')	9,810 9,695	Aleutian	7,679[b]
		Nansei Shoto (Ryuku)	7,507
		Java	7,450
Puerto Rico	9,200±20	New Hebrides (South)	7,070±20
New Hebrides (North)	9,165±20	Middle America	6,662±10

[a] Maximum sounding obtained in the vicinity of the Vitiaz Depth by French and Japanese vessels in connection with dives of the bathyscaph Archimède, July, 1962.
[b] Uncorrected, taken with nominal sounding velocity of 1500 m/sec.

* Dynamic m = $\dfrac{gd}{10}$, where g is acceleration of gravity in m/sec² and d water depth in m.

general rule of thumb to allow an increase of 1 atm for every 10 m or 33 feet of water depth.

Marine organisms inhabit all depths of the sea. While subsurface-living forms are exposed to pressures of little more than 1 atm, organisms inhabiting deep-sea floors are subjected to pressures in excess of 1000 atm (Table 8-7). At the greatest known oceanic depth—11,034 m in the Marianas Trench (Challenger Deep; HANSON and co-authors, 1959)—the pressure amounts to about 1160 atm.

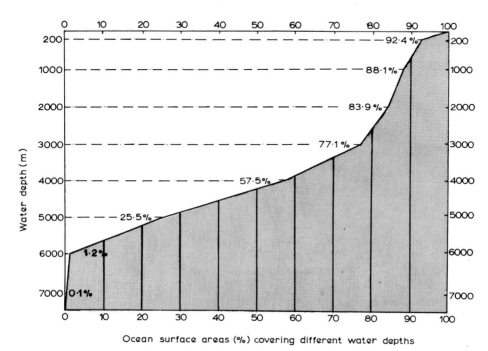

Fig. 8-19: Percentages of the total surface of all oceans (including small seas) covering the different depths indicated. Example: 83·9% of the total surface area covers depths in excess of 2000 m. (After HARDY, 1956; modified.)

About 88% of the total bottom area of all oceans carries water masses deeper than 1000 m. Some 84% of the surface areas of the oceans cover water depths in excess of 2000 m; 25.5%, depths greater than 5000 m (Fig. 8–19). The average depth of all oceans has been estimated to be about 3800 m; this is equivalent to a bottom pressure of about 380 atm. In terms of the earth's total surface area, more than half is covered by waters deeper than 2000 m, and about one-third lies buried under waters more than 4000 m deep.

'How does life on land—on that less than a third of the earth's surface—compare with this in quantity? Quite an appreciable portion of the land is either icecap, desert or high mountain with very little life; indeed, the whole antarctic continent is a barren frozen waste. The atmosphere has no permanent fauna of its own; at best the dry land has a zone of life from treetop height to but a few feet below the surface of the soil. We think again of the oceans—

thousands of feet deep, inhabited at all levels and stretching over the greater part of the earth's surface—and as we do so we begin to grasp the truth . . .' (HARDY, 1956, p. 219).

In the absence of pronounced variations and gradients of other environmental factors in the deep sea, pressure tends to become an ecological master factor. While lack of light, low temperature and diminished water movement—as well as scarcity and increasing uniformity of biologically useful energy sources—may also be found in other marine habitats, high pressure is a specific attribute of deep waters. For organisms occupying the free water of the oceans, pressure gradients and gravity represent the basic denominators providing orientation clues for vertical migrations and for positioning in the water column. Whereas, in water depths down to several hundred m, planktonic forms may also employ changes in light intensity or quality, motile bathypelagic organisms must rely primarily on variations in pressure and on the directional force of gravity for orientation, migration and positioning.

8. PRESSURE

8.1 BACTERIA, FUNGI AND BLUE-GREEN ALGAE*

R. Y. Morita

(1) Introduction

Although bacteria have been isolated from all depths of oceans and coastal waters, knowledge on the intimate relationship between hydrostatic pressure and life processes in the oceans still suffers from lack of investigations concerned with hydrostatic effects on biological systems. In addition, marine biochemists or chemists have paid little attention to hydrostatic pressure as a variable in their studies. Johnson's (1957b) statement still stands:

> 'Even now, the importance of pressure as a variable of both fundamental theoretical interest and ecological importance remains to be fully appreciated, in as much as contemporary books and extended discussions pertaining to ecology, comparative physiology, oceanography, deep-sea life, etc., often include the influence of temperature on living organisms with scarcely more than a passing reference to pressure'.

In other words, most biologists ignore the Ideal Gas Law ($PV = nRT$) as it applies to biological systems.

Since the pressure ranges from 1 atm to approximately 1100 atm in the ocean (Chapter 8.0), this chapter deals with this range of pressure only. Several studies have been performed on proteins, bacteria and viruses above the kilobar range. Results from such studies may aid in the interpretation of some of the data obtained from studies dealing with pressures that occur in the oceans (Morita, 1967).

Pressure also affects the chemical and physical environments of micro-organisms in the sea. For instance, the ionization of water is greater with increasing pressures (Owen and Brinkley, 1941; Hamann, 1963c). Pressure also modifies the viscosity (Horne and Johnson, 1966), pH (Buch and Gripenberg, 1932; Distèche, 1959; Pytkowicz and Conners, 1964), chemical reaction rates (Hamann, 1964), ionization of various substances (Owen and Brinkley, 1941; Ewald and Hamann, 1956; Distèche and Distèche, 1965; Hamann and Strauss, 1965), as well as the hydrophobic bonding of proteins (Kettmann and co-authors, 1965). According to Horne and Johnson (1966), the application of pressure tends to destroy the structured regions in liquid water. For instance, at 4·009°C, the relative viscosity of water is less at 984 kg/cm² than at 70 kg/cm² (0·9985 versus 0·9745 respectively). When electrolytes are added, the effect of pressure on the viscosity of water is less pronounced; but when the amount of NaCl added approaches 1·4 molar, the relative viscosity of water increases with increased pressure. At 4·5 molar, the relative viscosity increases with increased pressure; consequently, the addition of sodium chloride aids the structured regions of water (Horne and Johnson, 1967). The partial specific volume of various non-biological compounds

* Published as technical paper No. 2350, Oregon Agricultural Experiment Station.

has been measured under pressure by ANDERSON (1963). The apparatus employed could be adapted to determine the molecular volume changes of certain biological substances under pressure but there would be difficulty adapting the apparatus for enzymatic reactions. Unfortunately, protein chemistry is not sufficiently advanced; hence the effect of pressure on conformational changes of intracellular proteins cannot be studied readily.

The area in which pressure exerts most influence in the marine environment is below the thermocline, where the temperature is 5°C or lower. It should be recognized that very few hydrostatic pressure studies are performed at 5°C or less. If more meaningful data are to be obtained, then studies should be done with microbial cells grown at 5°C, since there is a difference between cells grown at 5°C and the optimum temperature for growth (HAIGHT and MORITA, 1966). In the depths of the oceans, temperature is more or less constant, while pressure represents the main ecological variable.

It should also be realized that many problems dealing with pressure effects on micro-organisms remain unsolved, and that a myriad of questions arises as to why certain organisms cannot withstand increased pressures, while other forms which live in the hadal deep sea cannot live without pressure. Since the environmental factor pressure does not generally enter into the studies of researchers dealing with terrestrial forms, marine ecologists must begin to investigate the biological consequences of variations in pressure, in order to appreciate fully the importance of pressure as an ecological variable in the sea. Furthermore, aid is needed from the marine chemists so that we will be able to understand the effects of pressure on the non-biological systems previously mentioned.

Hydrostatic pressure investigations, employing blue-green algae or fungi (MORITA, 1965) as test organisms, are limited in number and, as a result, this chapter will deal mainly with the effects of pressure on bacteria. For more information concerning this subject matter, JOHNSON and co-authors (1954), JOHNSON (1957b), ZoBELL (1964), MORITA, (1967) and ZIMMERMAN (1970a) should be consulted.

One of the main tasks yet to be undertaken in marine ecology is the study of pressure effects on organisms taken from the deeper zones of the oceans. This task must be performed with all the tools of modern biology so that basic mechanisms, characteristic of life in the deep sea, can be unravelled. Only in this way can progress be made in determining how pressure affects the physiology, biochemistry, distribution and morphology of micro-organisms. As can be seen from the following pages, practically all pertinent laboratory research has been done with shallow water or terrestrial micro-organisms.

(2) Functional Responses

(a) Tolerance

When micro-organisms are exposed to pressures above the kilobar range, death usually results (e.g. ROGERS, 1895; CHLOPIN and TAMMANN, 1903; HITE and co-authors, 1914; LARSON and co-authors, 1918; LUYET, 1937; ZIMMERMAN, 1970a). The duration of exposure is an important factor in the organism's ability to withstand pressure. However, ZoBELL and JOHNSON (1949) demonstrated that the suddenness, with which cultures were compressed or decompressed during short periods of exposure to high pressure, has no appreciable effect on the viability of the test organism. The temperature during pressure exposure may significantly modify the degree of tolerance. In view of the resulting, rather complex picture, no general rule can be laid down concerning the tolerance of bacteria to elevated pressures. Pressure tolerance varies greatly in different micro-organisms and is a function of intensity and patterns of pressure and temperature applied, as well as of the duration of exposure. The rate of compression or decompression may be of great importance when either (i) pressure shock, (ii) appreciable adiabatic heating or cooling, or (iii) compressed gas is involved (ZoBELL, 1970). The causes of lethal effects due to moderate pressures remain unknown; death caused by high pressures above the kilobar range seems to be due primarily to inactivation of cellular proteins.

ZoBELL and JOHNSON (1949) coined the term barophilic to designate bacteria which possess the ability to grow well under pressures higher than 400 or 500 atm (see also ZoBELL, 1970). The occurrence of bacteria on pelagic substrates was demonstrated by MORITA and ZoBELL (1955) during the Mid-Pacific Expedition of 1950. Most of these forms were cultured at 1 atm. Since that time, bacteria which are tolerant to or which require elevated pressure have been examined by ZoBELL (1952) and ZoBELL and MORITA (1957, 1959); see also MORITA and BECKER (1970) and ZoBELL (1970). It should further be mentioned that growth of bacteria from open ocean sediments is very slow, requiring 3 or 4 months incubation. Even after this incubation period, growth should be determined microscopically.

Sediment obtained from the hadal zones of the Pacific and Indian Oceans was examined for the presence of bacteria (Table 8-8). In many instances, bacteria

Table 8-8

MPN of different physiological types of bacteria detected per gram of wet sediment from the Philippine Trench, incubated in selective media at different pressures and 3° to 5°C. ‘Galathea’ Deep-Sea Expedition (After ZoBELL and MORITA, 1957; modified)

‘Galathea’ Station:	No. 418		No. 419		No. 420		No. 424	
Latitude	10°13′ N		10°19′ N		10°24′ N		10°28′ N	
Longitude	126°43′ E		126°39′ E		126°40′ E		126°39′ E	
Water depth	10,190 m		10,210 m		10,160 m		10,120 m	
Incubation pressures:	1 atm	1000 atm	1 atm	1000 atm	1 atm	1000 atm	1 atm	1000 atm
Total aerobes	10^3	10^6	10^3	10^5	10^4	10^5	10^4	10^6
Total anaerobes	10^3	10^5	10^4	10^5	10^3	10^5	10^4	10^5
Starch hydrolyzers	10^2	10^3	10	10^2	10	10^2	10^2	10^3
Nitrate reducers	10^2	10^5	10^2	10^4	10^3	10^5	10^2	10^5
Ammonifiers	10^3	10^5	10^3	10^4	10^3	10^5	10^3	10^5
Sulphate reducers	0	10^2	0	10	0	10^2	0	0

capable of existing at 1 atm were observed along with those which had the ability to grow at elevated pressures (ZoBELL and MORITA, 1957). Bacteria which grow at 1 atm and also at elevated pressures are termed 'barotolerant' or 'baroduric'. The barotolerant bacteria may be dormant forms from shallower depths, or passive mutants; it is also possible that not all of their cultural requirements have been met in the laboratory.

Growth of barophilic bacteria under conditions, isobaric and isothermic to those from which they were obtained, is slow (ZoBELL and MORITA, 1957). Although various physiological types were isolated from sediments taken from the depth of various hadal zones of the oceans, it still remains to be seen in what way these micro-organisms are different from their surface-dwelling counterparts. This question also applies to the other organisms dredged up during the 'Galathea' Deep-Sea Expedition from various trenches and deeps. In the top 1-cm layer of the deep-sea sediment, the 'standing crop' of bacteria was estimated by ZoBELL (1954) to be between 0·001 and 0·005 g of organic carbon per m². Since many of the animals inhabiting the ocean floor are mud eaters, they may obtain much of their energy by consuming sediment-living bacteria.

KRISS (1963) reported the existence of barophilic and baroduric bacteria from various oceanic regions. He also mentions the ability of soil bacteria to exist at pressures up to 1500 atm. Some of the data obtained by KRISS are not in agreement with those of ZoBELL and MORITA (1959). ZoBELL (1964) attributes this discrepancy to the fact that KRISS did not (i) allow sufficient incubation time for the cultures to develop (usually 3 days); (ii) provide oxygen to the cultures under pressure. A 3-day incubation period is insufficient because barophilic bacteria appear to be slow growers (ZoBELL and MORITA, 1959). Furthermore—as ZoBELL also points out—a hydrogen acceptor must be added to the medium, since it is impossible to supply oxygen to a closed pressurized system.

In order to conduct investigations on barophilic and barotolerant bacteria, certain technical difficulties must be surmounted. Some of the major problems are (i) the difficulty to isolate micro-organisms in pure culture, (ii) the inability to culture large quantities of individuals for physiological and biochemical studies, (iii) the problem of providing continuous aeration and of removal of carbon dioxide under pressure.

(b) Metabolism and Activity

Growth

In many respects, terrestrial bacteria respond to the effects of hydrostatic pressure similarly to micro-organisms isolated from the sea. As a result, many studies which have been made with terrestrial bacteria have contributed greatly to the understanding of pressure effects on bacteria in the sea.

In *Escherichia coli*, the rate of population growth is logarithmic at both 28·5° and 39·9°C at $66\frac{2}{3}$ atm, but when the pressure is raised to 333 atm, multiplication rate is greatly retarded. In cultures exposed to relatively low pressures, the rate of multiplication is slightly higher than in the atmospheric controls (JOHNSON and LEWIN, 1946b). In 1949, ZoBELL and JOHNSON subjected a number of terrestrial micro-organisms in various media to different levels of pressure (Table 8-9). All cultures are able to reproduce at 1 atm, but when the pressure is raised to 300

Table 8-9

Pressure effects on multiplication or destruction of terrestrial micro-organisms after 48 hrs incubation at 30°C at different hydrostatic pressures: + : degree of turbidity relative to the control culture at 1 atm; − : no apparent multiplication; d: culture has lost its ability to multiply following decompression (After ZoBELL and JOHNSON, 1949)

Organism	Hydrostatic pressure (atm)				
	1	300	400	500	600
Alkaligenes viscosus	+ + + +	+ + +	+ +	− d	− d
Bacillus alvei	+ + + +	+ + +	+	−	−
Bacillus brevis	+ + + +	+ + +	+	−	− d
Bacillus cereus	+ + + +	+ + + +	−	−	−
Bacillus circulans	+ + + +	+ + + +	−	−	−
Bacillus megatherium	+ + + +	+ + + +	+	−	− d
Bacillus mesentericus	+ + + +	+ + + +	+ + +	+ +	− d
Bacillus mycoides	+ + + +	+ + + +	+	−	− d
Bacillus subtilis	+ + + +	+ +	+	−	−
Clostridium chauvei	+ + + +	+ + + +	+ +	−	−
Clostridium histolyticum	+ + + +	+ + + +	+ + +	−	− d
Clostridium putreficum	+ + + +	+ + + +	+ +	−	−
Clostridium scpticum	+ + + +	+ +	+	−	−
Clostridium sporogenes	+ + + +	+ + +	+ +	−	−
Clostridium welchii	+ + + +	+ + + +	+	−	−
Escherichia coli	+ + + +	+ + + +	+ + +	+ +	−
Micrococcus lysodeikticus	+ + + +	+ +	−	− d	− d
Mycobacterium phlei	+ + + +	+ + +	+ +	− d	− d
Mycobacterium smegmatis	+ + + +	+ +	+	−	− d
Proteus vulgaris	+ + + +	+ +	−	−	− d
Pseudomonas fluorescens	+ + + +	+ + + +	+ + +	−	−
Sarcina lutea	+ + + +	+ + +	+	−	− d
Serratia marcescens	+ + + +	−	−	−	−
Staphylococcus albus	+ + + +	+ + +	+ + +	−	−
Staphylococcus aureus	+ + + +	+ +	−	− d	− d
Streptococcus lactis	+ + + +	+ + + +	+ + + +	+ +	−
Hansenula anomala	+ + + +	+ +	− d	− d	− d
Saccharomyces cerevisiae	+ + + +	+ + + +	− d	− d	− d
Saccharomyces ellipsoideus	+ + + +	+	− d	− d	− d
Schizosaccharomyces octosporus	+ + + +	−	− d	− d	− d
Sporobolomyces salmonicolor	+ + + +	+	− d	− d	− d
Torula cremoris	+ + + +	+ + +	+ +	− d	− d

atm, quite a few bacteria are adversely affected. At 600 atm, many of the cultures expire, while others remain viable but do not multiply. As can be seen in Table 8-9, a variety of responses occur with the different cultures tested at various pressures; as a general trend, population death rates increase with increasing pressures. Oxygen is not a limiting factor in these studies, since the amount of oxygen was the same in all tubes at the beginning of the experiments for the aerobic organisms, while no oxygen was present in the tubes containing anaerobes.

Table 8-10 illustrates the interrelationship between pressure and temperature on population growth of various micro-organisms. At 300 atm, most cultures grow

Table 8-10

Relative turbidity of cultures of micro-organisms in nutrient medium after 4 days' incubation at 20°, 2 days at 30°, or 1 day at 40°C at different pressures. All cultures listed show four-plus ($++++$) growth in the controls incubated at normal atmospheric pressure (After ZoBell and Johnson, 1949)

Micro-organism	300 atm			400 atm			500 atm			600 atm		
	20°C	30°C	40°C	20°C	30°C	40°C	20°C	30°C	40°C	20°C	30°C	40°C
Alkaligenes viscosus	++	+++	++++	++	++	++	−	−	−	−	−	−
Bacillus brevis	−	+++	++	−	+	+	−	−	−	−	−	−
Bacillus megatherium	−	++++	+++	−	++	++	−	−	−	−	−	−
Bacillus mesentericus	−	++++	+++++	−	+++	+++++	−	++	++++	−	+	+++
Bacillus subtilis	−	+++	+++++	−	++	+++++	−	++	+++	−	−	++
Clostridium bifermentans	++	+++++	+++++	−	−	+++++	−	−	−	−	−	−
Clostridium chauvei	−	+++++	+++++	−	++	+++	−	−	−	−	−	−
Clostridium histolyticum	−	+++++	+++++	−	−	+++	−	−	−	−	−	−
Clostridium putreficum	−	+++++	+++++	−	++	+++	−	−	−	−	−	−
Clostridium septicum	−	+	++	−	−	+	−	−	−	−	−	−
Clostridium sporogenes	−	+++++	+++++	−	++	+++	−	−	−	−	−	−
Clostridium welchii	−	+++++	+++++	−	+++	+++	−	++	+++++	−	−	+
Escherichia coli	++	+++++	+++++	−	+++	+++++	−	−	+++++	−	+	+
Mycobacterium phlei	−	+++	+++	−	++	+	−	−	−	−	−	−
Mycobacterium smegmatis	−	++	+++	−	+	+	−	−	−	−	−	−
Pseudomonas fluorescens	++	+++	+++++	−	++	+++++	−	−	+++	−	−	−
Sarcina lutea	++	++	+++	−	++	+++	−	−	−	−	−	−
Staphylococcus albus	++	++	+++	−	++	+++	−	−	−	−	−	−
Staphylococcus aureus	−	+++	+++++	−	++	+++	−	−	−	−	−	−
Streptococcus lactis	+++++	+++++	+++++	+	+	+++++	−	++	+++++	−	+	+++

at 30° and 40°C, but at 20°C some fail to multiply. When the pressure is raised to 600 atm, most organisms fail to multiply; population growth continues, in three cases, at 40°C only, but not at 20° or 30°C.

A comparison of Tables 8-9 and 8-10 with Table 8-11, suggests that marine bacteria are more resistant to pressure than their terrestrial counterparts. OPPEN-HEIMER and ZoBELL (1952) exposed 63 species of marine bacteria to elevated pressures and found that 10 fail to multiply when subjected to 200 atm at 27°C, while 5 are killed (k) within 4 days of incubation (Table 8-12). Of the 28 species which fail to multiply at 400 atm, 11 are killed, and of the 56 which fail to grow at 600 atm, 23 are killed. ZoBELL and BUDGE (1965) tested other marine bacteria for their ability to multiply under elevated pressures and found significant differences, depending upon the species in question. *Vibrio marinus* MP-1, an obligately psychrophilic marine bacterium isolated from an environment of 120 atm and 3·24°C, is capable of multiplication at 300 atm at either 3° or 15°C, but at higher pressures (400, 500, 600 atm) the number of cells, when compared to the inoculum, decreases. At 800 atm, the cells are killed within 24 hrs (MORITA and ALBRIGHT, 1965). Zoospores and mycelia of *Allomyces macrogynus* are rendered non-viable when incubated at 27°C at 600 or 1000 atm (HILL, 1962).

Population growth of *Escherichia coli* is retarded by pressures ranging from 1 to 500 atm when incubated in nutrient medium (ZoBELL and COBET, 1962). There exists a pronounced pressure–temperature relationship; for example, at 40°C, an initial decrease in multiplication rates (number of cells) of *E. coli* occurs at 400 atm compared to 1 or 200 atm; however, further incubation brings about a different picture: rates of population growth at 400 atm are higher than at 200 atm, and at 200 atm are higher than at 1 atm. The effect of pressure on growth and reproduction is less pronounced at 30°C than at 40°C or 20°C. The application of increased pressures also augments the lag phase of the culture, and at 400 to 1000 atm, death rates are accelerated—a response which becomes accentuated at the higher temperatures employed, i.e., 20°, 30° and 40°C. Studies by ZoBELL and COBET (1964) concerning filament formation in several strains of *E. coli* (see also p. 1387) indicated that increased pressure produces filament formation, while the biomass of cellular material contains the same amount of protein and nucleic acids. However, at higher pressures, the type of nucleic acid changes, resulting in appreciably more ribonucleic acid (RNA) than desoxyribonucleic acid (DNA). The data suggest that the DNA fails to replicate at increased pressures and, as a result, may be responsible for a repression of cell division.

Employing a system whereby air could be pulsed into culture tubes under pressure, HEDÉN and MALMBORG (1961) demonstrated that population growth of *Escherichia coli* B and *Staphylococcus aureus* is retarded. The retardation effect is not clearly noticeable until approximately 6 hrs after pressurization. Growth curves (35 atm) reveal that *E. coli* B probably divides once before a stationary phase is reached. If the pressure is released from the system, the culture under pressure gradually exhibits a regular growth pattern again. Although not mentioned by HEDÉN and MALMBORG, it appears that, if synthetic processes in the cells have been initiated, they will go on to completion before a cessation of growth occurs. However, the cells do not appear to be injured permanently when pressurized under this system for 3 hrs, since after a normal lag period growth again occurs.

Table 8-11

Relative turbidity caused by the multiplication of marine bacteria in nutrient broth for 6 days at 20°, 4 days at 30°, or 1 day at 40°C at different pressures. All cultures listed – except those marked with an asterisk, which fail to grow at 40°C – exhibit four-plus (+ + + +) growth at normal pressure (After ZoBell and Johnson, 1949)

Micro-organism	300 atm			400 atm			500 atm			600 atm		
	20°C	30°C	40°C	20°C	30°C	40°C	20°C	30°C	40°C	20°C	30°C	40°C
Achromobacter fischeri	+ + + +	+ +	*	+ +	–		–	–	–	–	–	
Achromobacter harveyi	+ + + +	+ + + + +	+ + + +	+	+ + + +	–	–	–	–	–	–	–
Achromobacter thalassius	–	+ + + + +	+ +	–	–	+ + +	–	–	–	–	–	–
Bacillus abysseus	+	+ + + + +	+ + + + +	–	+ + + + +	+ + + + +	–	+	+ + + +	–	–	+ +
Bacillus borborokoites	+ +	+ + + + +	+ + + + +	–	+ + +	+ + + + +	–	+ +	+ + + +	–	–	+ +
Bacillus cirroflagellosus	+ +	+ + +	*	–	+ +		–	–		–	–	
Bacillus submarinus	+ +	+ + + + +	+ + +	+	+ + + + +	+ + +	–	+ + + + +	+ + +	–	+ + + + +	+ + +
Bacillus thalassokoites	+ + +	+ + + + +	+ + +	+ +	+ + + + +	+ + +	+ +	+ + + + +	+ + + + +	–	+ + + + +	+ + +
Flavobacterium okeanokoites	+ + + + +	+ + + + +	+ + + + +	+ + + + +	+ + + + +	+ + + + +	+ + + + +	+ + + + +	+ + + + +	–	+ + + + +	+ + +
Flavobacterium uliginosum	+ + + + +	+ + + + +	+ + + + +	+ + + + +	+ + + + +	+ + + + +	–	+ + + + +	+	–	–	–
Micrococcus infimus	+	+ + + + +	*	–	+		–	–				
Photobacterium splendidum	+ + + + +	+ + + + +	*	+ +	+ + +		–	+				
Pseudomonas pleomorpha	+ +	+ + + + +	*	+ +	+ + +		–	+ +				
Pseudomonas vadosa	+ + +	+ + + + +	+ + + +	+ + + + +	+ + + + +	+ + +	+ +	+ + + + +	+ + +	–	–	+
Pseudomonas xanthochrus	+ + + + +	+ + + + +	*	+ + +	+ + + +	+	–	+ + +	+ +	+	+	+ +
Vibrio hyphalus	–	–	*	–	–		+	–		–	–	–
Mixed microflora from mud	+ + + +	+ + + + +	+ + + + +	+ + + + +	+ + + + +	+ + + + +	+ + + + +	+ + + + +	+ + + + +	+ + + + +	+ + + + +	+ + +

Table 8-12

Multiplication of marine bacteria as indicated by relative turbidity in sea-water broth after 8 days' incubation at different pressures at 27°C
(After OPPENHEIMER and ZOBELL, 1952)

Micro-organism*	1 atm	200 atm	400 atm	600 atm
Achromobacter stenohalis	+ + + +	−	−	k
Achromobacter aquamarinus	+ + + +	+ + + +	+ + +	−
Achromobacter stationis	+ + + +	+ + + +	+ + + +	−
Achromobacter thalassius	+ + + +	k	k	k
Actinomyces halotrichis	+ +	−	−	−
Actinomyces marinolimosus	+ + + +	+ + + +	−	−
Bacillus imomarinus	+ + + +	−	−	−
Bacillus cirroflagellosus	+ + + +	+ +	−	−
Bacillus epiphytus	+ + +	+ +	+ +	k
Bacillus submarinus	+ + + +	+ + +	+ + +	−
Bacillus thalassokoites	+ + + +	+ + +	+	−
Bacillus filicolonicus	+ + + +	−	k	k
Bacillus abysseus	+ + + +	+ + + +	+ + +	−
Bacillus borborokoites	+ + + +	+ + + +	+ + + +	+ + +
Flavobacterium marinotypicum	+ + + +	+ + + +	+ + + +	−
Flavobacterium marinovirosum	+ + + +	+ + + +	+ +	−
Flavobacterium neptunium	+ +	+	−	−
Flavobacterium okeanokoites	+ + + +	+ + + +	+	−
Micrococcus aquivivus	+ + + +	+ + + +	+ + + +	+ + +
Micrococcus sedimenteus	+ + + +	+ + + +	+ + + +	k
Micrococcus maripuniceus	+ + + +	+ + + +	−	−
Micrococcus infimus	+ + +	+ + + +	−	−
Micrococcus sedentarius	+ + + +	+ + + +	+ + + +	−
Micrococcus euryhalis	+ + + +	k	k	k
Pseudomonas enalia	+ + + +	+ + +	−	k
Pseudomonas neritica	+ + + +	+ + + +	+ + +	k
Pseudomonas azotogena	+ + + +	k	k	k
Pseudomonas vadosa	+ + + +	+ + + +	+ + + +	−
Pseudomonas oceanica	+ + + +	+ + +	+ + + +	−
Pseudomonas felthami	+ + + +	+ + + +	+ +	−
Pseudomonas aestumarina	+ + + +	−	k	k
Pseudomonas membranula	+	−	k	k
Pseudomonas stereotropis	+ + + +	+ + +	+	k
Pseudomonas coenobios	+ + + +	+ + + +	+ + + +	−
Pseudomonas obscura	+ + + +	+ + + +	+ +	−
Pseudomonas pleomorpha	+ + + +	+ + + +	+ +	−
Pseudomonas marinopersica	+ +	k	k	k
Pseudomonas periphyta	+ +	−	k	k
Pseudomonas hypothermis	+ + + +	k	k	k
Pseudomonas perfectomarinus	+ + + +	+ + + +	+ + +	+ + +
Pseudomonas xanthochrus	+ + + +	−	−	−
Sarcina pelagia	+ + + +	+ + +	+ + +	−
Serratia marinorubra	+ + + +	+ + + +	+ + +	−
Vibrio marinopraesens	+ + + +	+ + + +	+ + +	−
Vibrio algosus	+ + + +	+ + +	+ +	+
Vibrio adaptatus	+ + + +	+ + +	+ + +	k
Vibrio marinoflavus	+ + +	+ + +	+	k
Vibrio ponticus	+ + + +	+ + +	+ +	k

Table 8-12—*Continued*

Micro-organism*	1 atm	200 atm	400 atm	600 atm
Vibrio phytoplanktis	+ + + +	+ + + +	+ + + +	+ + +
Vibrio haloplanktis	+ + + +	+ + + +	+ + + +	+
Vibrio marinovulgaris	+ + + +	+ + +	–	–
Vibrio marinofulvus	+ + + +	+ + + +	+ +	–
Vibrio marinagilis	+ + + +	+	–	–
Vibrio hyphalus	+ + + +	+ + + +	k	k
Number 516	+ + + +	+ + + +	+ +	k
Number 549	+ + + +	+ + + +	+ + +	k
Number 595	+ +	–	–	–
Number 623	+ +	+ +	–	–
Number 632	+ + + +	–	–	–
Number 633	+ + + +	+ + + +	–	–
Number 639	+ + + +	+ +	k	k
Number 643	+ + + +	+ + +	+ + + +	+ + + +
Number 689	+ + + +	+ + +	–	k

* All except the numbered cultures have been described by ZoBELL and UPHAM (1944).

+ + + +	good multiplication
+ + +	fair multiplication
+ +	less multiplication
+	poor multiplication
–	no multiplication, but organisms not killed
k	all bacteria in culture killed

The growth inhibition was attributed to toxic oxygen effects (over-oxygenation); under pressure, oxygen tends to react with any available free radical present in the cells (cysteine, glutathione, etc.).

Growth is a complicated process which involves many subprocesses. Our knowledge on the subprocesses involved is insufficient, and hence a detailed assessment of pressure effects is, at present, impossible. Some specific aspects of growth have been studied under a variety of pressure and temperature conditions. Most of the information available is related to responses of enzymes, employing washed cells, cell-free extracts, or partially purified enzymes as test systems.

Molecular volume changes

Important studies on pressure effects on the metabolism of micro-organisms have been performed with the bioluminescent system by JOHNSON and co-authors (1954). The results obtained by these authors on the luciferin–luciferase system, under the influence of pressure and temperature, were analyzed in terms of the absolute reaction rate theory (JOHNSON and co-authors, 1954; JOHNSON, 1957b). LAMANNA and MALLETTE (1959) contend that the absolute reaction rate theory has not been successful when applied to biological systems other than biolumin-escence, and that this fact may reflect deficiencies in the theory, or experimental difficulties in making the necessary precise measurements.

Pressure effects on biological systems can be interpreted in terms of the amount of volume change that the internal cellular components undergo. Volume changes accompanying the formation of ion pairs, amino-acid dipoles, and peptide dipoles have been calculated from pressure effects on reaction rates (LINDERSTRØM-LANG and JACOBSEN, 1941). Although the precise way in which pressure affects the change in molecular structure is still obscure, some initial studies have been undertaken (RIFKIND and APPLEQUIST, 1964; GILL and GLOGOVSKY, 1965; KETTMAN and co-authors, 1965).

Molecular volume (partial molar volume in chemistry) may be defined as the volume occupied by a molecule. All cellular constituents undergo a molecular volume decrease or increase, depending upon pressure, temperature, ionic environment, and pH. Most of the research has dealt with the effect of pressure on enzymes, since these proteins have a high molecular weight and can be analyzed much more readily than other systems. Because of their high molecular weight, the changes that occur in proteins are easily influenced by both pressure and temperature. Increased pressure results in a decrease of the molecular volume, resulting in a negative ΔV. Increased temperature on the other hand, causes the molecular volume to increase and results in a positive ΔV; a decrease in temperature results in a negative ΔV. As a result, variations in pressure and temperature exhibit an interrelationship as would be expected from the Ideal Gas Law ($PV = nRT$). However, the exact way in which pressure and temperature affect the secondary, tertiary, or—possibly—the quaternary structure of proteins is not as yet known. As a general rule, reactions with a negative ΔV are accelerated by pressure increase, while those with a positive ΔV are retarded or inhibited. In terms of enzyme reaction rates, the formation of the enzyme–substrate complex is generally the limiting step (LAIDLER, 1951). When the enzyme unfolds (activated state) to accept the substrate to form the enzyme–substrate complex, the unfolding process involves a positive ΔV (see also Chapter 8.0).

Protein conformation and configuration is affected by the presence of water. Since all cells are bathed in an aqueous environment, water structure due to pressure becomes an important factor, as well as the way in which the water molecules are grouped around a particular molecule or ion. Physical biochemistry will play an important role in analysis by biologists of the data already at hand, and in a more detailed interpretation of the effects of pressure on cellular constituents. Presumably, pressure variations do not significantly affect the primary structure of proteins (sequence of amino acids, and disulphide cross linkages), but probably play an important role in regard to secondary (hydrogen bonds), tertiary (non-covalent bonds such as electrostriction, hydrophobic bonds, van der Waals interactions, etc.) and quaternary structures (non-covalent bonds) of proteins. Changes in secondary, tertiary and quaternary structures of proteins should result in volume changes—a situation which is, indeed, observed under elevated pressures.

How do proteins of micro-organisms living in the deep sea, especially in the hadal zone, differ from those of organisms living at lesser depths? This question opens up a wide new area of research for marine biochemists interested in proteins.

Analysis of previously pressurized cells

The effect of pressure on washed cells of *Escherichia coli* was studied by MORITA and ZOBELL (1956) and MORITA (1957a). Although these studies employed cells which were previously exposed to different pressures and temperatures, they

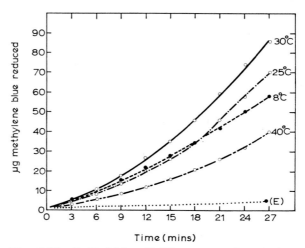

Fig. 8-20: *Escherichia coli*. Micrograms of methylene blue reduced at 30°C after different periods of time by 2 mg of lyophilized (10-day) cells which 30 mins previously had been subjected to a pressure of 600 atm for 4 hrs at 8°, 25°, 30° or 40°C. Upper four curves: succinic dehydrogenase activity with succinate as the hydrogen donor; dotted line (E): endogenous activity with no succinate added. (After MORITA and ZOBELL, 1956; modified.)

demonstrate the effects of pressure on certain enzymes within the cell. The amount of inactivation of succinic dehydrogenase increases with time of compression, and inactivation is more pronounced at temperatures above or below the temperature at which the organisms were grown (MORITA and ZOBELL, 1956; Fig. 8-20). The amount of ammonia produced from various amino acids by previously pressurized cells depends upon the amino acid employed as substrate. Cells pressurized at 600 atm and 30°C for 3 hrs produce more ammonia from alanine and glutamic acids than the controls at 1 atm; but, in the case of cysteine and histidine, less ammonia is produced (Table 8-13). These results indicate that, within cells, pressure acts differentially on enzymes; this could result in unbalanced, metabolic activities of the cell. Such metabolic disharmonization may cause death of the cell if it is subjected to prolonged pressurization.

Enzyme reactions under elevated pressures

Temperatures above optimal cause a decrease in enzyme activity, since a change from the active to the inactive states of the enzyme results in a molecular volume increase. If pressures of a few hundred atm are applied to the system, reaction rate may increase. This has been shown in bacterial luminescence (JOHNSON and co-authors, 1942a, b; STREHLER and JOHNSON, 1954) as well as in enzymes not of microbial origin (EYRING and co-authors, 1946; FRASER and JOHNSON, 1951; SCHNEYER, 1952). Pressure can also reduce the rate of protein inactivation resulting from heat or other denaturants (JOHNSON and CAMPBELL, 1945, 1946; JOHNSON and co-authors, 1948a).

Table 8-13

Ammonia production from various substrates by washed cells of *Escherichia coli* as a function of pressure. Incubation was for 15 mins at 30°C (After MORITA, 1957a)

Substrate	Cells previously treated at 1 atm	Cells previously treated at 600 atm
	μmoles	μmoles
L-Alanine	0·29	0·35
L-Glutamic acid	0·48	0·69
L-Cysteine hydrochloride	0·42	0·18
Histidine	0·13	0·00

At temperatures above the optimum for bioluminescence, augmented pressure causes increased bioluminescence, while at temperatures below optimum luminescence, pressure decreases the amount of luminescence (BROWN and co-authors, 1942). These results were interpreted as follows: (i) at low temperatures the major effect of pressure is the limiting of the enzyme reaction, and (ii) at higher temperatures the amount of reversibly denatured enzyme becomes rapidly more important as a limiting factor in the over-all rate. In the latter situation, under pressure the equilibrium shifts in favour of the denatured state, thereby increasing the amount of active enzyme. The amount of bioluminescence, as influenced by pressure, depends also upon the bacterial species tested, the temperature, and the medium on which the organism was grown (JOHNSON and EYRING, 1948). Thermal inactivation of the luciferase in *Photobacterium phosphoreum* can be retarded at 34°C (optimum for activity is 21°C) at 330 atm (JOHNSON and EYRING, 1948).

BERGER (1958) studied the effect of pressure on phenylglycosidase, an enzyme produced when a strain of *Streptomyces griseus* is grown on mineral salts supplemented with chitin as carbon and energy source. The function of this enzyme in the decomposition of chitin has been discussed by BERGER and REYNOLDS (1958). Pressure applied to the enzyme in the presence of its substrate decreases the rate of reaction, as well as the rate of thermal inactivation of the enzyme. If substrate is absent, thermal inactivation of the enzyme is much more rapid. The same relation was found by HAIGHT and MORITA (1962); but if inorganic pyrophosphatase is pressurized, the cofactor, rather than the substrate, protects the enzyme from thermal inactivation (MORITA and MATHEMEIER, 1964).

To further emphasize the temperature–pressure relationship, HAIGHT and MORITA (1962) studied the aspartase system in *Escherichia coli*, employing cell-free extracts and washed wells. Cell-free extracts at 1 atm increase with temperature, but at 56°C a decrease occurs due to thermal inactivation of the enzyme (Fig. 8-21). However, when pressure is applied to the system, the reaction rate at 37° and 45°C decreases with increased pressure (Fig. 8-22); the reaction rate at 50°C increases with increased pressure. Although the enzyme is slightly inactivated at 1 atm and 56°C, when pressure is applied there is an increased enzymic activity

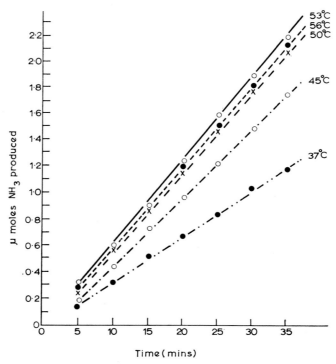

Fig. 8-21: *Escherichia coli*. Rate of aspartase activity as a function of temperature. The mixture consisted of equal volumes of the aspartase preparation ($11\cdot2$ μg N/ml) and L-aspartic acid (800 μmoles/ml in $0\cdot2$M phosphate buffer at pH $6\cdot9$) and was incubated at the various temperatures indicated. Ammonia was determined at 5-min intervals for each temperature. The values were corrected for residual ammonia in the enzyme and substrate preparation. (After HAIGHT and MORITA, 1962; modified.)

until a pressure of 600 atm is reached. Above 600 atm, the aspartase activity decreases at 56°C, and this is probably due to the fact that pressure becomes the more important factor in bringing about a retardation in the reaction rate. In order to test the protective effect of elevated pressure against damages due to heat, the enzyme system was subjected to various pressures and temperatures, and a portion of the mixture assayed after 35 mins (Fig. 8-23). It can be seen from the figure that $1\cdot2$ μmoles of ammonia are produced when the system is subjected to 1 atm and 37°C; $0\cdot8$ μmoles of ammonia are produced at 1000 atm and 37°C; $2\cdot0$ μmoles at 1 atm and 56°C; $2\cdot8$ μmoles at 700 atm and 56°C; 2.5 μmoles at 1000 atm and 56°C. Further incubation of the same mixture for 35 mins at 1 atm and 37°C results in the formation of $1\cdot2$ μmoles of ammonia in each case, with the exception of the reaction mixture which was held previously at 1 atm and 56°C. The mixture without substrate does not produce any ammonia when exposed to 1 atm and 56°C or 1000 atm and 56°C for 35 mins, and further incubation does not

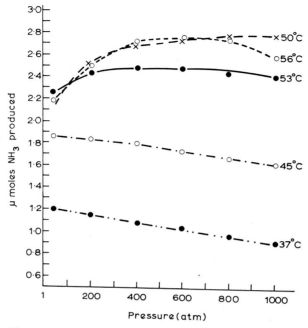

Fig. 8-22: *Escherichia coli.* Aspartase activity as a function of temperature and pressure. Reaction mixtures identical to those in Fig. 8-21. (After HAIGHT and MORITA, 1962; modified.)

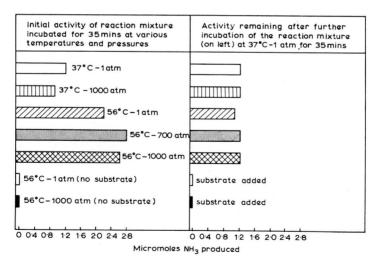

Fig. 8-23: *Escherichia coli.* Activity of aspartase remaining after treatment at various pressures and temperatures in the presence and absence of L-aspartic acid. Reaction mixtures identical to those in Fig. 82-1, except where no substrate was used initially. (After HAIGHT and MORITA, 1962; modified.)

produce any ammonia. Substrate was not limiting during the entire 70-min incubation period. The results indicate that elevated pressure decreases the aspartase activity at 37°C and that at 1 and 1000 atm and 37°C the enzyme is not denatured. Activity increases at 1 atm and 56°C, but at this temperature part of the enzyme becomes thermally denatured; this was obvious in cases in which the reaction mixture was further incubated for 35 mins. Although the enzyme denatures at 1 atm and 56°C, application of pressure does not allow it to undergo inactivation. As a result, at elevated pressures and 56°C, the enzyme remains in the undenatured state; the faster reaction rate can be attributed to the increased temperature or to the exposure of more reactive sites of the enzyme; without the added pressure the enzyme denatures.

Aspartase activity of washed cells also displays a temperature–pressure relationship (HAIGHT and MORITA, 1962). The 1 atm curve in Fig. 8-24 represents the activity as a function of temperature alone. An increase in temperature from 37°C to 45°C increases the rate of aspartase activity in whole cells, as one would expect according to van't Hoff's Law, at all pressures employed. When the temperature is raised from 45° to 50°C, thermal denaturation of the enzyme becomes noticeable at 1100, 200, 300, 400 and 500 atm. At 50°C, activity of the aspartase increases at 600, 700, 800, 900 and 1000 atm. At 53°C, there appears to be a crossover of all curves and, as a result, the 56°C curve indicates the highest amount of aspartase activity at 1000 atm and the smallest amount of activity at 1 atm.

Utilizing a cell-free extract of *Bacillus stearothermophilus* NCA 2184, MORITA and HAIGHT (1962) demonstrated that malic dehydrogenase is completely inactivated

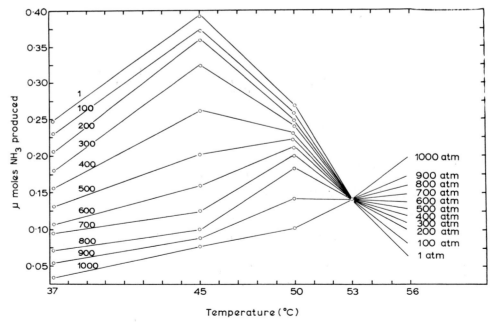

Fig. 8-24: Effect of pressure and temperature on the deamination of L-aspartic acid by washed cells of *Escherichia coli*. Values corrected for endogenous ammonia production. (After HAIGHT and MORITA, 1962; modified.)

at 1 atm and 78°C. Application of pressures from 1 to 700 atm did not produce any evidence of protection of the malic dehydrogenase at 101°C. However, when the pressure was elevated to 900 atm and more (1700 atm was highest employed), pressure protected the enzyme system from the destructive effects of heat so that malic dehydrogenase activity occurred (Fig. 8-25). Although reaction rate is much slower than at 1 atm and 56°C, there is a definite increase of activity with time at 1300 atm and 101°C (Fig. 8-26).

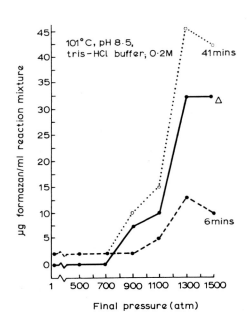

Fig. 8-25: *Bacillus stearothermophilus.* Activity of malic dehydrogenase at 101°C under various pressures. Δ: difference between 6-min and 41-min incubation times. Values are corrected for controls. (After Morita and Haight, 1962; modified.)

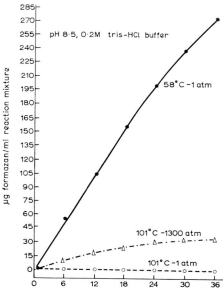

Fig. 8-26: *Bacillus stearothermophilus.* Rate of malic dehydrogenase activity. The curve obtained at 101°C and 1300 atm is corrected for the 6-min period required for initial pressure to reach final pressure. Values are corrected for controls. (After Morita and Haight, 1962; modified.)

Employing the same methods with inorganic pyrophosphatase, Morita and Mathemeier (1964) were able to demonstrate enzyme activity at 1700 atm and 105°C (Fig. 8-27). Morita and Mathemeier interpreted the data in terms of molecular volume changes: a rise in temperature produces a molecular volume increase and this increase is countered by the application of pressure. However, it should not be axiomatically assumed that pressure and temperature affect the structure of molecules in the same way.

A co-ordinated effort to study the effect of pressure on the various enzymic pathways in bacteria is still lacking. Most studies are concerned with dehydrogenases (especially of the Tricarboxylic Acid Cycle). Succinic, formic and malic dehydrogenases in cells of *Escherichia coli* were studied by Morita (1957b). The

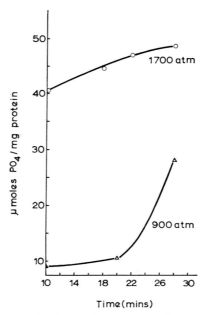

Fig. 8-27: *Bacillus stearothermo-philus*. Rate of pyrophosphatase activity at 105°C and two pressures. The reaction mixture contained in 1·8 ml medium: 200 μmoles of tris-HCl (pH 7·7) at 60°C, 0·26 mg of enzyme, 7·2 μmoles of $Na_4P_2O_7$, and 9·0 μmoles of CoCl: Values corrected for controls. (After MORITA and MATHEMEIER, 1964; modified.)

three dehydrogenases, measured during the period of compression, reveal that increased pressure decreases enzyme reaction rates, as can be seen in Table 8-14 and Figs 8-28 and 8-29. When the slopes of the reaction rates at 1, 200, 600 and 1000 atm on the oxidation of the substrates are compared, it is noted that the pressure applied does not affect each enzyme reaction in the same way (Table 8-15). The endogenous metabolism, as evidenced by methylene blue reduction, increases regularly with increased pressure. The reason(s) for the high endogenous metabolism under pressure is (are) not known.

HILL (1962) demonstrated that mould *Allomyces macrogynus* from shallow fresh water is rendered non-viable when incubated at 600 or more atm at 27°C. In an attempt to explain why death occurred when the mould was pressurized, HILL and MORITA (1964) tested the enzymatic activity of the mitochondria. Isolated mitochondria were used in order to rule out cell permeability factors. The activity of the dehydrogenases (succinic, alpha-ketoglutarate, oxalosuccinic and isocitric) decreases with increased pressure. The rate of the enzyme activity was measured during the period of compression by use of an optical pressure cylinder. At 500

Table 8-14

Micrograms of methylene blue reduced in the presence of malate by resting cells of *Escherichia coli* (0·5 ml, Klett turbidimetric reading of 250 employing a red filter) at various pressures (After MORITA 1957b; modified)

Reaction time (mins)	1 atm Control 26·7°C	1 atm Added malate 26·8°C	200 atm Control 27·5°C	200 atm Added malate 27·5°C	600 atm Control 27·5°C	600 atm Added malate 27·2°C	1000 atm Control 27·2°C	1000 atm Added malate 27·6°C
0	0·0	0·0	0·0	0·0	0·0	0·0	0·0	0·0
3	0·4	2·7	0·3	2·5	0·2	1·5	1·5	2·2
6	0·7	8·7	0·9	6·7	0·4	6·3	3·1	4·2
9	1·8	12·6	2·5	12·0	1·2	12·1	3·8	5·6
12	1·9	16·0	2·7	16·5	2·2	15·8	5·6	7·5
15	2·2	20·5	2·9	20·1	5·0	20·5	6·7	8·6

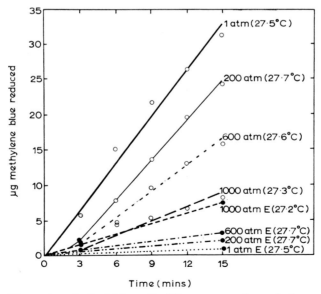

Fig. 8-28: *Escherichia coli.* Micrograms of methylene blue reduced in the presence of succinate by resting cells (0·5 ml, Klett turbidimetric reading of 250 employing red filter) at various pressures. (After MORITA, 1957b; modified.)

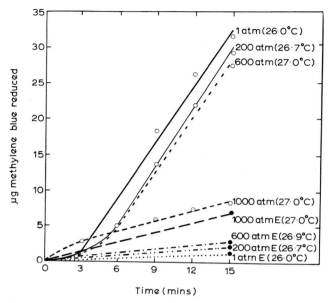

Fig. 8-29: *Escherichia coli*. Micrograms of methylene blue reduced in the presence of formate by resting cells (0·5 ml, Klett turbidimetric reading of 250 employing a red filter) at various pressures. (After MORITA, 1957b; modified.)

Table 8-15

Slope of the line of methylene blue reduction in Figs 8-28 and 8-29 and Table 8-14. Corrected for endogenous methylene blue reduction (Based on data of MORITA, 1957b)

Pressure (atm)	Malate*	Succinate**	Formate†
1	1·27	1·93	2·67
200	1·26	1·70	2·60
600	1·20	1·00	2·43
1000	0·17	0·04	0·01

* Slope calculated from data in Table 8-14 for the 3 to 15-min period.
** Slope calculated from data in Fig. 8-28 for the 3 to 15-min period.
† Slope calculated from data in Fig. 8-29 for the 6 to 15-min period.

atm, the rate of succinic dehydrogenase is very low, while no alpha-ketoglutarate and isocitric dehydrogenase activities can be demonstrated at 1000 atm. It was postulated that one of the reasons why death occurs in cells of *A. macrogynus* exposed to high pressure for a prolonged period of incubation, is a creation of a biochemical lesion—mainly the inoperativeness of the TCA cycle from which the main energy is derived in an aerobic organism.

An important microbial activity in oceans and coastal waters is the regeneration of primary nutrients. MORITA and HOWE (1957) demonstrated that the phosphatase activity of various marine forms varies with the organism in question. Generally, bacteria isolated from shallow waters show a decrease in phosphatase activity with increased pressure, although in a few forms the phosphatase activity remains rather unchanged. Thirty species of bacteria, tested by ZoBELL and BUDGE (1965), proved to be capable of reducing nitrate at 300 atm, but not as rapidly as at 1 atm. The enzyme nitrate reductase is inactivated more effectively at increased pressures combined with low temperatures, than with high temperatures such as 30°C. *Achromobacter stationis, Actinomyces halotrichis, Micrococcus euryhalis* and *Micrococcus sedimenteus* fail to grow at 600, 400, 200 and 1000 atm; the urease of the cells remains active at 1000 atm (ZoBELL ,1964).

The regeneration of ammonia from proteinaceous material is also a very important activity of marine bacteria. In the reviewer's laboratory, WEIMER (1967) studied proteolytic activity of the obligately psychrophilic marine *Vibrio* designated as MP-41 and employed gelatinase as proteolytic enzyme. The gelatinase activity of MP-41 cells decreases with increased pressure at 15°C. Although obtained from an obligately psychrophilic bacterium, the gelatinase was not abnormally thermolabile and had an optimum activity at 1 atm at approximately 43°C. When the activity of the partially purified gelatinase was tested at 40°C and 25°C under various pressures, it was found that the rate of its activity is not noticeably lowered between 200 and 600 atm; most of the activity occurs at pressures between 1 and 200 atm, but decrease with pressure is not great. At 25°C, enzyme rate reveals maximum reduction between 1 and 100 atm, while above 100 atm (to 600 atm) it does not change significantly. It appears that pressure does not hinder the formation of the enzyme-substrate complex or the number of active sites functioning; therefore, the internal conformational changes due to the application of pressure are such that the active sites remain unchanged.

In the sea, the area below the thermocline is fairly constant as far as temperature is concerned. The main variable is the pressure. HAIGHT and MORITA (1966) demonstrated that cells of *Vibrio marinus* MP-1 grown at 15°C (its optimum temperature for growth) and at 4°C (organisms isolated from a culture kept at 3·24°C) are physiologically different. Cells grown at 15°C are more heat stable and utilize glucose at a faster rate than those grown at 4°C. This difference was further analyzed by ALBRIGHT and MORITA (1965) and ALBRIGHT (1968) with pressure as additional parameter. All amino acids tested were deaminated. Deamination of serine by washed cells of *V. marinus* MP-1 grown at 15°C decreased with increased pressure at 4°C, but when tested at 15°C the optimum pressure for deamination was approximately 300 atm. In cells grown at 4°C, maximum serine deamination occurred at approximately 300 atm at 4°C and decreased nearly linearly with increased pressure. Cells grown at 4°C and tested at 15°C displayed an optimum serine deamination peak at slightly less than 200 atm. These results indicate that more meaningful laboratory data can be obtained from cells grown at low temperatures which parallel those found in the natural habitat. Only in these cases does it seem permissible to extrapolate directly from laboratory data to the situation met in the marine environment.

Synthesis of macromolecules

The growth of any cell depends on synthesis of protein, RNA and DNA. We must investigate the effect of pressure on the coding mechanism of macromolecular synthesis if we are to understand some of the basic phenomena taking place in cells exposed to different pressures. If pressure does not permit the synthesis of DNA, then growth and multiplication of the cell will cease. In other words, the fidelity of information transfer during gene function is an important aspect that must be investigated further. The terms 'fidelity', 'miscopy' and 'miscoding' were designated by WEINSTEIN and co-authors (1966) to express the degree of precision of the translation process and the finite mistakes in gene transcription. Any change in the conformation of DNA may change the transcription from DNA; hence 'miscopy' will occur in the RNA, thereby affecting translation so that a 'miscode' occurs, resulting in the formation of protein which may not be compatable with the cells. Likewise, change in the conformation of DNA may result in 'misreplication' (or mutation) in the new DNA. Pressure may also affect the already formed RNA in the cell, so that the translation process is a 'miscode'. Because of this situation, the effect of pressure on the synthesis of macromolecules such as protein and nucleic acid has received some attention lately. For instance, GILL and GLOGOVSKY (1965) demonstrated that pressure enhances the unfolding of ribonuclease.

Some evidence concerning the ability to synthesize protein (coat of phage) and DNA has been presented by HEDÉN (1964). A pressure of 740 atm employed for 2 mins on *Escherichia coli* infected with T_2 phage, reduces the burst size; the latency period is rather constant, regardless of whether or not the pressure was applied 7 or 12 mins after the cells were infected. It appears that pressure stops phage multiplication, which is resumed after the pressure is released; whether pressure affects primarily the synthesis of protein or of DNA is not known. HEDÉN further postulates that pressure may act on the initiator. *E. coli* cells, previously pressurized, have the capacity to support T_2 phage multiplication, indicating that pressure affects the intracellular material, eliminating the capacity for phage multiplication. HEDÉN also speculates on some other possible targets within cells affected by pressure.

The incorporation of ^{14}C-glycine into the protein of cells of *Escherichia coli* decreases with increasing hydrostatic pressures at 37°C (LANDAU 1966; Fig. 8-30). If the pressure is released, the inhibiting incorporation of ^{14}C-glycine is resumed at approximately the same rate as in the 1-atm control (Fig. 8-31). ^{14}C-leucine is incorporated into protein at a faster rate in a system incubated at 4000 psi at temperatures from 27° to 37°C than in the atmospheric control; but at temperatures below 27°C, rate of incorporation is reduced. At 6000 psi, the rate of incorporation is less than in the atmospheric control, except at 37°C where the rates are approximately the same. Nucleic acid synthesis, as evidenced by ^{14}C-adenine, at 37°C

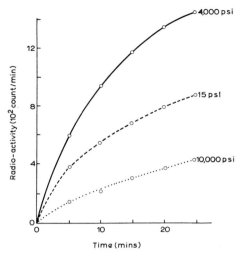

Fig. 8-30: *Escherichia coli*. Pressure effects on ^{14}C-glycine incorporation into protein at 37°C. (After LANDAU, 1966; modified.)

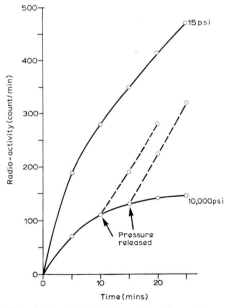

Fig. 8-31: *Escherichia coli*. The effects of release of inhibiting pressure (10,000 psi, 37°C) on ^{14}C-glycine incorporation. Rate of incorporation into protein resumes rapidly. (After LANDAU, 1966; modified.)

decreases as the pressure is elevated (Fig. 8-32). LANDAU (1966) does not postulate on the primary effect of pressure in the synthesis of nucleic acids and proteins.

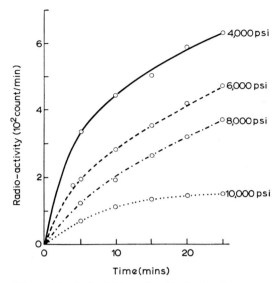

Fig. 8-32: *Escherichia coli.* ^{14}C-adenine incorporation into nucleic acids at various pressures
(37°C). (After LANDAU, 1966; modified.)

POLLARD and WELLER (1966) studied the incorporation of ^{14}C labelled valine, proline, uracil and thymine into protein and nucleic acids of *Escherichia coli.* No labelled valine or proline was found in the protein when *E. coli* cells were subjected to 1320 atm; incorporation is the same at 450 and 900 atm. Incorporation of labelled thymine into new DNA at 450 atm occurs at the same level as in the 1-atm control, but at 550 atm incorporation is only half that of the control value. At 900 atm, no labelled thymine is incorporated. On the other hand, incorporation of labelled uracil into RNA at 450 and 625 atm is equal to that in the control, and at 900 atm the amount of uracil incorporated is nine-tenths of the control value. From these data, it appears that protein synthesis is the process most sensitive to pressure, followed by DNA synthesis. RNA synthesis does not appear to be affected by pressures. It should be pointed out, however, that pressure studies were not of sufficient duration to offset the colony-producing ability of the cells.

The ability of a substrate to induce the formation of an enzyme was studied by POLLARD and WELLER (1966). When cells of *Escherichia coli* were placed in the presence of isopropylthiogalactophyranoside to induce β-galactosidase formation, application of 450 and 1320 atm suppressed formation, whereas controls showed normal amounts of β-galactosidase formation, as evidenced by its activity. If cells previously induced to form β-galactosidase are exposed to a pressure of 450 atm, activity of β-galactosidase ceases, but is sustained in the controls. The authors attribute this action to the cessation of the transcription process needed to form the enzyme. However, the data presented are not definitive enough to draw such a conclusion. In addition, the action of pressure on RNA and DNA in the cell is not known. If conformation of DNA is affected by pressure, its function as a template may be damaged completely or altered to such a degree that the messenger RNA is not correct. This latter situation could result in defective synthesis of proteins and could be one of the many reasons why certain cells expire under increased pressure.

In the authors's laboratory, ALBRIGHT (1968), working with *Vibrio marinus* isolated from a marine environment at 120 atm, found that syntheses of protein, RNA and DNA are completely inhibited at 1000 atm (see also Fig. 8-7). At 200 atm, the rate of protein synthesis decreases initially, but quickly resumes the 1-atm rate when the pressure is released, while the rate of DNA synthesis remains unaffected. At 500 atm, the rate of protein synthesis immediately drops to one-third of the 1-atm rate; RNA synthesis continues at the 1-atm rate for 30 mins, DNA synthesis for 60 mins, before both begin to diminish; after 150 mins, DNA synthesis stops completely. Within the time limits of the experiments, pressures of 500 to 600 atm on protein, RNA and DNA syntheses were reversible and resumed to 1-atm rate upon pressure release. Other data obtained suggest that pressures from 400 to 600 atm slow or stop RNA translation, which in turn may slow or stop DNA transcription or replication.

LANDAU (personal communication) studied induction, transcription and translation and found them to be inhibited by the application of increased pressures. In *Escherichia coli* ML-3, transcription seems the process least affected and still functions at 670 atm. Translation is totally inhibited at 670 atm, but appears unaffected at 265 atm; induction is inhibited above 275 atm. A larger volume change was calculated for translation than induction ($\Delta V = 100$ cm^3/mole and 55 cm^3/mole, respectively). One wonders just what makes the various organisms found in the hadal zones of the oceans different from their counterparts living near the surface. Two aspects that come immediately to mind are (i) functional differences due to long-term adjustments, and (ii) structural differences in proteins or in secondary and tertiary structures, due to evolutionary processes.

Again, we are in an unfortunate situation in that all the laboratory experiments are done with cells that grow best at 1 atm, instead of true barophilic or barotolerant forms.

(c) Reproduction

In microbiology (bacteriology) an increase in the number of cells (reproduction) is considered (population) growth and, as a result, bacteriologists are generally working with clones. No data exist in the literature concerning the effects of pressure on sexual reproduction in bacteria. Our present knowledge on pressure effects on sexual reproduction in fungi and blue-green algae is insufficient for critical assessment.

(d) Distribution

There exist numerous studies on the distribution of bacteria, fungi and blue-green algae in nearshore waters. They have been summarized by ZoBELL (1946), WOOD (1965, 1967), and others. The papers at hand indicate that pressure scarcely seems to play an important role.

In certain bacteria isolated from the sea, salinity variations may influence the temperature allowing maximum growth (STANLEY and MORITA, 1968; Chapter 4). In *Vibrio marinus* MP-1, an obligately psychrophilic marine bacteria, maximum growth temperatures change from 21°C to 10·5°C in salinities of 35‰ and 5‰, respectively. Recognizing this fact, ALBRIGHT and PALMER (personal communication) subjected cells of *V. marinus* MP-1 to various salinities under a range of pressures. They found that higher salinities support growth at higher pressures, whereas lower salinities do not support growth at high but only at very low pressures.

The horizontal and vertical distributions of bacteria in the open oceans have been investigated by KRISS (1963). The occurrence of bacteria in the hadal zones of the oceans has also been elucidated by ZoBELL and MORITA (1957, 1959). The biomass studies by KRISS (1963) only indicate the presence of certain micro-organisms but do not relate their presence to their activities at the geographical location and water depth from which they were isolated. In the deep sea, both low temperatures and high pressures act as additive in bringing about a decrease in molecular volume change and, therefore, the incubation of micro-organisms at room temperatures (or even 5°C) and at 1 atm does not tell us what their activities are under *in situ* conditions. It only suggests the potential of microbial activities provided the environmental conditions offered—such as temperature, pressure, pH, presence of organic matter, etc.—are sufficient for the micro-organism tested to metabolize and multiply. Marine microbiologists often recognize the dearth of bacteria in the open sea, but this situation only reflects the conditions from which they have been isolated (generally waters of low organic matter content). In the open oceans, bacteria regulate the amount of organic matter present to a great degree; if a sufficient amount develops (i.e., decomposition of dead organisms) they multiply and mineralize the organic matter until the micro-organisms, in turn, become a source of energy for other organisms (Chapter 10.1).

The vertical distribution of fungi in the ocean has not been investigated intensely. However, fungi have been demonstrated quite frequently in nearshore waters. SPARROW (1937) demonstrated the existence of saprophytic fungi at depths from 19 to 1127m, while HÖHNK (1956) has recovered fungi from sediments as far down as 4610 m. HÖHNK (personal communication), GAERTNER (1967, 1969, 1970) and MEYERS and co-authors (1967) have been able to isolate quite a number of various fungi from the North Sea. Yeasts have been demonstrated from water taken as deep as 4000 m (KRISS, 1963) and found to be numerous in the Indian Ocean by FELL (1967). Information on the horizontal distribution of marine fungi has been compiled by JOHNSON and SPARROW (1961).

The vertical distribution of blue-green algae has not been investigated to any great extent. The reason for the lack of data on pressure effects on their distribution is probably due to the fact that blue-green algae are ecologically important only in the photic zone where pressure plays a minor role as a factor controlling distributions.

(3) Structural Responses

The size and shape of the bacterium *Serratia marinorubra* was examined by phase microscopy after being subjected to elevated pressures (ZoBELL and OPPEN-

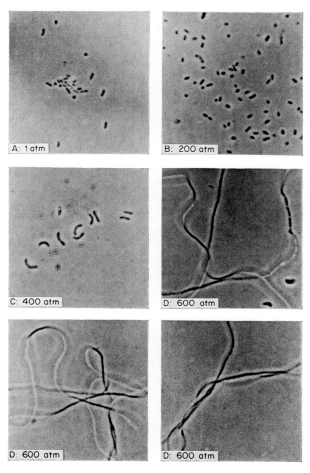

Fig. 8-33: *Serratia marinorubra.* Photomicrographs (wet preparation, Spencer phase contrast microscope with 97 × dark M objective) shortly after removal from pressure vessels in which cells had been incubated 4 days in nutrient broth at 23°C and the pressures indicated (× 1000). (After ZoBELL and OPPENHEIMER, 1950; modified.)

HEIMER, 1950). The cells appeared normal in size as well as in motility when subjected to 400 atm in nutrient broth after 6 to 8 hrs, but after 16 hrs they had lost their motility and long filaments began to appear (Fig. 8-33). Exposure to 400 and 600 atm caused the individual cells to grow into long filaments. After decompression, the long filaments fragmented into cells of normal size; at 25°C, the process of fragmentation was completed within 1 or 2 hrs. Not all bacteria tested respond in the same manner as *Serratia marinorubra*; formation of long filaments or increase in cell size were observed in 5 marine bacteria out of 63 (OPPENHEIMER and ZoBELL, 1952). Filament formation was also noted in *Escherichia coli* by ZoBELL and COBET (1962, 1964). In *E. coli*, increased pressures cause the cells to grow, without cell division, into long filaments. About 50 to 75% of the cells grown at 400 atm and 30°C were longer than 10 μ. In marine forms exposed to increased pressure, loss of motility and flagellation, formation of greater spores, granulation in cells and pleomorphism have been observed (ZoBELL and OPPENHEIMER, 1950; OPPENHEIMER and ZoBELL, 1952). The reasons why certain bacteria cells increase in size, or form filaments, under increased pressure are not known.

Electron micrographs of ultra-thin sections of *Allomyces macrogynus* kept under elevated pressures did not disclose any noticeable change in the internal structure of the cells (HILL, 1962).

Subjecting two *Bacillus* ssp. of marine origin to various pressures, BOATMAN (1967) found that culture No. 871 showed few detectable mesosomes but its striated cell wall was emphasized by pressure. At 6000 psi, a thin, single layer was present between cell wall and cytoplasmic membrane. At 1 atm, culture No. 874 revealed mesosomes; but when this culture was subjected to 4000 or 6000 psi for 3 days at 21°C, the mesosomes appeared at the periphery of the cells between the cell wall and the cytoplasmic membrane. According to BOATMAN (1967), the application of pressure may not only reveal additional structures (not seen at 1 atm), but may indicate the behaviour of certain bacterial organelles under pressure stress.

8. PRESSURE

8.2 PLANTS

W. Vidaver

(1) Introduction

Hydrostatic pressure is a significant factor in the ecology of plants. In order not to be exposed to any pressure, a plant would have to exist in a complete vacuum. In a strict sense, the term 'hydrostatic pressure' applies only to fluid systems (Chapter 8.0), but all metabolic processes take place in liquid phase. Consequently, it is convenient to consider pressure as an environmental factor for plants which possess liquid, solid, or gas phase components, or any combination of the three.

Pressure determines, in part, the environment of all plants. The biota of the upper atmosphere is no more independent of pressure than that of the extreme depths of the sea. This chapter, however, deals only with conditions where pressure is ordinarily in excess of 1 atm.

Historically, most biological observations have been made at a pressure of 1 atm: variations in pressure have been largely ignored as ecological parameters. Man, whose survival can be sustained only over a range of from about 1/2 to perhaps 2 atm, unless the earth's surface environment is at least partially recreated (Folk, 1966), has tended to be most interested in the organisms with which he shares this apparent limitation.

Having respiratory systems involving a gas phase, plants—like other organisms, including man—are limited to that part of the ecosphere where gases at the requisite concentrations for the maintenance of metabolic processes must pass in and out of aqueous solution (Renner, 1942). Lungs, spongy leaf mesophyll, and other organs specialized for the exchange of gases from liquid solutions to a gaseous atmosphere do not normally function in the exchange of gas molecules between internal and external liquid solutions.

In nature, pressures in excess of only a fraction over 1 atm are encountered in water bodies and in the depths of the earth. As the gas concentration can increase more or less linearly with pressure, organisms living under tens or even hundreds of atmospheres have no special problem in the exchange of dissolved gases, other than the transport of the appropriate molecules either into or out of their cells or tissues. Normally, gas exchange through cell membranes is accomplished easily by organisms at pressures of many atmospheres (MacDonald, 1965; Vidaver, 1969).

Photosynthetic plants, in common with other organisms not obligate anaerobes, are dependent on the transport of O_2 and CO_2 through their cell membranes. Transport of these gases is carried out by the barophilic organisms of the deepest ocean trenches (Oppenheimer and ZoBell, 1952). There is little reason to presume that such gas exchanges pose a particular problem for the deep-sea forms. Gas exchange through biological membranes is a physical process occurring in the direction of a concentration gradient. Plant cell membranes, therefore, should, like the

F

membranes of other cells, permit the passage of gases in the direction of the gradient, at least up to the limit imposed by disruptive pressure effects on the molecular structure of the membrane (KROGH, 1959). Some algae can carry out gas exchanges at pressures exceeding those of extreme marine depths (VIDAVER, 1969).

Pressure up to the limits occurring in the sea (1 to about 1100 atm) restricts the tolerance range of plants utilizing transitions between dissolved and molecular gas to a few atmospheres, but does not restrict those relying on the diffusion of dissolved gases. For plants with no molecular gas phase, pressure limitation presumably results from the effects of pressure on molecular structures and biochemical functions (JOHNSON and CAMPBELL, 1946; JOHNSON and co-authors, 1954; POLLARD and WELLER, 1966; ZIMMERMAN, 1970a).

There are innumerable examples of organisms which regularly sustain pressures of hundreds of atmospheres. In some cases, the exposure to pressure may be intermittent (diving animals, plankton and larger animals undertaking extended vertical migrations) while many kinds of micro-organisms and invertebrates spend their entire lives exposed to pressures which may exceed 1000 atm. Viable algae have been collected from depths of several thousand metres (BERNARD, 1963).

Pressure does not preclude the occurrence of algal species within the range of pressures encountered in the biosphere. Photo-autotrophic metabolism by plants is limited to the upper 200 m of the hydrosphere because insufficient light to support photosynthesis is transmitted beyond this depth. Plants surviving at greater depths must rely presumably on either organic or inorganic chemical energy sources. The 20-atm pressure range of the euphotic zone is probably a factor in determining the vertical distribution of marine plants. These rather low pressures would, however, appear to be of the same order of ecological significance as, for example, light quality, nutrient concentration and salinity. The vertical ranges of marine plants are controlled by environmental factors of which pressure, while significant, is not of paramount importance.

(2) Functional Responses

(a) Tolerance

Airborne plants may be confronted with atmospheric pressures of under 0·1 atm, but specific responses to such reduced pressures are not well defined. The effects of reduced pressures may be mostly on gas and water vapour exchanges.

There is a fairly extensive literature on tolerances of plants to elevated pressures. Tolerance to augmented pressures appears to be related to various parameters. Among these are the magnitude and duration of pressure exposure, presence or absence of light (GROSS, 1965), temperature (VIDAVER, 1969), and, possibly, the physiological state of the plant (GESSNER, 1955).

Different cellular processes appear to respond differently to elevated pressures. Some marine algae subjected to 1400 atm are unable to recover the ability to photosynthesize and die rather soon after exposure (VIDAVER, 1969). In *Elodea canadensis*, cytoplasmic streaming ceases at pressures between 400 and 500 atm, and resumes within a few minutes after pressure release if the exposure has not exceeded one-half hour (BYRNE and MARSLAND, 1965). GUYOT (1960) has shown that cytokinesis is generally abnormal in *Triticum vulgaris* seedling radicle meristems at pressures of 500 atm or above at ordinary temperatures. It seems unlikely that affected daughter cells are viable. GROSS (1965) found that pressures between 500 and 1000 atm induce pigment mutants in *Euglena viridis* which are heterotrophically viable but cannot photosynthesize. He also reported that 2 hrs at 1000 atm was lethal for 100% of the population of *E. viridis* tested. *Ulva lobata* exposed to 800 atm for 1 hr died within a few days (FONTAINE, 1929c). In another study, *U. lobata* survived a 5-min exposure to 1300 atm at 30°C, while the same pressure at 15°C was lethal (VIDAVER, 1969). *Porphyra perforata* withstood exposures to 730 atm for about 10 mins with no loss in growth potential; exposure to 1000 atm for a similar period reduced subsequent growth 50%, and death followed a similar exposure to 1300 atm (VIDAVER, 1969). GESSNER (1955) reports that 312 hrs of exposure to 400 atm inhibited subsequent growth rates of *Chlorella pyrenoidosa* populations by 50%.

For the small number of species that have been studied, it appears that the upper limit of tolerance to pressure lies around 1000 atm; this limit may be extended slightly by increased temperatures; it coincides closely with the pressure (about 1100 atm) occurring in the deepest marine trenches. The total range of tolerances of photosynthetic plants seems, then, to approximate the range of pressures existing in the biosphere.

The precise nature of the physiological effects of elevated pressures is far from being completely understood. More comprehensive studies of the relations between pressure, temperature and duration of exposure are needed for a better understanding of the limiting effects of pressure on plants.

(b) *Metabolism and Activity*

Photosynthesis

Like other metabolic systems, photosynthesis in plants is affected by variations in pressure. Under natural conditions, light rather than pressure limits the depths at which photosynthesis may occur, and there is no necessity for pressure resistant photosynthetic systems (other than to a few tens of atm) to have evolved. Nevertheless, laboratory observations indicate that photosynthetic systems of some algae, and possibly higher plants also, are surprisingly resistant to both reversible and irreversible inhibition by pressure.

VIDAVER (1963, 1969) studied the effects of pressure on photosynthesis and observed three general types of responses. The first response type, which appears around 100 atm and extends to about 600 or 700 atm, is immediately (within seconds or less) reversible upon the release of pressure. The second type of response begins at about 600 atm, persists to about twice this pressure, and becomes slowly reversible with time (minutes or more). The third response, which appears to mark the upper limits of tolerance, sometimes occurs at pressures as low as 1000 atm and is irreversible. The pressure ranges of these responses are not narrowly delineated; there is considerable overlap, and the limits may be shifted by concomitant changes in temperature. Duration of exposure also influences the responses; longer exposures, especially at higher pressures, usually result in slower recovery or in irreversible effects at less high pressures. Even when photosynthetic activity is recovered, in some cases there may be other, unobserved damage.

In our laboratory, relative rates of photosynthetic O_2 exchange were determined polarographically with the platinum electrode (BLINKS, 1950). The Pt electrode, with an AgCl reference electrode, was enclosed in a stainless steel pressure vessel. The vessel was fitted with a lucite window to permit the illumination of the photosynthetic sample held in contact with the Pt electrode. Thin thalloid marine algae and unicellular algae were used as experimental material. O_2 exchange rates were determined in light at various pressures and temperatures. Steady rates in continuous light, induction transients in continuous and intermittent light, and other phenomena such as enhancement (EMERSON and co-authors, 1957) were observed.

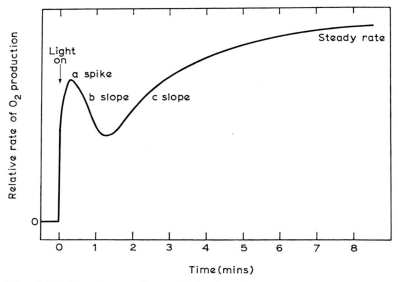

Fig. 8-34: *Porphyra perforata*. Induction time course of oxygen production. 1000 psi, 15°C. (After VIDAVER, 1969; modified.)

Fig. 8-34 represents the time course for O_2 production in white light of about one-half saturation intensity for the marine red alga *Porphyra perforata*. The response at a pressure of 1000 psi does not differ significantly from that at normal atmospheric pressure, but the increased pressure keeps the gases in the vessel in solution. The shape of the induction curve is influenced by light intensity and wavelength, duration of preceding light and dark intervals, temperature and pressure. There is no simple response to changes in these parameters, which suggests that the transients result from rate limitations by a series of overlapping reactions, shifting from one to another as substrate pools in the chloroplasts are diminished or incremented in the light. Variations in pressure alone drastically affect the shape of the response curve in different species, indication that some of these reactions are more pressure labile than others (Fig. 8-35).

The effect of pressure increase on steady O_2 production rates was different in the red alga *Porphyra perforata* and the green alga *Ulva lobata*. In both species, pressures of more than a few thousand psi were inhibitory, but the effect was greater in *U. lobata*.

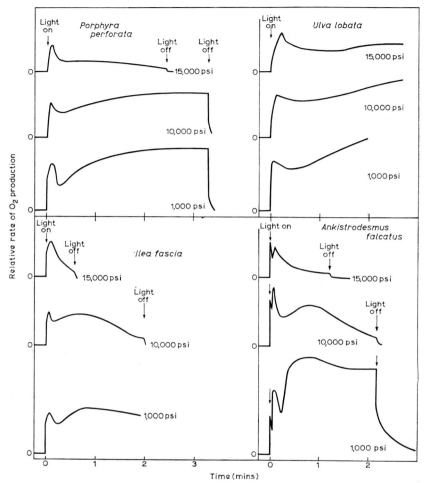

Fig. 8-35: Induction time courses of oxygen production in algae at three
different pressures (15°C). Intensity of white light was approximately
one-half that required for steady rate saturation. No pre a-spike transient
(see Fig. 8-34) is visible in the response of the marine algae *Porphyra
perforata*, *Ulva lobata* and *Ilea fascia*. A distinct transient appears on the
rise of the a-spike in *Ankistrodesmus falcatus*. While steady rates were
repressed more by pressure than maximum a-spike rates, the pre a-spike
transient was inhibited least. (After VIDAVER, 1969; modified.)

In *Ulva lobata* (Fig. 8-36), increased temperature tends to oppose the pressure-
induced suppression of O_2 production. The responses of *U. lobata* to variations in
pressure and temperature appear to be similar to the pressure–temperature relations of many enzymatic reactions observed *in vivo* (JOHNSON and co-authors,
1954) and to the responses of other biological systems (MURAKAMI, 1963; BYRNE
and MARSLAND, 1965; see also Chapters 8.0 and 8.1).

The responses of *Porphyra perforata* are more complex (Fig. 8-37). While increased pressure suppresses O_2 production, the tendency for increased tempera-

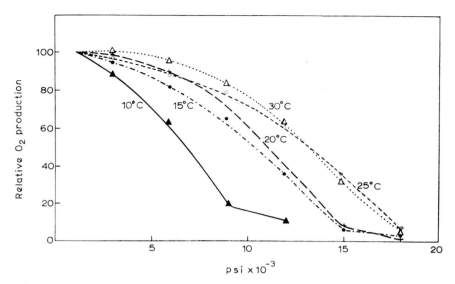

Fig. 8-36: *Ulva lobata*. Steady rate oxygen production under pressure at five different temperatures. (After VIDAVER, 1969; modified.)

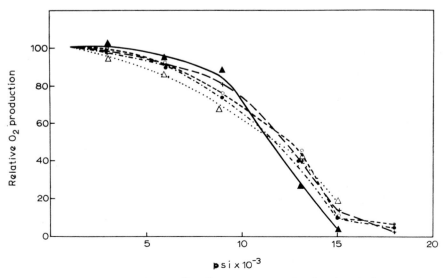

Fig. 8-37: *Porphyra perforata*. Steady rate oxygen production under pressure at five different temperatures. Symbols indicate the same temperatures as in Fig. 8-36. (After VIDAVER, 1969; modified.)

ture to oppose pressure inhibition is slight. This apparent insensitivity to temperature variations may be the result of individual differences in the samples. The alga was collected in the intertidal zone. No two thalli are likely to have the same genetic or environmental histories, hence, enzymatic differences in the samples can be expected. On the other hand, relative insensitivity of *P. perforata* to temperature changes may result from adaptive mechanisms. This species grows in the upper intertidal regions, where it is regularly exposed to extreme temperature fluctuations, as well as to direct sunlight and desiccation during the long intervals between high tides. The alga may well have evolved mechanisms which protect its metabolic enzymes from the extreme environmental changes to which it is subject.

We also compared the effects of increased pressure on the first O_2 transient (a-spike of Fig. 8-34) for several species of algae. Polarographic O_2-exchange determinations show a complex time course of O_2 production in light. Pressure increase or temperature change alters this induction time course during illumination. Time courses for the three marine algae *Porphyra perforata, Ulva lobata, Ilea fascia* (Fig. 8-35) show that both steady O_2 production and the a-spike are suppressed by increasing pressure, but steady rates decrease more than the a-spike. *Ankistrodesmus falcatus* is similarly affected by pressure (Fig. 8-35, lower right), but a small transient peak appears before the a-pike maximum. This transient remains almost unaffected by the highest pressures applied.

In view of this pre-a transient in the induction period of *Ankistrodesmus falcatus*, we looked for similar responses in other algae. High or low-temperature exposures do induce the appearance of the transient in the green alga *Ulva lobata* (Fig. 8-38) and in the brown alga *Ilea fascia* (Fig. 8-39); these transients are strikingly resistant to inhibition by increasing pressure. Investigations in our laboratory with higher electrode sensitivity and monochromatic light indicate that the transient commonly appears in the O_2 production of many algae at ordinary temperatures and atmospheric pressure (VIDAVER and FRENCH, 1965; VIDAVER, 1969). Yet the

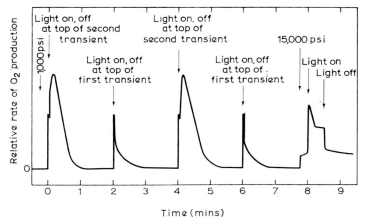

Fig. 8-38: *Ulva lobata.* Isolation of the pre a-spike transient by low temperature (4°C) and light flashes at two different pressures. Transient magnitude is almost undiminished by the highest pressure applied (15,000 psi). (After VIDAVER, 1969; modified.)

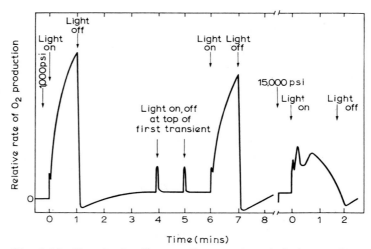

Fig. 8-39: *Ilea fascia*. Pre a-spike transient isolation at high temperature (30°C) and two different pressures. The transient persists at the higher pressure applied. (After VIDAVER, 1969; modified.)

transient has never been observed by us under any conditions in the responses of all the red algae examined.

We have interpreted these responses to pressure of algae showing the pre-a transient as indicating that O_2 must be produced by a process quite insensitive to pressures of at least 1000 atm. Kinetically, such a response differs from those usually associated with enzymatic reactions (JOHNSON and co-authors, 1954). This lack of pressure sensitivity may indicate the functioning of a non-enzymatic primary O_2 production mechanism. If the transient is evidence for the direct photo-chemical photolysis of water, then it may be assumed that, in plants having the transient, O_2 production results from photochemical reactions and does not require enzymatic activities. Assuming that a very early response to light is the production of O_2, and that O_2 release cannot continue unless the products formed during the reaction are somehow consumed (HILL reaction), it might be expected that only a pre-a transient would appear during illumination. Pressure may block the pathways by which the products are utilized. As a result, the transient persists under pressure, while steady rate O_2 production is inhibited.

The apparent lack of a pre-a transient, if real, in red algae suggests a process of O_2 production different from that in green or brown algae.

Within the range of pressures occurring in the biosphere, pressure effects on photosynthesis appear to manifest themselves primarily on the dark reactions of the process. Obligate photo-autotrophs would seem to be limited in their vertical ranges by the availability of light rather than by pressure effects on their metabolic processes.

Growth

In general, observations of the effects of pressure on plant growth have been made primarily on macro- or micro-algae exposed to relatively high pressures (hundreds of atmospheres). Higher green plants, including some species which are nominally aquatic, have most often been observed in relation to their responses to pressure increments ranging from less than 1 atm to seldom more than 10 atm.

GESSNER (1955, 1961) reported that pressures of more than 1·5 atm completely inhibit the growth of shoots and leaves of *Hippuris vulgaris*, while pressures below 1·5 atm are not inhibitory. *Cryptocoryne* is similarly retarded, but *Myriophyllum* is only slightly affected. GESSNER also observed low pressure stimulation of internode growth in submerged specimens of *Potamogeton densus*, but the effects differed according to the season and the developmental stage of the plants. Two atm was sufficient to block completely flower development in *P. densus*.

Pressure exposures inhibit root and hypocotyl elongation in *Helianthus* seedlings (GESSNER, 1961). Exposure to 100 atm for 5 mins causes about a 20% decrease in root length after 16 days. Inhibition of hypocotyls is more marked—about 40%—after 16 days, compared to the unpressurized controls.

Growth responses to small increments of pressure may be of adaptive value to aquatic plants. Assumptions that such growth responses of higher plants result directly from the effects of pressure on metabolism or are manifestations of endogenous control mechanisms appear problematical. Aquatic phanerogams obviously require such controls, but do small differences in pressure trigger the mechanisms? Some evidence points to the importance of other environmental factors—such as differences in light intensity, temperature and O_2 concentration—to growth and morphogenetic responses (BOSTRACK and MILLINGTON, 1962; BERG and co-authors, 1965). More research is needed in this area.

Sustained high pressure can affect the growth of algae. Responses to pressure may be observed either while the pressure is applied or following a previous exposure to elevated pressures. Most observations appear to be of the latter type and, unfortunately, have an *ex post facto* implication. GESSNER (1955) found that 308 hrs of exposure to 200 atm stimulates population growth in *Chlorella pyrenoidosa* over unpressurized controls, resulting in nearly three-fold increase in cell number after 15 days. In 30 days, the pressure-treated cultures still have 30% more cells. Exposure to 400 atm for 312 hrs, however, induces a decrease in the growth rate which results in 70% fewer cells after 15 days. Similar results were obtained by STURM (1941, *in*: GESSNER, 1961) using *C. pyrenoidosa* cultures maintained in organic media; STURM also found that there was usually a delayed action response to pressure exposure: exposures to 200 atm or more induce increases in cell numbers over the controls, but the increases do not begin for about 5 days following 10 to 13 days of exposure to pressure. STURM reported a small but significant increase of cell length in cells of *Hydrodictyon* over controls, induced by 5 hrs exposure to 100 atm at 20°C. Pressures of 200 to 400 atm for 5 hrs induce a somewhat more pronounced elongation, but 26 hrs at 40 atm cause a decrease in elongation of 25 to 35%. Lowering the temperature to 0°C causes a strong inhibition of elongation at 200 atm.

The post-treatment growth of the marine algae *Ulva lobata* and *Porphyra perforata* is affected by short exposures to pressure (VIDAVER, 1969). Growth in controls and pressure-treated discs of the algal thalli were compared 1 week after exposure to pressure. In *P. perforata*, 5 or 10-min exposures to pressures of 730 atm at 15°C or less do not inhibit growth. A similar exposure to 930 atm retards growth by

14%, and to 1100 atm by nearly 50%. At 30°C, pressure appears to be less in-hibitory, but more experimentation is necessary to determine the relationship between pressure and temperature. *Ulva lobata* subjected to 1350 atm at 30°C for 5 mins achieves almost 90% of the growth of the control after 1 week.

Mutagenic effects

Pressure variations have scarcely been examined as a mutagenic agent. In view of the rather well-known and relatively predictable effects of pressure on some biological macromolecules (MURAKAMI, 1961; WEIDA and GILL, 1966; Chapters 8.0, 8.1), pressure application may well serve as a more or less specific mutagen. Pressure-induced mutations then might—in contrast to random, radiation-induced mutations—be highly reproducible.

GUYOT (1960) appears to have observed a correlation between pressure treat-ment of root tip meristematic cells and the occurrence of certain cytokinetic aberrations (p. 1404).

Euglena viridis pigment mutants obtained by GROSS (1965) suggest an effect of pressure on chloroplast 'genes'. Cultures of *E. viridis* exposed to 1000 atm for 20 mins or 2 hrs produce about 20% colonies, which appear colourless on agar plates. Cultures exposed to 500 atm for $\frac{1}{2}$ hr and 2 hrs exhibit about 1% mutant colonies. These responses indicate that the phenomenon is pressure rather than time dependent. Mutants have been carried through as many as 14 serial subcultures in the light without resynthesis of chlorophyll. Two other chlorophyll-less mutants, differing from the first in carotenoid dry-weight ratios, were also obtained by pressure treatment.

BERG and co-authors (1965) have used O_2 under pressures up to 2000 psi ($\sim$135 atm) to induce mutations in barley seeds. A high frequency of root-tip cell chromo-somal aberrations is obtained by using very dry seeds (2·1 and 3·1% of water con-tent). In dry seeds, the occurrence of dicentric chromosomal bridges in root tip cells is about 1%. Seeds with an initial water content of 9·3% show a dicentric bridge frequency of about 0·1% under the same treatment. Dicentric bridge frequency in seeds exposed to 2000 psi of N_2 is approximately equal to that in controls (0·03%). Plants grown from the treated seeds reveal a variety of muta-tions. The most prevalent mutation which occurred in the 9·3% water-content seeds is albinism or failure to produce chlorophyll (90% of all mutations). In these experiments, the lack of a mutagenic effect of N_2 under pressure suggests that the role of pressure in the induction of mutations is an indirect one. As the quantity of the dissolved gas is proportional to pressure, O_2 concentration in the water in contact with the seeds would be very high (0·17 M compared to $1·25 \times 10^{-3}$M at atmospheric pressure). As suggested by the authors, the mutagenic agent is prob-ably O_2 itself at such high concentrations.

Membrane permeability

MURAKAMI (1963) observed pressure effects on plasmolysis and deplasmolysis in onion inner-epidermal cells using various concentrations of electrolytes and non-electrolytes in solution. Pressure chambers with transparent windows permitted observation of the cells while pressure was applied. Pressures up to 485 atm in-crease the plasmolysis time over that at atmospheric pressure in any test solution

Table 8-16

Pressure effects on membrane permeability. Permeability of onion inner-epidermal cells in various solutions is indicated by the differences in time (mins) required to bring about 50% plasmolysis under normal pressure and under 500 kg/cm²; 20°C (After MURAKAMI, 1963)

Solution	Concentration (M)	Difference (mins)
CaCl₂	0·2	7
Urea	0·5	8
Sucrose	0·5	10
NaCl	0·3	12
MgCl₂	0·2	15
KCl	0·3	17
Glycerin	0·5	25

(see Table 8-16). Increased plasmolysis time indicates greater membrane permeability. The degree of increased permeability varies according to the solution used. With 0·5M sucrose solution, permeability increase is maximal at 20°C. At 2°C, plasmolysis times at atmospheric pressure and 485 atm are nearly the same; raising the temperature to 30°C results in only slightly increased permeability under pressure. Deplasmolysis times in dilute solution following plasmolysis are increased by pressure, except when the tissue is placed in solutions of divalent cations (see Table 8-17). Differing from the plasmolytic response, deplasmolysis

Table 8-17

Pressure effects on permeability. Differences in time (mins) required to bring about 50% deplasmolysis in onion inner-epidermal cells at atmospheric and 500 kg/cm² pressure. In CaCl₂ and MgCl₂ solutions, deplasmolysis time is shorter by 15 and 3 mins under normal pressure (After MURAKAMI, 1963)

Solution	Concentration (M)	Difference (mins)
Urea*	0·3	0·5
NaCl*	0·1	1·0
KCl*	0·1	1·0
Glycerin	0·3	2·0
Sucrose	0·2	80·0
CaCl₂	0·05	− 15·0
MgCl₂	0·05	− 3·0

* membrane rupture

permeability is reported to be enhanced by higher temperatures and increased pressure. MURAKAMI interprets these observations as indicating that increased permeability under pressure affects the cells' sensitivity to the various ions and solute molecules.

Seed germination

Pressure is known to influence seed germination in many species of plants. DAVIES (1926, 1928a) reported enhancement of germination after pressure exposure, while in other works, DAVIES (1928b) and RIVERA and co-authors (1937) observed enhancement and repression depending on the plant species and the experimental conditions. VIDAVER and LUE-KIM (1967) studied the germination of lettuce seeds at different pressures and O_2 concentrations. Germination is stimulated by some combinations of pressure and O_2 and retarded by others. Germination occurs most rapidly under 2000 psi ($\sim$135 atm) and with an O_2 concentration of $1 \cdot 25 \times 10^{-3}$M (approximately equal to 1 atm of 100% O_2 in equilibrium with water at atmospheric pressure). Any further increase in pressure generally retards germination, but the effect is partially offset by added O_2. No germination occurs while pressure remains above 7000 psi at any O_2 concentration. Sustained pressures up to 15,000 psi retard germination after the release of pressure, but only when O_2 is present. Seeds to which pressure is applied in the absence of O_2 germinate as well after temporary pressure exposure as untreated seeds.

Life cycles

Little information is available about the effects of sustained pressure on life cycles of plants. The technical difficulties involved in the construction of pressure chambers in which environmental conditions are suitable for the completion of the whole life cycle are considerable, even in regard to unicellular algae. Not the least of the problems is the maintenance of relatively constant CO_2/O_2 ratios in a closed system under pressure, while respiration and photosynthesis are in process. This author's laboratory is presently engaged in an attempt to culture, synchronously, unicellular algae at pressures of up to 1000 atm. At the present time it is known that *Chlorella pyrenoidosa* will reproduce at sustained pressures of at least 5000 psi and that the daughter cells appear to be normally viable.

Inherent in experiments to determine the effects of small increments in atmospheric gas pressure is the problem of separating the effects of variations in pressure from changes in other factors such as water vapour concentrations, dissolved gas concentrations, internal plant water relations, transpiration rates and local temperature variations. Further experiments appear to be needed to ascertain the effects of small changes in atmospheric pressure on plants.

Pressure-induced (GROSS, 1965) or pressure–O_2 induced (BERG and co-authors, 1965) mutations, as discussed under *Mutagenic effects* (p. 1398), obviously affect the life cycle of the plant. Albinism (loss of chlorophyll) either in seed plants or algae requires a switch from photo-autotrophy to heterotrophy; consequently, the normally light-dependent phase of the life cycle in the mutants is replaced by a dark phase and organic energy requirements.

(c) Reproduction

The relationship between pressure and reproduction in plants is largely unknown. Experiments in the author's laboratory indicate that in a strain of *Chlorella pyrenoidosa*, in which the mother cells normally give rise to 4 daughter cells ($n = 4$), the n number of 4 has been maintained in cells growing under pressures up to 5000 psi.

In *Ulva lobata*, the release of gametes and zoospores from thalli has often been observed immediately upon decompression from pressures of up to 1000 atm (own, unpublished observations). However, in *U. lobata*, abrupt changes in illumination, temperature, or salinity, in some cases also induce such a response (PRINGSHEIM, 1951). Consequently, the role of pressure in gamete release may be no more than that of a metabolic shock which triggers the process.

It would be a relatively simple matter to examine the reproductive responses of unicellular algae, such as species of *Chlamydomonas* under pressure. Windowed pressure chambers and an inverted microscope are essentially all that would be required. It would be of considerable interest to know whether elevated pressures affect behavioural patterns of the various mating types (HOSHAW, 1965).

(d) Distribution

The role of pressure in governing the vertical distribution of plants remains poorly understood. Macro-algae appear to be limited, vertically, by the availability of light. The present author has made previously unreported observations on the red alga *Meripelta rotata* (HOLLENBERG and ABBOTT, 1966) collected from about 70 m in Monterey Bay (California, USA). Although normally exposed to several atm, this alga is less resistant to pressure-induced injury than any other species observed. Exposure to 600 atm for a few minutes is lethal; this contrasts markedly with the resistance of the intertidal red alga *Porphyra perforata*, which will survive exposures to 1000 atm or more. Perhaps greater pressure tolerance is a concomitant of adaptation to the rigorous intertidal environment.

BERNARD (1963) has reported on the occurrence of large populations of micro-algae at depths of the seas well below the euphotic zone (200 m). Heterotrophy is suggested as the only possible process of metabolism. The most common forms are coccolithophorids, chrysomonads, members of the Volvocales, blue-green algae (including species of *Nostoc*, *Dactyococcopsis*, and *Microcystis*), and various genera of dinoflagellates. Frequently, the density of aphotic populations, even to depths of 4000 m, is greater than in the euphotic zone. However, the deep-living species are always well represented in the euphotic zone and the number of species found in the depths is considerably smaller than that at or near the surface. The vertical range of these algae exposes them to variations in pressure from about 1 atm at the surface to at least 400 atm. It is not known if the deep-living algae spend their entire lives in the depths; their numbers are reported to be greater when the water contains large amounts of organic nutrients. Whether growth and reproduction occur only in the euphotic zone and the cells are later transported to the depths is likewise uncertain. To what extent pressure is an ecological variable in the vertical distribution of these algae remains, at present, a matter for conjecture.

Is the vertical distribution of higher plants limited by pressure? This question has two aspects. Firstly, there is the diminishing pressure in the atmosphere with increasing height above sea level. While the absence of pressure effects in the atmosphere cannot be demonstrated, the parallelism between the flora of high elevations and that of high latitudes would make it appear that changes in air pressure alone are not especially significant in determining plant distributions.

Secondly, there is the problem of submerged aquatic phanerogams. Several flowering plants are well adapted to an aquatic environment. However, in some members of the Ranunculaceae and of other families, flowering usually occurs only on branches extending above the water surface. The report of GESSNER (1955, 1961) that exposure to 2 atm inhibits flowering in *Potamogeton densus* may be considered in this light; flowering also does not occur on the submerged branches of this plant. GESSNER suggests that flowering is under the control of pressure. If this suggestion is correct, sensitivity to pressure, in ensuring that flowering occurs only in the air, aids in the vertical extension of the distributional range of these plants to the bottom of shallow water bodies. More experiments seem to be required in this area of research, as temperature, dissolved gas concentrations, and submersion itself would appear to be involved in such responses.

(3) Structural Responses

(a) Size

Micro-algae, collected from marine depths, have been reported to differ in size from those of the same species collected at or near the surface (BERNARD, 1963). As it is uncertain at what depths reproduction occurs, uniformity does not indicate the lack of pressure influence on the size of the cells of these algae. Presumably, a cell descending from one depth to another would decrease in volume by a very small amount, corresponding closely to the compressibility of water over the same pressure increment. No information appears to be available on the effects of pressure on the size of unicellular algae known to have divided while exposed to pressure. STURM (1941, *in*: GESSNER, 1961) reports that the inhibition of cell elongation in a species of *Hydrodictyon* following pressure exposure could be construed as a pressure effect on size. However, it is uncertain as to whether the cells would attain the size of the controls if given sufficient time. Responses to elevated pressures, either continuously or previously applied, in regard to the size of macro-algae appear equally unknown.

(b) External Structures

Exposure to small increments of atmospheric pressure has been shown by several authors (GESSNER, 1955, 1961) to have remarkable effects on the growth, differentiation, and morphogenesis of higher plants. GESSNER (1961) suggests that pressure of 1·5 to about 2 atm may control the development, morphogenesis, and flowering of aquatic phanerogams. These plants normally range to a depth of about 10 m at which pressures of approximately 2 atm are encountered. Shoots of *Potamogeton densus* and *Ranunculus circinatus* exposed to elevated pressures for 10 to 12 days produce far fewer leaves than do the controls (GESSNER, 1961).

Under 2 atm, development of the leaves of *P. densus* is about half of the controls, and 3 atm causes inhibition of nearly 90%, after 11 days under normal conditions. *R. circinatus* is slightly less sensitive in that inhibition amounts to about 25% at 2 atm, and to nearly 75% at 3 atm, after 20 days. GESSNER (1961) has interpreted these results as indicating a pressure-sensitive stage of the leaf primordia. If pressure is applied during the sensitive period no leaves form. On the other hand, in primordia which have developed beyond the sensitive stage when pressure is given, leaf development will occur. *Elodea canadensis* shoots, which have been exposed to 3 atm for 6 days and returned to atmospheric pressure, wither and die after 2 weeks, even though there are no visible signs of damage immediately after pressure treatment. Exposure of *E. canadensis* shoots to higher pressures for a short time yields inhibitory effects. Nineteen days after a 30-min exposure to 50 atm, the shoots develop 18% of the leaves of the controls. Only 11% of the control's leaf number develop after 30-min exposures to 200 atm. Micro-algae do not appear to respond to this type of temporary low pressure treatment (GESSNER, 1961). Presumably, pressure influences primarily the meristematic tissues of higher plants.

(c) Internal Structures

Most observations of pressure influences on plants have been limited to delayed effects of temporary exposures to elevated pressures. Plant tissues, with their generally simple internal organization usually do not show any conspicuous response to elevated pressures which may arise some time after the pressure treatment. Such responses no doubt exist, but little work has been done to elucidate them. On the other hand, some plant cellular structures, in common with animal cells, are subject to rapid and often drastic alterations upon the application of pressure. Pressure effects upon cellular organelles have frequently been observed on structural aspects of the nucleus, including the chromosomes and mitotic apparatus, the chloroplasts, and the plasma-gel layer, which may be involved in cyclosis.

MARSLAND (1939) reported the regular diminishing of protoplasmic streaming in cells of *Elodea canadensis* under elevated pressure. Streaming is completely inhibited between 400 and 500 atm. The effect is reversible if pressure is not maintained for more than 30 mins. Recovery can occur within a minute or two after the release of pressure. Centrifugation of cells under pressure and measurement of the time required for displacement of the chloroplasts to the ends of the cells indicates that the consistency of the protoplasm is reduced proportionately to the applied pressure. Inhibition of cyclosis due to increased pressure is also proportional to the protoplasmic consistency. MARSLAND suggests that these results indicate that cyclosis is motivated by a cycle of protoplasmic sol–gel reactions and is related to amoeboid movement.

In *Euglena gracilis*, protoplasmic-gel structures are also affected by pressure (BYRNE and MARSLAND, 1965). Solational weakening occurs with increasing pressure and decreasing temperature; the greatest effect (a rounding out of the cells) is observed with 15,000 psi at 15°C. Flagellatory movement continues at this pressure, but euglenoid movement ceases at about 10,000 psi. After 10 mins of exposure to 15,000 psi, normal form and activity is recovered within 10 mins of decompression.

Nuclear aberrations as a result of exposure to increased pressure have been observed in germinating seeds and in growing root tip cells (GUYOT, 1960). Chromosomal dicentric bridges of O_2 root tip cells of barley seedlings induced by high pressure have been discussed previously.

PEASE (1946) has shown differential pressure effects on spindle, chromosomes, and movement of chromosomes during division of *Tradescantia* pollen mother cells. The spindle of the first division appears normal up to 280 atm, but disappears at 480 atm. The second cell division spindle is more resistant and remains almost unaffected by 420 atm. At 500 atm, the spindle fibres are nearly invisible, and at 700 atm, fibrillar structure appears to be absent. Spindles of somatic cells, which are the most resistant, do not disappear until the pressure exceeds 560 atm. Chromosomes are more sensitive, and at pressures as low as 140 atm they may be bound together with chromatin bridges. Increased pressure accentuates the tendency towards fusion of the chloroplasts, and under 1050 atm, they are agglutinated into a mass of chromatin.

Pressures of from 280 to 420 atm induce a shortening and rounding up of the chromosomes. Fusion of the chromosomes interferes with their movement at pressures as low as 210 atm. Chromosomal movement ceases at 350 atm in the first division, 490 atm for the second division, and 630 atm for the somatic divisions. Complete reversibility of these pressure effects has not been observed, and it is uncertain as to whether new nuclei would arise from the pressurized cells. However, following the release of pressure, the spindle usually reappears, sometimes in a different plane of the cell in relation to the original position. During recovery, upon the release of pressure, the chromosomes may regain their original appearance, and, through the dissolution of the chromatin bridges, their initial separated condition.

In an extensive investigation into the interrelations of pressure, temperature and time relationships of cell division in the radicle meristems of *Triticum vulgaris* seedlings, GUYOT (1960) reported findings similar to those of PEASE (1946). Pressure of 200 atm at 20°C for 1 hr has little effect. Anomalies begin to appear at 300 atm, and, at pressures of 450 atm and above, all nuclear divisions are abnormal. Increased exposure times result in more anomalies at lower pressures. Decrease in temperature induces the appearance of anomalies at lower pressures, while increasing the temperature tends to oppose the effects of pressure. After exposure to pressures of 450 atm at 20°C, nuclear divisions appear to occur normally upon recovery after pressure release; but at this pressure, cytoplasmic division does not recover and resultant cells are binucleate. At pressures above 450 atm, nuclear divisions, when they occur, tend to be asymmetric. Persistent pressure effects after release are proportional to the exposure time.

(4) Conclusions

Plants respond to exposure to increased pressure in many different ways, a few of which have been discussed here. In a search for the general principles underlying these responses, certain points may be of value. Generally, pressures of a few hundred atmospheres appear to be tolerated by most mature plant cells, though there may be a reversible inhibition of some metabolic processes. Dividing cells, and especially meristematic cells appear to be highly sensitive to elevated pressure.

Growing populations of algal cells may be inhibited or stimulated by exposure to increased pressure, depending on the magnitude. Often a recognizable pressure effect may not appear for days or weeks. It is tempting to assume that such delayed responses result from the effects of pressure on cellular replicative (genetic?) processes.

Pressure can induce gross nuclear aberrations. Not surprising then, are the mutational responses of seeds and algal cells.

Morphogenetic alterations as a result of exposures to extremely small increments of pressure appear to represent a special case. Perhaps these low pressure effects result from the triggering of endogenous control mechanisms which are part of the plant's adaptation to its environment. Since low pressure acts mainly on meristematic cells, it is reasonable to presume that these cells are the ones capable of implementing morphogenetic control. This suggests a mechanism analogous to the phytochrome system, but responding to environmental factors other than light.

To achieve any understanding of the role of pressure as it influences plants, it is essential that the mechanisms of these widely varying responses be sorted out.

8. PRESSURE

8.3 ANIMALS

H. Flügel

(1) Introduction

In contrast to marine plants which depend to a great extent on the availability of light, both invertebrate and vertebrate animals have conquered all zones of the oceans. Near the water surface, the effects of pressure may overlap with effects of light (Chapter 2), temperature (Chapter 3), salinity (Chapter 4), and so forth. Nevertheless, there is evidence that animals are able to perceive, and to respond to, pressure variations within the range of 1 atm or fractions thereof.

On the floor of the deep-sea trenches, marine animals are exposed to hydrostatic pressures exceeding 1000 atm, or 14,696 psi (Chapter 8.0). Together with constant low temperatures, lack of light, shortage of food, and the specific, usually soft condition of the substratum, pressure seems to be responsible for the fact that the number of species which can cope successfully with these conditions decreases at great depths (Table 8-18). The distribution of echinoderms in relation to depth clearly demonstrates this phenomenon (EKMAN, 1954). High pressure, therefore, is one of the most important factors governing the distribution and evolution of life in the deep sea. Even pressures of hundreds of atmospheres do not in the slightest way injure animals once they have adapted to life in great depths. At 10,000 m depth, sea water—and the physico-chemical properties of body fluids of marine animals are closely related to those of sea water (Chapters 1 and 4.3) —becomes compressed by only 4·16%.

Table 8-18

Number of echinoderm species (with the exception of Crinoidea) found at different depths in the central Atlantic Ocean. Sea-stars of the western Atlantic Ocean south of Cape Cod and genuine shelf species are not included (After EKMAN, 1954; modified)

Depth ranges	Asteroidea	Ophiuroidea	Echinoidea	Holothuroidea except Elasipoda	Elasipoda
Total number of species considered:	134	173	99	65	42
200–1000 m	63	123	83	20	4
1000–2000 m	75	76	40	35	14
2000–3000 m	39	33	21	29	26
3000–4000 m	14	23	11	13	17
4000–5000 m	17	10	3	16	19
5000 m and more	7	4	1	2	4

Stimulated by the classical explorations of research vessels in the 19th century—especially the French '*Talisman*' and the British '*Challenger*'—the physiologists CERTES (1884a, b) and REGNARD (1884a, c, d, 1885, 1891) performed pioneering experiments analyzing pressure effects on micro-organisms and marine and fresh-water animals. In their very first reports to the Société de Biologique, the French scientists emphasized the ecological importance of their work. They soon extended their investigations to a wide variety of animals and isolated tissues, and recognized the general physiological aspects of elevated pressure. Since the days of REGNARD and CERTES, pressure devices have changed little in principle (Chapter 8.0). The interpretation of pressure experiments was handicapped by difficulties in differentiating between responses to rapid compression and decompression and those to constant high pressure. Furthermore, the experiments had to be carried out in a relatively small volume of sea water. The duration of such investigations was limited by the availability of oxygen in the pressure chamber. Only recently have efforts been made to study animals under pressure in freely circulating sea water.

Reviews on or including the effects of pressure on marine animals have been published by REGNARD (1891), CATTELL (1936), EBBECKE (1944), MARSLAND (1958), SCHLIEPER (1963a, 1968a), KNIGHT-JONES and MORGAN (1966) and FLÜGEL and SCHLIEPER (1970).

(2) Functional Responses

(a) Tolerance

The tolerance range of marine animals to variations in pressure is limited by the highest and the lowest pressure value at which the test organisms remain viable. As in micro-organisms (Chapter 8.1) and plants (Chapter 8.2), pressure tolerance of animals depends on, and may be modified by, variations of other simultaneously effective environmental factors, especially temperature, salinity (ionic composition), and pH of the medium. Furthermore, pressure tolerance may be a function of the physiological state of the test animal (developmental stage, breeding or non-breeding condition), as well as of the speed and duration of compression and decompression. To date, experiments on pressure tolerance have been carried out only with representatives of species collected from the seashore or shallow ocean regions. Investigations on the tolerance of deep-sea species to reduced pressure are still lacking.

REGNARD (1884a, c, 1891) exposed limnic and marine animals to pressures of up to 1000 atm. Echinoderms *Asterias rubens* and anthozoans *Actinia plumosa* survived at 1000 atm for only 30 mins. At decompression, the tested specimens had increased their body volume and almost doubled their initial weight. REGNARD concluded that, at such extreme pressure, water inflates the tissues and thereby creates the increase in size and weight observed. Unfortunately, details on his experimental conditions, such as test temperature and salinity, are not available. In his critical review, EBBECKE (1944) suggests that the swelling of specimens exposed to high pressure might be due to osmotic water uptake.

Teleosts exhibit distinctly lower pressure tolerances. Three small plaice *Pleuronectes platessa* survived 10 mins at 'fortes pressions' (apparently 600 atm). Prolongation of pressure exposure of up to 1 hr killed the fishes. Goldfish *Carassius auratus*, which, in contrast to *P. platessa*, have a swimbladder, were tested at 100, 200, 300 and 400 atm. They survived 100 atm without difficulty; at 200 atm, however, they seemed to be narcotized ('comme endormi'); and at 300 atm, the experimental specimen was dead (REGNARD, 1884a). At 400 atm, REGNARD noticed that rigor mortis had set in. Similar responses, beginning with retarded movements followed by death, were observed in tunicates, polychaetes (Nereidae, Serpulidae), and molluscs (*Mytilus edulis, Buccinum undatum, Cardium edule*).

REGNARD, the pioneer of modern pressure physiology, continually improved the original simple pressure apparatus. In 1885, he described, and later frequently used, a windowed pressure chamber. This chamber enabled him to observe the pressurized specimens of the genera *Daphnia, Gammarus* and *Cyclops* with the aid of a microscope. On July 25th, 1885, he reported to the 'Société de Biologique':

'Au delà de 1000 mètres, ils tombent lentement au fond de l'eau; leurs membres s'agitent avec rapidité, leurs appareils natatoires se raidissent et sont pris d'un tremblement très énergique. Les animaux demeurant à part cela immobiles au fond de l'eau. Ils semblent incapables de se mouvoir, ils sont tétanisés.'

REGNARD's work was continued by FONTAINE (1928), DRAPER and EDWARDS (1932) and EBBECKE (1935a, b). In his review, EBBECKE (1944) points out that the pressure tolerance of marine animals decreases in the order: Anthozoans, sea-stars, sea-urchins, jellyfish, gastropods, worms (apparently annelids), crustaceans, teleost fish.

NAROSKA (1968) tested the pressure tolerance of various invertebrates and three species of fish under controlled laboratory conditions. In order to reduce the degree of heterogeneity of his material NAROSKA selected specimens of nearly the same size, sex, and state of maturity. All individuals used in one experiment were collected from the same area (Baltic Sea or North Sea). Prior to the experiments, he acclimated the specimens to constant levels of temperature ($5°$, $10°$, $15°$ C) and salinity ($15‰$, $30‰$) for periods of 8 to 10 days. The test animals were then transferred into plastic jars. Two jars at a time were placed in an AMINCO-pressure device (Chapter 8.0) and exposed to pressures ranging from 100 to 800 atm for 1-hr periods. Thereafter, the test individuals were carefully removed and kept for 24 hrs in well-aerated water of identical salinity for determination of LD_{50} values. The resulting data reveal rather high pressure tolerances of isopods and amphipods (Fig. 8-40). Molluscs, polychaetes and sea-stars proved to be resistant to pressures rarely encountered in their natural environments. Decapods, mysids and teleosts did not tolerate such high pressures. The LD_{50}–24-hr values determined lay between 120 and 230 atm (*Eupagurus bernhardus, Neomysis vulgaris, Platichthys flesus, Pleuronectes platessa, Crangon crangon,* and *Carcinus maenas*). 50% of the tunicates *Ciona intestinalis* and the teleosts *Zoarces viviparus* survived 350 and 370 atm, respectively.

In situ experiments on pressure tolerance were carried out by MENZIES and WILSON (1961). While experimental conditions in the field are much more difficult to control than in the laboratory, there is no undesired interference of oxygen deficiency, accumulation of toxic metabolic by-products and pH changes, which may occur in small pressure chambers. On the other hand, results obtained during *in situ* experiments may be significantly affected by temperature changes during submersion and do not allow direct observation of the specimen tested. As emphasized by MENZIES and WILSON, neither in laboratory nor in *in situ* experiments is a separation of biological effects due to rapid compression or decompression possible. The two authors selected intertidal crabs *Pachygrapsus crassipes* and mussels *Mytilus edulis diegensis* as test animals. The experiments on crabs were conducted on groups of 5 to 8 adult *P. crassipes*. Each group was placed in a net-covered jar, attached to the hydrographic wire of the ship, and submerged into the ocean. The crabs were exposed for 1 min to each maximum depth. 100% of the crabs survived exposure to 88 atm, 60% exposure to 92 atm. Pressures in excess of 94 atm caused 100% mortality. *Mytilus edulis diegensis* exhibited a remarkably high tolerance to compression under *in situ* conditions. The mussels tolerated a hydrostatic pressure of 222 atm (2227 m water depth). At 357 atm, all specimens

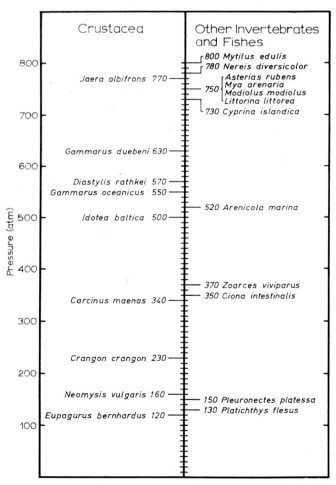

Fig. 8-40: Pressure tolerances (LD_{50} values in atm) of invertebrates and fishes from the Baltic Sea. 10° C; 15‰S. (After NAROSKA, 1968; modified.)

hauled up were dead, muscles relaxed and shells agape. These experiments confirm the observations of REGNARD (1891), EBBECKE (1944), NAROSKA (1968), and others that decapods exhibit low, but bivalves high pressure tolerances (Fig. 8-40).

Responses of decapods and teleosts to compression were thoroughly studied by EBBECKE and HASENBRING (1935b). Employing a windowed pressure chamber in connection with a microscope, EBBECKE was able to determine not only the lethal limits but also the behaviour of the specimens while under increased pressure. Shrimps (apparently *Leander adspersus*, *Crangon crangon* and *Pandalus montagui*; species not indicated) respond to 50 atm with temporarily increased excitement; upon decompression, they instantly behave normally again. Repetition of the experiment 10 to 20 times did not harm the shrimps. They became increasingly conditioned to the pressure changes, but augmentation of compression slightly beyond the original levels again induced excitement and accelerated movements of pleopods. Pressures of around 150 atm seemed to paralyze the shrimps; they slowly sank to the bottom of the pressure chamber. Further pressure increase to 200 atm resulted in brief excitement, but the shrimps became completely motionless if that compression was sustained. Surprisingly, the shrimps survived even at 500 atm; they became absolutely motionless and stiff; only their hearts continued to beat. Following decompression, the shrimps resumed normal activity after 1 hr. These experiments were conducted at the Marine Station of the 'Biologische Anstalt Helgoland'. Thus, we may conclude that the salinity of the sea water used was in the range of 32‰ to 33‰. Unfortunately, details as to the time course, temperature and light intensities during the experiments are not given (EBBECKE and HASENBRING, 1935b).

Small individuals of teleost fishes such as *Gobius* sp., *Pleuronectes platessa* and *Spinachia spinachia* responded to suddenly increased pressures of 50 and 100 atm with rapid movements. At 200 atm, sustained compression killed the fishes, while 500 atm caused immediately 100% mortality. EBBECKE (1944) concluded that surface-dwelling fishes are not able to exist at water depths greater than 2000 m (200 atm).

In order to reduce the shortcomings of experiments conducted in small pressure chambers, SCHLIEPER (1963a, b, 1966, 1968a, b) and his associates investigated the pressure tolerance of small, isolated surviving tissue pieces of marine animals. Responses of isolated tissues have been used successfully as criteria in general physiology and in pressure physiology (REGNARD, 1891; GRAY, 1928; PEASE and KITCHING, 1939; EBBECKE, 1944; SCHLIEPER, 1955; SCHLIEPER and KOWALSKI, 1956).

Under convenient environmental conditions (near-optimum temperatures, salinities, etc.), excised gill tissue of marine molluscs (total area about 2 by 4 mm) can survive for more than 8 days under normal atmospheric pressure. The general physiological conditions of the epithelia can be assessed in terms of the degree of ciliary activity exhibited. Normal, healthy cells show high ciliary activity (classified as 3); cell death leads to complete stand-still of cilia (classified as 0); intermediate stages are classified as 1 (about 25% surviving cells) and 2 (about 75% surviving cells), respectively (see also Chapters 3.31 and 4.31). SCHLIEPER and his associates realized that this method is far from being ideal; however, it yields reproducible results which are largely in keeping with comparable information obtained on whole, intact individuals, and provides a suitable tool for analyzing the effects of environmental stress. Occasionally, after excision of tissue pieces, the cilia cease to beat. This is, apparently, a phenomenon which is comparable to a surgical shock. Unfortunately, slightly injured tissues exhibit not only retarded ciliary activity but, in many cases, also loss of beat synchronization. Therefore, stroboscopic investigations are of no or little advantage.

PONAT (1967) examined the relative cellular activity of excised tissues of various bivalves, of the sea-anemone *Metridium senile* and of the sea-star *Asterias rubens* from the North Sea and the Baltic Sea. She kept two pieces of the tissue, each with an area of about 6 to 8 mm², in plastic jars filled with water of habitat salinity. One of the jars was then placed in the pressure chamber and pressures applied between 300 and 900 atm for 6 hrs. After 1-hr recovery periods at normal atmospheric pressure, the activity of the tissue was examined microscopically. The tissues of *Asterias rubens* survived compression up to 700 atm, those of *Mytilus edulis* pressures up to 600 atm. Lower pressure tolerances were found in excised tissue pieces of *Cyprina islandica* and *Metridium senile*. Maximum tolerance of *M. senile* and *C. islandica* amounts to about 300 atm (Fig. 8-41).

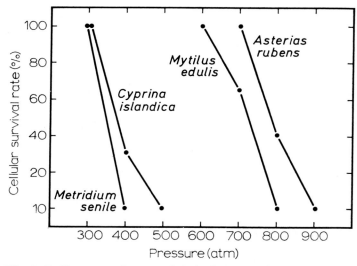

Fig. 8-41: Pressure tolerance of excised tissues of various marine invertebrates at 10°C. *Metridium senile*: isolated tentacle epithelium (30‰S); *Cyprina islandica* and *Mytilus edulis*: gill epithelium; *Asterias rubens*: papulae epithelium (15‰S). (After PONAT, 1967; modified.)

Pressure tolerance as a function of temperature

Detailed analyses of the combined effects of pressure and temperature on amoebae, ciliates, cleaving sea-urchin eggs and excised tissues of various aquatic invertebrates have been presented by MARSLAND (1950, 1957), LANDAU and MARSLAND (1952), LANDAU and co-authors (1954) and AUCLAIR and MARSLAND (1958). High pressure leads to a critical weakening of protoplasmic gel structures, particularly of the peripheral plasmagel layer of the cytoplasm (BROWN, 1934a; BROWN and

MARSLAND, 1936; MARSLAND and BROWN, 1936; Chapter 8.0). These detrimental effects of high pressure are counteracted by high temperatures within the physiological range of the species considered, while lowering of temperatures amplifies the pressure stress.

At 10° C, the cleavage of *Arbacia punctulata* eggs, for instance, becomes blocked by a pressure of 204 atm (3000 psi); at 30° C, a blockage occurs at 476 atm (7000 psi) (MARSLAND, 1950). A combination of high pressure (680 atm = 10,000 psi) and low temperature (12° C) killed 100% of the test population of the ciliate *Blepharisma undulans*; while the same pressure at 25° C killed only 80% (AUCLAIR and MARSLAND, 1958).

SCHLIEPER (1963a) and his associates paid much attention to the modifying effects of temperature on pressure tolerance of excised tissues and intact invertebrates. In excised gill tissue pieces of *Mytilus edulis*, the critical pressure which significantly reduces the ciliary activity is distinctly affected by temperature: it is 300 atm at 10° C and about 400 atm at 20° C (Fig. 8-42a). SCHLIEPER's findings are in accordance with those of LANDAU and MARSLAND (1952) on the heart beat rate of tissue culture explants from the tadpole of *Rana pipiens*. The pressure effects on the heart beat varies according to the temperature at which the experiments are performed. At 272 atm (4000 psi), the heart beat rate of the cultured tissue mounted in the microscope-pressure chamber decreases below 14° to 16° C and increases at higher temperatures if compared with the beat rate of tissue at atmospheric pressure.

PONAT (1967) acclimated bivalves from the Baltic Sea for at least 5 days to 5°, 10° or 15° C. She then excised small gill pieces and exposed them for 24 hrs to 400

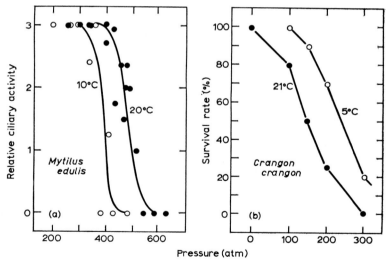

Fig. 8-42: Pressure tolerance as a function of temperature. (a) *Mytilus edulis*. Relative activity of terminal cilia of gill pieces in 32‰ S. 3: normal activity, 2: slightly reduced activity, 1: strongly reduced activity, 0: complete cessation of beating in all cilia. (After SCHLIEPER, 1963a; modified.) (b) *Crangon crangon*. Survival rate of whole intact individuals in 15‰ S. (After NAROSKA, 1968; modified.)

atm at the same three temperature levels. Her observations show that the ciliary rate is relatively high at the two test temperatures 10° and 15° C, independently of the acclimating temperature. The ciliary rate falls drastically, however, when the experiments are performed at 5° C. In other words, the pressure tolerance is significantly reduced when warm-acclimated tissue is tested at low temperatures (Table 8-19).

Table 8-19

Mytilus edulis. Relative ciliary activity of excised gill tissues as a function of pressure and temperature. Intact bivalves were acclimated to 5°, 10° and 15° C for at least 5 days. Excised gill pieces were examined after 24-hr exposure to 400 atm at test temperatures (After PONAT, 1967; modified)

Acclimation temperature	Test temperatures (° C); 400 atm		
(°C)	5	10	15
5	1·7	2·4	2·4
10	0·9	2·6	2·5
15	0·6	1·9	2·5

From the information presented above, SCHLIEPER (1968a) concluded that the combination of high pressures and low temperatures (hardly ever above 4° C) prevailing at the deep-sea floor severely challenges prospective immigrants; at higher temperatures, possibly a larger number of species would be able to cope successfully with deep-sea conditions. However, recent investigations demonstrate that the combination high (within the physiological range) temperature and high pressure by no means always increases the pressure tolerance of marine invertebrates (NAROSKA, 1968). *Crangon crangon*, fully acclimated to 5° or 20° C, for example, is less pressure resistant at 20° than at 5° C (Fig. 8-42b).

It must be kept in mind, that responses obtained on single cells or excised tissues may not be the same as those of intact individuals. Obviously, genetic adaptations of intact organisms to defined pressure–temperature conditions are most important. In marine deep-sea species, pressure tolerance apparently tends to decrease with rising temperatures. It is, however, too early to say whether temperature increase leads to increased pressure tolerances in all cases. The information at hand is still insufficient for large-scale generalizations. Critical laboratory experiments on the combined effects of pressure and temperature on healthy multicellular deep-sea animals have not yet been conducted.

Pressure tolerance as a function of salinity

The degree of pressure tolerance of excised gill tissues may be modified by salinity. PONAT (1967) acclimated *Mytilus edulis* from the Baltic Sea for 10 days to 15‰ and 30‰ S, respectively. She then tested the pressure tolerance of excised gill pieces, removed from 15‰ and 30‰S acclimated individuals in the respective acclimation salinity. Exposure to 500 atm leads to cessation of the ciliary activity after 20 mins in 15‰S, but after 28 mins in 30‰S. Cross-acclimation of North Sea *Mytilus edulis* (30‰) and Baltic Sea *Mytilus edulis* (15‰) revealed that the pressure tolerance of excised tissues increases as a function of salinity; it is higher in 30‰ than in 15‰ (Fig. 8-43, see also Fig. 8-40).

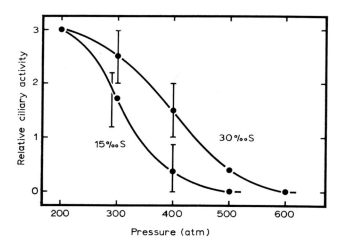

Fig. 8-43: Pressure tolerance as a function of salinity. Excised gill tissue of *Mytilus edulis* from the Baltic Sea, tested in 15‰S (left curve), and from the North Sea, tested in 30‰S, 24 hrs after pressure exposure at 5° C. All bivalves examined were acclimated to 10° C for at least 5 days prior to the experiment. Averages of 12 individuals in each case. (After PONAT, 1967; modified.)

Similar experiments on intact amphipods were carried out by NAROSKA (1968). *Gammarus oceanicus* from the Baltic Sea was acclimated to 5‰, 10‰, 30‰ and 50‰S, respectively for 9 to 12 days. 100% mortality within 1 hr (LD_{100}–1hr) was obtained in 5‰S at 550 atm, in 10‰S at 600 atm, in 30‰S at 650 atm, and in 50‰S at 700 atm (Fig. 8-44). Increases in pressure tolerance were also observed if the total osmoconcentration of the test medium was increased by adding glycerin.

Salinity may affect pressure tolerance not only via variations in total osmoconcentration but also through variations in ionic ratios. Specific ion effects have been demonstrated to exist both at the individual and at the subindividual level. Calcium has long been known to stabilize protoplasmic components (HÖBER, 1947; HEILBRUNN, 1952). Recently, PONAT (1967), SCHLIEPER and co-authors (1967) and NAROSKA (1968) have examined the pressure tolerance of marine invertebrates in the presence of excess Ca. Excised gill tissues of *Mytilus edulis* and intact whole crabs *Eupagurus zebra* revealed significantly increased pressure resistance in the presence of excess calcium. In sea water with normal ion ratios, *Eupagurus zebra* was killed by exposure to 250 atm for 1 hr; in sea water containing twice as much calcium, the test animals survived exposure to 270 atm.

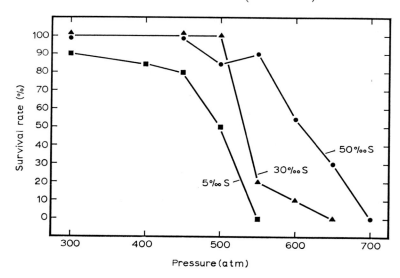

Fig. 8-44: Pressure tolerance as a function of salinity. *Gammarus oceanicus* from the Baltic Sea, tested at 10° C after 9 to 12 days acclimation to 10° C and the three salinity levels indicated. (After NAROSKA, 1968; modified.)

NAROSKA (1968) kept the polychaete *Arenicola marina* for 8 to 10 days in brackish water of 15‰S with additional Ca (15‰ + 100% Calcium). He then examined the lethal pressure limits and found only very small differences compared to the performances of comparable individuals kept in normal sea water.

Gammarus oceanicus exhibits pressure tolerances in calcium-enriched sea water which are similar to those found in sea water of normal ionic composition. NAROSKA points out that in osmoregulators such as *Gammarus oceanicus* external variations in ion ratios are rapidly and efficiently compensated for, resulting in a constant ion composition of body fluids. Thus, changes in external calcium amounts cannot affect internal protoplasmic components (see also Chapter 4.31).

Pressure tolerance as a function of pH values

pH values are pressure dependent. According to BUCH and GRIPPENBERG (1932), a pH value of 8 measured at the water surface is equivalent to a pH of 7·78 at a depth of 10,000 m. A surface pH of 7·5 shifts to pH 7·16 at a pressure of 1000 atm. It was for this reason that the cellular pressure tolerance of *Mytilus edulis* tissue was studied at various pH levels ranging from 3 to 10 (PONAT and THEEDE, 1967); the normal pH value of Baltic Sea water was adjusted accordingly by adding HCl or NaOH to the test medium. Small pieces of excised gill tissue were transferred into Baltic Sea water of abnormal pH and exposed to a constant pressure for 3 hrs. The ciliary rate was then determined at normal atmospheric pressure about 15 mins after decompression. Control experiments were conducted under conditions of normal atmospheric pressure. Maximum ciliary activities are found between pH 6 and pH 9. Under a pressure of 400 atm a well-marked maximum in ciliary activity was obtained at pH 8; at 600 atm the maximum shifted to pH 6 (Fig. 8-45).

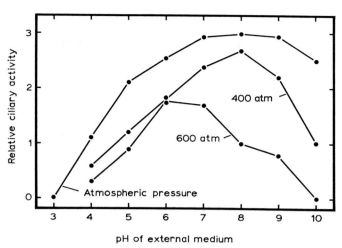

Fig. 8-45: Pressure tolerance as a function of pH. Excised gill tissues of *Mytilus edulis* from the Baltic Sea. Criterion: relative ciliary activity; for details consult legend to Fig. 8-42a. 20° C; 15‰S. (After Ponat and Theede, 1967; modified.)

The physiological mechanism underlying the combined effects of pressure and pH is not yet fully understood. Without doubt, pH changes affect the cellular tolerances to compression at the molecular level. According to Johnson and co-authors (1954), the strength of hydrogen bonds as well as the ionization of proteins are affected by changes both in pH and pressure. Increased pressure tends to retard the process of proteins unfolding, which is accompanied by volume increase. Processes which involve volume decrease, such as the ionization of molecules, are accelerated by pressure. Thus, pressure affects ionization and stability of hydrogen bonds in opposite ways. (For further pertinent information consult Chapter 8.0.)

Pressure tolerance as a function of oxygen tension

The oxygen content of the polar water masses decreases on the way from the surface to the abyss. Living organisms and the oxidation of dead organic material are responsible for this reduction in oxygen content. In fact, the oxygen content varies remarkably near the bottom of all oceans (Table 8-20). According to these data, abyssal marine animals exist in habitats with reduced oxygen tensions.

Theede and Ponat (1970) studied the pressure tolerance of the isopod *Idotea baltica* at oxygen tensions ranging from 0% to 120% saturation (Fig. 8-46). At 15 to 85% oxygen saturation, the test animals survive 500 atm for 1 hr; above 90% and below 15% saturation, pressure tolerance decreases. When exposed to 700 atm (15° C, 15‰S) for a period of 1 hr, 94% of the test population of the bivalve *Cyprina islandica* survive an oxygen tension of 39% saturation. Survival rates decrease progressively, both in media with lower and higher oxygen tensions (Table 8-21).

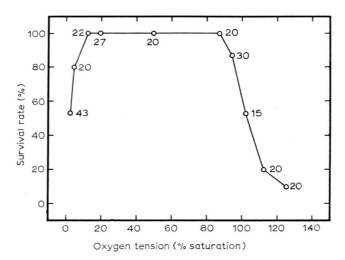

Fig. 8-46: Pressure tolerance as a function of oxygen tension. *Idotea baltica* exposed for 1 hr to 500 atm at 15° C and 15‰ S. After decompression, the test individuals were allowed to recover for 1 hr in well-aerated water and then examined for survivors. Numbers of individuals tested are given at the appropriate points. (After THEEDE and PONAT, 1970; modified.)

Table 8-20

Oxygen content of deep-sea waters (Based on data by various authors; after THEEDE and PONAT, 1970; modified)

Ocean area and position			O₂ content (% of saturation value)	Water depth (m)	Author
Atlantic Ocean	1° 45′ N	44° 50′ W	79·0	3488	METCALF and STALCUP (1969)
	7° 30′ N	79° 19′ W	28·9	2500	SCHMIDT (1929)
Indian Ocean	14° 24′ S	46° 08′ E	50·0	3035	I.I.O. Expedition
	4° 26′ N	85° 21′ E	38·1	3500	DANA Rep. 1937
Pacific Ocean	7° 10′ N	78° 15′ W	25·2	2000	DANA Rep. 1937
	18° 53′ S	163° 02′ W	56·2	5000	DANA Rep. 1937
China Sea	15° 22′ N	115° 20′ E	33·3	4000	DANA Rep. 1937

Table 8-21

Cyprina islandica. Pressure tolerance at different oxygen tensions.
The intact whole bivalves were exposed to 700 atm for 1 hr (15° C,
15‰S; Baltic Sea specimens) (After THEEDE and PONAT, 1970;
modified)

Oxygen tension (% saturation)	3·4	8·5	39	90	140	202
Survival (%)	75	83	94	92	67	58

THEEDE and PONAT (1970) also studied the tolerance of excised tissues at various oxygen tensions. At 700 atm, the ciliary rates of excised gill tissues of *Mytilus edulis* and *Cyprina islandica* decrease gradually with increasing oxygen tension. In fully saturated Baltic Sea water of 15‰S, 100% of the epithelial cells of *Cyprina islandica* are killed within 60 mins at 700 atm. The tissue of *Mytilus edulis* is somewhat more resistant when tested at 700 atm and oxygen tensions in excess of 100% saturation (Fig. 8-47).

The treatment of the tested tissues prior to pressure experiments is of great importance. If the gill pieces are acclimated for 18 hrs to sea water from the habitat (15‰S) of only 2–4% oxygen saturation and compressed in the same medium, the

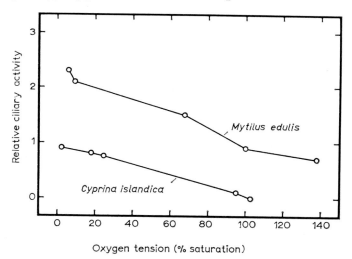

Fig. 8-47: Pressure tolerance as a function of oxygen tension. Excised gill tissues of *Cyprina islandica* and *Mytilus edulis* from the Baltic Sea exposed for 1 hr to 700 atm at 15° C and 15‰S. After decompression, the tissues were allowed to recover for 1 hr and then examined for their relative ciliary activities (see legend to Fig. 8-42a). Average values of 20 measurements in each case. Controls at normal atmospheric pressure survived at all oxygen levels applied with undiminished activites. (After THEEDE and PONAT, 1970; modified.)

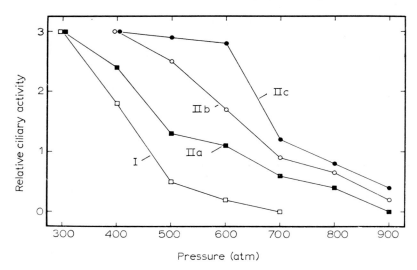

Fig. 8-48: Pressure tolerance of excised gill tissues of *Cyprina islandica*
in oxygen saturated and unsaturated (2 to 4% saturation) Baltic Sea
water of 15‰S at 15° C. I: acclimated to, and tested in a saturated
medium; IIa: acclimated to an unsaturated and tested in a saturated
medium; IIb: acclimated to a saturated and tested in an unsaturated
medium; IIc: acclimated to, and tested in an unsaturated medium. In
all cases, the tissues were exposed to the pressures indicated for 1 hr;
after decompression, they were allowed to recover in a well-aerated
medium and then examined for the relative activity (see legend to
Fig. 8-42a) of their terminal cilia. Average values of 12 individual
measurements in each case. (After THEEDE and PONAT, 1970;
modified.)

tolerance is noticeably higher than that of tissues acclimated to fully saturated
brackish water (15‰S). Further results obtained under combinations of varying
oxygen tension before and during compression are illustrated in Fig. 8-48.

Does the pressure tolerance of excised tissues depend on the amount of lactic
acid present? Lactic acid, the product of anaerobic glycolysis, is formed particularly under conditions of oxygen deficiency. THEEDE and PONAT (1970) could,
indeed, demonstrate that external concentrations of 1 or 2 mMol/l lactic acid
augmented the cellular pressure tolerance as compared to the tolerance of tissue
pieces compressed in absence of lactic acid in the external medium. However,
there is evidence that the gain in pressure tolerance is due primarily to the concomitant pH change of the external medium (see also PONAT and THEEDE, 1967).

The reduced pressure resistance of whole individuals and excised tissue pieces
at high ambient oxygen tensions may be interpreted in terms of oxidation and
inhibition of pressure sensitive SH-enzymes and of other protoplasmic components.
The reduced pressure tolerance at extremely low oxygen tensions, on the other
hand, may possibly be a result of retarded ATP-synthesis (Chapter 8.0).

Experiments on the pressure tolerances of marine animals have thus far been
carried out only on coastal species. Many of them turned out to be resistant to
pressures rarely encountered in their natural environments; they are eurybath.

Animals sensitive to pressure changes, such as the bivalve *Cyprina islandica*, are referred to as stenobath. The degree of species-specific tolerance to hydrostatic pressure seems to be related to the degree of tolerance to variations in intensities of other ecological factors such as temperature, salinity, etc. Thus the gill epithelia of *Mytilus edulis* are tolerant to extreme intensities of temperature, salinity and pressure.

There is urgent need for experiments on the pressure tolerance of deep-sea living animals. Such experiments require a high degree of technical sophistication; they have not yet been attempted. In normal sampling devices, the drastic temperature rise during the long journey from the sea floor to the deck of the research vessel kills the collected specimens long before they become available to the eagerly waiting marine ecologists. Research in arctic or antarctic regions, where temperature gradients are less pronounced, may lead sooner to the much needed breakthrough in pressure ecology.

(b) Metabolism and Activity

Pressure effects on respiratory rates

It is known that the frequency and amplitude of muscle contraction is augmented by slight increase in pressure. Thus, we may expect parallel changes in respiratory rates. Respiratory rates of fishes and decapods exposed to different pressures were determined for the first time by the French physiologist FONTAINE (1928; 1929a, b; 1930). He measured the oxygen consumption of *Pleuronectes platessa*, *Ammodytes lanceolatus*, *Gobius minutus*, *Crangon crangon* and *Palaemon serratus* before and during exposure to 100 atm. In all species, respiratory rates increased upon compression. The augmentation of oxygen consumption after 25 to 60 mins at elevated pressures varied between 27 and 114%.

In further experiments, FONTAINE (1929a) tested the influence both of the intensity of the pressure employed and of the duration of the compression on the respiratory rates of small specimens of *Pleuronectes platessa*. In the pressure range 25 to 125 atm, oxygen consumption rises significantly (Table 8-22); but exposure to 150 atm causes death.

Table 8-22

Respiratory rates of the plaice *Pleuronectes platessa* exposed to different pressures (After FONTAINE, 1929a; modified)

Pressure (atm)	Oxygen consumption (cm³)		Change (%)
	before experiment	under pressure	
25	0·232	0·299	+28
50	0·224	0·311	+38·8
100	0·176	0·279	+58
125	0·172	0·268	+54
150	0·184	0·112	−39

G

In complex animals such as teleosts, experiments in small pressure chambers may provide inadequate information, due to uncontrollable behavioural aspects and confinement in the narrow chambers, possibly also by stimulating effects of oxygen deficiency during prolonged experiments. With increasing duration of exposure to 100 atm, the oxygen consumption of plaice increased. The highest respiratory rates were obtained during a 90-min experiment; after maintaining the compression for 120 mins, respiratory rates decreased (FONTAINE, 1929b).

Recent experiments on pressure effects on respiratory rates of various marine invertebrates and fishes have been conducted by NAROSKA (1968). His experiments

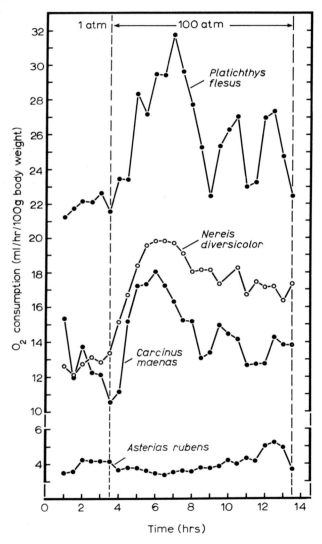

Fig. 8-49: Pressure effects on respiratory rates as a function of time. 14° C, 15‰S. (After NAROSKA, 1968; modified.)

revealed that the respiratory rates of sea-stars *Asterias rubens* and—to a lesser degree—the sea-urchins *Psammechinus miliaris* decrease with increasing pressure. For the first time, NAROSKA determined also the continuous rate of oxygen consumption in running sea water. The sea-stars *Asterias rubens* and *Henricia sanguinolenta*, the brittle star *Ophiura texturata* and the sea-urchin *Psammechinus miliaris* responded to stepwise pressure increases (100, 200, 300 atm) with an initial 'under-shoot', i.e. a sudden, temporary reduction in oxygen consumption. At 100 atm, the teleost *Platichthys flesus* and the crab *Carcinus maenas*, however, immediately accelerated their metabolic rates. After about 3 hrs under pressure, this acceleration was followed by a slow reduction. By the end of the experiment, the respiratory rate had returned to about the original level. During the same experiment, *Asterias rubens* showed no, or only very slight respiratory changes (Fig. 8-49). At 200 and 300 atm, the respiratory rate fell below the original level. However, this reduction is followed by a slow recovery, which has been interpreted by NAROSKA as acclimation to increased pressures. Since the pressure devices available at present do not allow pressure experiments to be conducted under adequate conditions for extended periods, the phenomenon of non-genetic adaptation to increased or decreased pressures can still not be studied appropriately. Until more suitable pressure devices for long-term experiments become available, *in situ* experiments aboard research vessels could possibly fill this gap.

Pressure effects on metabolic performance and activity

Windowed pressure chambers make it possible to study living animals under elevated pressures (REGNARD, 1885, 1891; DRAPER and EDWARDS, 1932; MARSLAND and BROWN, 1936; PONAT, 1967; NAROSKA, 1968; Chapter 8.0). Rhythmic movements of heart, umbrella (medusae), cilia, antennules, etc. provide useful criteria for the study of metabolic performances of animals exposed to elevated pressures. The rhythmic contractions of the medusae *Cyanea capillata*, for instance, were studied at pressures of up to 600 atm by EBBECKE and HASENBRING (1935b). Slight compression induces an increase in frequency and amplitude of the contractions. Under moderate pressures, umbrella movements became more regular. At 300 atm, umbrella movements cease. At pressures in excess of 300 atm, umbrella muscles firmly contract and reveal changes in shape. Medusae exposed to 600 atm for only a few minutes easily recover and, at normal pressures, exhibit again rhythmic contractions. Decompression leads to temporary standstill; pressures of 200 atm and less cause an increasing frequency of umbrella contractions. According to EBBECKE (1944), acceleration and retardation of contraction may be due to the responses of the nerve net, while muscular tetany seems to be induced by compression of the muscle cells.

Similar variations in the rate of performances due to changes in pressure have been observed in respect of the heart beats of intact marine invertebrates and fishes (*Fundulus* sp., embryo: DRAPER and EDWARDS, 1932; *Pandalus montagui*: EBBECKE and HASENBRING, 1935b; *Zoarces viviparus*, *Gammarus duebeni*, *G. oceanicus*, *Ciona intestinalis*: NAROSKA, 1968). Initial acceleration of heart beats may be a temporary phenomenon (over-shoot reaction) followed by retardation. A pressure of about 82 atm produced a reduction in heart beat rate of the embryos of *Fundulus* sp. within 2 mins. The average decrease, relative to controls, amounted to 9·9%. By the end of a 10-min compression period, the average reduction was 16·6% (DRAPER and EDWARDS, 1932). Initial acceleration, followed by a decline in heart beat rate, was also observed by NAROSKA (1968). He subjected larvae of *Zoarces viviparus* to 50 atm and 200 atm, respectively. At 50 atm, the rate returned to normal within 4 mins; at 200 atm, the heart came practically to a standstill at first, then beat activity gradually increased again to one-third of the initial rate. Initial temporary acceleration and subsequent decline in the heart beat rate occurs also in amphipods such as *Gammarus oceanicus* (Fig. 8-50). In the shrimp *Pandalus montagui*, an increase from normal to 100 atm reduces the heart rate from 200 beats/min to 134 beats/min; at 300 atm, the heart beats only 96 times/min. After decompression, recovery to normal rates occurs within 2 mins (EBBECKE and HASENBRING, 1935b).

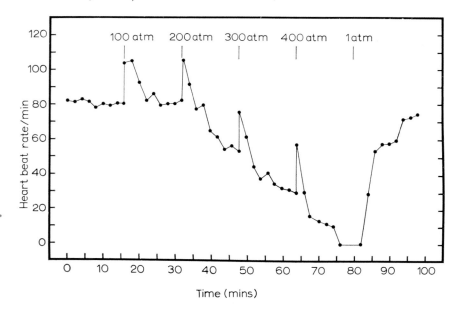

Fig. 8-50: Pressure effects on heart beat rate of *Gammarus oceanicus*. 5° C, 15‰S. (After NAROSKA, 1968; modified.)

The physiological responses to pressure described above can also be observed in excised cardiac muscle and heart preparations of frog, turtle and dogfish (EDWARDS and CATTELL, 1928); the differences recorded between the various species are of degree rather than kind. Muscle tissues can easily be removed from the organism and kept alive for long periods. As early as 1914, EBBECKE demonstrated successfully, under well-defined conditions, that brief exposure of the frog's gastrocnemius muscle to 200 to 300 atm yielded no response, but that pressures in excess of 300 atm caused contraction. If the pressure was still higher, or the exposure period prolonged, continued shortening resulted. Investigations into the effects of increased pressures on smooth, striated and cardiac muscles have been repeatedly reviewed (CATTELL, 1936; EBBECKE, 1944). CATTELL (1936, p. 466) writes:

'As the pressure is increased the first change to occur is the conversion of more energy, which is shown by an augmentation of the twitch tension and a corresponding increase in the initial heat production. Next there is evidence of an increased viscosity and decreased elasticity, and these factors adequately explain the smaller and slower twitch as well as the reduction in efficiency. Finally there is a gradual loss in the ability to contract under the higher pressures, which in the beginning is reversible, but if long continued results in the permanent loss of function. As already described, the pressure itself causes a development of tension or contracture, and this may be in part responsible for the falling off in twitch tension during the prolonged application of higher pressures.'

Investigations on the effects of pressure on isolated muscle tissue have been published also by CATTELL and EDWARDS (1928, 1929, 1930, 1931, 1932), BROWN (1931, 1934a, b, 1935, 1936, 1957), BROWN and EDWARDS (1932), EBBECKE (1935b, 1936b), EBBECKE and HASENBRING (1935a), EBBECKE and SCHÄFER (1935).

Very informative experiments have been carried out on the activity of compressed unicellular animals such as *Amoeba proteus* (BROWN and MARSLAND, 1936; MARSLAND and BROWN, 1936) and various flagellates and ciliates (EBBECKE 1935a; KITCHING, 1957a). As is well known, amoeboid movements consist of formation and reabsorption of pseudopodia. Two distinct layers of the protoplasm can easily be distinguished: the steadily flowing 'endoplasm' and the more rigid outer 'ectoplasm'. The latter provides a sheath, while the former advances to the tip of a forming pseudopodium. Pressures up to 136 atm lead to elongation of pseudopodia. The diameter of the pseudopodia decreases as the pressure increases. Sudden compression to about 250 atm prevents the streaming of the protoplasm within a pseudopodium. If the pressure is raised to 340 or 408 atm, the pseudopodia become shortened, finally forming terminal spheres. If the high pressure is not maintained for more than 30 mins, these effects are reversible. At 476 atm all pseudopodia collapse and the amoebae become spherical. These effects are explained on the basis of the liquifaction of the plasmagel which occurs above a certain critical pressure. This interpretation is supported by the observation that dense protoplasmic inclusions such as food vacuoles, nuclei and other relatively heavy cell components sink toward the lower parts of the amoebae indicating reduced viscosity of the cell protoplasm. Upon pressure release, a new ectoplasm is formed, pseudopodia begin to appear and the movement is re-established within a few minutes.

Investigations on various flagellates and ciliates demonstrate that even closely related species may respond very differently to compression (KITCHING, 1957a). In general, locomotory activity of protozoans ceases at pressures of 544 to 953 atm. Such effects are reversible unless the pressure is maintained beyond certain critical periods. The pressures required for complete depression of flagellar or ciliary movements are very close to the lethal dosis. In ciliates and flagellates, moderate pressures (68 to 204 atm) initially increase ciliary activities.

Accelerating and 'narcotizing' effects of pressure variations were also demonstrated by PEASE and KITCHING (1939), who used gills of *Mytilus edulis* as test objects. The beat of the lateral cilia of a single gill filament was measured stroboscopically. Compression of gill filaments causes an instantaneous rise in ciliary activity, but the rate returns gradually to the original level. When the pressure is raised by successive steps of 68 atm, the frequency of ciliary beatings rises each time. But at pressures exceeding 340 or 408 atm, the basic ciliary frequency falls and the tissues becomes injured irreversibly (Fig. 8-51).

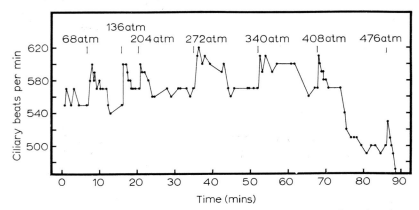

Fig. 8-51: *Mytilus edulis*. A single complete record of the ciliary beatings per min during which the pressure was raised rapidly by successive equal steps of 68 atm. 21° to 24° C; sea water. (After PEASE and KITCHING, 1939; modified.)

The experiments described on the previous pages have dealt primarily with changes in metabolism and activity of marine animals due to rather high pressures. Discoveries of HARDY and PATON (1947) and HARDY and BAINBRIDGE (1951), however, made it clear that pressures of less than 2 atm can be readily perceived by numerous marine animals. The copepod *Calanus finmarchicus*, for example, when lowered into the sea in vertical cylinders, tends to move upwards with increasing depth.

HARDY and PATON (1947) suggested that pressure affects the vertical migration of planktonic animals. In order to test this suggestion, decapod larvae (almost entirely zoea and megalopa stages of *Portunus* sp. and *Carcinus maenas*) were transferred into 20 inch long 'Perspex' tubes (about 25 larvae per tube), with a rectangular cross-section of $2 \times 1 \frac{1}{2}$ inch (HARDY and BAINBRIDGE, 1951). The pressure in the experimental tubes was then varied over a range of approximately 2 atm by raising or lowering a column of mercury connected to the tube. Poisoning of the sea water was prevented by a freshwater buffer and a rubber diaphragm separating the mercury and the freshwater buffer. Controls were kept in tubes at normal atmospheric pressure. During the experiments, the number of larvae found swimming in the upper half of the tube was recorded at intervals of 15 or 30 mins. All experiments were performed in the dark; only for counting were the larvae exposed to red light. If the pressure was raised to 2 atm, more than 50% of the larvae responded with upward swimming, while only about 20% of the controls occupied the upper half of the tube (Fig. 8-52a). Upon pressure release, the experimentals behaved exactly as the controls. If the pressure was maintained at a constant value for 3 hrs, a gradually increasing percentage of the larvae was found

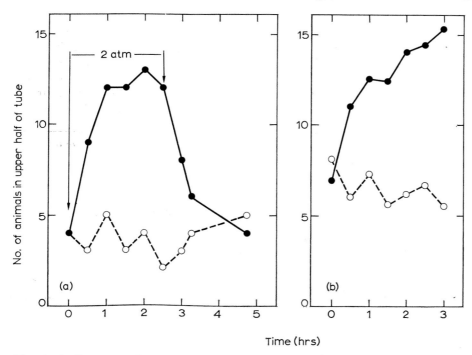

Fig. 8-52: Pressure effects on swimming activity of mixed samples of decapod larvae (almost entirely zoea and megalopa of species of *Portunus* and *Carcinus*). (a) The pressure was raised to 2 atm and, after 2½ hrs, reduced to normal atmospheric pressure again. Filled circles: numbers of larvae in the upper half of the experimental tube; open circles: the same in the controls. (b) The pressure was maintained at a constant value for 3 hrs. (After HARDY and BAINBRIDGE, 1951; modified.)

swimming in the upper half of the tube. The authors emphasized that the responses are very similar at pressures of 0·5, 1, 1·5 or 2 atm (Fig. 8-52b).

HARDY and BAINBRIDGE's (1951) experiments stimulated numerous investigations on low pressure effects on a wide variety of planktonic and benthonic marine animals. In 1955, KNIGHT-JONES and QASIM were able to demonstrate sensitivity to pressure changes in hydromedusae, ctenophorans, polychaetes, copepods, isopods, decapods and fishes (Table 8-23). The responses of marine animals to pressures amounting to fractions of 1 atm have been summarized by KNIGHT-JONES and QASIM (1955, p. 941) as follows:

'When the pressure was increased, they generally became more active and swam upwards; when it was decreased, they became less active, or completely inactive and allowed themselves to sink.'

Similar results have been obtained by DIGBY (1967). RICE (1961) reported that the mysids *Schistomysis spiritus*, *Siriella armata*, *Praunus flexuosus* and *P. neglectus* respond to small changes in low pressure. Also these mysids swim upwards as the pressure is increased and downwards as it is decreased; *Schistomysis spiritus* responds to changes as small as 15 mbar (corresponding to a water depth of 15 cm).

Table 8-23

Minimum pressure changes producing significant responses in various marine animals (After Knight-Jones and Qasim, 1955; modified)

Taxonomic group	Species	Minimum pressure changes (in mbar) producing a significant response
Hydromedusae	*Phialidium hemisphericum*, adults	(800)
	Gossea corynetes, adults	(800)
	Eutima gracilis, adults	(800)
Ctenophorans	*Pleurobrachia pileus*, adults	50
Polychaetes	*Poecilochaetus serpens*, larvae	(800)
	Autolytus aurantiacus, pelagic adults	(800)
Copepods	*Caligus rapax*, probably adults	(800)
Isopods	*Eurydice pulchra*, adults (pelagic during high water)	50
Decapods	*Carcinus maenas*, megalopa larvae	10
	Galathea sp., corresponding larval stages	10
Teleost fishes	*Blennius pholis*, larvae	5

Rice suggested that these responses assist in vertical orientation of *S. spiritus* as the water level rises and falls due to tidal rhythms. Further support of the hypothesis that small pressure changes may assist in local orientation and adjust activity rhythms associated with the tides has been produced in studies on the amphipod *Corophium volutator* (Morgan, 1965), the decapod *Carcinus maenas* (Williams and Naylor, 1969) and the isopod *Eurydice pulchra* (Jones and Naylor, 1970). Responses similar to those described by Hardy and Bainbridge (1951) and Knight-Jones and Qasim (1955) have also been reported from the chaetognath *Sagitta setosa* (Singarajah, 1966) and the freshwater cladoceran *Daphnia magna* (Lincoln, 1970). *S. setosa* uses gravity rather than light as orientation clue; but small changes in temperature, salinity and pH, as well as mechanical disturbances, may interfere.

A detailed study on pressure responses of the amphipod *Synchelidium* sp. has been presented by Enright (1961, 1962). Increase or decrease in pressure causes sudden 'rapid scrambling' and 'darting'. Under conditions of slow linear pressure increase (2·8 mbar/sec), 25 to 50 mbar are required to evoke a significant reaction. The response threshold lies between 5 and 10 mbar in the case of rapid pressure increases; it rises to 10 to 15 mbar after the amphipods have been exposed to a series of pressure changes of approximately 30 mbar.

While responses to pressure changes are rather common among marine crustaceans, the isopod *Excirolana chiltoni* and the mysid *Archaeomysis maculata* do not reveal appreciable reactions to pressure increases of up to 100 mbar (Enright, 1962).

Changes in locomotory activity, opposite to those normally observed, have been reported in the pycnogonid *Nymphon gracile* (Morgan and co-authors, 1964). Cyclical pressure changes (800 mbar) of tidal range and frequency led to active swimming under conditions of decreasing pressures or during late ebb. The authors suggest that *N. gracile* may be able to discriminate between slow tidal pressure undulations and more rapid pressure changes such as would be encountered during vertical upward or downward movements (KNIGHT-JONES and MORGAN, 1966).

The responses to small pressure changes have prompted the question: what kind of pressure receptors are available to marine animals and where are they located? In teleost fishes, the swimbladder wall, the Weberian ossicles and the exteroceptors act as pressure receptors (VASILENKO and LIVANOW, 1936; DIJKGRAAF, 1940; MÖHRES, 1940; ABRAHAM and STAMMER, 1954; QUTOB, 1962).

In marine and freshwater invertebrates without gas-containing organs of detectable size, the mechanisms and sites of pressure receptors remain obscure. However, the small changes in pressure eliciting a reaction in many planktonic animals suggest the presence of some kind of gas phase (DIGBY, 1967). As is well known, gas may accumulate on the body surfaces of several aquatic insects. In crustaceans, such as the shrimps *Leander serratus* and *Palaemonetes varians*, minute vesicles or films of gas may occur on the lipid-covered cuticle surfaces. Mechanical compression of such tiny amounts of gas is believed to be insufficient for producing adequate stimulation; however, it may alter substantially ionic diffusion and electrical properties of the crustacean cuticle (DIGBY, 1961, 1965, 1967). Variations in pressure produce, indeed, changes in the potentials across the body surfaces of decapods. DIGBY studied the behaviour of metallic electrodes and found that the electrical properties of such a system, when exposed to variations in pressure, are similar to those of animal body surfaces. His experiments suggest that pressure-sensitivity of the crustaceans examined is due to the properties of a thin layer of electrolytic hydrogen produced at the cuticle's surface. However, in spite of DIGBY's fascinating experiments, many problems remain unsolved. From the early experiments of BROWN (1934b) employing the pressure centrifuge, it is well established that significantly elevated pressures cause reductions in the viscosity of the cytoplasm. These and related changes at the subcellular level may well account for the perception of pressure changes, even small ones, in the absence of specific pressure receptors.

(c) Reproduction

Cell division

Animal reproduction involves a number of processes, such as gamete maturation, gonad growth, and gamete release (Chapters 3.31, 4.31). All these processes are based on cell division. Since detailed field observations and long-term laboratory experiments regarding pressure effects on animal reproduction are lacking, studies on cleaving cells must suffice to elucidate how reproduction may be affected by variations in pressure.

MARSLAND (1936, 1938) succeeded in arresting cleavage in fertilized eggs of the sea-urchin *Arbacia punctulata* by exposing them quickly to 450 atm. Under sustained compression, the furrow slowly receded. As soon as the pressure was released, cleavage continued. At 1 atm, cleavage was delayed for about 3 mins, at 333 atm, for 15 mins. Eggs of the parasitic nematod *Ascaris megalocephala* var. *univalens* continue to cleave up to a pressure of 800 atm, a condition hardly ever encountered in their natural environment; furrowing continues up to 600 atm. However, if

600 atm are applied more than 20 mins prior to the expected cleavage, the cleavage becomes delayed and the rate of furrowing slows down. Compression of the eggs at 270 atm for 24 hrs causes only a slight retardation in cleavage rate. The resistance to high pressures of cleavage processes in eggs of *Ascaris megalocephala* is unique and remarkably higher than in any of the other species listed in Table 8-24. In most of the marine animals examined, 200 to 400 atm are required to block egg cell division. In other words, the marine species listed in Table 8-24 would not be able to reproduce successfully at water depths exceeding 2000 to 4000 m.

Table 8-24

Elevated pressures required to block cell division in eggs of various animals (After PEASE and MARSLAND, 1939; modified)

Animals	Pressure (atm)	Author
Protozoa		
Amoeba dubia	400	MARSLAND and BROWN
Amoeba proteus	400	MARSLAND and BROWN
Echinodermata		
Arbacia punctulata	330–400	MARSLAND, PEASE
Arbacia lixula	330	MARSLAND
Echinarachnius parma	330	PEASE
Paracentrotus lividus	330	MARSLAND
Psammechinus microtuberculosis	330	MARSLAND
Sphaerechinus granularis	400	MARSLAND
Nematoda		
Ascaris megalocephala	800	PEASE
Annelida		
Chaetopterus pergamentaceus	220–270	PEASE
Mollusca		
Cumingia tellenoides	270–330	PEASE
Planorbis sp.	270	PEASE
Solen siliqua	230	MARSLAND
Insecta		
Drosophila melanogaster (pole cell division)	330–400	MARSLAND
Tunicata		
Ciona intestinalis	200	MARSLAND
Vertebrata		
Fundulus heteroclitus	330–400	MARSLAND
Rana pipiens	330–400	MARSLAND

The mechanism involved in blocking egg cleavage seems to be comparable to pressure effects on protozoans. Elevated pressures tend to weaken the more rigid cortical portions of the cell and finally cause solation, thus inhibiting amoeboid movement as well as the intrusion of the cleavage furrow. In addition to weakening the cortical ooplasm, high pressure affects the mitotic apparatus or spindle-aster complex of dividing cells (ZIMMERMAN and MARSLAND, 1964). At about 140 atm, chromosomal movements in cleaving eggs of *Arbacia punctulata* become retarded, at about 280 atm, completely arrested. Retardation and arrest are, in general, entirely reversible. However, exposure to pressures exceeding 544 atm for more than 5 mins leads to drastic disorganization of the mitotic apparatus beyond recovery when subsequently decompressed.

MARSLAND (1950, 1970) has shown that increasing temperatures (within the tolerable range) strengthen the cortical gel structures of various dividing marine eggs, while decreasing temperatures amplify the blocking effect of pressure. The information at hand seems to indicate that, in the deep sea, successful reproduction of numerous marine animals may be severely limited by the low temperatures prevailing in that habitat.

Breeding cycles

Very small pressure changes, associated with the tides, appear to play an important role in phasing (timing) the breeding cycles of a variety of marine animals. According to KORRINGA (1947, 1957), the European flat oyster *Ostrea edulis* shows marked spawning maxima during spring tides and releases its larvae mostly at neap tides about 8 days later; it is suggested that the pressure differences between spring and neap tides assists in synchronizing spawning.

QASIM and co-authors (*in:* KNIGHT-JONES and MORGAN, 1966) reported that the emission of the larvae of the polychaete *Spirorbis borealis* can be delayed experimentally for a few hours by employing pressures in the range of high water spring tide. In their natural environment, the larvae are shed mostly at the moon's quarters. The ecological significance of this behaviour is possibly the protection of the larvae against desiccation (low water). The larvae, attached to *Fucus serratus* growing below low water neap tide, can thus add to the length of their tubes for several more days.

Since weather-induced variations in water temperature or light (turbidity) are frequently rather irregular, pressure changes associated with the tides seem to provide much more reliable clues for synchronization of spawning in many marine animals.

(d) Distribution

About 84% of the total surface area of the oceans and adjacent seas lies above depths exceeding 2000 m, and about 75% covers abyssal depths between 3000 and 6000 m. The abyssal zone constitutes the largest habitat on earth (see also Chapter 8.0). It is characterized by an unusual degree of uniformity of environmental factor intensities. The temperature reveals no seasonal changes of biological significance and remains always below 4° C; from 50°N to 58°S it varies only from 3·6° C to − 0·6° C. In the Atlantic Ocean, at 50°N, 3·32° C were measured at 2000 m and 2·38° C at 4000 m (MENZIES, 1965). The salinity is close to 34·8‰ and varies only about ± 0·2‰. The oxygen content (Chapter 9) suffices to support life even in oceanic trenches with depths in excess of 10,000 m, since azoic trench areas have not yet been found. The sediments in the abyssal zone consist of soft siliceous and calcareous oozes, and, to a great extent, of red clay. None of these environmental

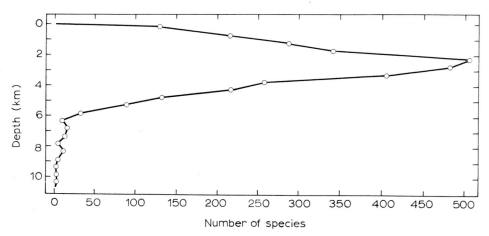

Fig. 8-53: Total number of deep-water bottom invertebrate species as a function of depth. (After Vinogradova, 1952; modified.)

factors seem to be responsible for the fact that the number of species decreases with increasing depth (Fig. 8-53, Table 8-18). According to Bruun (1956, p. 1107),

'between 6000 m and 7000 m a considerable number of major groups disappear. This is the transition zone leading to the depths of the trenches, which at 10 000 m in the Kurile Trench and the Philippine Trench contain only six and eight species, respectively.'

This decrease in species number, the food shortage due to decreases in biomass (82% of the total biomass of the oceans is produced in shallow coastal waters, only 0·8% at depths exceeding 3000 m), together with pressures up to 1100 atm characterize the basic ecological conditions of the hadal zone (from 'hades', Greek, the underworld). However, food shortage should not be over-emphasized. Bruun (1956, p. 5) summarized the availability of food in the abyssal zone as follows:

'On the whole, the food supply in the abyssal zone may not be termed plentiful; but it is no more scanty than in any other benthic zone of the deep ocean below the layers in the proximity of the photosynthetic zone.'

According to Wolff (1960), bottom samples taken from deep-sea trenches reveal biomasses as high as 10 to 40 g/m².

It is beyond the scope of this chapter to review the vast number of investigations concerned with the distribution of deep-sea animals. They contain a surprisingly small amount of information pertaining directly to pressure effects on the penetration of marine abyssal species into the hadal zones of the oceans.

Spärck (in: Ekman, 1954) and other authors doubt the ecological significance of pressure as a controlling factor in vertical distributions of multicellular marine animals. To them, temperature and, particularly, lack of food are of greater importance. In the previous sections of this review, we have learned that pressure may influence functional properties of organisms quite differently. Even closely related species may respond very differently. The fact that the bottom fauna of the deep trenches consists of species which represent the same taxonomic groups as the shallow-water fauna by no means contradicts the assumption that pressure acts as major environmental factor controlling the distribution of animals. Long-term laboratory experiments on (i) pressure-acclimated shallow-water inhabitants and (ii) animals recovered from the deep sea are necessary to elucidate further the ecological significance of pressure.

(3) Structural Responses

(a) Size

At present, long-term cultivation of marine multicellular animals under sustained elevated pressures is impossible. Hence pertinent experimental information on possible effects of increased pressure on body size, external and internal structures is not available.

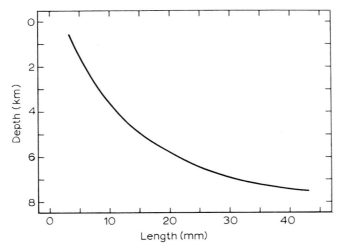

Fig. 8-54: Body length of species of the abyssal isopod genus *Storthyngura* as a function of water depth. The coefficient of correlation between increasing length and depth is 0.62 ± 0.15. (After BIRSHTEIN, 1957; modified.)

WOLFF (1956), who studied the isopods collected during the '*Galathea*' Expedition, emphasized that most hadal species 'are larger than almost all hitherto known species within the same genus'. This observation has since been confirmed by several authors. BIRSHTEIN (1957), for example, noted increased size in the abyssal isopod genus *Storthyngura* and termed it 'abyssal gigantism' (Fig. 8-54). Evidence for abyssal gigantism in deep-sea animals, other than crustaceans, is scarce. Abyssal gigantism may be due to the effect of elevated pressure on the metabolism (WOLFF, 1956 and BIRSHTEIN, 1957).

MADSEN (1961) challenged this view and suggested that the increased size of hadal animals is a consequence of their greater longevity due to fewer predators and stable ecological conditions. However, WOLFF (1962) remains convinced that pressures of several hundred atmospheres play a significant role in abyssal gigantism (overgrowth):

'It is probable that the overgrowth is determined by a combination of several factors, such as low temperature, large supply of food in restricted areas, and the effect of pressure resulting in an increased metabolic rate or, alternatively, furthering the excess in size, e.g. by retarded sexual maturity and/or greater longevity' (WOLFF, 1962, p. 237).

(b) External Structures

Also in regard to external structures, we must carefully distinguish between *in situ* pressure effects under deep-sea conditions and responses to rather short pressure exposures under laboratory conditions.

We do not yet know if certain peculiarities in regard to the external structures of hadal animals are a direct consequence of living under significantly increased pressures.

In laboratory experiments, elevated pressures cause a reduction in protoplasmic gel strength of *Amoeba proteus* (LANDAU and co-authors, 1954). Direct observation (microscopic pressure chamber) has revealed that the pseudopodia become slender and reduced in number at pressures between 68 and 102 atm. At 139 and 204 atm, locomotory movement slows down and pseudopodia length decreases. At about 260 atm, pronounced pseudopodia can no longer be formed. At 272 to 408 atm, the *A. proteus* rounds up (if no pseudopodial plasm is pinched off), or forms motionless spheres (if pseudopodial material is lost upon pressure application).

Table 8-25

Cessation of movement and onset of structural damage in various flagellates and ciliates exposed to elevated pressures. Standard stepwise experiments (After KITCHING, 1957a; modified)

Species	Pressures at which movement ceases (atm)	Pressures at which structural damage becomes evident (atm)	Nature of structural damage
Polytoma uvella	680–953	1089–1228	Minute blebs on some individuals
Chlamydomonas pulsatilla	749–885	—	—
Astasia longa	544–680	544–680	Rounded or irregular shape
Paramecium aurelia	476 or over	476 or over	Distorted and irregular shape
P. caudatum	272–476	272–408	Blisters develop
Tetrahymena pyriformis	749–817	817 or over	Bulbously deformed or rounded
Colpidium campylum	612–817	680–953	Rounded or spherical
Colpoda cucullus	408–476	408–476	Distorted or blistered
Spirostomum ambiguum	680–749	408 and upwards	Bent, contorted, or otherwise deformed
Stentor polymorphus	612–680	544 and over	General distortion, especially bulging of foot; some bulging of disk even at lower pressures

In the suctorian *Discophrya piriformis*, a creasing of the body surface area was observed a few seconds after application of 139 atm (KITCHING, 1954). This response is accompanied by an expansion of the pellicle. At 204 atm, the protoplasm begins to separate from the pellicle. These structural effects are completely reversible after pressure release. At 272 atm and still higher pressures, the protoplasm separates suddenly from the pellicle, and recovery under normal pressures requires several hours.

KITCHING (1957a) reported that the pressures required to cause visible structural damage to various flagellates and ciliates are close to those which stop ciliary movement (Table 8-25). Similar pressure effects on the ciliates *Blepharisma undulans* and *Paramecium caudatum* have been described by AUCLAIR and MARS-LAND (1958). At pressures less than 476 atm, the ciliates retain their slender form. But at pressures exceeding 476 atm, they become more spherical. In both species, structural consequences of variations in pressure depend on temperature. The ciliates are more resistant to deformation at 25° C than at 12° C.

In the heliozoan *Actinophrys sol*, form stability of body and axopods depends on pressure and temperature (KITCHING, 1957b). At 204 atm, the axopods shorten in most cases and a few are withdrawn. At 272 atm, some of the axopods appear to be 'beaded' and subsequently disappear. At 15° C, the critical value for axopod distortion is about 238 atm; this value shifts to 320 atm at 25° C. Upon pressure release, normal axopods are formed again.

(c) Internal Structures

TILNEY and MARSLAND (1965) and TILNEY and co-authors (1966) performed interesting electron microscopic investigations on internal fine structures of axopods in pressurized heliozoans *Actinosphaerium nucleofilum*. They used a device which permits the fixation of the specimen, while being compressed. Thus it was possible to compare the internal structures before, during and after exposure to pressures between 272 and 544 atm. Most of the cell organelles exhibit a rather high resistance to disintegration; however, the microtubular elements of the axopods become unstable. The disintegration of tubular elements is correlated with the 'beading' and retraction of the axopodia. If exposure to 272 atm is extended to 20 mins, reorganization of microtubules and axopodia occurs. At 408 atm, such reorganization is reduced to 'knoblike' axopodial remnants. But at 544 atm, no regeneration takes place. After pressure release, the microtubuli reappear as soon, or sooner, as the axopodia are reformed. These observations are in complete accordance with, and confirm, the previous light microscopic studies of KITCHING (1957b) on the axopods of *Actinophrys sol*.

Highly organized microtubuli occur also in the mitotic apparatus of dividing cells and in germ cells. Exposure of metaphase *Arbacia punctulata* eggs to 680 atm for 1 min causes complete disorganisation of the cytoplasmic microtubuli (ZIM-MERMAN, 1970b).

The effects of increased pressures on internal fine structures of *Amoeba proteus* were studied by LANDAU and THIBODEAU (1962) and ZIMMERMAN and RUSTAD (1965). At 544 atm, plasmalemma, mitochondria and other cell organelles exhibit no distinguishable differences if compared with cells fixed at atmospheric pressure.

The existence of numerous vesicles in close proximity to the cell membrane is interpreted as disintegrated pinocytosis channels. While in non-pressurized amoebae the Golgi complex is readily observable, it is not evident in pressurized specimens. These effects have been confirmed by ZIMMERMAN and RUSTAD (1965), who found no detectable disintegration on pinocytosis at 68 atm; at 139 atm, most of the pinocytosis channels disappear, but small pseudopodia, as well as some new channels are distinguishable. No pinocytosis occurs at 204 atm. About 30 mins after decompression, pinocytosis becomes normal again.

FLÜGEL and FRITSCH (unpublished) studied the effects of increased pressures on the ciliated gill epithelia of *Mytilus edulis* and *Modiolus modiolus*. These marine bivalves exhibit a striking resistance of their cell organelles against structural disintegration. The basal lamina and adjacent cells restrict pressure-induced

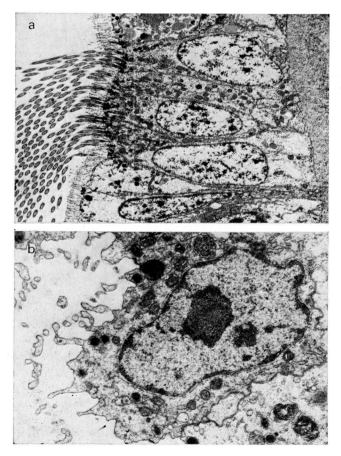

Fig. 8-55: Pressure effects on internal structures of the gill epithelium of *Modiolus modiolus*. (a) normal atmospheric pressure; ×5400; (b) 300 atm; ×18,000. 30 mins after pressure elevation, the distal cell membrane is thrown into irregular pseudopodia-like processes. (After FLÜGEL and FRITSCH, unpublished.)

modifications in epithelial cell structures largely to the free, distal cell surface. As is well known, these cells bear at their distal surface numerous microvilli and cilia. After exposure to 300 atm for 30 mins, the distal cell membrane of *Modiolus modiolus* forms many irregular pseudopodia-like processes (Fig. 8-55a, b). At pressures in excess of 600 atm, the finger-shaped processes disappear and the microvilli become reduced to short knob-like structures. The inner-microtubular elements of the cilia are relatively stable. But at 600 atm, they begin to disintegrate.

On the whole, littoral marine bivalves are apparently much more resistant to pressure-induced structural disintegration than limnic protozoans. Presumably, this pronounced form stability represents an adaptation which is part of a high over-all tolerance to environmental stress. The stability of epithelial cell organelles may also be the result of a fast recovery under sustained low pressures, since fixation was carried out after 30-min exposures to test pressures. It seems possible that 'adaptational' reorganization of cell organelles occurs during prolonged pressure experiments (see also TILNEY and co-authors, 1966).

(4) Conclusions

In spite of numerous investigations on the effects of variations in pressure on marine animals, the ecological significance of high pressures remains largely obscure. Our present knowledge is based on short-term experiments conducted on shallow-water species. As long as it is technically not possible to acclimate the test animals to sustained elevated pressures in running sea-water systems, the information obtained is hardly adequate for interpreting or analyzing the situation actually met in the seas.

There is considerable need for the construction of pressure apparatus which allows the measurement of functional and structural properties of well-acclimated animals from different oceanic depths. Another important 'bottle neck' is our present inability to obtain healthy stenobathic deep-sea animals. Studying their performance under reduced pressures is likely to improve considerably our present assessment of pressures as an ecological variable. It is encouraging that efforts are being made to attack some of these problems (NAROSKA, 1968; SCHLIEPER, 1968a; MacDONALD and GILCHRIST, 1969).

Field investigations, preferably in arctic or antarctic waters where a critical rise in temperature may be avoided, will also add to our knowledge. Biochemists and cell biologists have always considered pressure a major factor influencing chemical and biological systems. It is now the turn of the marine ecologists to extend the fundament provided and to deepen our knowledge on pressure as an environmental variable in oceans and coastal waters.

Literature Cited (Chapter 8)

ABRAHAM, A. and STAMMER, A. (1954). Pressorezeptoren in der Wand der Schwimmblase. *Annls biol. Univ. hung.*, **2**, 345–360.

ADAMS, K. T. (1942). *Hydrographic Manual*. U.S. Government Printing Office, Washington, D.C. (Spec. Publ. No. 143.)

ALBRIGHT, L. J. (1968). The effect of temperature and hydrostatic pressure on protein, ribonucleic acid and deoxyribonucleic acid synthesis by *Vibrio marinus*, an obligate psychrophile. Ph.D. thesis, Oregon State University, Corvallis.

ALBRIGHT, L. J. and MORITA, R. Y. (1965). Temperature–hydrostatic pressure effects on deamination of L-serine by *Vibrio marinus*, an obligate psychrophile. *Bact. Proc.*, **65**, 20.

ALBRIGHT, L. J. and MORITA, R. Y. (1968). Effect of hydrostatic pressure on synthesis of protein, ribonucleic acid, and deoxyribonucleic acid by the psychrophilic marine bacterium, *Vibrio marinus*. *Limnol. Oceanogr.*, **13**, 637–643.

ANDERSON, G. R. (1963). A study of the pressure dependence of the partial specific volume of macromolecules in solution by compression measurements in the range of 1–8,000 atm. *Ark. Kemi*, **20**, 513–571.

ANSEVIN, A. T. and LAUFFER, M. A. (1959). Native tobacco mosaic virus protein of molecular weight 18,000. *Nature, Lond.*, **183**, 1601–1602.

ARONSON, M. H. (1967). *Pressure Handbook*, Rimbach, Pittsburgh, Pa.

AUCLAIR, W. and MARSLAND, D. A. (1958). Form stability of ciliates in relation to pressure and temperature. *Biol. Bull. mar. biol. Lab., Woods Hole*, **115**, 384–396.

BALCHAN, A. S. and DRICKAMER, H. G. (1961). High pressure electrical resistance cell, and calibration points above 100 kilobars. *Rev. scient. Instrum.*, **32**, 308–313.

BECKER, R. R. and SAWADA, F. (1963). Enzymatic properties of polypeptidyl-ribonuclease. *Fedn Proc. Fedn Am. Socs exp. Biol.*, **22**, 419.

BERG, C. C., NILAN, R. A. and KONZAK, C. F. (1965). The effect of pressure and seed water content on the mutagenic action of oxygen in barley seeds. *Mutation Res.*, **2**, 263–273.

BERGER, L. R. (1958). Some effects of pressure on a phenylglycosidase. *Biochim. biophys. Acta*, **30**, 522–528.

BERGER, L. R. and REYNOLDS, D. M. (1958). The chitinase of a strain of *Streptomyces griseus*. *Biochim. biophys. Acta*, **29**, 522–534.

BERNARD, R. (1963). Density of flagellates and Myxophyceae in the heterotrophic layers related to environment. In C. H. Oppenheimer (Ed.), *Symposium on Marine Microbiology*. C. C. Thomas, Springfield, Ill. pp. 215–228.

BIRSHTEIN, J. A. (1957). Certain peculiarities of the ultra-abyssal fauna as exemplified by the genus *Storthyngura* (Crustacea, Isopoda, Asellota). *Zool. Zh.*, **36**, 961–985.

BLINKS, L. R. (1950). Photosynthetic action spectra of marine algae. *J. gen. Physiol.*, **3**, 389–442.

BOATMAN, E. S. (1967). Electron microscopy of two marine *Bacillus* spp. subjected to hydrostatic pressure during growth. *Bact. Proc.*, **67**, 25.

BOSTRACK, J. M. and MILLINGTON, W. F. (1962). On the determination of leaf form in an aquatic heterophyllous species of *Ranunculus*. *Bull. Torrey bot. Club.* **98**, 1–20.

BOYD, F. R. and ENGLAND, J. L. (1960). Apparatus for phase-equilibrium measurements at pressures up to 50 kilobars and temperatures up to 1750° C. *J. geophys. Res.*, **65**, 741–748.

BRADLEY, R. S. (Ed.) (1963). *High Pressure Physics and Chemistry*, Vols I and II. Academic Press, New York.

BRANDT, B., EDEBO, L., HEDÉN, C. G., HJORTZBERG-NORDLUND, B., SELIN, I. and TIGERSCHIOLD, M. (1962). The effect of submerged electrical charges on bacteria. *TVF*, **33**, 222–229.

BRANDTS, F. (1964). The thermodynamics of protein denaturation. I. The denaturation of chymotrypsinogen. *J. Am. chem. Soc.*, **85**, 4291–4301.

BRIDGMAN, P. W. (1949). *The Physics of High Pressures*, 2nd ed., G. Bell and Sons, London.

BROWN, D. E. S. (1931). Pressure and the dynamics of cardiac muscle. *Am. J. Physiol.*, **97**, 508.

BROWN, D. E. S. (1934a). The effect of rapid changes in hydrostatic pressure upon the contraction of skeletal muscle. *J. cell. comp. Physiol.*, **4**, 257.

BROWN, D. E. S. (1934b). The pressure coefficient of 'viscosity' in the eggs of *Arbacia punctulata*. *J. cell. comp. Physiol.*, **5**, 335–346.

BROWN, D. E. S. (1935). The liberation of energy in the contracture and simple twitch. *Am. J. Physiol.*, **113**, 20.

BROWN, D. E. S. (1936). The sequence of events in the isometric twitch at high pressure. *Cold Spring Harb. Symp. quant. Biol.*, **4**, 242–251.

BROWN, D. E. S. (1957). Temperature–pressure relation in muscular contraction. In F. H. Johnson (Ed.), *Influence of Temperature on Biological Systems*. American Physiological Society, Washington, D.C. pp. 83–110.

BROWN, D. E. S. and EDWARDS, D. J. (1932). A contracture phenomenon in cross-striated muscle. *Am. J. Physiol.*, **101**, 15.

BROWN, D. E. S., GUTHE, K. F., LAWLER, H. C. and CARPENTER, M. P. (1958). The pressure, temperature, and ion relations of myosin and ATP-ase. *J. cell. comp. Physiol.*, **52**, 59–77.

BROWN, D. E. S., JOHNSON, F. H. and MARSLAND, D. A. (1942). The pressure–temperature relations of bacterial luminescence. *J. cell. comp. Physiol.*, **20**, 151–168.

BROWN, D. E. S. and MARSLAND, D. A. (1936). The vicosity of Amoeba at high hydrostatic pressure. *J. cell. comp. Physiol.*, **8**, 159–165.

BRUNS, E. (1962). *Ozeanologie*, Vol. II: Ozeanometrie, 1. Deutscher Verlag der Wissenschaften, Berlin.

BRUNS, E. (1968). *Ozeanologie*, Vol. III: Ozeanometrie, 2. Teubner, Leipzig.

BRUUN, A. F. (1956). The abyssal fauna: Its ecology, distribution and origin. *Nature, Lond.*, **177**, 1105–1108.

BRUUN, A. F. (1957). Deep sea and abyssal depths. In J. W. Hedgpeth (Ed.), *Treatise on Marine Ecology and Paleoecology*, Vol. I, Ecology. *Mem. geol. Soc. Am.*, **67**, 641–672.

BUCH, K. and GRIPENBERG, S. (1932). Über den Einfluß des Wasserdruckes auf pH und das Kohlensäuregleichgewicht in größeren Meerestiefen. *J. Cons. perm. int. Explor. Mer*, **7**, 233–245.

BUNDY, F. P., HIBBARD, W. R. and STRONG, H. M. (Eds) (1961). *Progress in Very High Pressure Research*, Wiley, New York.

BYRNE, J. and MARSLAND, D. A. (1965). Pressure–temperature effects on form stability and movement of *Euglena gracilis* variety, Z. *J. cell. comp. Physiol.*, **65**, 277–284.

CANTON, J. (1762). Experiments to prove that water is not incompressible. *Trans. Phil. Soc., London*, **52**, 640–643.

CATTELL, McK. (1936). The physiological effects of pressure. *Biol. Rev.*, **11**, 441–476.

CATTELL, McK. and EDWARDS, D. J. (1928). The energy changes of skeletal muscle accompanying contraction under high pressure. *Am. J. Physiol.*, **86**, 371–382.

CATTELL, McK. and EDWARDS, D. J. (1929). The influence of pressure on the refractory period and rhythmicity of the heart. *Am. J. Physiol.*, **90**, 308.

CATTELL, McK. and EDWARDS, D. J. (1930). The influence of hydrostatic pressure on the contraction of cardiac muscle in relation to temperature. *Am. J. Physiol.*, **93**, 87–104.

CATTELL, McK. and EDWARDS, D. J. (1931). Epinephrin action in relation to the hydrostatic pressure effect on the contraction of cardiac muscle. *Am. J. Physiol.*, **96**, 657.

CATTELL, McK. and EDWARDS, D. J. (1932). Conditions modifying the influence of hydrostatic pressure on striated muscle, with special reference to the role of viscosity changes. *J. cell. comp. Physiol.*, **1**, 11–36.

CERTES, A. (1884a). Note relative à l'action des hautes pressions sur la vitalité des microorganismes d'eau douce et d'eau de mer. *C. r. Séanc. Soc. Biol.*, **36**, 220.

CERTES, A. (1884b). Sur la culture, à l'abri des germes atmosphériques, des eaux et des sédiments rapportés par les expéditions du 'Travailleur' et du 'Talisman', 1882–1883. *C. r. hebd. Séanc. Acad. Sci., Paris*, **98**, 690.

CHLOPIN, G. W. and TAMMANN, C. E. (1903). Über den Einfluß hoher Drucke auf Mikroorganismen. *Z. Hyg. InfektKrankh.*, **45**, 171–204.

COMINGS, E. W. (1956). *High Pressure Technology*, McGraw-Hill, New York.

DAVIES, P. A. (1926). Effect of high pressure on the germination of seeds (*Medicago sativa* and *Melilotus alba*). *J. gen. Physiol.*, **9**, 805–809.

DAVIES, P. A. (1928a). High pressure and seed germination. *Am. J. Bot.*, **15**, 149–156.

DAVIES, P. A. (1928b). The effect of high pressure on the percentage of soft and hard seeds of *Medicago sativa* and *Melilotus alba*. *Am. J. Bot.*, **15**, 433–436.

DIGBY, P. S. B. (1961). Mechanism of sensitivity to hydrostatic pressure in the prawn, *Palaemonetes varians* LEACH. *Nature, Lond.*, **191**, 366–368.

DIGBY, P. S. B. (1965). Semi-conduction and electrode processes in biological material. I. Crustacea and certain soft-bodied forms. *Proc. R. Soc.* (*B*), **161**, 504–525.

DIGBY, P. S. B. (1967). Pressure sensitivity and its mechanism in the shallow marine environment. *Symp. zool. Soc. Lond.*, **19**, 159–188.

DIJKGRAAF, S. (1940). Über die Bedeutung der Weberschen Knöchel für die Wahrnehmung von Schwankungen des hydrostatischen Druckes. *Z. vergl. Physiol.*, **28**, 389–401.

DISTÈCHE, A. (1959). pH measurements with a glass electrode withstanding 1500 kg/cm² hydrostatic pressure. *Rev. scient. Instrum.*, **30**, 474–478.

DISTÈCHE, A. (1962). Electrochemical measurements at high pressure. *J. electrochem. Soc.*, **109**, 1084–1092.

DISTÈCHE, A. and DISTÈCHE, S. (1965). The effect of pressure on pH and dissociation constants from measurements with buffered and unbuffered electrode cells. *J. electrochem. Soc.*, **112**, 350–354.

DRAPER, J. W. and EDWARDS, D. J. (1932). Some effects of high pressure on developing marine forms. *Biol. Bull. mar. biol. Lab., Woods Hole*, **63**, 99–107.

EBBECKE, U. (1914). Wirkung allseitiger Kompression auf den Froschmuskel. *Pflügers Arch. ges. Physiol.*, **157**, 79–116.

EBBECKE, U. (1935a). Das Verhalten von Paramäcien unter der Einwirkung hohen Druckes. *Pflügers Arch. ges. Physiol.*, **236**, 658–661.

EBBECKE, U. (1935b). Kompressionsverkürzung und idiomuskuläre Kontraktion und die Beziehung zwischen elektrischer und mechanischer Reizung. *Pflügers Arch. ges. Physiol.*, **236**, 662–668.

EBBECKE, U. (1936b). Über das Verhalten der Querstreifung und des Muskelspektrums bei der Kompressionsverkürzung. *Pflügers Arch. ges. Physiol.*, **238**, 749–757.

EBBECKE, U. (1944). Lebensvorgänge unter der Einwirkung hoher Drücke. *Ergebn. Physiol.*, **45**, 34–183.

EBBECKE, U. and HASENBRING, O. (1935a). Über die Kompressionsverkürzung des Muskels bei Einwirkung hoher Drücke. *Pflügers Arch. ges. Physiol.*, **236**, 405–415.

EBBECKE, U. and HASENBRING, O. (1935b). Über die Wirkungen hoher Drücke auf marine Lebewesen. *Pflügers Arch. ges. Physiol.*, **236**, 648–657.

EBBECKE, U. and SCHÄFER, H. (1935). Über den Einfluß hoher Drücke auf den Aktionsstrom von Muskeln und Nerven. *Pflügers Arch. ges. Physiol.*, **236**, 678–692.

EDWARDS, D. J. and CATTELL, McK. (1928). The stimulating action of hydrostatic pressure on cardiac function. *Am. J. Physiol.*, **84**, 472–484.

EKMAN, S. (1954). Betrachtungen über die Fauna der abyssalen Ozeanböden. *Un. int. Sci. biol.* (*B*), **16**, 5–11.

EMERSON, R., CHALMERS, R. and CEDERSTRAND, G. (1957). Some factors influencing the long-wave limit of photosynthesis. *Proc. natn. Acad. Sci. U.S.A.*, **43**, 133–143.

ENRIGHT, J. T. (1961). Pressure sensitivity of an amphipod. *Science, N.Y.*, **133**, 758–760.

ENRIGHT, J. T. (1962). Responses of an amphipod to pressure changes. *Comp. Biochem. Physiol.*, **7**, 131–145.

EWALD, A. H. and HAMANN, S. D. (1956). The effect of pressure on complex ion equilibria. *Aust. J. Chem.*, **9**, 54–60.

EYRING, H., JOHNSON, F. H. and GENSLER, R. L. (1946). Pressure and reactivity of proteins, with particular reference to invertase. *J. phys. Chem., Ithaca*, **50**, 453–464.

FAIRBRIDGE, R. W. (Ed.) (1966). *The Encyclopedia of Oceanography*, Reinhold, New York.

FELL, J. (1967). Distribution of yeasts in the Indian Ocean. *Bull. mar. Sci.*, **17**, 454–470.

FERLING, E. (1957). Die Wirkungen des erhöhten hydrostatischen Druckes auf Wachstum und Differenzierung submerser Blütenpflanzen. *Planta*, **49**, 235–270.

FISHER, R. L. and HESS, H. H. (1963). Trenches. In M. N. Hill (Ed.), *The Sea*, Vol. III. Wiley, New York. pp. 411–436.

FLÜGEL, H. and SCHLIEPER, C. (1970). The effects of pressure on marine invertebrates and fishes. In A. M. Zimmerman (Ed.), *High Pressure Effects on Cellular Processes*. Academic Press, New York. pp. 211–234.

FOLK, G. E. (1966). *Introduction to Environmental Physiology*, Lea and Febiger, Philadelphia, Pa.

FONTAINE, M. (1928). Les fortes pressions et la consommation d'oxygène de quelques animaux marins. Influences de la taille de l'animal. *C.r. Séanc. Soc. Biol.*, **99**, 1789–1790.

FONTAINE, M. (1929a). De l'augmentation de la consommation d'O₂ des animaux marins sous l'influence des fortes pressions. Ses variations en fonction de l'intensité de la compression. *C.r. hebd. Séanc. Acad. Sci.*, *Paris*, **188**, 460–461.

FONTAINE, M. (1929b). De l'augmention de la consommation d'oxygène des animaux marins sous l'influence des fortes pressions. Ses variations en fonction de la durée de la compression. *C.r. hebd. Séanc. Acad. Sci.*, *Paris*, **188**, 662–663.

FONTAINE, M. (1929c). De l'influence comparée de la pression sur la respiration et la photosynthese des algues. *C.r. Séanc. Soc. Biol.*, **100**, 912–914.

FONTAINE, M. (1930). Recherches expérimentales sur les réactions des êtres vivants aux fortes pressions. *Annls Inst. océanogr.*, *Monaco*, **8**, 1–99.

FRASER, D. and JOHNSON, F. H. (1951). The pressure–temperature relationship in the rate of casein digestion by trypsin. *J. biol. Chem.*, **190**, 417–421.

GAERTNER, A. (1967). Ökologische Untersuchungen an einem marinen Pilz aus der Umgebung von Helgoland. *Helgoländer wiss. Meeresunters.*, **15**, 181–192.

GAERTNER, A. (1969). Marine niedere Pilze in Nordsee und Nordatlantik. *Ber. dt. bot. Ges.*, **82**, 287–306.

GAERTNER, A. (1970). Einiges zur Kultur mariner niederer Pilze. *Helgoländer wiss. Meeresunters.*, **20**, 29–38.

GERBER, B. R. and NOGUCHI, H. (1967). Volume change associated with the G-F transformation of flagellan. *J. molec. Biol.*, **26**, 197–210.

GERTHSEN, H. O. (1966). *Physik. Ein Lehrbuch zum Gebrauch neben Vorlesungen*, 9th ed. (Bearbeitet und ergänzt von H. O. KNESER) Springer, Berlin.

GESSNER, F. (1955). *Hydrobotanik*, Vol. I. Energiehaushalt. Deutscher Verlag der Wissenschaften, Berlin.

GESSNER, F. (1961). Hydrostatischer Druck und Pflanzenwachstum. *Handb. PflPhysiol.*, **16**, 668–690.

GILL, S. J. and GLOGOVSKY, R. L. (1964). Apparatus for the measurement of optical rotation of the solution at high pressure. *Rev. scient. Instrum.*, **35,**, 1281–1283.

GILL, S. J. and GLOGOVSKY, R. L. (1965). Influence of pressure on the reversible unfolding of ribonuclease and poly-y-benzyl-L-glutamate. *J. phys. Chem.*, *Ithaca*, **69**, 1515–1519.

GILL, S. J. and RUMMEL, S. D. (1961). Teflon and sapphire cell for optical absorption studies under high pressure. *Rev. scient. Instrum.*, **32**, 752.

GILL, S. S. (Ed.) (1970). *The Stress Analysis of Pressure Vessels and Pressure Vessel Components*, Pergamon Press, Oxford.

GILLEN, R. G. (1971). The effect of pressure on muscle lactate dehydrogenase activity of some deep-sea and shallow-water fishes. *Mar. Biol.*, **8**, 7–11.

GRAY, J. (1928). *Ciliary Movement*, Cambridge University Press, Cambridge.

GROH, G. (1963). Über den Druckeinfluß auf das elektrische Leitvermögen von Meerwasser. *Z. angew. Physik*, **15**, 181–184.

GROSS, J. A. (1965). Pressure-induced color mutation of *Euglena gracilis. Science*, *N.Y.*, **147**, 741–742.

GUYOT, M. (1960). Analyse par les haute pressions de quelques functions physiologiques des végétaux et des animaux. Thèse, Université de Poitiers.

HAIGHT, J. J. and MORITA, R. Y. (1966). Some physiological differences in *Vibrio marinus* grown at environmental and optimal temperatures. *Limnol. Oceanogr.*, **11**, 470–474.

HAIGHT, R. D. and MORITA, R. Y. (1962). Interaction between the parameters of hydrostatic pressure and temperature on aspartase of *Escherichia coli. J. Bact.*, **83**, 112–120.

HAMANN, S. D. (1957). *Physico-Chemical Effects of Pressure*, Academic Press, New York.

HAMANN, S. D. (1963a). Chemical equilibria in condensed systems. In R. S. Bradley (Ed.), *High Pressure Physics and Chemistry*, Vol. II, Academic Press, New York. pp. 131–162.

HAMANN, S. D. (1963b). Chemical kinetics. In R. S. Bradley (Ed.), *High Pressure Physics and Chemistry*, Vol. II. Academic Press, New York. pp. 163–207.

HAMANN, S. D. (1963c). The ionization of water at high pressures *J. phys. Chem., Ithaca*, **67**, 2233–2235.

HAMANN, S. D. (1964). High pressure chemistry. *A. Rev. phys. Chem.*, **15**, 349–370.

HAMANN, S. D. and STRAUSS, W. (1965). The chemical effects of pressure. 3. Ionization constants at pressures up to 1200 atm. *Trans. Faraday Soc.*, **51**, 1684–1690.

HAMON, B. V. (1956). The effect of pressure on the electrical conductivity of sea-water. *J. mar. Res.*, **16**, 83–89.

HANSON, P. P., ZENKEVITCH, N. L., SERGEEV, U. V. and UDINTSEV, G. B. (1959). Maximum depths of the Pacific Ocean. (Russ). *Priroda*, **6**, 84–88.

HARDY, A. C. (1956). *The Open Sea: Its Natural History*, Pt 1—The World of Plankton. Collins, London.

HARDY, A. C. and BAINBRIDGE, R. (1951). Effect of pressure on the behaviour of decapod larvae. *Nature, Lond.*, **168**, 327–328.

HARDY, A. C. and PATON, W. N. (1947). Experiments on the vertical migration of plankton animals. *J. mar. biol. Ass. U.K.*, **26**, 467–526.

HATSCHEK, E. H. (1958). Viscosity. In *Encyclopaedia Britannica*, Vol. XXIII. William Benton, Chicago. pp. 194–199.

HEDÉN, C. -G. (1964). Effects of hydrostatic pressure on microbial systems. *Bact. Rev.*, **28**, 14–29.

HEDÉN, C. -G. and MALMBORG, A. S. (1961). Aeration under pressure and the question of free radicals. *Sci. Rep. Ist. Super. Sanita*, **1**, 213–221.

HEILBRUNN, L. V. (1952). *An Outline of General Physiology*, Saunders, Philadelphia, Pa.

HERMOLIN, J. and ZIMMERMAN, A. M. (1969). The effect of pressure on synchronous cultures of *Tetrahymena*: a ribosomal study. *Cytobios*, **1**, 247–256.

HILL, E. P. (1962). Some effects of hydrostatic pressure on growth and mitochondria of *Allomyces macrogynus*. Ph.D. thesis, University of Nebraska, Lincoln.

HILL, E. P. and MORITA, R. Y. (1954). Dehydrogenase activity under hydrostatic pressure by isolated mitochondria obtained from *Allomyces macrogynus*. *Limnol. Oceanogr.*, **9**, 243–248.

HIRSCHFELDER. J. O., CURTISS, C. F. and BIRD, R. B. (1954). *Molecular Theory of Gases and Liquids*, Wiley, New York.

HITE, H., GIDDINGS, N. J. and WEAKLEY, C. E. (1914). The effect of pressure on certain microorganisms encountered in the preservation of fruits and vegetables. *Bull. W. Va Univ. agric. Exp. Stn*, **146**, 1–67.

HOCHACHKA, P. W., SCHNEIDER, D. E. and KUZNETSOV, A. (1970). Interacting pressure and temperature effects on enzymes of marine poikilotherms: Catalytic and regulatory properties of FDPase from deep and shallow-water fishes. *Mar. Biol.*, **7**, 285–293.

HÖBER, R. (1947). *Physikalische Chemie der Zellen und Gewebe*, Stämpfli, Bern.

HÖHNK, W. (1956). Studien zur Brack- und Seewassermykologie VI. Über die pilzliche Besiedlung verschieden salziger submerser Standorte. *Veröff. Inst. Meeresforsch. Bremerh.*, **4**, 195–213.

HOLLENBERG, G. and ABBOT, I. (1966). *Supplement to Smith's Marine Algae of the Monterey Peninsula*, Stanford University Press, Stanford, California.

HORNE, R. A. (1969). *Marine Chemistry*, Wiley, New York.

HORNE, R. A. and FRYSINGER, G. R. (1963). The effect of pressure on the electrical conductivity of sea water. *J. geophys. Res.*, **68**, 1967–1973.

HORNE, R. A. and JOHNSON, D. S. (1966). The viscosity of water under pressure. *J. phys. Chem., Ithaca*, **70**, 2182–2190.

HORNE, R. A. and JOHNSON, D. S. (1967). The effect of electrolyte addition on the viscosity of water under pressure. *J. phys. Chem., Ithaca*, **71**, 1147–1149.

HOSHAW, R. W. (1965). Mating types of *Chlamydomonas* from the collection of Gilbert M. Smith. *J. Phycol.*, **1**, 194–196.

JOHNSON, F. H. (Ed.) (1957a). *Influence of Temperature on Biological Systems*, American Physiological Society, Washington, D.C.

JOHNSON, F. H. (1957b). The action of pressure and temperature. In R. E. O. Williams and C. C. Spicer (Eds), *Microbial Ecology*. Universities Press, Cambridge. pp. 134–167.

JOHNSON, F. H., BAYLOR, M. B. and FRASER, D. (1948a). The thermal denaturation of tobacco mosaic virus in relation to hydrostatic pressure. *Archs Biochem.*, **19**, 237–245.

JOHNSON, F. H., BROWN, D. E. and MARSLAND, D. A. (1942a). A basic mechanism in the biological effects of temperature, pressure and narcotics. *Science, N.Y.*, **95**, 200–203.

JOHNSON, F. H., BROWN, D. E. and MARSLAND, D. A. (1942b). Pressure reversal of the action of certain narcotics. *J. cell. comp. Physiol.*, **20**, 269–276.

JOHNSON, F. H. and CAMPBELL, D. H. (1945). The retardation of protein denaturation by hydrostatic pressure. *J. cell. comp. Physiol.*, **26**, 43–46.

JOHNSON, F. H. and CAMPBELL, D. H. (1946). Pressure and protein denaturation. *J. biol. Chem.*, **163**, 689–698.

JOHNSON, F. H. and EYRING, H. (1948). The fundamental action of pressure, temperature, and drugs on enzymes, as revealed by bacterial luminescence. *Ann. N.Y. Acad. Sci.*, **49**, 376–396.

JOHNSON, F. H. and EYRING, H. (1970). The kinetic basis of pressure effects in biology and chemistry. In A. M. Zimmerman (Ed.), *High Pressure Effects on Cellular Processes*. Academic Press, New York. pp. 1–44.

JOHNSON, F. H., EYRING, H. and POLISSAR, M. J. (1954). *The Kinetic Basis of Molecular Biology*, Wiley, New York.

JOHNSON, F. H., EYRING, H., STEBLAY, R., CHAPLIN, H., HUBER, C. and GHERARDI, G. (1945). The nature and control of reactions in bioluminescence with special reference to the mechanism of reversible and irreversible inhibitions by hydrogen and hydroxyl ions, temperature, pressure, alcohol, urethane, and sulfanilamide in bacteria. *J. gen. Physiol.*, **28**, 463–537.

JOHNSON, F. H. and FLAGLER, E. A. (1951). Activity of narcotized amphibian larvae under hydrostatic pressure. *J. cell. comp. Physiol.*, **37**, 15–25.

JOHNSON, F. H., KAUZMANN, W. J. and GENSLER, R. L. (1948b). The urethan inhibition of invertase activity in relation to hydrostatic pressure. *Archs Biochem.*, **19**, 229–236.

JOHNSON, F. H. and LEWIN, I. (1946a). The disinfection of *E. coli* in relation to temperature, hydrostatic pressure and quinine. *J. cell. comp. Physiol.*, **28**, 23–45.

JOHNSON, F. H. and LEWIN, I. (1946b). The influence of pressure, temperature and quinine on the rates of growth and disinfection of *E. coli* in the logarithmic growth phase. *J. cell. comp. Physiol.*, **28**, 77–97.

JOHNSON, T. W., JR. and SPARROW, F. K., JR. (1961). *Fungi in Oceans and Estuaries*, J. Cramer, Weinheim.

JONES, D. A. and NAYLOR, E. (1970). The swimming rhythm of the sand beach isopod *Eurydice pulchra*. *J. exp. mar. Biol. Ecol.*, **4**, 188–199.

KAO, C. Y. and CHAMBERS, R. (1954). The internal hydrostatic pressure of the *Fundulus* egg. I. The activated egg. *J. exp. Biol.*, **31**, 139–149.

KENNEDY, G. C. and LaMORI, P. N. (1961). Pressure calibration about 25 kilobars. In F. P. Bundy, W. R. Hibbard and H. M. Strong (Eds), *Progress in Very High Pressure Research*, Wiley, New York. pp. 304–313.

KENNINGTON, G. S. (1961). The influence of temperature and atmospheric pressure on the rate of oxygen uptake in *Tribolium confusum*. *Ecology*, **42**, 212–215.

KETTMAN, M. S., NISHIKAWA, A. H., MORITA, R. Y. and BECKER, R. R. (1965). Effect of hydrostatic pressure on the aggregation reaction of poly-L-valyl-ribonuclease. *Biochem. biophys. Res. Commun.*, **22**, 262–267.

KITCHING, J. A. (1954). The effects of high hydrostatic pressure on a suctorian. *J. exp. Biol.*, **31**, 56–57.

KITCHING, J. A. (1957a). Effects of high hydrostatic pressures on the activity of flagellates and ciliates. *J. exp. Biol.*, **34**, 494–510.

KITCHING, J. A. (1957b). Effects of high hydrostatic pressures on *Actinophrys sol* (Heliozoa). *J. exp. Biol.*, **34**, 511–517.

KNIGHT-JONES, E. W. and MORGAN, E. (1966). Responses of marine animals to changes in hydrostatic pressure. *Oceanogr. mar. Biol. A. Rev.*, **4**, 267–299.

KNIGHT-JONES, E. W. and QASIM, S. Z. (1955). Responses of some marine plankton animals to changes in hydrostatic pressure. *Nature, Lond.*, **175**, 941–942.

KORRINGA, P. (1947). Relations between the moon and periodicity in the breeding of marine animals. *Ecol. Monogr.*, **17**, 347–381.

KORRINGA, P. (1957). Lunar periodicity. In J. W. Hedgpeth (Ed.), *Treatise on Marine Ecology and Paleoecology*, Vol. I, Ecology. *Mem. geol. Soc. Am.*, **67**, 917–934.

KRISS, A. E. (1963). *Marine Microbiology (Deep Sea)*. (Transl. by J. M. Shewan and Z. Kabata) Oliver and Boyd, London.

KROGH, A. (1959). *The Comparative Physiology of Respiratory Mechanisms*, University of Pennsylvania Press, Philadelphia.

KUHN, O. and STROTKOETTER, E. (1967). Untersuchungen von Drucken auf den tierischen Körper. I. Druckreception bei Fischen und ihre Mitwirkung bei der Orientierung im Raum. *Forsch. Ber. Landes N Rhein-Westf.*, **1857**, 1–57.

LAIDLER, K. J. (1951). The influence of pressure on the rates of biological reactions. *Archs Biochem.*, **30**, 226–236.

LAMANNA, C. and MALLETTE, M. F. (1959). *Basic Bacteriology*, 2nd ed., Williams and Wilkins, Baltimore, Maryland.

LANDAU, J. V. (1966). Protein and nucleic acid synthesis in *Escherichia coli*: Pressure and temperature effects. *Science, N.Y.*, **153**, 1273–1274.

LANDAU, J. V. (1970). Hydrostatic pressure on the biosynthesis of macromolecules. In A. M. Zimmerman (Ed.), *High Pressure Effects on Cellular Processes*. Academic Press, New York. pp. 45–70.

LANDAU, J. V. and MARSLAND, D. A. (1952). Temperature–pressure studies on the cardiac rate in tissue culture explants from the heart of the tadpole (*Rana pipiens*). *J. cell. comp. Physiol.*, **40**, 367–382.

LANDAU, J. V. and PEABODY, R. A. (1963). Endogenous adenosine triphosphate levels in human amnion cells during application of high hydrostatic pressure. *Expl Cell Res.*, **29**, 54–60.

LANDAU, J. V. and THIBODEAU, L. (1962). The micromorphology of *Amoeba proteus* during pressure-induced changes in the sol-gel cycle. *Expl Cell Res.*, **27**, 591–594.

LANDAU, J. V., ZIMMERMAN, A. M. and MARSLAND, D. A. (1954). Temperature–pressure experiments on *Amoeba proteus*: Plasmagel structure in relation to form and movement. *J. cell. comp. Physiol.*, **44**, 211–232.

LARSON, W. P., HARTZELL, T. B. and DIEHL, H. S. (1918). The effect of high pressures on bacteria. *J. infect. Dis.*, **22**, 271–279.

LAWSON, A. W. and HUGHES, A. J. (1963). High pressure properties of water. In R. S. Bradley (Ed.), *High Pressure Physics and Chemistry*, Vol. I, Academic Press, New York. pp. 207–225.

LAWSON, A. W., LOWELL, R. and JAIN, A. L. (1959). Thermal conductivity of water at high pressure. *J. chem. Phys.*, **30**, 643–647.

LETTS, P. J. (1969). *In vitro* studies of protein synthesis in *Tetrahymena pyriformis*: A pressure study. M.Sc. thesis, University of Toronto.

LINCOLN, R. J. (1970). A laboratory investigation into the effects of hydrostatic pressure on the vertical migration of planktonic Crustacea. *Mar. Biol.*, **6**, 5–11.

LINCOLN, R. J. and GILCHRIST, I. (1970). An observational pressure vessel for studying the behaviour of planktonic animals. *Mar. Biol.*, **6**, 1–4.

LINDERSTRØM-LANG, K. and JACOBSEN, C. F. (1941). The contraction accompanying enzymatic breakdown of proteins. *C.r. Trav. Lab. Carlsberg (Ser. Chim.)*, **24**, 1–46.

LOWE, L. (1968). The effects of hydrostatic pressure on synchronized *Tetrahymena*. Ph.D. thesis, University of Toronto.

LUYET, B. (1937). Sur le méchanisme de la mort cellulaire par les hautes pressions; l'intensité et la durée des pressions lethales pour la levure. *C.r. hebd. Séanc. Acad. Sci., Paris*, **204**, 1214–1215.

McCutcheon, F. H. (1958). Swimbladder volume, buoyancy, and behaviour in the pinfish, *Lagodon rhomboides* (Linn.). *J. cell. comp. Physiol.*, **52**, 453–480.

MacDonald, A. G. (1965). The effect of high hydrostatic pressure on the oxygen consumption of *Tetrahymena pyriformis*. *Expl Cell Res.*, **40**, 78–84.

MacDonald, A. G. (1967). The effect of high hydrostatic pressure on the cell division and growth of *Tetrahymena pyriformis*. *Expl Cell Res.*, **47**, 569–580.

MacDonald, A. G. and Gilchrist, I. (1969). *The Physiological Problems and Equipment for the Recovery and Study of Deep Sea Animals*, Oceanology International, Brighton, England.

Madsen, F. J. (1961). On the zoogeography and origin of the abyssal fauna in view of the knowledge of the Porcellanasteridea. *Galathea Rep.*, **4**, 177–218.

Marsland, D. A. (1936). The cleavage of *Arbacia* eggs under hydrostatic compression. *Anat. Rec.*, **67**, 38.

Marsland, D. A. (1938). The effects of high hydrostatic pressure upon cell division in *Arbacia* eggs. *J. cell. comp. Physiol.*, **12**, 57–70.

Marsland, D. A. (1939). The mechanism of protoplasmic streaming. The effects of hydrostatic pressure upon cyclosin in *Elodea canadensis*. *J. cell. comp. Physiol.*, **13**, 23–30.

Marsland, D. A. (1950). The mechanisms of cell division : temperature–pressure experiments on the cleaving eggs of *Arbacia punctulata*. *J. cell. comp. Physiol.*, **36**, 205–227.

Marsland, D. A. (1957). Temperature–pressure studies on the role of sol-gel reactions in cell division. In F. H. Johnson (Ed.), *The Influence of Temperature on Biological Systems*. American Physiological Society, Washington, D.C. pp. 111–126.

Marsland, D. A. (1958). Cells at high pressure. *Scient. Am.*, **199**, 36–43.

Marsland, D. A. (1970). Pressure–temperature studies on the mechanisms of cell division. In A. M. Zimmerman (Ed.), *High Pressure Effects on Cellular Processes*. Academic Press, New York. pp. 259–312.

Marsland, D. A. and Brown, D. E. S. (1936). Amoeboid movement at high hydrostatic pressure. *J. cell. comp. Physiol.*, **8**, 167–178.

Marsland, D. A. and Brown, D. E. S. (1942). The effects of pressure on sol-gel equilibria, with special reference to myosin and other protoplasmic gels. *J. cell. comp. Physiol.*, **20**, 295–305.

Mazia, D. and Zimmerman, A. M. (1958). SH compounds in mitosis. II. The effect of mercapto-ethanol on the structure of the mitotic apparatus in sea urchin eggs. *Expl Cell Res.*, **15**, 138–153.

Menzies, R. J. (1965). Conditions for the existence of life on the abyssal sea floor. *Oceanogr. mar. biol. A. Rev.*, **3**, 195–210.

Menzies, R. J. and Wilson, J. B. (1961). Preliminary field experiments on the relative importance of pressure and temperature on the penetration of marine invertebrates into the deep sea. *Oikos*, **12**, 302–309.

Metcalf, W. G. and Stalcup, M. C. (1969). Current meter and hydrographic station data from *Crawford* cruise no. 165 in the tropical Atlantic Ocean, February–April 1968. Unpubl. MS Ser. Woods Hole, Ref. No. 69–72.

Meyers, S. P., Ahearn, D. G., Gunkel, W. and Roth, F. J., Jr. (1967). Yeasts from the North Sea. *Mar. Biol.*, **1**, 118–123.

Miyagawa, K. and Suzuki, K. (1963a). Pressure inactivation of enzyme : Some kinetic aspects of pressure inactivation of trypsin. *Rev. phys. Chem. Japan*, **32**, 43–50.

Miyagawa, K. and Suzuki, K. (1963b). Pressure inactivation of enzyme: Some kinetic aspects of pressure inactivation of chymotrypsin. *Rev. phys. Chem. Japan*, **32**, 51–56.

Miyatake, T. (1957). Studies on effects of high hydrostatic pressure on blood cells. II. On the exchange of ions in the erythrocyte. (Japan.) *J. Okayama med. Soc.*, **69**, 461–471.

Möhres, F. P. (1940). Untersuchungen über die Frage der Wahrnehmung von Druckunterschieden des Mediums (Versuche an Bodenfischen). *Z. vergl. Physiol.*, **28**, 1–42.

Morgan, E. (1965). The activity rhythm of the amphipod *Corophium volutator* (Pallas) and its possible relationship to changes in hydrostatic pressure associated with the tides. *J. Anim. Ecol.*, **34**, 731–746.

Morgan, E., Nelson-Smith, A. and Knight-Jones, E. W. (1964). Responses of *Nymphon gracile* (Pycnogonida) to pressure cycles of tidal frequency. *J. exp. Biol.*, **41**, 825–836.

MORITA, R. Y. (1957a). Ammonia production from various substrates by previously pressurized cells of *Escherichia coli*. *J. Bact.*, **74**, 231–233.

MORITA, R. Y. (1957b). Effect of hydrostatic pressure on succinic, formic, and malic dehydrogenases in *Escherichia coli*. *J. Bact.*, **74**, 251–255.

MORITA, R. Y. (1965). The physical environment for fungal growth. 2. Hydrostatic pressure. In G. C. Ainsworth and A. S. Sussman (Eds), *The Fungi*, Vol. I. Academic Press, New York. pp. 551–556.

MORITA, R. Y. (1967). Effects of hydrostatic pressure on marine microorganisms. *Oceanogr. mar. Biol. A. Rev.*, **5**, 187–203.

MORITA, R. Y. (1970). Application of hydrostatic pressure to microbial cultures. In J. R. Norris and D. W. Ribbons (Eds), *Methods in Microbiology*. Academic Press, London. pp. 243–257.

MORITA, R. Y. and ALBRIGHT, L. J. (1965). Cell yields of *Vibrio marinus*, an obligate psychrophile, at low temperature. *Can. J. Microbiol.*, **11**, 221–227.

MORITA, R. Y. and BECKER, R. R. (1970). Hydrostatic pressure effects on selected biological systems. In A. M. Zimmerman (Ed.), *High Pressure Effects on Cellular Processes*. Academic Press, New York. pp. 71–83.

MORITA, R. Y. and HAIGHT, R. D. (1962). Malic dehydrogenase activity at 101° C under hydrostatic pressure. *J. Bact.*, **83**, 1341–1346.

MORITA, R. Y. and HOWE, R. A. (1957). Phosphatase activity by bacteria under hydrostatic pressure. *Deep Sea Res.*, **4**, 254–258.

MORITA, R. Y. and MATHEMEIER, P. F. (1964). Temperature–hydrostatic pressure studies on partially purified inorganic pyrophosphatase activity. *J. Bact.*, **88**, 1667–1671.

MORITA, R. Y. and ZOBELL, C. E. (1955). Occurrence of bacteria in pelagic sediments collected during the Mid-Pacific Expedition. *Deep Sea Res.*, **3**, 66–73.

MORITA, R. Y. and ZOBELL, C. E. (1956). Effect of hydrostatic pressure on the succinic dehydrogenase system in *Escherichia coli*. *J. Bact.*, **71**, 668–672.

MUNRO, D. C. (1963). Production and measurement of high pressures. In R. S. Bradley (Ed.). *High Pressure Physics and Chemistry*, Vol. I. Academic Press, New York. pp. 11–49.

MURAKAMI, T. H. (1958). Hydrostatic pressure effects on the adenosinetriphosphatase activities. (Japan.) *Symp. cell. Chem.*, **8**, 71–77.

MURAKAMI, T. H. (1960). The effects of high hydrostatic pressure on cell division. (Japan.) *Symp. cell. Chem.*, **10**, 233–244.

MURAKAMI, T. H. (1961). The effects of high hydrostatic pressure on cell division-synthesis of nucleic acids. *Symp. cell. Chem.*, **11**, 223–233.

MURAKAMI, T. H. (1963). Effect of high hydrostatic pressure on the permeability of plasma membranes at various temperatures. (Japan.) *Symp. cell. Chem.*, **13**, 147–156.

MURAKAMI, T. H. (1970). Japanese studies on hydrostatic pressure. In A. M. Zimmerman (Ed.), *High Pressure Effects on Cellular Processes*. Academic Press, New York. pp. 131–138.

MURAKAMI, T. H. and ZIMMERMAN, A. M. (1970). A pressure study of galvanotaxis in *Tetrahymena*. In A. M. Zimmerman (Ed.), *High Pressure Effects on Cellular Processes*. Academic Press, New York. pp. 139–153.

NAROSKA, V. (1968). Vergleichende Untersuchungen über den Einfluß des hydrostatischen Druckes auf Überlebensfähigkeit und Stoffwechselintensität mariner Evertebraten und Teleosteer. *Kieler Meeresforsch.*, **24**, 95–123.

NEUMANN, G. (1968). *Ocean Currents*, Elsevier, Amsterdam.

NEWITT, D. M. (1940). *High Pressure Plant and Fluids at High Pressure*, Oxford University Press, London.

NOGUCHI, H. and YANG, J. T. (1962). Dilatometric and refractometric studies of the helix-coil transition of poly-L-glutamic acid in aqueous solution. *Biopolymers*, **1**, 359–370.

OKADA, K. (1954a). Effects of high hydrostatic pressure on the permeability of plasma membranes. I. Resting current of muscle. (Japan.) *J. Okayama med. Soc.*, **66**, 2071–2075.

OKADA, K. (1954b). Physiological studies of hydrostatic high pressure. Supplement. II. On pH found in a neutral red solution. (Japan.) *J. Okayama med. Soc.*, **66**, 2105–2110.

OKADA, K. (1954c). Effects of high hydrostatic pressure on the permeability of plasma membranes. III. On salt current. (Japan.) *J. Okayama med. Soc.*, **66**, 2083–2088.

OKADA, K. (1954d). Effects of high hydrostatic pressure on the permeability of plasma membranes. IV. On electric conductivity. (Japan.) *J. Okayama med. Soc.*, **66**, 2089–2094.

OKADA, K. (1954e). Effects of high hydrostatic pressure on the permeability of plasma membranes. V. On plasmolysis. (Japan.) *J. Okayama med. Soc.*, **66**, 2095–2099.

OPPENHEIMER, C. H. and ZOBELL, C. E. (1952). The growth and viability of sixty-three species of marine bacteria as influenced by hydrostatic pressure. *J. mar. Res.*, **11**, 10–18.

OWEN, B. B. and BRINKLEY, S. R. (1941). Calculation of the effect of pressure upon ionic equilibria in pure water and salt solutions. *Chem. Rev.*, **29**, 461–473.

PEASE, D. C. (1946). Hydrostatic pressure effects on the spindle figure and chromosome movement. II. *Biol. Bull. mar. biol. Lab., Woods Hole*, **91**, 145–169.

PEASE, D. C. and KITCHING, J. A. (1939). The influence of hydrostatic pressure upon ciliary frequency. *J. cell. comp. Physiol.*, **14**, 135–142.

PEASE, D. C. and MARSLAND, D. A. (1939). The cleavage of *Ascaris* eggs under exceptionally high pressure. *J. cell. comp. Physiol.*, **14**, 407–408.

PLANCK, M. A. (1887). Über das Prinzip der Vermehrung der Entropie. *Ann. Phys. Chem.*, **32**, 462–503.

POLLARD, E. C. and WELLER, P. K. (1966). The effect of hydrostatic pressure on the synthetic processes in bacteria. *Biochim. biophys. Acta*, **112**, 573–580.

PONAT, A. (1967). Untersuchungen zur zellulären Druckresistenz verschiedener Evertebraten der Nord- und Ostsee. *Kieler Meeresforsch.*, **23**, 21–47.

PONAT, A. and THEEDE, H. (1967). Die pH-Abhängigkeit der zellulären Druckresistenz bei *Mytilus edulis*. *Helgoländer wiss. Meeresunters.*, **16**, 231–237.

PRINGSHEIM, E. G. (1951). Methods for the cultivation of algae. In G. S. Smith (Ed.), *Manual of Phycology*. Chronica Botanica, Waltham, Mass.

PYTKOWICZ, R. M. and CONNERS, D. N. (1964). High pressure solubility of calcium carbonate in sea water. *Science, N.Y.*, **144**, 840–841.

QUTOB, Z. (1962). The swimbladder of fishes as a pressure receptor. *Archs néerl. Zool.*, **15**, 1–67.

REGNARD, P. (1884a). Note sur les conditions de la vie dans les profondeurs de la mer. *C.r. Séanc. Soc. Biol.*, **36**, 164–168.

REGNARD, P. (1884b). Note relative à l'action des hautes pressions sur phénomènes vitaux (mouvement des cils vibratiles, fermentation). *C.r. Séanc. Soc. Biol.*, **36**, 187–188.

REGNARD, P. (1884c). Effect des hautes pressions sur les animaux marins. *C.r. Séanc. Soc. Biol.*, **36**, 394–395.

REGNARD, P. (1884d). Recherches expérimentales sur l'influence des très hautes pressions sur les organismes vivants. *C.r. hebd. Séanc. Acad. Sci., Paris*, **98**, 745–747.

REGNARD, P. (1885). Phénomènes objecktifs que l'on peut observer sur les animaux soumis aux hautes pressions. *C.r. Séanc. Soc. Biol.*, **37**, 510–515.

REGNARD, P. (1891). *Recherches Expérimentales sur les Conditions Physiques de la Vie dans les Eaux*, Masson, Paris.

RENNER, O. (1942). Hydrostatic pressure and gas content. *Ber. dt. bot. Ges.*, **60**, 292–294.

RICE, A. L. (1961). Responses of certain mysids to changes in hydrostatic pressure. *J. exp. Biol.*, **38**, 391–401.

RIFKIND, J. and APPLEQUIST, J. (1964). The helix interruption for poly-L-glytamic acid from the pressure dependence of optical rotation. *J. Am. chem. Soc.*, **86**, 4207–4208.

RIVERA, R., POPP, H. W. and DOW, R. B. (1937). The effect of high hydrostatic pressures upon seed germination. *Am. J. Bot.*, **24**, 508–513.

ROBERTSON, W. W. (1963). Spectral measurement in high pressure systems. In H. B. Jonassen and A. Weissberger (Eds), *Techniques of Inorganic Chemistry*, Vol. I. Wiley, New York. pp. 157–172.

ROGERS, H. (1895). Action des hautes pressions sur quelques bactéries. *Archs physiol. norm. pathol. (Ser. 5)*, **7**, 12–17.

SCHERBAUM, O. and ZEUTHEN, E. (1954). Induction of synchronous cell division in mass cultures of *Tetrahymena pyriformis*. *Expl Cell Res.*, **6**, 221–227.

SCHLIEPER, C. (1955). Über die physiologischen Wirkungen des Brackwassers. *Kieler Meeresforsch.*, **11**, 22–23.

SCHLIEPER, C. (1963a). Biologische Wirkungen hoher Wasserdrücke. Experimentelle Tiefsee-Physiologie. *Veröff. Inst. Meeresforsch. Bremerh.* (Sonderbd), **3**, 31–48.

SCHLIEPER, C. (1963b). Neuere Aspekte der biologischen Tiefseeforschung. *Umschau*, **15**, 457–461.

SCHLIEPER, C. (1966). Genetic and nongenetic cellular resistance adaptation in marine invertebrates. *Helgoländer wiss. Meeresunters.*, **14**, 482–502.

SCHLIEPER, C. (1968a). High pressure effects on marine invertebrates and fishes. *Mar. Biol.*, **2**, 5–12.

SCHLIEPER, C. (1968b). Ökologisch-physiologische Untersuchungsmethoden. B. Tiere. In C. Schlieper (Ed.), *Methoden der Meeresbiologischen Forschung*. G. Fischer, Jena. pp. 290–303.

SCHLIEPER, C., FLÜGEL, H. and THEEDE, H. (1967). Experimental investigations of the cellular resistance ranges of marine temperate and tropical bivalves: Results of the Indian Ocean Expedition of the German Research Association. *Physiol. Zool.*, **40**, 345–360.

SCHLIEPER, C. and KOWALSKI, R. (1956). Über den Einfluß des Mediums auf die thermische und osmotische Resistenz des Kiemengewebes der Miesmuschel *Mytilus edulis* L. *Kieler Meeresforsch.*, **12**, 37–45.

SCHMIDT, J. (1929). Introduction to the oceanographical reports including list of stations and hydrographical observations. *Oceanogr. Rep. 'Dana' Exped.*, **1**, 1–87.

SCHNEYER, L. H. (1952). Effects of hydrostatic pressure and selected sulfhydryl inhibitors on salivary amylase. *Archs Biochem. Biophys.*, **41**, 345–363.

SHULYNDIN, A. A. (1967). Some characteristics of high hydrostatic pressure effects on biological objects. (Russ.; Engl. summary.) *Tsitolgiya*, **9**, 1328–1345.

SIE, H., CHANG, J. J. and JOHNSON, F. H. (1958). Pressure-temperature-inhibitor relations in the luminescence of *Chaetopterus variopedatus* and its luminescent secretion. *J. cell. comp. Physiol.*, **52**, 195–225.

SINGARAJAH, K. V. (1966). Pressure sensitivity of the chaetognath *Sagitta setosa*. *Comp. Biochem. Physiol.*, **19**, 475–478.

SMITH, A. H. and LAWSON, A. W. (1954). The velocity of sound in water as a function of temperature and pressure. *J. chem. Phys.*, **22**, 351–359.

SPARROW, F. K., JR. (1937). The occurrence of saprophytic fungi in marine muds. *Biol. Bull. mar. biol. Lab.*, Woods Hole, **73**, 242–248.

SPRING, W. and HOFF, J. H. VAN'T (1887). Über einen Fall durch Druck bewirkter chemischer Zersetzung. *Z. phys. Chem.*, **1**, 227–330.

STANLEY, S. O. and MORITA, R. Y. (1968). Salinity effect on the maximum growth temperature of some bacteria isolated from marine environments. *J. Bact.*, **95**, 169–171.

STEELE, W. A. and WEBB, W. (1963a). Compressibility of liquids. In R. S. Bradley (Ed.), *High Pressure Physics and Chemistry*, Vol. I. Academic Press, New York. pp. 145–162.

STEELE, W. A. and WEBB, W. (1963b). Transport properties of liquids. In R. S. Bradley (Ed.), *High Pressure Physics and Chemistry*, Vol. I. Academic Press, New York. pp. 163–176.

STREHLER, B. L. and JOHNSON, F. H. (1954). The temperature-pressure-inhibitor relationships of bacterial luminescence *in vitro*. *Proc. natn. Acad. Sci. U.S.A.*, **40**, 606–617.

SUZUKI, C. and SUZUKI, K. (1962). The protein denaturation by high pressure. Changes of optical rotation and susceptibility of enzymatic proteolysis with ovalbumin denaturated by pressure. *J. Biochem.*, Tokyo, **52**, 67–71.

SUZUKI, K. and KITAMURA, K. (1960a). Denaturation of hemoglobin under high pressure. I. *Rev. phys. Chem. Japan*, **29**, 81–85.

SUZUKI, K. and KITAMURA, K. (1960b). Denaturation of hemoglobin under high pressure. II. *Rev. phys. Chem. Japan*, **29**, 86–91.

SUZUKI, K., MIYOSAWA, Y. and SUZUKI, C. (1963). Protein denaturation by high pressure. Measurements of turbidity of isoelectric ovalbumin and horse serum albumin under high pressure. *Archs Biochem. Biophys.*, **101**, 225–228.

THEEDE, H. and PONAT, A. (1970). Die Wirkung der Sauerstoffspannung auf die Druckresistenz einiger mariner Wirbelloser. *Mar. Biol.*, **6**, 66–73.

TILNEY, L. G., HIRAMOTO, Y. and MARSLAND, D. A. (1966). Studies on the microtubules in Heliozoa. III. A pressure analysis of the role of these structures in the formation and maintenance of the axopodia of *Actinosphaerium nucleofilum* (BARRETT). *J. Cell Biol.*, **29**, 77–95.

TILNEY, L. G. and MARSLAND, D. A. (1965). The role of microtubules in the formation and maintenance of the axopodia of *Actinosphaerium nucleofilum*; a pressure analysis. *Biol. Bull. mar. biol. Lab., Woods Hole*, **129**, 393.

TIMASHEFF, S. N. (1966). Turbidity as a criterion of coagulation. *J. Colloid Sci.*, **21**, 489–497.

TOKUMOTO, H. (1962). Effects of high hydrostatic pressure on respiration of liver tissue. (Japan.) *J. Okayama med. Soc.*, **74**, 639–648.

TONGUR, V. S. and KASATOCHKIN, V. I. (1950). Reversibility of thermal denaturation of protein under pressure. *Dokl. Akad. Nauk SSSR*, **74**, 553–556.

TONGUR, V. S. and KASATOCHKIN, V. I. (1954). Protein regeneration under pressure. The kinetics and thermodynamics of the process. *Trudȳ vses. Obshch. Fiziol. Biokhim. Farmak.*, **2**, 166–176.

TROSKOLANSKI, A. T. (1960). *Hydrometry. Theory and Practice of Hydraulic Measurements.* (Transl. from Polish.) Pergamon Press, Oxford.

VASILENKO, Th. D. and LIVANOV, M. N. (1936). Oscillographic studies of the reflex function of the swimming bladder in fish. *Bull. Biol. Med. Exp. U.S.S.R.*, **2**, 264–266.

VIDAVER, W. E. (1963). Effects of hydrostatic pressure on induction transients of oxygen evolution. In *Photosynthetic Mechanisms of Green Plants.* National Academy of Sciences Washington, D.C. pp. 726–732. (*Publs natn. Res. Coun., Wash.* 1145.)

VIDAVER, W. E. (1969). Hydrostatic pressure effects of photosynthesis. *Int. Revue ges. Hydrobiol.*, **54**, 697–747.

VIDAVER, W. E. and FRENCH, C. S. (1965). Oxygen uptake and evolution following monochromatic flashes in *Ulva* and an action spectrum for System I. *Pl. Physiol.*, **40**, 7–12.

VIDAVER, W. E. and LUE-KIM, H. (1967). Interactions between hydrostatic pressure and oxygen concentration on the germination of lettuce seeds. *Pl. Physiol.*, **42**, 243–246.

VINOGRADOVA, N. G. (1962). Vertical zonation in the distribution of deep-sea benthic fauna in the ocean. *Deep Sea Res.*, **8**, 245–250.

VODAR, B. and SAUREL, J. (1963). The properties of compressed gases. In R. S. Bradley (Ed.), *High Pressure Physics and Chemistry*, Vol. I. Academic Press, New York. pp. 51–143.

VOIGT, K. (1964). Zum Einfluß des Druckes auf die elektrische Leitfähigkeit des Meerwassers. *Wiss. Z. Karl-Marx-Univ. Lpz.*, **13**, 421–423.

WEALE, K. E. (1967). *Chemical Reactions at High Pressures*, Spon, London.

WEIDA, B. and GILL, S. J. (1966). Pressure effect on DNA transition. *Biochim. biophys. Acta*, **12**, 179–181.

WEIMER, M. S. (1967). Purification and kinetics of gelatinase obtained from an obligately psychrophilic marine vibrio. M.S. thesis, Oregon State University, Corvallis.

WEINSTEIN, I. B., FRIEDMAN, S. M. and OCHOA, M., JR. (1966). Fidelity during translation of the genetic code. *Cold Spring Harb. Symp. quant. Biol.*, **31**, 671–681.

WENTORF, R. H., JR. (Ed.) (1962). *Modern Very High Pressure Techniques*, Butterworth, London.

WESTPHAL, W. H. (1963). *Physik. Ein Lehrbuch*, 22nd–24th ed., Springer, Berlin.

WILLIAMS, B. G. and NAYLOR, E. (1969). Synchronization of the locomotor tidal rhythm of *Carcinus. J. exp. Biol.*, **51**, 715–725.

WISEMAN, J. D. H. and OVEY, C. D. (1955). Proposed names of features on the deep-sea floor. *Deep Sea Res.*, **2**, 93–106.

WOLFF, T. (1956). Isopoda from depths exceeding 6000 meters. *Galathea Rep.*, **2**, 85–157.

WOLFF, T. (1960). The hadal community, an introduction. *Deep Sea Res.*, **6**, 95–124.

WOLFF, T. (1962). The systematics and biology of bathyal and abyssal *Isopoda itsellota. Galathea Rep.*, **6**, 7–319.

WOOD, E. J. F. (1965). *Marine Microbial Ecology*, Rheinhold, New York.

WOOD, E. J. F. (1967). *Microbiology of Oceans and Estuaries*, Elsevier, Amsterdam.

WYLLIE, P. J. (1963). Applications of high pressure studies to the earth sciences. In R. S. Bradley (Ed.), *High Pressure Physics and Chemistry*, Vol. II. Academic Press, New York. pp. 1–89.

YAMATO, H. (1952a). Effects of high hydrostatic pressure on the action of erythrocytes. III. On potassium contents. (Japan.) *J. Okayama med. Soc.*, **64**, 874–881.

YAMATO, H. (1952b). Effects of high hydrostatic pressure on the action of erythrocytes. V. On lysis. (Japan.) *J. Okayama med. Soc.*, **64**, 888–900.

YUYAMA, S. and ZIMMERMAN, A. M. (1969). Temperature–pressure effects on RNA synthesis in synchronized *Tetrahymena*. *Biol. Bull. mar. biol. Lab., Woods Hole*, **137**, 384.

ZENKEVITCH, L. A. and BIRSHTEIN, J. A. (1956). Studies of the deep water fauna and related problems. *Deep Sea Res.*, **4**, 54–64.

ZIMMERMAN, A. M. (1969). Effects of high pressure on macromolecular synthesis in synchronized *Tetrahymena*. In G. M. Padilla, G. L. Whitson and I. L. Cameron (Eds), *The Cell Cycle*. Academic Press, New York. pp. 203–225.

ZIMMERMAN, A. M. (Ed.) (1970a). *High Pressure Effects on Cellular Processes*. Academic Press, New York.

ZIMMERMAN, A. M. (1970b). High pressure studies on synthesis in marine eggs. In A. M. Zimmerman (Ed.), *High Pressure Effects on Cellular Processes*. Academic Press, New York. pp. 235–257.

ZIMMERMAN, A. M. and MARSLAND, D. A. (1964). Cell division: Effects of pressure on the mitotic mechanisms of marine eggs (*Arbacia punctulata*). *Expl Cell Res.*, **35**, 293–302.

ZIMMERMAN, A. M. and RUSTAD, R. C. (1965). Effects of high pressure on pinocytosis in *Amoeba proteus*. *J. Cell Biol.*, **25**, 397–399.

ZIMMERMAN, S. B. and ZIMMERMAN, A. M. (1970). Biostructural, cytokinetic, and biochemical aspects of hydrostatic pressure on Protozoa. In A. M. Zimmerman (Ed.), *High Pressure Effects on Cellular Processes*. Academic Press, New York. pp. 179–210.

ZOBELL, C. E. (1946). *Marine Microbiology*, Chronica Botanica, Waltham, Mass.

ZOBELL, C. E. (1952). Bacterial life at the bottom of the Philippine Trench. *Science, N.Y.*, **115**, 507–508.

ZOBELL, C. E. (1954). The occurrence of bacteria in the deep sea and their significance for animal life. *Publs Un. int. Sci. biol.*, (*Ser. B*), **16**, 20–26.

ZOBELL, C. E. (1964). Hydrostatic pressure as a factor affecting the activities of marine microbes. In *Recent Researches in the Fields of Hydrosphere, Atmosphere and Nuclear Geochemistry—Ken Sugawara Festival Volume*. Maruzen, Tokyo. pp. 83–116.

ZOBELL, C. E. (1970). Pressure effects on morphology and life processes of bacteria. In A. M. Zimmerman (Ed.), *High Pressure Effects on Cellular Processes*. Academic Press, New York. pp. 85–130.

ZOBELL, C. E. and BUDGE, K. M. (1965). Nitrate reduction by marine bacteria at increased hydrostatic pressures. *Limnol. Oceanogr.*, **10**, 207–214.

ZOBELL, C. E. and COBET, A. B. (1962). Growth, reproduction, and death rates of *Escherichia coli* at increased hydrostatic pressures. *J. Bact.*, **84**, 1228–1236.

ZOBELL, C. E. and COBET, A. B. (1964). Filament formation by *Escherichia coli* at increased hydrostatic pressure. *J. Bact.*, **87**, 710–719.

ZOBELL, C. E. and JOHNSON, F. H. (1949). The influence of hydrostatic pressure on the growth and viability of terrestrial and marine bacteria. *J. Bact.*, **57**, 179–189.

ZOBELL, C. E. and MORITA, R. Y. (1957). Barophilic bacteria in some deep sea sediments. *J. Bact.*, **73**, 563–568.

ZOBELL, C. E. and MORITA, R. Y. (1959). Deep-sea bacteria. *Galathea Rep.*, **1**, 139–154.

ZOBELL, C. E. and OPPENHEIMER, C. H. (1950). Some effects of hydrostatic pressure on the multiplication and morphology of marine bacteria. *J. Bact.*, **60**, 771–781.

ZOBELL, C. E. and UPHAM, H. C. (1944). A list of marine bacteria including descriptions of sixty new species. *Bull. Scripps Instn Oceanogr.* (*non-techn. Ser.*), **5**, 239–292.

9. DISSOLVED GASES

9.0 GENERAL INTRODUCTION

K. Kalle

(1) General Aspects of Dissolved Gases

Of the gases dissolved in oceans and coastal waters, nitrogen, oxygen, carbon dioxide and hydrogen sulphide play a significant physical as well as biochemical role. All four differ from each other considerably in their behaviour. Whilst the first three gases are atmospheric components, which pass into the sea from the atmosphere, hydrogen sulphide is formed in the sea itself by chemico-bacteriological transformations.

The composition of the atmosphere at sea level is as follows:

77·0	vol.%	Nitrogen
20·6	,,	Oxygen
1·47	,,	Water vapour (on the average)
0·9	,,	Argon
0·03	,,	Carbon dioxide
0·0024	,,	Trace gases

(a) Physical Aspects

In regard to the physics of the gases, two regularities exist which are of fundamental importance to the understanding of the behaviour of gases in sea water.

The first has to do with the concept of **gas solubility**. This is a function of temperature, salinity and pressure. With pressure, the situation is rather simple, because gaseous and dissolved phases are in a physical equilibrium: solubility in the liquid phase is proportional to the pressure in the gaseous phase. Physically, these relations may be represented most unequivocally by the **Ostwald absorption coefficient**, α', which equals the ratio of the amount of gas present in identical volumes in the liquid and gaseous phase, respectively.

On closer consideration of these ratios, it is seen that, just as with many other basic physical phenomena, water exhibits an anomalous behaviour (DIETRICH and KALLE, 1963). This is already shown by the fact that water possesses an appreciably smaller capacity to dissolve gases than do other liquids under normal temperature and pressure conditions (Fig. 9-1). On the other hand, we find in the temperature-gas solubility relations, intermediary minima and maxima, of a sort familiar to us from numerous other physical aspects of water behaviour, which are recognizable in the diagram by parabolic function curves. In the gas solubility function, there are minima, which in pure water (0‰S) lie at about 37°C for hydrogen, and in the neighbourhood of 80° and 75°C for oxygen and nitrogen, respectively (Fig. 9-2).

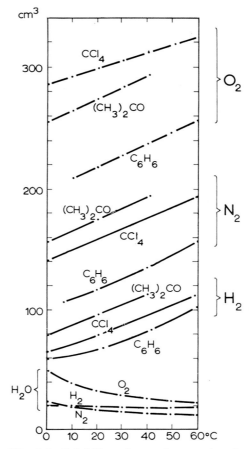

Fig. 9-1: Solubility of gases in organic sol-
vents, CCl_4, $(CH_3)_2CO$, C_6H_6, and in
water, H_2O, as a function of temperature.
O_2: oxygen, N_2: nitrogen, H_2: hydrogen.
(Original.)

The second concept, essential for understanding the behaviour of atmospheric gases in the sea, is that of gas saturation. Theoretically, there are three possibilities in defining this term. Firstly, one could start from the 'true physical saturation'; one must then take as a basis the appropriate pressure value for each depth in the sea, as would pertain to the corresponding level of an 'ideal' atmosphere (KALLE, 1945). An ocean under equilibrated conditions would approach this ideal state, if no vertical or horizontal currents and exchange processes were to disturb this equilibrium. Consequently, this concept of gas saturation, which has only theoretical significance, cannot be used in practice. Secondly, one could start from the hydrostatic maximum pressure in the different water depths, a pressure increase of approximately 1 atm corresponding to an actual depth increase of 10 m. Theoretically one could, therefore, bring considerable amounts of gas into solution in the deep sea, although only for limited periods of time. Both vertical mixing of the

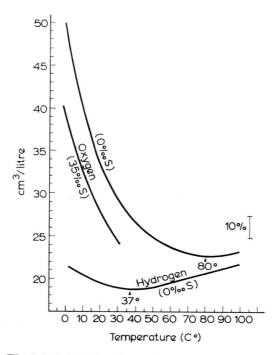

Fig. 9-2: Solubility of gases in pure water of 0‰S
and in sea water of 35‰S, as a function of
temperature. (After KALLE, 1945; modified.)

water masses and gas diffusion would cause excess of gases to escape gradually at the sea surface into the atmosphere. There remains, therefore, only the third possibility: to refer the gas saturation to the physical equilibrium conditions at the boundary sea-atmosphere, at which gas exchange alone takes place. It has been agreed, in this case, to assume a normal pressure of 760 mm of a mercury column at the sea surface, and to ignore regional and daily pressure fluctuations between about 740 and 770 mm; this is justified because the adjustment of equilibrium between atmosphere and sea proceeds so sluggishly that short-term air pressure fluctuations are without appreciable influence (see also Chapter 8.0).

(b) Biochemical Aspects

Let us now consider some important aspects of the chemistry of the gases. All gases present in the atmosphere are in exchangeable equilibrium with their dissolved phase in sea water, in accordance with recognized physical principles. However, whilst nitrogen (together with the noble gas argon), apart from a scarcely measurable portion, does not participate in chemical reactions within the water masses, the oxygen and carbon dioxide in sea water undergo constant and intensive biological and chemical transformations. In the case of carbon dioxide, the solution conditions are still more complicated, because, by the molecular addition of water, it becomes free carbonic acid (H_2CO_3), which, in turn, enters into

H

chemical interaction with the cations present in sea water (DIETRICH and KALLE, 1963; SKIRROW, 1965). Thus the CO_2 economy and the calcium carbonate budget in the sea are intimately linked with each other (**carbon dioxide-calcium carbonate system**). The operation of this system may be best understood if we consider the sea as a bicarbonate solution in equilibrium with atmospheric carbonic acid. The three basic quantities of this closed system are the carbon dioxide pressure, the hydrogen ion concentration (pH) and the alkalinity ($CaCO_3$ content). If two of these basic quantities in sea water are determined (for practical reasons, usually hydrogen ion concentration and alkalinity), the third quantity (carbon dioxide pressure) automatically becomes known also (Chapter 1).

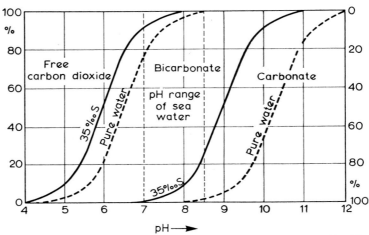

Fig. 9-3: Percentage distribution of the three forms of carbon dioxide (free carbon dioxide, bicarbonate, carbonate) in sea water of 35‰S and in pure water as a function of pH. (After KALLE, 1945; modified.)

The percentage proportions of the three forms of carbon dioxide (free carbon dioxide, bicarbonate, carbonate) as a function of pH, both in sea water of 35‰S and in pure water, are illustrated in Fig. 9-3. Below pH 4, practically free carbon dioxide only occurs; at pH 7·5, the percentage of bicarbonate reaches its maximum; at still higher pH values, bicarbonate is increasingly replaced by carbonate ions. The adjustment of the CO_2 system is of great importance in affecting the solubility of calcium carbonate in sea water and hence for the formation of marine sediments on the sea bottom. In general, it may be said that increased bicarbonate content leads to a solution of calcium carbonate, while increased carbonate content induces precipitation of the dissolved calcium carbonate.

Just as the carbon dioxide–calcium carbonate system is closely associated with oxygen transformations in sea water and regulates the level of hydrogen ion concentration, the **oxygen–hydrogen sulphide system** is responsible for the development of the oxidation-reduction potential (rH). This system begins to operate when the oxygen supply becomes depleted, mostly due to the presence of unusually large amounts of organic matter associated with effective vertical separation of the water masses. Under anaerobic conditions, bacteria develop which use the oxygen

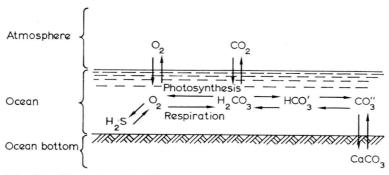

Fig. 9-4: The carbon dioxide–oxygen–hydrogen sulphide system in the ocean. (After KALLE, 1945; modified.)

bound in sulphate for oxidation of their organic nutrients, with concomitant formation of gaseous hydrogen sulphide, which dissolves in the sea water. As hydrogen sulphide is a powerful biological poison, normal plant and animal life is no longer possible in such regions. Fig. 9-4 shows in a schematic presentation the connections of the two systems, carbon dioxide–calcium carbonate and oxygen–hydrogen sulphide.

(2) Measuring Dissolved Gases: Methods

Of primary importance are chemical analytical methods; these are used in the determinations of oxygen, hydrogen sulphide and alkalinity; for the determination of pH, colorimetric and electrometric methods have been developed (for details consult BARNES, 1959).

As a basis for oxygen saturation, the temperature and salinity dependent values calculated by Fox (1907) are still in use, although 'improved' tables have been published by TRUESDALE and co-authors (1955). Apparently, these too cannot claim absolute correctness; the determination of correct gas saturation values is associated with considerable experimental difficulties (RICHARDS, 1965).

(3) Dissolved Gases in Oceans and Coastal Waters

The primary gases and their concentrations in the atmosphere are listed in Table 9-1, their saturation values in fresh and sea water in Table 9-2.

The ratio of oxygen to nitrogen amounts to about 1:4 in the atmosphere; in sea water, this ratio is approximately 1:2. The volume per cent of carbon dioxide amounts to only 0·03% in the atmosphere, but increases 40 to 60 times in sea water. The solubility of biologically important gases, therefore, leads to considerably higher percentage values in the sea than in the atmosphere.

The total amounts of water and atmospheric gases, as distributed between oceans and atmosphere, are listed in Table 9-3. Whilst the proportion of atmospheric water vapour amounts to only 1/100,000 of the water in the oceans, the proportions of atmospheric gases in sea water are much higher. In the case of nitrogen, the proportion of concentration in oceans to atmosphere amounts to

Table 9-1

Concentration of primary gases in
the atmosphere (reduced to dry
atmosphere of 1013 mbar and 0°C)
(After Fox, 1907)

	cm³/l	vol.%
Nitrogen	780·9	78·09
Oxygen	209·5	20·95
Argon	9·3	0·93
Carbon dioxide	0·3	0·03
Total	1000·0	100·00

Table 9-2

Saturation values of atmospheric gases in fresh and sea water
(After Fox, 1907)

	Fresh water (0‰ S)				Sea water (35‰ S)			
	0°C		30°C		0°C		30°C	
	cm³/l	vol.%	cm³/l	vol.%	cm³/l	vol.%	cm³/l	vol.%
Nitrogen	18·10	61·4	10·98	63·8	14·04	61·2	9·08	65·1
Oxygen	10·29	35·0	5·57	33·2	8·04	35·1	4·50	32·2
Argon	0·54	1·8	0·30	1·8	0·41	1·8	0·21	1·5
Carbon dioxide	0·52	1·8	0·20	1·2	0·44	1·9	0·18	1·2
	29·45	100·0	17·05	100·0	22·93	100·0	13·97	100·0

Table 9-3

Total quantities of water and atmospheric gases in oceans and
atmosphere, related to 1 cm² of the surface of the earth (Original)

	Oceans		Rel. proportion	Atmosphere		Rel. proportion
Water	269·000	kg/cm²	(500,000)	2·6	g/cm²	(5)
Oxygen (gaseous)	1·924	g/cm²	(4)	231·37	g/cm²	(503)
Nitrogen (gaseous)	4·412	g/cm²	(10)	754·81	g/cm²	(1640)
Carbon dioxide	29·266	g/cm²	(64)	0·46	g/cm²	(1)

1/165, or oxygen 1/126, and of carbon dioxide—if one restricts oneself to the 'free carbonic acid' dissolved in the gaseous state in sea water—to 1/2. If one considers the total carbonic acid, including the carbonic acid present in sea water in a chemically bound form, however, then the proportion turns in favour of the oceans, reaching the value 64/1 (Table 9-3). The oceans are, therefore, to be regarded as a gigantic reservoir of carbonic acid. As the carbonic acid contents in the atmosphere and in the oceans are in exchange equilibrium with each other, it must be assumed that the oceans act, over geological periods of time, as a regulator of the carbon dioxide content of the atmosphere, exercising a damping influence on large-scale climatic variations.

At the surface of the open oceans, oxygen saturation values amount to about 100% and vary, in general, only slightly. In coastal waters, considerable deviations from oxygen saturation may occur, depending on whether normal, polluted or nutrient-rich waters with high rates of primary production are considered. In areas with high primary production, super-saturations up to 120% are not unusual during periods of intensive solar radiation. In shallow waters, such temporary super-saturations may reach values up to 500%.

The water masses with oxygen minima—extending in all three oceans on both sides of the Equator, at about 500 m depth, from east to west—are a peculiarity; particularly in the eastern half of the Atlantic Ocean, oxygen contents may fall below 1 cm³/l, in the Pacific Ocean even below 0·5 cm³/l. Another special situation is represented by sea regions without significant vertical water exchanges due to surface waters with reduced salinities. Examples are the Black Sea and several Norwegian fjords. In these areas, the oxygen supplies of the deeper water layers may be completely consumed; subsequently, large amounts of free hydrogen sulphide are formed.

9. DISSOLVED GASES

9.1 BACTERIA, FUNGI AND BLUE-GREEN ALGAE

G. Rheinheimer

(1) Introduction

Marine bacteria, fungi and blue-green algae depend greatly upon the presence of several dissolved gases. Obligate aerobic bacteria and fungi require a certain amount of oxygen; photo-autotrophic purple and green bacteria and blue-green algae need carbon dioxide; heterotrophs assimilate carbon dioxide and require, in addition, organic carbon. Some highly specialized bacteria use hydrogen, hydrogen sulphide, methane or carbon monoxide as energy sources, and some other bacteria are able to assimilate free nitrogen. On the other hand, oxygen is toxic for obligate anaerobes, and hydrogen sulphide is lethal for many micro-organisms, as well as for plants and animals.

In spite of the great importance of dissolved gases for micro-organisms in sea water and marine sediments, our pertinent knowledge is still very limited. Little is known about the influence of dissolved gases on bacteria and blue-green algae, nearly nothing on marine fungi. More information is urgently required.

(2) Functional Responses

(a) Tolerance

Oxygen

On the basis of their responses to oxygen, four groups of micro-organisms may be distinguished: (i) **obligate aerobes**, which can grow (establish populations) only in the presence of oxygen; (ii) **micro-aerophilic organisms**, with optimum growth at low oxygen concentrations; (iii) **facultative aerobes** or **facultative anaerobes**, which can grow in both the presence and absence of oxygen; (iv) **obligate anaerobes** which grow only if oxygen is completely absent, oxygen being toxic for them.

Most of the bacteria and fungi isolated from sea water or marine mud are facultative aerobes. They show better population growth in the presence than in the absence of oxygen under ordinary conditions of laboratory cultivation (ZoBell, 1946). For these bacteria and fungi, no tolerance limits exist in regard to oxygen concentration.

In addition to the great majority of facultative aerobes, there exists a relatively small number of obligate aerobes in oceans and coastal waters; they are of great importance for the life processes in the sea. Obligate aerobes are, for example, the chemolithotrophic nitrifiers and some Thiobacilli, which use dissolved oxygen, not only for respiration, but also for the oxidation of ammonia, nitrite or hydrogen sulphide. These bacteria require certain amounts of oxygen for the oxidation processes mentioned. According to Schöberl and Engel (1964), the minimum concentrations of ambient oxygen for substrate oxidation by *Nitrosomonas europaea* and *Nitrobacter winogradskyi* are rather low. These two bacteria show complete inhibition of both ammonia and nitrite oxidation below 0·2 mg O_2/l. In *N. europaea*, activity decreases below 1 mg O_2/l at 30°C. Longer periods of total oxygen deficiency seem to be lethal for both these nitrifiers. Hence they are generally absent in anaerobic habitats.

For obligate anaerobes, such as *Desulfovibrio desulfuricans*, *Chlorobium limicola* and some purple bacteria, even small amounts of oxygen may be toxic. Some other anaerobes—for example, species of *Bacteroides* and *Clostridium*—tolerate low oxygen concentrations. But there are no exact data available on tolerance limits of marine anaerobes. Spores of spore-forming anaerobic bacteria, such as the different species of the genus *Clostridium*, are not affected by high oxygen concentrations; but they grow only under anaerobic conditions. Some micro-aerophilic bacteria, e.g. species of *Achromatium*, are found in marine environments (BREED and co-authors, 1957), but there is no information on their tolerance limits.

Carbon dioxide

All micro-organisms, autotrophic as well as heterotrophic, require carbon dioxide or bicarbonate. However, normal CO_2 concentrations in sea water and in marine sediments are usually within the tolerance limits of bacteria, fungi and blue-green algae, as the pH values of sea water range from 8·1 to 8·3; or, in extreme cases, from 7·5 to 8·5 (ZoBELL, 1946). With increasing pH values, the concentration of free carbon dioxide and of bicarbonate decreases (Chapter 9.0), and at pH 9·4, photosynthesis of marine green plants ceases even in bright sunlight, because there is no more carbon dioxide or bicarbonate available. Such a situation may occur only in heavily grassed estuarine flats with large amounts of periphyton (WOOD, 1967). Only under these extreme conditions may the photosynthesis of marine micro-organisms be limited, due to lack of CO_2 and bicarbonate. Neither in the free water, nor on the sea floor may the concentration of CO_2 be high enough to inhibit photosynthesis of blue-green algae or photosynthetic bacteria; such inhibition is possible in some other special habitats only.

Hydrogen sulphide

Hydrogen sulphide is toxic for most aerobic and facultative anaerobic micro-organisms. Small concentrations of H_2S already cause growth inhibition. Higher concentrations are lethal. Therefore, neither obligate aerobes nor many of the facultative aerobes occur in waters or sediments containing hydrogen sulphide. Nitrifying bacteria seem to survive for only a short time in media containing hydrogen sulphide. On the other hand, besides sulphur bacteria, some anaerobic bacteria, mainly of the orders Eubacteriales and Spirochaetales, may tolerate hydrogen sulphide (DURNER and co-authors, 1965). However, exact data are still lacking. Some blue-green algae also are tolerant to hydrogen sulphide, for example, many species of the orders Chroococcales and Hormogonales.

Hydrogen, methane, nitrogen

It is not known whether increased concentrations of hydrogen, methane and nitrogen can affect marine micro-organisms.

(b) Metabolism and Activity

Oxygen

For aerobic respiration, micro-organisms use molecular oxygen as the final hydrogen (electron) acceptor. At the end of their respiratory chain, where reaction with oxygen takes place, they have cytochrome pigments. Some of these are able to react rapidly with molecular oxygen. While facultative aerobic bacteria are also able to respire anaerobically if oxygen is absent, obligate aerobic bacteria depend on the presence of molecular oxygen. According to JOHNSON, F. H. (1936), some heterotrophic marine bacteria consumed, within 1 hr, 2·8 to 185×10^{-12} mg oxygen per 'resting' cell in sea water at 25°C; about one-fifth of this quantity was consumed at 5°C. Addition of 0·04% glucose resulted in an increase in oxygen uptake of 10 to 378%. ZOBELL (1940) estimated that marine bacteria multiplying in sea water at 22°C consume, on average, $20·9 \times 10^{-12}$ mg oxygen per cell and hour. The rate of oxygen consumption increases about four times when the water is enriched with 0·05% of glucose or asparagine. The rate of consumption is independent of the ambient oxygen tension between 0·43 and 17·84 mg O_2/l.

For some chemolithotrophic bacteria, oxygen is necessary, not only for respiration but also for oxidation of anorganic compounds such as ammonia, nitrite, hydrogen sulphide, molecular sulphur, thiosulphate, as well as for ferro- and mangano-compounds. The nitrifying bacteria *Nitrosomonas europaea* and *Nitrobacter winogradskyi* use a minimum of 0·2 mg O_2/l for oxidation of ammonia and nitrite (SCHÖBERL and ENGEL, 1964). Most species of the genus *Thiobacillus* can oxidize hydrogen sulphide or other reduced sulphur compounds only in the presence of oxygen. *Thiobacillus denitrificans* is the only known chemolithotrophic micro-organism which is capable of sulphur oxidation under anaerobical conditions also, as long as nitrate and organic matter are present. Ferro- and mangano-oxidizers use molecular oxygen also for respiration, as well as for oxidation of ferro- and mangano- compounds. On the other hand, molecular oxygen is toxic for many obligate anaerobic micro-organisms. Most of these lack the enzyme catalase. They are thus unable to destroy the poisonous hydrogen peroxide, which is formed in the final phase of aerobic respiration by transformation of hydrogen to oxygen. Hydrogen peroxide has to be transformed immediately into water and oxygen by the enzyme catalase—otherwise the cells will be poisoned.

Carbon dioxide

Carbon dioxide is required by all green plants—including blue-green algae and photosynthetic green and purple bacteria—and by the small group of chemosynthetic bacteria. These carbon autotrophic organisms reduce carbon dioxide and transform it into carbohydrates. For this process, they use sun energy (photosynthesis) or chemical energy derived from reactions catalyzed by the bacteria concerned (nitrification, sulphurication, mangano- and ferro-oxidation). During photosynthesis, water is used as hydrogen donator for the reduction of carbon dioxide; thus, oxygen is released. During chemosynthesis, oxygen is consumed. Consequently, the oxygen content of the water is increased by photosynthesis but decreased by chemosynthesis.

Some blue-green algae are facultative photo-autotrophic, some iron and sulphur bacteria facultative chemo-autotrophic. They may live heterotrophically also, i.e., they are able to use organic compounds (Chapter 10.1) for nutrition. Micro-organisms exhibiting such characteristics are called mixotrophs or amphitrophs (SCHWARTZ and SCHWARTZ, 1960). However, in the last three decades, it has become known that heterotrophic bacteria and fungi, which require some form of organic carbon as nutrients, use small amounts of carbon dioxide too. Until 1935, CO_2 was believed to be completely inert in heterotrophic organisms (WOOD and STJERNHOLM, 1962). In that year, the utilization of carbon dioxide by heterotrophs was discovered.

In the following years, biochemical research (mainly tracer experiments) revealed many reactions within the heterotrophic metabolism of bacteria and fungi, which combine CO_2 with different compounds. Some bacteria, for example, were shown to be able to form malate from pyruvate and carbon dioxide. Di- and tri-carboxylic acids also can be formed by CO_2 fixation. These reactions require reduced pyridine nucleotides as energy source. Carbon dioxide, therefore, is necessary for all micro-organisms, in marine habitats as well as in other biotopes.

Hydrogen sulphide

The photosynthetic purple sulphur bacteria (Thiorhodaceae) and green bacteria (Chlorobacteriaceae) use hydrogen sulphide as hydrogen donator for the reduction of carbon dioxide. They lack the enzymes catalase and peroxydase; hence they cannot—as do algae and higher green plants—use water as hydrogen donator. Some blue-green algae of the genus *Oscillatoria* only are able to use both H_2O and H_2S.

Among the purple sulphur bacteria are some facultative photosynthetic forms which may use organic material also (e.g. the Athiorhodaceae), when no hydrogen sulphide is present. While purple or green bacteria are obligate anaerobic or micro-aerophilic organisms, colourless sulphur bacteria (*Thiobacillus, Thiothrix, Beggiatoa, Thioploca, Thiospirillopsis, Achromatium, Thiovulum* and *Thiospira*) are obligate or facultative aerobes. They are chemolithotrophic forms, able to oxidize hydrogen sulphide (and other reduced sulphur compounds) to sulphur or sulphate. They accomplish this with the help of molecular oxygen or—in the absence of oxygen—by reduction of nitrate, for example, *Thiobacillus denitri-ficans*. Colourless sulphur bacteria obtain the energy necessary for the reduction of carbon dioxide by oxidation of hydrogen sulphide.

In addition to obligate chemolithotrophic organisms, there are facultative chemolithotrophic forms (for example, *Thiobacillus novellus* and some Beggia-toales), which may also live heterotrophically from organic substances (Chapter 10.1). However, for most aerobic micro-organisms, hydrogen sulphide is toxic and even low concentrations will kill them.

Hydrogen sulphide (like potassium cyanide and carbon monoxide) inhibits the enzyme cytochrome oxidase, the final link of the respiratory chain (BALDWIN, 1957). It is in this way that hydrogen sulphide inhibits respiratory processes in most aerobic organisms.

Hydrogen, methane, carbon monoxide

Under anaerobic conditions, a few specialized bacteria are able to produce molecular hydrogen, methane or carbon monoxide from various organic compounds.

Other bacteria—chemolithotrophic ones as well as heterotrophs—oxidize these gases with free oxygen or, in anoxic habitats, with nitrate or sulphate. A few bacteria grow only in the presence of one of the gases mentioned (for example, *Carboxydomonas* in the presence of carbon monoxide), while others can use them in addition to other organic compounds. Thus, *Hydrogenomonas* is capable of oxidizing molecular hydrogen as well as living heterotrophically on gelatine or other organic substances. Carbon monoxide, like hydrogen sulphide, is toxic for many aerobic micro-organisms; it inhibits their respiration.

Nitrogen

Several bacteria and blue-green algae are able to fix molecular nitrogen. In the presence of insufficient amounts of bound nitrogen, they can use molecular nitrogen as the sole source of this element. For reduction of free nitrogen, organic substances are necessary, in order to provide the hydrogen donators.

(c) Reproduction

Although the reproductive activities of bacteria, fungi and blue-green algae are greatly influenced by the occurrence and concentration of dissolved gases, exact information is scarce.

Oxygen

Obligate aerobic micro-organisms multiply only if sufficient oxygen is present, obligate anaerobes only if oxygen is completely absent. Reproduction of micro-aerophilic forms is restricted to habitats with low oxygen concentrations.

Carbon dioxide

The fact that most micro-organisms will multiply only in the presence of carbon dioxide is more of theoretical than practical interest, because carbon dioxide is not usually absent in marine environments.

Hydrogen sulphide

Hydrogen sulphide is required for reproduction by several sulphur bacteria. However, multiplication of most aerobic micro-organisms is inhibited even by low concentrations of hydrogen sulphide. In addition to these two groups, there are also some forms which can withstand rather high hydrogen sulphide concentrations; some heterotrophic bacteria, as well as a number of blue-green algae, belong to these H_2S tolerant organisms, probably several fungi also. Although these forms multiply well in hydrogen sulphide-containing media, we do not know yet if all of them reproduce in the normal way under these conditions.

Hydrogen, methane, carbon monoxide

The presence of molecular hydrogen, methane and carbon monoxide is a prerequisite for reproduction in very few micro-organisms which oxidize one or more of these gases.

Nitrogen

Molecular nitrogen seems to exert little effect on the reproductive capacity of bacteria, fungi and blue-green algae. This gas may promote multiplication of nitrogen-fixing bacteria and blue-green algae in habitats lacking nitrogen compounds which are acceptable to these organisms.

(d) Distribution

Dissolved gases may influence the horizontal and vertical distribution of bacteria, fungi and blue-green algae in sea water as well as in marine sediments.

Oxygen

Obligate aerobic bacteria require habitats with oxygen, while obligate anaerobes grow only in habitats without oxygen. Hence the species composition of the microbial flora in aerobic areas is different from that of anaerobic areas, and localities with periodically or aperiodically changing oxygen concentrations reveal concomitant changes in their microbial populations. Such changes are small at oxygen saturations ranging from 100% (or more) to about 5%. However, further decrease in oxygen saturation causes death in many obligate aerobic bacteria, fungi and blue-green algae and, at the same time, population increases in micro-aerophilic and obligate anaerobic species. Therefore, in waters or sediments with high oxygen deficits, small changes in oxygen concentrations may be followed by rather large changes in the composition of microbial populations. In oxygen-containing waters with high turbidity, populations of aerobic and anaerobic micro-organisms may occur together, since the latter can thrive in the anoxic microzones surrounding the detritus particles (Chapter 6.1). Aerobic and anaerobic populations may also occur close together in some sediments.

Carbon dioxide

Dissolved CO_2 exerts only little influence on the distribution of bacteria, fungi and blue-green algae, because sufficient amounts of CO_2 (and bicarbonate) are usually available.

Hydrogen sulphide

Under anaerobic conditions, hydrogen sulphide is chemically inert; hence the H_2S concentration of the water may increase as a result of microbial activity. Proteolytic bacteria release hydrogen sulphide mainly from the sulphur-containing amino acids cystine, cysteine and methionine. Sulphate-reducing bacteria, such as *Desulfovibrio desulfuricans* and *Clostridium nigrificans*, are able to produce considerable amounts of H_2S, as long as sulphate and organic material (as energy sources) are present. In this case, the species composition of the microflora and fauna changes completely within a short period of time; only micro-organisms tolerant to hydrogen sulphide survive. Habitats containing large amounts of hydrogen sulphide are called 'sulphureta' (BAAS-BECKING, 1925). There are different types of sulphureta, for example, light and dark sulphureta.

In a model light sulphuretum, DURNER and co-authors (1965) found 2 species of purple bacteria, 1 green bacterium and 14 blue-green algae; they also observed 4 diatoms and 2 green algae. In addition to these photosynthetic organisms, chemo-lithotrophic and heterotrophic bacteria and protozoans were present. In dark sulphureta, photosynthetic bacteria and algae are absent.

DURNER and co-authors also established that plate counts (aerobic and anaero-bic forms) of natural sapropel mud samples yield higher values without than with sodium sulphide (Table 9-4).

Table 9-4

Bacteria counts on sapropel mud extract agar with and without Na$_2$S, after 4 days (aerobic) and 8 days (anaerobic), at 27°C. The body of the Table gives numbers of bacteria per g dry mud (After DURNER and co-authors, 1965; modified)

Bacteria	Without Na$_2$S	With Na$_2$S
Aerobic	$2 \cdot 7 \times 10^6$	$2 \cdot 3 \times 10^5$
Anaerobic	$2 \cdot 7 \times 10^5$	$5 \cdot 9 \times 10^4$

Observations in the Kiel Fjord (West Germany) repeatedly revealed heavy decreases in total bacterial counts on ZoBELL'S medium (2216 E), and sometimes also decreases in numbers of coliform bacteria, when hydrogen sulphide was present in the deep water (RHEINHEIMER, unpublished). On the other hand, the numbers of sulphate-reducing and of sulphur-oxidizing bacteria increased in the H$_2$S containing zones (Fig. 9-5). A similar situation may be found in other waters with hydrogen sulphide zones, e.g. Black Sea and some Norwegian fjords. In the Black Sea, the hydrogen sulphide zone generally lies below 200 m water depth. Since there is not enough light to allow photosynthesis, mainly colourless sulphur bacteria are present (ISSATSCHENKO and JEGOROWA in: KUSNEZOW, 1959). However, in shallow lagoons, beach lakes and small brackish-water ponds along the coasts of the Black Sea, the hydrogen sulphide zones lie near the water surface; hence, they often contain large numbers of purple and green bacteria. The green bacteria tend to prefer waters with higher hydrogen sulphide concentrations, while many purple bacteria also live in zones with very low H$_2$S concentrations; some of them require small amounts of free oxygen (ZoBELL, 1946), which they obtain from the co-existing, photosynthesizing algae.

In some Russian lakes, the border area between oxygen and hydrogen sulphide containing waters lies in about 5 m depth and contains maximum cell numbers of *Chlorobium limicola* (KUSNEZOW, 1959). In the Belowod Lake, where this border is located in about 14 to 15 m depth, an impressive maximum of *Chromatium* cells has been reported (Fig. 9-6). Green bacteria, such as *Chlorobium limicola*, generally prefer places with more light than purple bacteria (PFENNIG, 1965). Accordingly, in small brackish-water ponds, vertical zonations may be found which consist of a surface zone with algae, a zone of green bacteria and, below this, a zone of purple bacteria. GIETZEN (1931) recorded extensive populations of purple sulphur

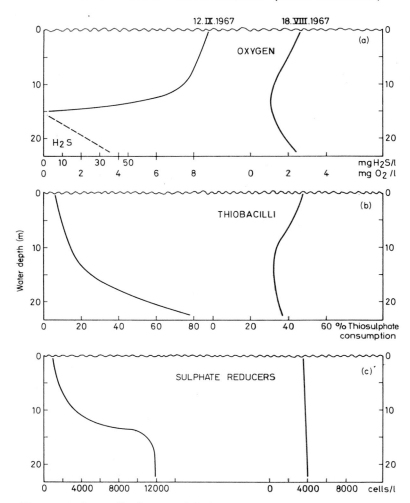

Fig. 9-5: Effects of hydrogen sulphide on the numbers of Thiobacilli and
sulphate-reducing bacteria in Kiel Bay (West Germany). (a) H$_2$S and
O$_2$ concentrations (mg/l) on August 18 and September 12, 1967;
(b) thiosulphate consumption (% within 7 days) of Thiobacilli;
(c) most probable numbers of sulphate-reducing bacteria (cells/l). As
can be seen from the left sides of (b) and (c), the numbers of both
groups of bacteria increase, in the presence of H$_2$S, with increasing
water depth. (Original.)

bacteria along the coast of Holstein (West Germany), growing together with
decomposing algae, jellyfish, etc., and colouring the sea distinctly red. In such
places, three factors allow maximum development of purple bacteria: micro-
aerophilic or anaerobic conditions, the presence of hydrogen sulphide, and sunlight.
In the Dreckee mud swamps along the Danish coast, UTERMÖHL (1925) found
several thousand *Chromatium* and *Thiopedia* cells and some hundred cells of
Thiocystis per ml of water.

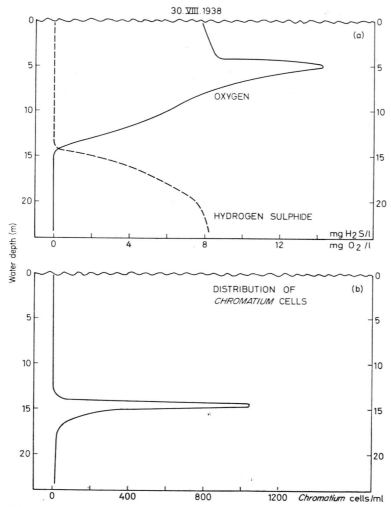

Fig. 9-6: Vertical distributions of oxygen and hydrogen sulphide (a),
and of *Chromatium* cells (b) as a function of water depth in the
Belowod Lake (USSR). Maximum abundance is attained at 14 to
15 m, the border zone between O_2 and H_2S containing water bodies.
(After KUSNEZOW, 1959; modified.)

BAVENDAMM (1924) listed 16 species of Thiorhodaceae in sunlit brackish or
marine waters containing hydrogen sulphide. In calcareous mud near the Bahama
Islands, BAVENDAMM (1932) recognized 4 species of the genus *Chromatium*. In
addition to sulphur bacteria, there was a large heterotrophic bacterial population,
including sulphate reducers; associated with these bacteria, blue-green algae,
especially of the genus *Oscillatoria*, and some diatoms and protozoans were observed. BAVENDAMM regarded this association as being ideal for sulphur bacteria;
sulphate reducers provide hydrogen sulphide and algae supply free oxygen.

Colourless sulphur bacteria, such as species of *Thiothrix* and *Beggiatoa* live mainly on the floor of the seas if hydrogen sulphide is available. They may also be found on decomposing algae (MOLISCH, 1912) and in sea water (BAVENDAMM, 1924). Representatives of the genera *Achromatium*, *Thiospira*, *Thioploca* and *Thiophysa* occur also in marine habitats; they oxidize hydrogen sulphide and are able to deposit sulphur granules intracellularly (ZoBELL, 1946).

According to KRISS (1961), the numbers of threadlike, colourless sulphur bacteria, as well as of spherical cells containing sulphur granules, increase considerably in the hydrogen sulphide zone of the Black Sea. Bacteria of the genus *Thiobacillus* have been found in sea water and in marine sediments too. Besides hydrogen sulphide, they oxidize elementary sulphur and other reduced sulphur compounds. The end-product of their oxidative metabolism is sulphate; but some of them often deposit sulphur extracellularly. *Thiobacillus thioparus* has repeatedly been isolated from sea water. Large numbers of this species have been noted in the Black Sea, and RAWITCH-SCHTERBO (1930) holds *T. thioparus* responsible for the absence of hydrogen sulphide in certain zones of the Black Sea. This aerobic micro-organism seems to occur mainly in the euphotic zone of the Black Sea where oxygen is present (KRISS, 1961). *Thiobacillus denitrificans* is often found in marine habitats also; it is facultative aerobic and can use nitrate as hydrogen acceptor while oxidizing hydrogen sulphide. According to KRISS and RUKINA (1949), large numbers of *Thiobacillus denitrificans* live in the anaerobic zones of the Black Sea, especially in the mud above the sea floor.

Observations in the Kiel Fjord of the Baltic Sea indicate that the numbers of *Thiobacillus* increase significantly in hydrogen sulphide containing zones, i.e. from the water surface to the ground (Fig. 9-5). Sometimes, maximum density of *Thiobacillus* cells was observed in the border zone between oxygen and hydrogen sulphide containing water bodies.

Hydrogen, methane, carbon monoxide

There is hardly any information available on the influence of dissolved hydrogen, methane or carbon monoxide on the distribution of bacteria, fungi and blue-green algae in oceans and coastal waters. We know only that these gases may be formed by highly specialized micro-organisms in anaerobic marine habitats. Other bacteria are able to oxidize the gases mentioned. HUTTON and ZoBELL (1949) found species of *Methanomonas* and *Hydrogenomonas* in marine environments; the first-named organism is able to oxidize methane, the latter to oxidize hydrogen (some strains, carbon monoxide also). According to LUKINS and FOSTER (1963), *Mycobacterium marinum* also is a chemolithotrophic bacterium capable of oxidizing molecular hydrogen in marine habitats. It does this, either in the presence of molecular oxygen or under anaerobic conditions, in the presence of nitrate or sulphate. Consequently, the numbers of bacteria oxidizing hydrogen, methane or carbon monoxide will tend to increase in the presence of one of these gases.

Nitrogen

It is unlikely that the different concentrations of molecular nitrogen which may occur in oceans and coastal waters have any influence on the distribution of bacteria, fungi or blue-green algae. Several bacteria, yeasts and blue-green algae are capable of using free nitrogen. They occur in places without bound nitrogen, as long as free nitrogen and organic hydrogen donators are present. Different species of the blue-green algae genera *Nostoc, Anabaena* (ZoBELL, 1946), *Trichodesmium* (WOOD, 1965) and *Calothrix* (ALLEN, 1963) are able to fix nitrogen. Among the marine bacteria, species mainly of the anaerobic genus *Clostridium* can fix nitrogen. However, the aerobic *Azotobacter* was also isolated from marine waters (ZoBELL, 1946). PSHENIN (1963) recorded *Azotobacter* cells in all depths of the Black Sea, and RHEINHEIMER (1965) found *Azotobacter chroococcum* and *Azotobacter agilis* in the brackish water of the Elbe estuary (West Germany).

According to ALLEN (1963), pink yeasts of the genus *Rhodotorula*, which are able to fix nitrogen, were found in the Gulf of California and near Friday Harbor (Washington, USA); and off the Florida coast, up to 6000 cells of *Rhodotorula* per litre have been observed.

(3) Structural Responses

Size and shape of bacteria, fungi and blue-green algae may certainly be expected to be influenced by some dissolved gases; however, there is hardly any conclusive information available. Cell size of some bacteria seems to depend on the ambient oxygen concentration, and the size of several blue-green algae depends on carbon dioxide concentration. Hydrogen sulphide or carbon monoxide may influence not only the size but also the shape of micro-organisms.

Detailed information on possible effects of dissolved gases on size and shape of marine micro-organisms, as well as their internal structures, may open up new avenues for assessing their morphological and ecological potentials.

9. DISSOLVED GASES

9.2 PLANTS

W. VIDAVER

(1) Introduction

Of the atmospheric gases dissolved in the relatively shallow waters of the euphotic zone of oceans, lakes or streams, only two—carbon dioxide and oxygen—are of primary significance for submerged plants. Carbon dioxide is the major substrate for photosynthesis. Oxygen is necessary to plants, as it is to other aerobic organisms, because of its role in oxidative metabolism.

The gas which occurs at the highest concentrations, both in the atmosphere and dissolved in natural waters, is nitrogen. Since N_2 is, metabolically, virtually inert, except in some species of nitrogen-fixing blue-green algae (Chapter 9.1), it is of little significance to most aquatic plants. Some atmospheric gases, such as argon, neon and helium, are present in the sea, but in such low concentrations as to have negligible influence on plants; they also tend to be metabolically inert.

Gases sporadically present in the atmosphere at concentrations sufficient to influence, when dissolved, growth or distribution of submerged plants would nominally be considered as pollutants. Except in the polluted or otherwise abnormal environments, variations in the concentrations of carbon dioxide and oxygen only appear to have major influence on tolerance, metabolism, reproduction, distribution and morphology of aquatic plants.

In waters devoid of organisms or oxidizable detritus, the source of dissolved gases is primarily the atmosphere. Under abiotic conditions, gases dissolved in water tend to come to equilibrium with the atmosphere. The concentrations of dissolved gases are determined, to a large extent, by the solubility of each gas, its partial pressure in the atmosphere, the degree to which it reacts with water, the barometric pressure, the temperature and pH of the water, and the concentrations of dissolved salts and other solutes (Chapter 9.0).

In waters which contain organisms, on the other hand, the dissolved gases are seldom in equilibrium with those of the atmosphere. Metabolic gas exchange of organisms precludes the establishment of an equilibrium. In animals and many plants, oxidative metabolism requires the uptake of O_2 and culminates in the production of CO_2. Many, if not all non-photosynthetic eucaryotic cells have some capacity to take up CO_2. In addition, plants consume large quantities of CO_2 through photosynthesis, which utilizes light energy to carry out the photolysis of water and the production of O_2. Consequently, plants exposed to light have the power to increase the ratio of dissolved O_2 to CO_2 in their aquatic environment. Non-photosynthetic aerobic organisms at all times, and plants in the dark, increase the CO_2 concentration at the expense of O_2.

Largely as a consequence of metabolic exchanges, concentrations of dissolved gases in oceans and coastal waters can vary over wide ranges. O_2 may, at times, be entirely absent or be present in concentrations exceeding equilibrium values by three or four fold (HUTCHINSON, 1957). Both marine and limnic plants may, if their numbers are sufficient and other factors are not limiting, raise the concentration of dissolved O_2 well beyond that which occurs through equilibration with the atmosphere. Freshwater plants sometimes lower the concentration of CO_2 to levels at which photosynthesis does not occur. However, in the marine environment, the buffering action of sea water tends to maintain a fairly constant CO_2 concentration, despite the uptake by plants (Chapter 9.0).

In this chapter, CO_2 can only be considered in meaningful terms as 'metabolic CO_2' present in the environment in chemical forms which may participate in the metabolism of organisms. Only small amounts of CO_2 actually occur in the ambient water as dissolved gas; most CO_2 is present in the form of carbonic acid (H_2CO_3), bicarbonate (HCO_3), and carbonate (CO_3) ions. The ratios of these dissolved forms of CO_2 depend, to a great extent, on the pH of the water. Detailed examination of all factors controlling quantitative aspects of dissolved CO_2 (and O_2) in the sea is outside the scope of this chapter; the reader is referred to the works of SVERDRUP and co-authors (1942), HUTCHINSON (1957), RILEY and SKIRROW (1965) and SILLEN (1967).

Among the gases which could be termed pollutants are sulphur dioxide, hydrogen sulphide, ozone, carbon monoxide, fluorine, hydrogen fluoride, chlorine, and ozonated hexenes. It is possible that the ambient concentration of any of these gases could become high enough to influence the metabolism or species distribution of submerged plants. Almost certainly, adverse effects on marine or limnic plants caused by such gases may be observed in waters contaminated with industrial and sewage wastes. CARR (1961) has reviewed the effects of some of these gases on various plants. Hydrogen sulphide does occur naturally at very high concentrations, mostly as a product of organic decompositions. In the Black Sea, hydrogen sulphide concentrations can be so high in deeper water layers as to preclude the existence of aerobic organisms including plants.

A few other gases, such as ethylene and related hydrocarbons, could possibly influence submerged plants by affecting hormonal growth and differentiation; but such gases are not normally present in the atmosphere, and hence will not be considered here.

GAFFRON and BISHOP (1962) have shown that some algae can utilize dissolved H_2 for the photoreduction of CO_2; in this process, which is inhibited by small amounts of O_2 or high light intensity, no O_2 is produced. Since the conditions allowing photoreduction by H_2 would seldom, if ever, occur in natural environments, this capacity appears to be of little ecological importance. The topic has been reviewed by SPRUIT (1962).

Many Cyanophyta can fix dissolved N_2 (FOGG, 1962). Most species with this capacity belong to the genera *Anabaena*, *Cylindrospermum* and *Nostoc*. The process of N_2 fixation is carried out photosynthetically, and the algae produce more O_2 when using N_2 than when using combined forms of nitrogen. Presumably, the overall reaction is:

$$N_2 + 3H_2O \xrightarrow{\text{light}} 2NH_3 + 1\tfrac{1}{2}O_2$$

but interactions between N_2 fixation and photosynthetic CO_2 assimilation, along with the probable participation of organic hydrogen carriers, complicates actual determinations. It is possible that much of the fixed nitrogen, essential to all forms of life, originates from the photoreduction of N_2 by these algae.

The extent to which many of the environmental gases contribute to functional and structural aspects of plant life is, at best, uncertain and requires clarification. Most of the information available refers to the effects of variations in O_2 and CO_2. This chapter will thus be restricted to a consideration of the interactions of normally submerged or periodically emerged photosynthetic plants with variations in the concentrations of these two gases in their water-dissolved phase. Topics and examples are selected mainly from those impinging on this author's own investigations.

Evidence for the effects of variations in dissolved gas concentrations on plants comes generally from two sources: uncontrolled field observations and laboratory experiments. While the ecological significance of variations of gases in the natural environment is far from clear, except in some obvious cases, it seems nonetheless useful to presume that distributions of plant populations are, in part, determined by the gas concentrations of the milieu. In laboratory experiments, the effects of varying a single gas have been tested; most of our knowledge about gas metabolism results from such work.

(2) Functional Responses

(a) Tolerance

Plants can survive only when the concentrations of gases in their environment remain within acceptable limits. High or low CO_2 and O_2 concentrations can inhibit metabolic processes and hence lead to death; critically high concentrations of toxic gases also can be lethal.

Tolerance to carbon dioxide

As CO_2 is the substrate for photosynthesis, no photo-autotrophic plant can exist in its absence except in those cases, usually under laboratory conditions, where an organic acid or similar substance may substitute for it. Minimum critical CO_2 concentrations for autotrophic plant survival are highly variable (interspecific differences) and affected by such factors as O_2 concentration, light intensity, temperature, and pH. The minimum CO_2 requirement for growth and photosynthesis is often raised by increases in O_2 concentration, light intensity, and temperature. Except for this statement, it is impossible to generalize about minimum gas requirements of plants. The reader is referred to reviews or discussions by RABINOWITCH (1951, 1956), GESSNER (1959) and STEEMANN NIELSEN (1960).

Effects of high CO_2 concentrations are difficult to observe because of the close relationship between dissolved CO_2 and pH. According to EMERSON and GREEN (1938), dissolved CO_2 only exceeds 50% of the total CO_2 in water when the pH is below 6. At pH 6, water in equilibrium with air will contain about $2 \cdot 5 \times 10^{-4}$ mol/l CO_2 at 0°C. Any further increase in dissolved CO_2 would only increase the acidity. Since most plants, including algae, are injured by high hydrogen ion concentrations, the biological effects of CO_2 in high concentrations remain largely unknown.

Tolerance to oxygen

Low O_2 concentrations can become critical for photosynthetic plants only in the dark or at very low light intensities, since plants produce O_2 in the light. Complete lack of O_2 is just as lethal for plants as it is for obligate aerobes. Most marine algae succumb rapidly in the dark if given insufficient amounts of O_2. A few plants, however, are able to ferment sugars anaerobically in the dark (TURNER and BRITTAIN, 1962) and thus may withstand extended periods of exposure to O_2 free media (FOGG, 1953).

High O_2 concentrations appear to have a somewhat toxic effect on many plants (TURNER and BRITTAIN, 1962). High light intensities combined with high O_2 concentrations can damage irreversibly the photosynthetic apparatus of plant cells (MYERS and BURR, 1940).

(b) Metabolism and Activity

In contrast to non-photosynthetic organisms, which depend on the breakdown of previously synthesized organic substances for their energy needs (chemolitho-autotrophs are a minor exception), photosynthetic plants in light synthesize organic substances while exchanging O_2 for CO_2. The capacity to utilize light energy for simultaneous photolysis of water and CO_2 assimilation is present only in plants possessing both chlorophyll a and some other photosynthetic pigment system. Consequently, there exist two basic, relatively independent metabolic systems in these plants. One system manifests itself through the functioning of the photosynthetic apparatus, the other functions through the pathways of glycolysis (fermentation in some cases) and oxidative metabolism. Both systems are sensitive, but quite differently so, to variations in the concentrations of atmospheric gases. For example, photosynthesis appears to reach maximum efficiency in complete absence of O_2, while under the same conditions oxidative metabolism will have completely stopped. Both photosynthesis and oxidative metabolism can take place at rates high enough to alter the concentrations of O_2 and CO_2 in the ambient water to the extent that the processes may themselves be drastically affected. It is necessary, therefore, to consider effects of dissolved gases on both photosynthesis and dark metabolism in relation to how these two processes, as they take place, may alter the ambient concentrations of the dissolved gases (HUTCHINSON, 1957).

Generally, the ratio of gas exchange via photosynthesis amounts to one O_2 molecule produced to each CO_2 molecule taken up ($O_2/CO_2 = 1$); for oxidative metabolism the reverse is true ($CO_2/O_2 = 1$). At any given instant, the gas exchange rate of a photosynthesizing plant is equivalent to the algebraic sum of the two processes.

Metabolic responses to carbon dioxide

Rates of photosynthesis. Photosynthetic rates may be limited by the availability of CO_2, especially when other factors such as temperature or light are not limiting. On the other hand, respiration can be affected by variations in gas concentration.

While CO_2 is soluble in pure water, the quantity in solution depends on the partial pressure of the gas in the atmosphere (about 0.033% by volume on average), the temperature and the atmospheric pressure. At $0°C$ and 1 atm there is about

1·10 mg/l CO_2 in solution; at 20°C the amount is only half of this. Upon solution, some of the CO_2 reacts with the water to produce carbonic acid (H_2CO_3), which in turn dissociates to hydrogen, bicarbonate, and carbonate ions (H, HCO_3, and CO_3). At neutrality, approximately 21% of the CO_2 in dilute solution is present as CO_2 or H_2CO_3, nearly all the remainder is in the form of HCO_3, and there is a small amount of CO_3. Lowering the pH to 5 results in almost all the CO_2 being present in form of the undissociated acid; at pH 11, most is in the form of carbonate ion.

The presence of carbonates in the water greatly increases CO_2 solubility. The most important carbonate is $CaCO_3$. A suspension of $CaCO_3$, coming into equilibrium with air (0·033% CO_2), will dissolve about 54 mg/l of the carbonate and result in a solution of 65 mg/l of HCO_3, about 50% of which came from $CaCO_3$ and the other 50% from air. Any increase in atmospheric CO_2 results in further solution of the carbonate. The pH of the water is obviously related to the CO_2, HCO_3, CO_3, and cationic concentrations in solution.

In marine waters, high carbonate concentrations and the buffering action of sea water ($\sim$pH 8) tend to maintain stable amounts of CO_2, resulting in CO_2 levels more than adequate for plant metabolic requirements.

On the other hand, fresh waters can undergo large variations in the amounts of available (metabolic) CO_2. The pH of lake water may vary from as low as 1·7 in volcanic lakes containing sulphuric acid to 12·0 and above in closed alkaline lakes, where evaporation concentrates dissolved carbonates. Open lakes tend toward pH values between 6 and 9, depending mainly on the kind of soil or rocks. The buffer (i.e. CO_3 concentration) capacity of fresh waters may be very small, and dissolved CO_2 can vary over a wide range due to biological activities, and to some extent temperature changes. Under such conditions, plants may be exposed to sub- or possibly supra-optimal concentrations of CO_2. In natural waters, wide variations in pH can occur, shifting the ionic equilibrium to a point where CO_2 may be predominately present in only one of its dissolved forms.

Since aquatic plants do survive under a wide range of conditions, they must have mechanisms for utilization of various forms of CO_2. Some plants appear to be restricted to the use of CO_2 only, while others may utilize HCO_3 or even CO_3. The capacity to use the different CO_2 forms does not appear to be immutable.

OSTERLIND (1951) reported that 10-day old *Scenedesmus quadricauda*, and all his *Chlorella pyrenoidosa* cultures are unable to utilize bicarbonate, but 5-day old *S. quadricauda* cultures are capable of photosynthesizing at maximum rate in 10^{-5}m $NaHCO_3$ solution (pH 8·1), which contains negligible free CO_2. GAFFRON (*in:* RABINOWITCH, 1956) observed that the ability to utilize CO_2 depends on the previous history of the cells. *Chlorella pyrenoidosa* cells exposed to high CO_2 concentration (50% in air) are unable to take up bicarbonate upon transfer to CO_2 deficient medium, but are able to do so after a period of adaptation. According to STEEMANN NIELSEN (1960), higher plants can take up bicarbonate through their leaf surfaces. STEEMANN NIELSEN's results on *Potamogeton lucens* show a flow from lower to upper leaf surfaces, and indicate that about 15% of the bicarbonate diffuses through the leaf, 50% undergoes the reaction:

$$2HCO_3^- \longrightarrow CO_3^{--} + CO_2 + H_2O$$

and 35% is split by the reaction:

$$HCO_3^- \longrightarrow CO_2 + OH^-.$$

The CO_2 produced in the leaf is used in photosynthesis, while the hydroxyl and carbonate ions diffuse out of the leaf. The uptake rate of bicarbonate is proportional to the concentration gradient and therefore a passive process. An active transport system is involved in the elimination of hydroxyl ions. Whether the upper or lower surface of the leaf was exposed to light made no difference. No such direct transport of bicarbonate occurred through the thallus of *Ulva lactuca*.

The relationship of CO_2 to plant metabolism is complex. In addition to pH, variations in such parameters as light (Chapter 2.2), temperature (Chapter 3.2), and O_2 tension may influence the utilization of CO_2. For example, leaf stomates close as the internal CO_2 tension increases, causing an apparent decline in photosynthesis when CO_2 concentration is high (CHAPMAN and co-authors, 1924).

The CO_2 requirements of plants appear to vary from species to species. The experimental conditions under which CO_2 assimilation rates are determined also affect results. Table 9-5 lists results obtained by several authors on various plant species. Reported CO_2 requirements range from $\cdot 25 \times 10^{-5}$ mole/l to 2×10^{-3} mole/l for half saturation of photosynthesis. For full saturation, the range is from $0\cdot5 \times 10^{-5}$ mole/l to 4×10^{-3} mole/l. Such large variations can more likely be attributed to availability of CO_2 for photosynthesis arising from experimental conditions than to a wide range of species-specific assimilatory capacities. Large, apparent saturation requirements for CO_2 may be due to depletion in the water immediately surrounding the plants, limiting photosynthesis to the diffusion of CO_2 according to concentration gradients. Photosynthesis by plants in very still water could be limited in this way, especially in poorly buffered lakes or ponds. In the laboratory, adequate stirring or buffering should avoid such effects.

For reasons already stated in the section on *Tolerance*, effects of very high CO_2 concentrations on CO_2 uptake are difficult to test. At low concentrations, photosynthetic rates are limited by the availability of CO_2 down to the minimum concentration. This minimum, at which the CO_2 uptake and CO_2 production by the plant are equal, is known as the CO_2 compensation point. For most plants, the compensation point varies in accordance with changes in O_2 concentration, temperature and light intensity (DOWNTON and TREGUNNA, 1968).

Other effects. Other effects of variations in the concentration of CO_2 are also difficult to observe in plants. In the presence of O_2, plant cells respire and produce CO_2 in the dark. Except for those plants which can meet metabolic needs through fermentation, there is no respiration in the dark if O_2 is absent. Plants unable to ferment organic substrates produce no CO_2 in the dark without O_2, but the lack of O_2 itself can cause rapid cell damage or death. The consequences of low CO_2 without adequate O_2 are masked by the effects of anaerobiosis; on the other hand, in the presence of O_2, there will always be CO_2 production. This situation may explain the dearth of experimental data on the effects of diminished CO_2 on respiration. More research is needed in this area, especially since most cells appear to be able to assimilate CO_2 in the dark (OCHOA, 1948).

At least two effects have been observed in plants deprived of, or given only small amounts of CO_2, which may be more or less related to respiration. (i) There is an increase in the CO_2 compensation point (the minimum CO_2 concentration at which the gas can still be assimilated) with increasing O_2 concentration in the light (FORRESTER and co-authors, 1966). (ii) Photo-oxidation (indicated by greater O_2 consumption in the light than in the dark) may appear (reviewed by RABINO-WITCH, 1945).

Upward shifting in the CO_2 compensation point under some conditions may reflect a higher rate of photorespiration, i.e. of CO_2 turnover (BULLEY, 1969). Photo-oxidation proceeds with the consumption of cellular reserve materials (RABINOWITCH, 1945). These phenomena, which resemble respiration in some respects, may actually result from the functioning of the photosynthetic apparatus (RABINOWITCH, 1945; BULLEY, 1969).

Metabolic responses to oxygen

Rate of photosynthesis. Photosynthetic plants must exchange O_2 for CO_2 in the light, and this exchange may itself be affected by the availability of O_2. Production of O_2 in the light may be depressed if the plants are incubated anaerobically in the dark. According to BOUSSINGGAULT and PRINGSHEIM (*in*: RABINOWITCH, 1945), plants lose their capacity for photosynthesis after exposures to atmospheres of H_2, N_2, or methane. WILLSTÄTTER and STOLL (*in*: RABINOWITCH, 1945) observed that photosynthesis recovers at O_2 partial pressures much lower than required for dark respiration. Anaerobic incubation can lengthen the induction period for O_2 production in *Ulva* and other algae (RABINOWITCH, 1945; VIDAVER, 1964). *Ulva* will also produce less O_2 under N_2 at low light intensities than in air; however, at saturating light intensities, rates of production are about equal in N_2 and air (VIDAVER, 1965). OLSON and BRACKETT (*in*: RABINOWITCH, 1956) noted total inhibition of O_2 production lasting up to 2 mins in a *Chlorella* species after a 3-hr dark anaerobic incubation; after a longer or shorter incubation period (<1 hr or 18–24 hrs), no inhibition occurred. Similar effects of anaerobiosis on photosynthesis have been observed by other workers (RABINOWITCH, 1956; TURNER and BRITTAIN, 1962).

While the mechanism is poorly understood, it is clear that lack of O_2 has an inhibitory effect on photosynthesis. Several hypotheses have been put forward. One of these states that O_2 actually participates in photosynthesis, and supporting evidence for this has been found by some workers, including HILL and WHITTING-HAM (1957), WARBURG and co-authors (1959) and WARBURG and KRIPPAHL (1966). Another hypothesis, initiated by GAFFRON (*in*: RABINOWITCH, 1945), received considerable support; it suggests that during anaerobic incubation certain fermentation products accumulate and thus affect the enzymes of an O_2-liberating system. According to WILLSTÄTTER and STOLL (*in*: RABINOWITCH, 1945), small quantities of O_2 are believed to have an autocatalytic effect on photosynthesis.

There is evidence that, regardless of the duration of a dark anaerobic incubation, O_2 is produced immediately in response to light (FORK, 1963; VIDAVER, 1965). If this is so, it is unlikely that O_2 is actually required for photosynthesis. On the other hand, no endogenous chemical inhibitor of photosynthesis has ever been found. Anaerobiosis appears to prevent the transfer of electrons in the presumed electron transport system of chloroplasts; this has been reported by FORK and URBACH (1965), who used the technique of absorbance-change spectrophotometry. A block

Table 9-5

Determination of carbon dioxide curves during photosynthesis in aquatic plants (After Rabinowitch, 1951; modified)

Species	Medium	Light intensity (in klux if not otherwise designated)	(CO_2) in 10^{-5} mole/l required for		Author
			Half saturation	Full saturation	
Fontinalis antipyretica	CO_2 in water	Gas burner	200	400	Blackman and Smith (1911)
Elodea canadensis	CO_2 in water	Gas burner	100	200	
Fontinalis antipyretica	KHCO₃ soln	0·611	2	>32	Harder (1921)
		2	4		
		18	15 ⎫		
		18	7·7 ⎭		
Cladophora	KHCO₃ soln	20 rel units	12 ⎫	>80	James (1928)
Fontinalis antipyretica	KHCO₃ soln	40 rel units	40 ⎭		
Cabomba caroliniana	Carbonate buffers	~1·4	1·5	≃10	Smith (1937)
		21·9	3	>10	
		282	8	>30	
Cabomba caroliniana	Carbonate buffers	0·41	<1	≃0·5	Smith (1938)
		1·74	1	7·5	
		6·31	2	≃20	
		21·9	3·5	≃30	
		123·0	4	>30	
Myriophyllum spicatum	Acid CO_2 soln	32	17	50	Steemann Nielsen (1946)
	Alkaline soln (pH 8·4)	32	1	10	
Fontinalis antipyretica	Acid CO_3 soln	15	17	50	
	Alkaline soln (pH 8·4)	15	10	35	

Table 9-5—*Continued*

Species	Medium	Light intensity (in klux if not otherwise designated)	(CO_2) in 10^{-5} mole/l required for		Author
			Half saturation	Full saturation	
Chlorella sp.	Carbonate buffers	300 W lamp (20 cm away)	0·4	9	WARBURG (1919)
Chlorella pyrenoidosa	Phosphate buffers (pH 4·6)	Saturating	0·35	0·7	EMERSON and GREEN (1938)
Chlorella pyrenoidosa	Carbonate buffer No. 9	Saturating	~1	~10	EMERSON and ARNOLD (1932)
Hormidium flaccidum	CO_2-air stream over wet cells	2·0 rel units	0·25	2	VAN DER HONERT (1930)
Hormidium flaccidum	CO_2-air stream over wet cells	6·2 rel units	0·6	2	
Hormidium flaccidum		150 W lamp (13 cm away)	0·8	1·6	VAN DER PAAUW (1932)
Gigartina harveyana (red; thick tissue)	Carbonate buffers	60 W lamp (8 cm away)	8	35	EMERSON and GREEN (1934)
Gelidium cartilagineum (red)	Bicarbonate solutions	5 or 8 fluorescent lamps	~5	≥11	TSENG and SWEENEY (1946)
Nitzschia palea (diatom)	Carbonate buffers	75 W lamp (9 cm away)	0·1	0·5	BARKER (1935)

in the transport of electrons would inhibit non-cyclic photophosphorylation and pyridine nucleotide reduction, with subsequent inhibition of photosynthesis. Either increased illumination or added O_2 quickly overcame the block, making it unlikely that O_2 is actually required at this step.

Anaerobic inhibition of photosynthesis is readily overcome by supplying the plants with O_2 or by increasing light intensity. As O_2 is produced only in the light, it seems certain that, if O_2 is essential to the photosynthetic process, plants must retain some residual pool of O_2 or an O_2 precursor which becomes available upon illumination despite anaerobic incubation. If O_2 is not essential, it must be assumed that lack of O_2 deactivates the photosynthetic apparatus only partially and residual activity of the system produces enough O_2 during the induction period in the light to overcome the anaerobic inhibition. Further study seems necessary to resolve this apparent paradox.

The most obvious response to high concentrations of O_2 is an inhibition of photosynthesis, discovered by WARBURG in 1920 and subsequently named the 'Warburg Effect' (TURNER and BRITTAIN, 1962; Table 9-6). High O_2 concentrations inhibit both O_2 production and CO_2 uptake by plants in the light. This inhibition is most apparent under conditions of high light intensities and limiting concentrations of CO_2. Consequently, high O_2 concentrations may reduce the photosynthetic efficiency of plants under natural conditions which would otherwise be optimal.

Inhibition of photosynthesis by O_2 does not appear to be due to stimulated respiration in the light. While an increase of O_2 from 2 to 100% inhibits photosynthesis in *Chlorella pyrenoidosa*, between 41 and 73%, dark respiration is unchanged (WARBURG, 1920). In *Cladophora* spp., increase from 2 to 100% augments dark respiration by 21%, while photosynthesis decreases 25 to 40% (TURNER and co-authors, 1956). A 20% increase in dark respiration has an almost negligible effect on the net gas exchange as the photosynthetic rate can easily be 20 or 30 times the dark respiration. It is generally believed that increase of O_2 from 0 to 20% has little or no effect and that inhibition appears mainly in the 20 to 100% range. However, LUE-KIM (1969) found a strong inhibition in *Chlorella* by even 20% O_2 and nearly complete inhibition at concentrations of 60 to 70% O_2.

TURNER and BRITTAIN (1962) have speculated that the depressing effect of O_2 on photosynthesis is an evolutionary consequence of increased O_2 content of the atmosphere resulting from photosynthesis itself; thus photosynthesis could ultimately be a self-limiting process. The assumption that the presence of gaseous O_2 in the atmosphere is due entirely to the photosynthetic process may not be entirely valid. According to a recent hypothesis by LLOYD (1967), there exists at least a partial equilibrium between atmospheric O_2, dissolved oceanic O_2 and the sulphates in sea water. The existence of such an equilibrium may indicate a limitation of the role of plants in maintaining the partial pressure of O_2 in the atmosphere, to the extent that photosynthetically produced O_2 interacts with a natural chemical equilibrium. Perhaps it is in order to re-examine the concepts according to which the primeval atmosphere was devoid of O_2 and biochemical evolution took place in a reducing atmosphere.

Rate of respiration. Variations in the amount of O_2 dissolved in water appear to affect the respiration of plants growing in it. Water in equilibrium with air contains

Table 9-6

Percentage inhibition of photosynthesis (Warburg Effects, W.E.) in green algae exposed to high carbon dioxide concentrations and high light intensities (After TURNER and BRITTAIN, 1962; modified)

Species	CO_2 (mole/l)	O_2 (%)	Light intensity (lux)	Temperature (°C)	W.E. Range	W.E. Mean	No. of expts	Author
Chlorella pyrenoidosa	91×10^{-6}	2–100	10,000–20,000	25	57–71	65	5	WARBURG (1920)
Chlorella pyrenoidosa (Emerson's strain)	91×10^{-6}	21–100 / 0–21	40,000			30 / 0		BRIGGS and WHITTINGHAM (1952)
Chlorella pyrenoidosa	91×10^{-6}	0–100 / 20–100 / 2·5–20	30,000	25	22–100 / 20–42 / 0–9	35 / 30	12 / 7 / 5	TURNER and co-authors (1956)
Chlorella vulgaris	38×10^{-6}	20–100	30,000	25	24, 27	25	2	TURNER and co-authors (1956)
Chlorella vulgaris (var. *viridis*)	91×10^{-6}	0–100 / 0–20	3,500		33, 47	40 / 0	2	WASSINK and co-authors (1938)
Chlorella ellipsoidea	91×10^{-6}	0–100 / 0–20	25,000	25	17–25 / 0–2	21 / 2	5	TAMIYA and HUZISIGE (1949)
Cladophora spp. (freshwater)	91×10^{-6}	2·5–100	High, near saturation	25	5–74	30 (S.D.15)	56	TURNER and co-authors (1956)
Cladophora spp. (marine)	36×10^{-6}	20–100	17,000	20		43	1	TURNER and co-authors (1956)
Caulerpa spp. (marine)	36×10^{-6}		17,000	20		62	1	TURNER and co-authors (1956)
Enteromorpha spp. (marine)	36×10^{-6}		17,000	20		18	1	TURNER and co-authors (1956)
Bryopsis spp. (marine)	36×10^{-6}		17,000	20		5	1	TURNER and co-authors (1956)
Scenedesmus 'D$_3$'	4%	0–100	10,000	20·5	11, 43	27	2	GAFFRON (1940)

30 to 35 times less O_2 than is present in air itself. While the O_2 concentration in air remains quite constant at about 21%, the O_2 content of water can vary between 0 and more than 100% of saturation. In addition, the diffusion rate in air is some 4 orders of magnitude greater in air than in water. This means that respiration can be limited by the diffusion rate of O_2 both inside the plant and in the surrounding water. These facts can be extremely important to plants growing in waters with large animal populations. Submerged plants are more or less independent of a supply of O_2 while in light; however, they become entirely dependent on the dissolved O_2 in darkness.

Conditions which favour photosynthesis are probably of ecological advantage to plants, as they obtain their materials and energy from growth and reproduction this way. Respiration, on the other hand, consumes reserve materials and biologically usable energy; any condition causing respiratory rates exceeding those required for maintenance would appear, therefore, to be ecologically disadvantageous.

GESSNER and PANNIER (in: GESSNER, 1959) examined the relationship between dark respiration and O_2 in many aquatic plants. Most plants have little control over respiration in response to variations in dissolved O_2. In five species of freshwater algae, decreasing O_2 content of the ambient water was accompanied by a nearly proportional decrease in respiratory rates, suggesting that the intensity of respiration depends on rates of O_2 diffusion. Species of two green alga genera, *Cladophora glomerata* and *Oedogonium* sp., collected in still water (which could be low in O_2 at times), were able to maintain a somewhat higher respiration rate at a reduced O_2 content than the chrysophyte *Hydrurus foetidus*, the chlorophyte *Pithophora oedogonium* and the red alga *Batrachospermum monoliforme*, taken from rapidly flowing water, which presumably is always at least saturated with O_2. GESSNER and PANNIER suggest that still-water plants possess some capacity to adapt to waters with diminished O_2 concentrations. However, no evidence of an ability to sustain an O_2 debt, to be repaid on return to O_2-rich water, was observed in any of the plants.

In water over-saturated with O_2, respiration rates rose with increasing O_2 concentrations (GESSNER, 1959). In 300% supersaturation, a maximum rate of about 4 times that in water at equilibrium with air was measured in *Pithophora oedogonium*; 400% supersaturation reduced respiration to about $2\frac{1}{2}$ times the rate in air. Diatoms collected from surface waters of a eutrophic lake, where O_2 was already in high concentration, nevertheless showed increases in respiration up to 3 times the rate in air when placed in 400% supersaturated water. Maximum respiration rates tended to be reached within an hour after transfer of *Pithophora oedogonium* from air-equilibrated water to that with a different O_2 concentration, and persisted at the new rate with only minor fluctuations for up to 5 hrs (GESSNER, 1959).

Respiratory rates of marine algae also depend on the O_2 content of the water. GESSNER and PANNIER (in: GESSNER, 1959) found considerable interspecific variations in algal responses to decreased O_2 concentration. The red alga *Laurentia papillosa* was scarcely affected in water with an O_2 content 20% less than air equilibrium. At the same O_2 concentration, *Ulva lactuca* respired at only 50% of the air rate; several other species, including *Cystoseira barbata*, *Stypocaulon scoparium*, *Padina pavonia*, *Colpomenia sinuosa*, *Sargassum hornschuii* and

Gracilaria armata, were intermediate in response to these extremes. However, with diminishing O_2 concentrations, respiratory rates decreased in all the plants tested.

The capacity to recover from exposures to critically low O_2 concentrations also differs significantly in different algae. According to GESSNER (1959), the green alga *Ulva lactuca* exhibits the lowest recovery capacity, while *Laurentia papillosa* recovers fairly completely. The brown alga *Cystoseira barbata* shows the smallest degree of irreversible damage due to O_2 lack; other species were intermediate.

The respiratory rates of three *Anabaena* species (*Anabaena inequalis*, *Anabaena flos aquae* and *Anabaena spirodes*), collected from fresh water, are only moderately sensitive to variations in dissolved O_2 (GESSNER, 1959). In 10% air-saturated water, respiratory rate still amounts to 90% of that in 100% air-saturated water. At 5 times air saturation, respiratory intensity doubles. At intermediate concentrations, respiration rates are nearly proportional to the concentration of O_2. In contrast to the green, red and brown algae examined, these blue-greens appear to be able to maintain a relatively constant respiration intensity despite large fluctuations in O_2 supply.

For some species of aquatic angiosperms, GESSNER and PANNIER (*in*: GESSNER, 1959) report responses similar to those shown by algae. *Ranunculus flaccidus*, taken from standing water with occasionally low O_2 contents, maintained much higher respiration rates with diminished O_2 concentrations than individuals of the same species collected from running streams with continuously high O_2 contents. Recovery from exposure to critically low O_2 concentrations is similarly greater in still-water plants than in those from running water. Generally, respiration rates of higher plants respond less sensitively to low O_2 levels than those of any algae hitherto tested. High O_2 concentrations stimulate respiration in angiosperms and algae. The extent of stimulation varies considerably in the various species examined.

(c) *Reproduction*

The effects of dissolved gases on rate and mode of reproduction of aquatic plants are poorly documented and not well understood. Since growth and reproduction are end points of a variety of metabolic processes, and metabolic rates are known to be highly sensitive to variations in ambient CO_2 and O_2 concentrations, it is probable that these gases interact primarily with general metabolism rather than with specific reproductive functions. Interactions between variations in the CO_2 and O_2 content of the water and assimilated light energy are to be expected in plants growing or germinating in the light. The effects of dissolved gases vary with the life-cycle stage and the physiological condition of the plant; a germinating seed does not show the same responses as an adult plant, and a resting algal cell responds differently from a cell undergoing fission.

Attempts to correlate plant reproduction with dissolved gas concentrations of open waters are few. Measurements of total carbon content, carbon assimilation rates and O_2 exchange, as commonly carried out in primary productivity determinations, are of little use in relating gas availability to reproductive rates or functions. Most of the information available is related to laboratory experiments and may not reflect accurately plant responses in natural environments.

Reproductive responses to dissolved gases have been observed in laboratory cultures of unicellular algae (OSTERLIND, 1948a; TAMIYA and co-authors, 1953; BRITTAIN, 1957; GESSNER, 1959). As with other metabolic functions, sub- or supranormal concentrations of either CO_2 or O_2 may retard or inhibit reproduction.

Because they permit the experimental separation of cell division, which is dependent on CO_2 in plants, from growth processes, synchronous cultures of unicellular algae are useful in determining the effects of dissolved gases on cell division. In these cultures, the cells have identical life histories and all are in the same stage of the life cycle at any given time. Experimental treatments should have the same effects on all the cells of a culture, thereby vastly simplifying the interpretation of results.

Carbon dioxide requirements for reproduction

SOROKIN (1962b) reported some effects of CO_2 concentration on cell division (asexual reproduction) in *Chlorella pyrenoidosa* (7–11–05, high temperature strain). His cells were previously grown 8 hrs in the light in medium bubbled with 5% CO_2, washed, then transferred to other media of various bicarbonate concentrations and air-CO_2 mixtures for a 16-hr dark period. In these experiments, growth takes place autotrophically in the light and cell division occurs during the dark period. Therefore, a complete separation of cell division or asexual reproduction from growth processes in the cells is evident. The ability of the cells to divide during the dark period depended on the relationship between the percentage of CO_2 in the air and the HCO_3 concentration of the medium. Adjusting the ratio of CO_2 to HCO_3 permitted more than 90% of the cells to divide in all cultures. More than 90% cell division occurred in cultures given air only ($\sim$0·033% CO_2) with HCO_3 concentrations in the range of from 10^{-5} to $10^{-2\cdot5}$ M; few cells divided when HCO_3 exceeded 10^{-2} M. Division by cells transferred to 1% CO_2 was over 90% only with HCO_3 between about 10^{-3} and $10^{-1\cdot5}$ M. Inhibition was nearly complete with less than $10^{-3\cdot5}$ M and more than 10^{-1} M HCO_3.

Increased CO_2 concentration to 5% extended the higher limiting concentration for cell division slightly, but 50% inhibition appeared again when HCO_3 was decreased to slightly under 10^{-3} M. When CO_2 was raised to 25%, cell division occurred at still higher HCO_3 concentrations (50% division at 10^{-1} M HCO_3), but there was nearly 40% inhibition at $10^{-2\cdot5}$ M HCO_3. Cell division was greatly impeded by 50%, and completely blocked by 90% CO_2 at any HCO_3 concentration. Clearly, however, the inhibitory effect of CO_2 is offset by increasing the HCO_3 concentration. Variations in pH must be an important, but not the only, factor in affecting cell division in these experiments. Examination of the results in terms of variations in pH show that, in air, more than 50% of the cells divided at pH 9·5; giving 1% CO_2 lowered the pH of 50% division to 8, and 5% CO_2 still further to nearly 7·5. In 1 and 5% CO_2 cultures, cell division was reduced to 50% at pH 6·5 and 6·2 respectively. SOROKIN (1962a) suggests that CO_2 might be a natural regulator of cell division, but the ecological advantage for such regulation as well as its mechanism remain uncertain.

No CO_2 requirement was found for mating in the gametes of *Chlamydomonas moewusii* (LEWIN, 1956). However, half the male gametes of *Chlamydomonas eugametos* cultures were reported by STIFTER (*in*: COLEMAN, 1962), to lose their activity within an hour after removal of CO_2. On return to air ($0\cdot033\%$ CO_2), full activity is resumed within a few minutes. Involvement of CO_2 in gametic activity is suggested by the reported lack of a requirement for added CO_2 by gametes of darkened cell suspensions. The gametes which have previously received low CO_2 levels are more active on initial exposure to dim light than those which have received the gas in the dark.

Sexual activity of *Chlamydomonas eugametos* is drastically curtailed within minutes after giving 5% CO_2. This inhibition suggests an optimal CO_2 concentration for gamete activity which is probably close to the endogenous level of the gas in gametes of the dark cell suspensions; upon illumination, a high incidence of activity results, but when CO_2 is added, the optimal concentration may be exceeded. It would not be surprising if a CO_2–bicarbonate interaction were discovered to be involved in the mating response.

Oxygen requirements for reproduction

Since uni-algal cells do divide in darkness, a large metabolic CO_2 requirement seems unlikely for reproduction. Presumably, energy requirements for cell division may be met by respiration. O_2 should then be necessary for reproduction of algal cells; ample evidence exists that this is so. As light and O_2 are in many respects equivalent in photosynthetic plants, effects of low concentrations of dissolved O_2 might appear only in darkness or under conditions of light limitation. On the other hand, effects of high light intensity or high O_2 concentrations on plant cells may sometimes be indistinguishable. These considerations make it difficult to ascertain exactly what influence variations in O_2 availability have on cell division or reproduction of algae.

There is still a further difficulty. Synchronous cultures which should yield much information about reproductive responses are normally grown with an air-CO_2 mixture bubbled through the cell suspension, which remains unchanged throughout the entire light-dark cycle. In natural environments, however, gas concentrations are highly variable and dependent on gas exchanges by whatever organisms are present. In synchronous cultures, therefore, one would not expect cell division rates to reflect any limitation of dark respiration by the availability of O_2 and to show secondary effects resulting from interactions of all the gases present (which vary only slightly in concentration with time) with all cellular activities.

Very different conditions exist in natural waters with dense plant populations; animals are usually present and affect the gas contents of the water. Here, O_2 concentrations may be several times the air equilibration value after periods of high photosynthetic activity, and drop to nearly 0 at night, with concomitant high CO_2 values. Profuse algal blooms, resulting from extremely high cell division rates, do occur under such conditions (FOGG, 1966). Definitive experiments on the effects of variations in single gases, including O_2, on reproduction appear to have yet to be performed.

BRITTAIN (1957) found that rates of reproduction in cultures of *Chlorella* spp., in light and with CO_2 present, decrease on raising the O_2 concentration from air equilibrium (21%) to 95%. Rates of reproduction in fact, decrease almost linearly with increasing O_2 concentrations from $0\cdot5\%$ upwards. These results resemble the depressing effect of O_2 on photosynthesis and probably are due ultimately to a

I

decreased photosynthetic output, rather than to any direct effect of O_2 on cell division.

PANNIER (*in*: GESSNER, 1959) reported an unsuccessful attempt to assess the effects of various O_2 concentrations on cell division rates in *Chlorella pyrenoidosa*. Cultures were kept 11 days in the dark at 100%, 13% and 442% of atmospheric O_2 equilibrium values, but a tendency of the cells to clump together during treatment made it impossible to determine actual cell numbers.

In *Chlorella ellipsoidea*, cell division rates were determined in synchronous cultures by LUE-KIM (1969). Some cultures were transferred after a 14-hr light period from air-5% CO_2 bubbled suspensions to closed vessels containing various amounts of dissolved O_2. All cells remaining in the bubbled suspensions divided into 4 daughter cells within the first 2 hrs of a subsequent 10-hr dark period. In the closed vessels, any cell division which occurred during 10 hrs of darkness depended on added O_2. At an initial concentration of 6·6 ml O_2/l (these vessels were closed and no light was available; hence O_2 decreased while CO_2 increased), the cell number approximately doubled, and at lower concentrations no divisions occurred. The normal 4-fold increase was attained with 13·2 ml O_2/l. Increasing O_2 to 26·4 ml/l still yielded 4 daughter cells per cell, but at higher concentrations the increase in cell number began to drop. In these experiments, at the higher O_2 concentrations not all of the gas dissolves normally. It is nevertheless possible to observe the effects of very high O_2 concentrations by applying hydrostatic pressure to the vessels (gas concentration is approximately proportional to hydrostatic pressure; Chapter 8.0). Up to 26·4 ml O_2/l, pressure did not affect cell division, but at higher concentrations, division was repressed and completely inhibited by 50 ml O_2/l. Pressure of up to 200 atm had no effect on cell division at optimal O_2 concentrations. Calculation reveals that about $2·6 \times 10^{-10}$ ml O_2 (4×10^{-4} mg) per mother cell is required to double (it is not known if all the cells divided into 2, or if some divided into 4, while others did not divide) and $5·2 \times 10^{-10}$ ml O_2/cell to quadruple the initial cell number. Inhibition began to appear at about $10·4 \times 10^{-10}$ ml O_2/cell (about equal to 1 atm of 100% O_2) and was complete with 20×10^{-10} ml O_2/cell. On a dry weight basis, the O_2 requirement was approximately 35 mg O_2/g dry weight for some division to appear: 70 to 140 mg O_2/g permitted normal division into 4 daughter cells and greater O_2 amounts were inhibitory.

These O_2 requirements for cell division appear rather high, especially in view of O_2 consumption rates reported by GESSNER (1959) of about 1 to 3 mg O_2/hr/g dry weight for several non-synchronized freshwater algae species. In LUE-KIM's (1969) experiments, it is possible that inhibition of cell division by O_2 concentrations is a result of O_2 stimulated respiration (GESSNER, 1959) in which materials required by daughter cells are consumed instead by oxidation. Unfortunately, these closed vessel experiments yield no information about actual O_2 consumption or the effects of the CO_2 produced by respiration on cell division.

(d) Distribution

Any discussion of plant distribution in relation to the important dissolved gases must, at this time, remain conjectural. Light (Chapter 2), rather than dissolved CO_2 or O_2, is by far the most significant determinant of the vertical distribution of marine (and perhaps many limnic) plants. O_2 is probably most important in densely populated waters, especially in the dark.

Plants able to float on surface waters may exhibit adaptations enabling them to avoid O_2 deficiencies of subsurface waters. As pointed out by GESSNER (1959), a capacity for algae to maintain respiration rates despite decreases in dissolved O_2 concentrations (p. 1482) is probably an ecological advantage for growth below the surface of standing waters; plants without this ability are restricted to the surface or flowing, well-aerated waters. Of the higher plants, only those with efficient O_2 transport systems are likely to survive when rooted in soils deficient in O_2, a common situation in many ponds and lakes. The satisfaction of metabolic energy requirements in the dark through fermentation would also appear to be of adaptive advantage in many circumstances. Seeds of some plants germinate anaerobically, while some require O_2, and still others appear to be influenced by CO_2. Similarly, variations in dissolved gas concentrations may affect gametogenesis and sporulation, as well as the germination of the reproductive cells.

All of the above characteristics occur in some plants, yet this author knows of no systematic effort clearly relating any of them to influences of dissolved gases on species distribution. Fertile fields lie waiting here for the marine plant ecologist.

Water eutrophication, a process becoming more and more associated with pollution of sea areas due to human activities (Volume V), may often cause upset in the previous patterns of fluctuations of CO_2 and O_2 concentrations, together with the production of high levels of such gases as H_2S or methane. The eutrophic environment has the ecological effect of selecting a few plants and other organisms which then tend to establish populations with high individual numbers. Poorly understood in this regard, is whether eutrophic plants survive because they are especially well adapted to such environments or because reduced competition with — or predation by — other species enhances their development. A systematic approach, involving step-by-step alterations of environmental parameters, including dissolved gases, should be valuable in obtaining answers to this question.

(3) Structural Responses

The difficulties which beset attempts to relate changes in a single environmental factor to plant responses are especially evident in any consideration of the effects of dissolved gases on plant structures. Few observations lend themselves to detailed understanding of the mechanisms by which a specific plant structure might be directly affected by the concentration of a metabolic gas. Reports relating plant structure to environmental gases have been published by GESSNER (1959) and CARR (1961).

(a) Size

It is obvious that photosynthetic plants cannot achieve normal body size if growing under CO_2-deficient conditions. Size increase in plants is related to dry weight increase, and with insufficient CO_2 they are unable to sustain the synthesis, in light, of their constituent materials.

In what ways size of plants may be affected by excess CO_2 is much less certain; little investigation appears to have been done of this problem. Presumably, the size attained by a plant is related to its general metabolism. For this reason, adverse effects of high CO_2 concentrations on final size, resulting in part from high hydrogen concentrations, might be expected. There could be important exceptions, however.

(b) External Structures

An elegant study by PAASCHE (1964) could serve as a model for investigators who wish to obtain basic information about gas effects on external structures of plants. PAASCHE examined the relationship between photosynthesis and coccolith formation in the marine planktonic coccolithophorid *Coccolithus huxleyi*. Coccolithophorids form geometrically complex skeletal structures within their cytoplasm. These coccoliths migrate to the external surface of the cells, remaining there in an organized pattern, to form a kind of armour around the cell. Since the coccoliths are composed of $CaCO_3$, their formation might be expected to be involved with photosynthesis and CO_2 assimilation.

PAASCHE (1964) demonstrated that this expectation is correct. Some of his conclusions are (i) coccolith formation depends on light and CO_2, none are formed in the dark; (ii) CO_2 concentrations above 0·3mM depress rates of photosynthesis and coccolith formation; (iii) photosynthesis and coccolith formation are probably distinct processes in that HCO_3 is the main carbon source for coccoliths, and photosynthesis utilizes either CO_2 or HCO_3; (iv) both photosynthesis and coccolith production decrease under conditions of excess CO_2 and this repression depends on pH. For a similar relationship between pH and cell division consult SOROKIN (1962b) and the section on *Reproduction* in this chapter.

Several investigators have studied the effects of variations in dissolved gas concentrations on the size and shape of plant cells. Among these is NEEB (1962), who used the water net alga *Hydrodictyon reticulatum* as object. In this plant, zoospores produce cells which begin to form little nets within a few days of spore release. In diffuse light, the final size of individual cells depends on the O_2 concentration of the medium. After 14 days, cells in air-saturated water (21% O_2) reach a larger size than those in either 63% or 4% O_2. In air-saturated water, the average cell size is $1280 \times 80\mu$; in the high O_2 concentration the dimensions are $350 \times 40\mu$, in the low concentration only $194 \times 40\mu$. No injury to the cells was discovered as a result of any of the treatments.

The interpretation offered for the small size of cells grown in high O_2, which is in accordance with GESSNER'S (1959) view, is that plant respiration under these conditions is greater than the capacity of the cells to meet metabolic needs through photosynthesis. While such an interpretation may be partially correct, direct inhibition of photosynthesis by O_2 must also be involved. Retardation of growth in low O_2 may result from the inability of dark respiration, when supplied with too little O_2 to metabolize photosynthates produced in the light. It would be of interest to determine whether the effects of O_2 deficiency could be overcome by offering additional light.

(c) Internal Structures

In a study by PANNIER (*in*: GESSNER, 1959), sections of single thalli of *Ulva lactuca* were maintained in sea water at various O_2 concentrations. Cells of the tissue grown in water in equilibrium with air (21% O_2) maintained the normal size and appearance (about $12 \times 14 \mu$). With 93% O_2, cells measured only $5 \times 9 \mu$, while the cells of tissue aerated in sea water with about 5% O_2 became enlarged to an average of $16 \times 19 \mu$. Other than these differences in size and shape, no adverse effects of either the normal air (21% O_2) or higher O_2 concentrations are suggested. On the other hand, sections of tissue receiving 6% O_2, producing the extraordinarily large cells, showed signs of disintegration. This disintegration appeared to result from plasmolysis of some of the cells. GESSNER'S interpretation of these results is that the high O_2 concentration induced more frequent cell divisions, and hence smaller cells, than the air-saturated controls, and that low O_2 retarded division, giving rise to larger cells. While these interpretations may be correct, others are certainly possible. For instance, it is well known that environmental shocks can give rise to gametogenesis or sporogenesis in members of the Chlorophyta, including the Ulvales. Perhaps tissue exposed to high O_2 concentrations produced numerous cell divisions because of a partial activation of a reproductive response. Production of large cells, which occurred with less than normal O_2, suggests, well enough, a retardation of cell division, but does not account for a tendency toward cell plasmolysis. This last effect could be explained more easily as resulting from a weakened cell wall structure. Bacteria without cell walls (SALTON, 1964) may be produced by growing them in the presence of lysozyme (an enzyme which breaks down cell wall material) or in penicillin, which interferes with the synthesis of cell wall materials (Chapter 10.1). The resulting protoplasts are spherical in shape, larger than normal cells, and very susceptible to plasmolysis. Possibly, O_2 deficiency has interfered with cell wall synthesis in the *Ulva lactuca* cells, thereby inducing tissue disintegration, large cell size, and cell plasmolysis. For other interpretations, the review by CARR (1961) should be consulted. Possibly, in this instance also, additional light might make up for the lack of O_2.

9. DISSOLVED GASES

9.3 ANIMALS

F. J. Vernberg

(1) Introduction

Dissolved gases of principal importance in animal ecology are oxygen, carbon dioxide, hydrogen sulphide, nitrogen and pollutant gases. Although one of the more obvious roles of dissolved gases is in the respiratory metabolism of animals, other functions have been described. For example, in siphonophores, gases are found in floating structures. Earlier it was thought these gases were of atmospheric origin. However, Wittenberg (1960) demonstrated the presence of carbon monoxide in addition to atmospheric gases; this gas originates in the gas gland tissue as a product of serine metabolism. Additional studies on the dynamics of this process have been published by Lane (1960) and Larimer and Ashby (1962). The swimbladder of fishes has received wide attention. Only a few references are given here as this subject is tangential to the main theme of this chapter (Jones and Marshall, 1953; Scholander and co-authors, 1956; Kanwisher and Ebeling, 1957; Denton, 1961).

The concentrations of the various dissolved gases may vary markedly, either when comparing different habitats or in the same habitat at different times. Oxygen may be used as an example. Certain habitats are more or less anaerobic, e.g. mud flats or the deeper regions of the Black Sea; others contain reduced levels of oxygen, e.g. the oxygen minimum zone (Richards, 1957) or intestinal tracts of organisms inhabited by parasites. Turbulent waters on rocky shores or waters inhabited by a dense vegetation, on the other hand, may be supersaturated with oxygen (Shelford and Powers, 1915). On a temporal basis, the concentration of gases in one habitat may vary regularly, depending on storm action and pollution (ZoBell, 1962). Broekhuysen (1935) reported the absence of oxygen in *Zostera* beds at 5.00 a.m., while during daylight hours a peak of 260% saturation was reached at 3.00 p.m. A similar situation had been reported earlier for the eel-grass vegetation in Puget Sound, Washington, USA (Powers, 1920). Diurnal fluctuations in oxygen content also have been observed in salt marsh pools (Nicol, 1935), mangrove swamps, and waters associated with the Great Barrier Reef (Orr and Moorhouse, 1933). Consult Harvey (1945), Sverdrup and co-authors (1942), Marshall (1954) and Chapter 9.0 for further discussions in regard to variations in concentration of dissolved gases in marine environments.

Marine animals exhibit varying degrees of dependence on the presence or absence of dissolved gases. Typically, there is a relationship between the physiological capacity of an animal and its ecological requirements. The diversity in marine animals ranges from free-living to parasitic species, from forms occupying rather constant environmental conditions in the deep sea to those inhabiting temperate zone intertidal areas marked by extreme daily and seasonal fluctuations. Accompanying the diversity in habitat conditions is a marked divergence in physiological capacity. Some forms require high ambient oxygen concentrations while others appear to survive in the absence of oxygen. Since, in some habitats, the concentration of dissolved gases may vary as a function of time, animals found there may have to adjust to the alternating stresses of high and low concentrations. In addition to such interspecific differences, the requirements for dissolved gases may vary in individuals, with life-cycle stage or physiological condition.

In general, a given intensity of an environmental factor is considered to be lethal to a particular population if a certain percentage, usually 50% of the individuals examined, do not survive indefinite exposure. As will be illustrated later, percent mortality and survival duration may be influenced by a number of factors. The critical portion of the factor gradient, reaching to lethal conditions, is called the zone of tolerance or resistance. Typically, the lower and upper zones of tolerance include a region of minimal and maximal intensities. The intensities in the mid-section of the gradient are compatible with continued life and hence called the zone of compatibility. This chapter reviews the principal papers dealing with the influence of dissolved gases on marine animals, and attempts to correlate physiological and ecological data.

(2) Functional Responses

(a) Tolerance

The ability of an animal to survive critical concentrations of dissolved gases varies as a function of endogenous properties, e.g. life-cycle stage, body size, sex or rate of acclimation. In nature, an animal may not be subjected to the stress of one environmental factor only at any given time. Therefore, assessment of the lethal effects of dissolved gases must take into consideration multifactorial influences (e.g. DOUDOROFF, 1957; FRY, 1964; Chapter 12).

Oxygen

The influence of low levels of oxygen is not easily determined under natural conditions. Anoxic habitat conditions are always accompanied by other alterations in the environment, such as the accumulation of CO_2 and/or H_2S. Laboratory experiments may provide the means for controlled variation of a single, or a few, environmental factor(s), but there is reasonable doubt concerning the investigator's ability to duplicate adequately 'natural' conditions for maintenance, growth and reproduction.

In the sea, oxygen has not been thought to be a limiting factor. It is absent only in isolated areas with poor circulation or unusually high rates of oxygen utilization. However, BRONGERSMA-SANDERS (1957) reported eight cases of mass mortality in oceanic waters, apparently caused by oxygen deficiency and/or by poisonous gases, particularly H_2S. TULKKI (1965) described the disappearance of the benthic fauna from the southern Baltic Sea as a result of oxygen deficiency. In general, survival limits of marine animals to extreme concentrations of dissolved gases are poorly known when the number of species and the diversity of habitats are considered.

Different criteria of survival must be kept in mind when discussing the limiting nature of an environmental factor. At the individual level, death resulting from rapid loss of protoplasmic integrity may be immediate or result from prolonged stress preventing the individual from continuing one or more of its vital processes, such as feeding. At the population level, death is identical with complete cessation of reproductive processes. Hence, the lower and upper lethal limits lie closer together in the case of population survival than for the survival of individuals.

Survival experiments, performed by suspending the polychaete *Capitella capitata* in plastic cages in various field test sites of different oxygen content, yielded results which illustrate this point. In regions of little or no dissolved oxygen, the test individuals ceased feeding and survived for several days only; with a median dissolved oxygen content of 2·9 ppm, feeding ability was restored; when the median oxygen content rose to 3·5 ppm, the experimentals were able to reproduce and complete their life cycle (REISH and BARNARD, 1960).

In a number of species of salmonoid fishes, information on the lower lethal oxygen concentrations has been summarized in tabular form by TOWNSEND and EARNEST (1940). The 50% mortality level for juvenile American shads *Alosa sapidissima* occurs at dissolved oxygen values of less than 2 ppm (TAGATZ, 1961). These results are similar to those of DAVISON and co-authors (1959), who worked with juvenile coho salmon *Oncorhynchus kisutch*. A very active species of fish, such as the scup *Stenotomus chrysops*, dies quickly when the oxygen tension falls below 16 mm Hg, while the sluggish toadfish *Opsanus tau* lives for 24 hrs with oxygen tensions between 0 and 1 mm Hg (HALL, 1929). DAS (1934) reported species differences in the ability of air-breathing fish to survive when prevented from reaching the surface of the water; *Anabas* sp. dies within 35 to 40 mins, while *Pseudapocryptes lanceolatus* survives for 15 to 20 hrs.

The critical level of oxygen content for survival of the cod *Gadus callarias* is 2·7 ml O_2/l. This value is in agreement with the field observation that the cod is not usually found in water which has so little dissolved oxygen (SUNDNES, 1957). For an exposure period of 4 days, the median tolerance limit of sculpin *Cottus perplexus* is estimated to be about 1·46 mg/l (DAVISON and co-authors, 1959).

Artemia salina survives equally well if exposed for 14 days to water with 100%, 21% and 4% oxygen saturation (FOX and TAYLOR, 1955). The brackish-water copepod *Eurytemora hirundoides* is extremely tolerant to low oxygen levels and survives in polluted areas of the River Tyne (UK), except where the oxygen has entirely disappeared (BULL, 1931). However, harpacticoid copepods from the bottom muds of the Clyde Sea survive only 24 hrs in the absence of oxygen (MOORE, 1931). Additional experimental data on the oxygen tolerance of various species of mysids and gammarids have been cited by CASPERS (1957).

In laboratory studies, the mud-dwelling polychaete *Scoloplos armiger* survives better in poorly aerated water (4% oxygen) than in fully aerated water (21% oxygen). Another polychaete, *Arenicola marina*, which lives in sandy mud, survives equally well at 4%, 10% and 21% oxygen; but 100% oxygen gradually becomes detrimental. A third tube-dwelling polychaete, *Sabella pavonina*, survives exposure to high oxygen concentration, but cannot live in water which is only 10% oxygen saturated (FOX and TAYLOR, 1955).

Field observations in different habitats of a harbour demonstrated that a polychaete species was dominant in a given type of bottom and occurred in other bottoms only in reduced numbers. These different bottom types had different oxygen regimes (REISH, 1955).

Differential tolerance to reduced oxygen tensions has been observed in cercarial stages of various trematodes. Cercariae of *Schistosoma mansoni*, which are very active and rapid swimmers, are affected within 1 hr by oxygen lack, and death results within 4 hrs (OLIVIER and co-authors, 1953). The more robust, but also quite active, cercariae of *Himasthla quissetensis* survive anaerobic conditions for 6 to 8 hrs, but very low oxygen tensions will cause these larvae to become quiescent. In contrast, the tailless, non-swimming cercariae of *Zoogonus lasius* do not die until after 12 hrs of anaerobiosis (VERNBERG, 1963).

Stage of life cycle. Different life-cycle stages may exhibit different degrees of tolerance to extreme levels of dissolved gases. Stage V of *Calanus finmarchicus* is more resistant to oxygen deficiency than the adult (MARSHALL and co-authors, 1935). Adults died within 1 hr at $1 \cdot 4$ ml O_2/l; stage V individuals, however, were not moribund until after about 2 hrs at this concentration; they died after 1 hr at an oxygen level of $0 \cdot 7$ ml/l. In species where the various life-cycle stages occupy different habitats, tolerance differences may be even more distinct.

In salmonoid fishes, the response to low oxygen is dependent on age (BISHAI, 1962). Alevins (1 to 4 weeks after hatching) are not responsive to abnormally low oxygen tensions. Fry (5 to 16 weeks after hatching) avoid water which has a concentration of less than $4 \cdot 6$ mg O_2/l. Older individuals (26 weeks after hatching) are less sensitive; they avoid water with still lower oxygen concentrations ($3 \cdot 0$ mg O_2/l).

Body size. Large-sized prawn *Penaeus indicus* are less sensitive to oxygen deficiency than are smaller individuals (SUBRAHMANYAN, 1962). For example, the lethal oxygen level for 10 g individuals is $3 \cdot 80$ ml/l; for $0 \cdot 6$ g specimens, it is $1 \cdot 49$ ml/l. Large *Fundulus parvipinnis* are more tolerant than smaller individuals (KEYS, 1931). But JOB (1955, 1957) reported that, in certain fishes, body size is not a variable in determining the lethal oxygen level.

Sex. Females of *Calanus finmarchicus* appear to be slightly more resistant to anoxia than males (MARSHALL and co-authors, 1935). In contrast, males of the fish *Fundulus parvipinnis* are more resistant to asphyxiation than are females during the breeding season (KEYS, 1931).

Parasitism. Parasitized and diseased *Fundulus parvipinnis* die more quickly than 'healthy' fish when exposed to oxygen-deficient waters (KEYS, 1931). OLIVIER and co-authors (1953) reported that individuals of the snail *Australorbis glabratus* infected with *Schistosoma mansoni* are more susceptible to oxygen lack than are uninfected specimens.

Moulting stage. Tolerance to anoxic conditions varies with the stage of moulting in the American lobster *Homarus americanus*. Hard-shelled lobsters survive exposure to sea water with an oxygen content of 0·3 mg O_2/l for an average time of 1040 mins, while the average for moulting individuals is only 750 mins (McLEESE, 1956).

Temperature. At 31°C, *Mya arenaria* survives in oxygen-free media for about 24 hrs, while at 14°C survival time is 8 days (COLLIP, 1921). This response may reflect differences in metabolic requirements. However, copepods are slightly more resistant to low oxygen content at 5°C than at 15°C, even though the metabolic rate is similar at these two thermal points (MARSHALL and co-authors, 1935). Survival of juvenile coho salmon *Oncorhynchus kisutch* to hypoxia is relatively independent of temperature over the range of 12° to 20°C (DAVISON and co-authors, 1959); but with the approach of high sublethal temperatures, this species becomes very sensitive to low dissolved oxygen levels.

Anaerobiosis. The reviews by VON BRAND (1945) and BEADLE (1961) have dealt, in detail, with the general problem of animal survival under anoxic conditions. There exists a considerable diversity in the responses of marine animals to oxygen lack: some aerobic species can tolerate only short periods of oxygen lack, while obligatory anaerobes die when exposed to normal levels of oxygen. Between these extremes, some animals may live under anaerobic conditions for long periods of time, but can utilize oxygen when it becomes available (facultative anaerobes), whereas others live under aerobic conditions but can survive rather prolonged periods of oxygen lack. In general, there is a positive correlation between the oxygen content in the habitat and the response of the animal to anaerobiosis.

During prolonged periods of exposure to anaerobic conditions, marine animals may show an adaptive response by utilizing metabolic processes which do not need oxygen and thus may incur an oxygen debt. Subsequently, when they are exposed to an oxygen-enriched medium, this debt is repaid, as reflected by initially increased rates of oxygen uptake.

Marine invertebrates inhabiting vegetations of large brown seaweeds show differences in surviving the anaerobic conditions which may be encountered at low tide. Slow-moving, non-swimming animals, which are more or less permanent residents of these vegetations, can survive anaerobiosis for approximately 16 hrs at 25°C. In contrast, species capable of leaving this habitat are much more sensitive to oxygen deficiency. Representatives of one of these die within minutes, specimens of two others survive half an hour to 3 hrs of anaerobiosis (WIESER and KANWISHER, 1959).

Annelids vary in their tolerance to anaerobic conditions. PACKARD (1905) states that *Amphitrite* sp. and *Nereis* sp. could survive for 1 day in the absence of oxygen; JACUBOWA and MALM (1931), working with a number of annelid species, report survival times of 1 to 10 days; *Arenicola marina* lives 9 days (HECHT, 1932); *Owenia fusiformis* tolerates anoxia for 21 days (VON BRAND, 1927).

Certain molluscs have the ability to survive anaerobiosis for long periods of time: Examples are species of the genera *Mya* (BERKELEY, 1921; COLLIP, 1921; RICKETTS and CALVIN, 1948), *Saxidomus* and *Paphia* (BERKELEY, 1921), *Ostrea* (GALTSOFF and WHIPPLE, 1930) and *Littorina* (PATANE, 1946 a, b). COLLIP (1921) found, for example, that *Mya arenaria* can survive anaerobic conditions for weeks when kept at low temperatures; at 14°C it survives for 8 days, and at 31°C for about 24 hrs.

The fiddler crab *Uca pugnax* can survive 24 hrs of anoxia at 21°C (TEAL and CAREY, 1967). The response pattern to anoxia is typical, in that lactic acid is accumulated, glycogen apparently decreases, and an oxygen debt is incurred. These physiological properties enable *Uca pugnax* to remain burrowed in salt marshes over long periods of time.

Adaptive metabolic differences in the ability to incur oxygen debts have been reported in two species of snails from different habitats by BANNISTER and co-authors (1966): *Gibbula divaricata*, a subtidal species, is usually exposed to anaerobic conditions longer than the midlittoral species *Monodonta turbinata*. Since pyruvate is not completely oxidized, the ability of tissue homogenates from *G. divaricata* to consume more pyruvate than a similar tissue preparation from *M. turbinata* was considered by these authors to demonstrate a better capacity to withstand an oxygen debt.

The annelid *Arenicola marina*, when exposed to anaerobic conditions, consumes glycogen; but, unlike the fiddler crab and the snails mentioned, does not accumulate lactate or pyruvate, which could account for the apparent absence of an oxygen debt (BORDEN, 1931; DALES, 1958). Another worm, *Owenia fusiformis*, which can withstand anaerobiosis for a longer period of time (21 days), does not deplete its glycogen stores and appears to survive by becoming quiescent (DALES, 1958). In 1955, ELIASSEN re-investigated the earlier suggestion that, during periods of anoxia, the haemoglobin of *Arenicola marina* could function as a reservoir of oxygen (BORDEN, 1931). Based on analyses of water samples, taken in the vicinity of this worm's burrow, and blood samples, ELIASSEN concluded that, at least in the Waddensea, during the summer haemoglobin should never need to function as an oxygen reservoir. Also, Fox (1955) found no increase in haemoglobin of *A. marina* when subjected to low oxygen tensions.

In preliminary experiments, MANWELL and co-authors (1966) established that *Siboglinum atlanticum*, a pogonophorian, survives anaerobiosis for 24 hrs without apparent harmful effects. During exposure to anoxia, its tentacle is not protruded and body movements are reduced. The haemoglobin was not completely deoxygenated during the 24-hr period.

Pollution. High tolerance to anoxic conditions may enable certain species to inhabit coastal regions affected by pollution. WALDICHUK and BOUSFIELD (1962) found a number of amphipod species in low oxygen, marine waters adjacent to a sulphite pulp mill. According to TOWNSEND and EARNEST (1940), the lower lethal oxygen level of salmonoid fishes is raised markedly when the test individuals are exposed to 1:1000 waste sulphite liquor. Also, juvenile shad *Alosa sapidissima* are less tolerant of low oxygen levels when exposed simultaneously to sublethal concentrations of petroleum products (TAGATZ, 1961; Volume V).

ZoBELL (1962) has summarized much of the literature on the influences of oil pollution on marine organisms. There may be a direct effect on organisms by an active chemical compound, or the presence of oil may upset the existing oxygen regimen, thus indirectly exerting a profound biological effect. An example of the direct influence is the report of GALTSOFF and co-authors (1936) that a water-soluble fraction in crude oil narcotizes the ciliated epithelium of oyster gills and inhibits feeding (see also GUNKEL, 1967; KINNE and AURICH, 1968).

Fish maintained at low, but sublethal concentrations of oxygen are killed when stressed further by exposure to oil pollutants (TAGATZ, 1961). SINGH PRUTHI (1927a) suggested that CO_2 increase or pH change, rather than oxygen depletion, is of prime importance in fish kills in polluted waters. Mineral oils may undergo microbial oxidation or autoxidation; they have a very high biochemical oxygen demand. Thus, it is not uncommon to find anaerobic conditions in regions subjected to continuous or excessive oil pollution.

Acclimation. Homarus americanus acclimated to 5°C are more tolerant to anoxia than individuals acclimated to 15°C and these, in turn, are more tolerant than those acclimated to 25°C (MCLEESE, 1956). *H. americanus* acclimated to high salinities survive low oxygen levels better than specimens acclimated to low salinities. While acclimation to different oxygen concentrations does not have a significant effect on the lethal oxygen level, groups of *H. americanus* acclimated to various oxygen levels exhibit different lethal temperatures and salinities.

Acclimation to different ambient concentrations of oxygen appears to take place, at least in fishes. TAGATZ (1961) reported that, in juvenile American shad *Alosa sapidissima*, the low critical oxygen level necessary to produce 50% mortality varies with the rate at which the oxygen content is reduced during experimentation.

In view of the extremely large number of variables, acting independently or in combination, which influence the responses of marine animals to dissolved oxygen, it is apparent that detailed, long-term studies designed to take the variables into consideration are needed. In the past, many investigations have been confined to a few summer months; it is well known, however, that tolerances to oxygen stress may vary seasonally, hence summer research alone will not give a complete picture.

Carbon dioxide

The amount of CO_2 dissolved in sea water may influence the survival of marine animals (although BRUCE, 1928, stated that excessive CO_2 is never a limiting factor to animals associated with sandy beaches).

Tolerances to CO_2 narcosis differ among freshwater arthropods: species of *Gammarus* die after 1 hr, of *Asellus* after 5 hrs, while those of *Sialis* and *Corethra* survive 9 days' exposure (BEADLE and BEADLE, 1949). The euryhaline crab *Carcinus maenas*, which may live in isolated shallow pools, can tolerate higher CO_2 concentrations than many other decapod crustaceans (ARUDPRAGASAM and NAYLOR, 1964). Bottom-dwelling fishes, *Solea solea* and *Cottus bubalis*, survive longer in sea water with a high CO_2 content (20 cm³/l) than does the herring *Clupea harengus* (SHELFORD and POWERS, 1915).

Free CO_2 in the water, pH and salinity influence the toxicity of ammonium salts to fish (LLOYD and HERBERT, 1960; HERBERT and SHURBEN, 1965). When the dissolved oxygen level approaches the lethal point for silver salmon *Oncorhynchus kisutch*, a further stress of lowered pH will produce the same lethal effect as further lowering of oxygen tension (TOWNSEND and CHEYNE, 1944). Similar results were obtained for four other fishes when the pH was decreased by addition of H_2SO_4

(SHELFORD, 1918). SINGH PRUTHI (1927b) suggested that the pH decrease acts directly to kill fish.

Sulphides

The production of sulphides on sandy beaches has been described by BRUCE (1928). When the uppermost surface of the sulphide-producing sand region comes into contact with the free atmosphere, a black layer is formed which represents a region of complex oxidative activity. The boundary layer between H_2S and O_2 fluctuates, depending on various factors, especially water circulation. If the boundary rises closer to the surface, the result is mass mortality of benthic organisms (BRONGERSMA-SANDERS, 1957). In laboratory experiments, SHELFORD and POWERS (1915) found that a population sample of the herring *Clupea harengus*, died after 6 mins exposure to H_2S ($7 \cdot 6 \, \mathrm{cm^3/l}$); soles *Solea solea* were dead by 24 mins. In contrast, a cottid fish was still alive at the end of 3 hrs. A similar progression of tolerance was noted in these fishes when subjecting them to combinations of H_2S and CO_2. The differences in tolerance may be correlated with the different habitats occupied. Unlike the herring, soles and cottids live on the sea bottom, which is more apt to be a site of temporary increases in H_2S than the free water. The latter two fishes can detect and, where possible, avoid dangerous concentrations of H_2S. *Clupea pallasii* also swims away from sublethal levels of H_2S (SHELFORD and POWERS, 1915).

Marine invertebrates also exhibit interspecific variations in the degree of tolerance to high concentrations of H_2S; *Arenicola marina*, for example, survive in sea water through which H_2S has been bubbled for 24 hrs (PATEL and SPENCER, 1963); in contrast, the gastropods *Bullia laevissima* and *B. digitalis*, treated in the same manner as *A. marina*, die within 30 mins to 2 hrs (BROWN, 1964). In mammals and certain other vertebrates, H_2S combines with haemoglobin and inhibits the heavy metal enzyme system, but the haemoglobin of *A. marina* does not combine with H_2S, nor does the haemocyanin of the two *Bullia* species. Like fishes, *B. laevissima* and *B. digitalis* are able to detect small quantities of H_2S.

Factor combinations

Although examples of animal responses to factor combinations have been presented in the previous pages and will be discussed in detail in Chapter 12, special attention is drawn here to this important aspect. Marine animals live in environments where different factors may vary simultaneously. While one factor may attain a critical level without being lethal, the combined critical intensities of two or more factors—each one sublethal *per se*—may easily cause death (e.g. McLEESE, 1956; HOOD and co-authors, 1960; TAGATZ, 1961; ZOBELL, 1962).

(b) Metabolism and Activity

Within the zone of compatibility, metabolism and activity of marine animals are greatly influenced by dissolved gases. The general statement by VON LEDEBUHR (1939), that marine animals can withstand large variations in oxygen tension and only at rather low levels reveal variations in respiratory rate, reflects the information available at that time. In the last few decades, our pertinent knowledge has increased considerably and revealed a multiplicity of responses.

Oxygen

Oxygen consumption of adult individuals. The relation between oxygen tension and the rate of oxygen uptake has received much attention. Generally, the responses to varying oxygen tension fall into two categories: (i) oxygen-dependent responses, where the oxygen uptake by an animal is proportional to the amount of oxygen present; and (ii) oxygen-independent responses, where the rate of oxygen uptake is relatively constant over a wide range of oxygen tensions until some critical value (the 'critical oxygen tension' or Pc) is reached, and the rate declines rapidly (see also Chapter 3.31). TANG (1933), KROGH (1941), VON BRAND (1945) and ZEUTHEN (1955) have published reviews dealing with metabolic responses of many groups of animals to oxygen tension. Difficulties arise in comparing data because different experimental techniques and procedures are used. For example, BOSWORTH and co-authors (1936) emphasized that it is valid to compare the relation between oxygen tension and oxygen consumption only when animals are allowed to reduce tension at a rapid rate. If too much time elapses, adjustments may occur.

Early in this century, HENZE (1910) postulated that the respiration of simpler, bulkier invertebrates is oxygen dependent, while the rate of oxygen uptake of the 'higher' invertebrates is independent of oxygen concentration. The later work of others, listed below, demonstrates many exceptions to this generalization.

Echinoderms. *Caudina* species and *Strongylocentrotus purpuratus* are oxygen dependent (NOMURA, 1926; HYMAN, 1929), while species of *Asterias*, *Echinus* and *Thyone* are oxygen independent (MOORE and co-authors, 1912; HIESTAND, 1940). According to MEYER (1935), *Asterias rubens* does not regulate its metabolic rate in decreased oxygen tension; the degree of dependence on oxygen tension decreases with body size. However, MALOEUF (1937a) noted that *Asterias forbesii* is oxygen-dependent up to a critical point, and then becomes independent. The rate of oxygen uptake by the sea-star *Patiria miniata* is highly affected by ambient oxygen tension until the oxygen content is increased to a high level (10 to 11 cm³/l). At these high concentrations, the oxygen-consumption rate increases initially, then declines markedly. This respiratory response demonstrates a relation to the sea-star's habitat (HYMAN, 1929); *P. miniata* lives in rocky surf waters always saturated with oxygen.

Oxygen consumption of *Holothuria forskali* stops when the oxygen level falls below 60 to 70% air saturation (NEWELL and COURTNEY, 1965). At lower concentrations, there is no exchange through the respiratory tree and no corresponding increase in absorption through the general body surface. The pumping rhythm increases as the ambient oxygen content drops until the 60 to 70% air-saturation level is reached. Oxygen stored in coelomic fluid is used by individuals during the period in which they actively move towards a source of oxygen when initially in a medium of low concentration. In contrast, *Thyone briareus* is oxygen independent (HIESTAND, 1940). NEWELL and COURTNEY suggested that perhaps the muscles of *T. briareus* are less sensitive than those of *Holothuria forskali* to hypoxic conditions.

Earlier, LUTZ (1930) had reported that lack of oxygen influences the activity of the pumping muscles of the holothurian *Stichopus*.

Arthropods. The metabolic rate of some marine arthropods, such as *Carcinus maenas*, *Scyllarus latus* and *Eriocheir sinensis*, is oxygen independent (HENZE, 1910; VAN HEERDT and KRIJGSMAN, 1939), while that of *Homarus americanus*, *Limulus polyphemus*, *Callinectes sapidus* and *Penaeus indicus* is oxygen dependent (AMBERSON and co-authors, 1924; MALOEUF, 1937b; THOMAS, 1954; SUBRAHMANYAN, 1962).

Differences in oxygen consumption of seven species of Hawaiian crabs were reported by VAN WEEL and co-authors (1954). In four species (*Calappa hepatica*, *Metopograpsus messor*, *Pseudosquilla ciliata* and *Pseudozius caystrus*), percentage oxygen utilization from sea water passing at a constant rate through the respirometer remains relatively unchanged over the range 6·2 to 2·6 ml O_2/l; this indicates oxygen consumption in proportion to tension. In two species (*Phymodius ungulatus* and *Platypodia granulosa*), the utilization percentage decreases over this range of concentration, which indicates an even steeper decline of oxygen intake with declining oxygen tension. The metabolism of a mud-dwelling species (*Podophthalmus vigil*) is independent of ambient oxygen tension over part of this range, but at tensions between 3 to 2·5 ml O_2/l it is dependent. Below 2·5 ml O_2/l, in all seven species, a similar response is noted: a steep increase in the percentage utilization. This response is interpreted as demonstrating that, below a critical tension, these animals begin to regulate in order to insure a minimum amount of oxidation.

SUBRAHMANYAN (1962) reported that large-sized (10 g) *Penaeus indicus* show a greater metabolic dependency on ambient oxygen tension below 100 mm Hg than small-sized (0·6 g) prawns. In *Calanus finmarchicus*, respiration is independent of oxygen content until the concentration drops below 3 ml/l (MARSHALL and co-authors, 1935). However, unlike the response of the adult, oxygen consumption of stage V is little affected by the external oxygen level, even when it drops to 2·2 ml/l. (Additional earlier papers on crustaceans may be found in WOLVEKAMP and WATERMAN, 1960.)

The rate of oxygen consumption of amphipods, determined at various levels of oxygen tension in the external media, indicates that the critical tension is higher in semiterrestrial species and lower in brackish and freshwater species (WALSHE-MAETZ, 1956).

Recently, TEAL and CAREY (1967) investigated the metabolic responses to reduced oxygen of three species of marsh crabs. Although much variation is noted in their data, they conclude that, at a critical oxygen pressure of 1 to 3% of an atmosphere, respiratory rate of inactive *Uca pugilator* and *U. pugnax* decreases; when the crabs are active, the critical oxygen pressure is about 3 to 6%. No mention is made of the method of observing locomotory behaviour. These values differ from those reported earlier by TEAL (1959), which are probably too low since the oxygen pressure of the respiring media was calculated rather than determined by direct measurement. *U. pugilator* and *U. pugnax* consume oxygen down to a level of 0·4% of an atmosphere while *Sesarma cinereum* stop respiring at a somewhat higher value.

Annelids. *Nereis diversicolor* consumes oxygen at a rate independent of ambient oxygen tension until a low pressure of 40 mm Hg is reached (JÜRGENS, 1935).

According to AMBERSON and co-authors (1924) and HYMAN (1932), *Nereis virens* may be classified as being oxygen dependent. However, LINDROTH (1938) reported that the critical pressure for *N. virens* is 100 mm Hg, with little change in Q_{O_2} over the range of oxygen tension equivalent to 150 mm to 300 mm Hg. At low oxygen levels, *Arenicola marina* suspends external respiration with only occasional sampling of water by pumping (WELLS, 1949; WELLS and DALES, 1951). In an extensive review, LINDROTH (1941) described the ventilation mechanisms in many polychaete families and attempted a classification of respiratory types.

Molluscs. The metabolism of many molluscs is independent of ambient oxygen tension until some low oxygen tension is reached. *Pecten grandis* and *P. irradious* maintain a relatively constant rate of oxygen consumption until the concentration of oxygen is about 1·0 to 0·5 ml O_2/l (VAN DAM, 1954). It is not known if more water is pumped or if the oxygen utilization percentage is increased. *Ostrea gigas* is metabolically independent until a low value of 1·5 cm³ O_2/l is reached (ISHIDA, 1935); for *O. circumpicta*, this low critical value is 1 cm³ O_2/l (NOZAWA, 1929); for *Ostrea edulis* it is 2·5 cm³ O_2/l (GALTSOFF and WHIPPLE, 1930). Species of *Buccinium*, *Aplysia* and *Fusus* are also independent, down to a critical low oxygen level (MOORE and co-authors, 1912); in *Aplysia*, oxygen consumption remains constant even at increased oxygen tensions (HENZE, 1910). *Loligo peali* is independent until oxygen saturation of the water drops below 30% (AMBERSON and co-authors, 1924).

The Q_{O_2} of *Mytilus edulis* is strictly proportional to ambient tension, even at a level 4 times the air saturation value at 25°C (MALOEUF, 1937a). In contrast, in *M. californianus*, respiration is unaffected by external oxygen tension until a low level is reached (WHEDON and SOMMER, 1937).

Many intertidal zone molluscs withdraw within their shells during periods of environmental stress. This response may result in approaching anaerobic conditions in the intrashell water. Depending on the period of confinement, the intertidal mollusc may incur an oxygen dept which has to be paid off later by increased removal of oxygen (increased pumping rates) from the ambient water (VAN DAM, 1935; SCHLIEPER, 1957). In contrast, *Lasaea rubra*, exposed to air for 3 hrs, consumes oxygen at the pre-stress rate when re-immersed in sea water (MORTON and co-authors, 1957).

The rate of oxygen consumption of the heart ventricle of the clam *Tivella stultorum* is proportional to oxygen tension over the range of 25 to 100 mm Hg and reduced at 160 mm Hg (BASKIN and ALLEN, 1963).

Fishes. As in other marine animals, the rate of oxygen consumption of various fish species is independent of the ambient concentration of dissolved oxygen until a critical low level is attained. This fact has been established, for example, for species of the genera *Coris* and *Sargus* (HENZE, 1910), *Blennius* (MOORE and co-authors, 1912), *Cyprinus* (GAARDER, 1918), and *Fundulus* (AMBERSON and co-authors, 1924). In *Clupea pallasi*, *Oncorhynchus kisutch* and *Cymatogaster aggregatus*, metabolic rates decrease when the ambient oxygen content is lowered; the influence of intermediate oxygen tensions was not studied adequately (POWERS and SHIPE, 1928). Of the three species mentioned, *Clupea pallasi* seems to be most sensitive to reduced tensions of dissolved oxygen. In the Atlantic cod *Gadus morhua*, reduction of the ambient oxygen level from 10 to 3 mg O_2/l lowers the rate of oxygen consumption, but increases the respiratory volume (SAUNDERS, 1962).

In *Fundulus heteroclitus*, oxygen consumption is a hyperbolic function of oxygen pressure (MALOEUF, 1937b). In *Fundulus parvipinnis*, respiration is independent of small deviations from usual oxygen tensions of the medium (KEYS, 1930 a, b); when the oxygen consumption rate is independent of external oxygen pressure, the intracellular oxidation-reduction system is oxygen saturated. At critical oxygen pressure, oxygen becomes a limiting factor. For a more comprehensive discussion including both adult freshwater and marine fishes, consult BLACK (1951) and FRY (1957).

The correlation between the swimming activity of fishes and the influence of ambient oxygen tension on their oxygen consumption was studied by HALL (1929). The metabolic rate of the sluggish toadfish *Opsanus tau* is dependent on the oxygen concentration of the surrounding water. In contrast, oxygen uptake of the rather active scup *Stenotomus chrysops* is independent over a wide range of oxygen tension (40 to 120 mm Hg). Apparently, the puffer *Spheroides maculatus* occupies an intermediate position between these two species (VERNBERG and VERNBERG, 1966). In 1930, HALL reported on the locomotory activity of 11 species of marine fishes and the minimum oxygen tensions at which they can remove dissolved oxygen from the surrounding water. The minimum tension is 0·0 mm Hg for the sluggish *Opsanus tau*, while—at the other extreme—the very active mackerel *Scomber scombrus* fails to withdraw oxygen when the tension drops below 70 mm Hg.

Oxygen consumption of fishes, maintained at a constant rate of locomotory activity by electrical stimulation, was determined by BASU (1959). At a constant oxygen content of the water, the rate of oxygen consumption decreases with increased CO_2 levels. In the absence of CO_2, the oxygen concentration determines the level of oxygen consumption. BEAMISH (1964) pointed out that, above 80 mm Hg, the standard metabolism of certain fishes is unchanged at different partial pressures of oxygen. Below 80 mm Hg, the standard rate first increases and then decreases as the partial pressure of oxygen is reduced.

GRAHAM (1949), working with the freshwater speckled trout *Salvelinus fontinalis*, found that the Q_{O_2} value of active fish becomes dependent on oxygen tension of the water at a critical level which changes with temperature; at 5°C, the critical point is 100 mm Hg and at 24·5°C, 150 mm Hg.

Changes in the structure of large fish-schools can be correlated with declining oxygen content (MCFARLAND and MOSS, 1967). Individual fish of the school may sense marked reductions in dissolved oxygen or increases in carbon dioxide. These individuals alter their swimming behaviour and thus affect the structure of the school. In one case, schooling broke down after the oxygen level had been reduced by about 25%.

Other marine animals. At different ambient concentrations of dissolved oxygen, the rate of oxygen uptake by *Branchiostoma lanceolatum* varies considerably. In sealed vessels, some individuals utilize oxygen linearly down to less than 20% of air saturation; others cease to absorb oxygen at high saturation while some display two different rates in a sequence (COURTNEY and NEWELL, 1965).

BRAFIELD and CHAPMAN (1965) found the Q_{O_2} values of the coelenterates *Pennatula rubra* and *Calliactis parasitica* to be largely proportional to the external oxygen concentration; this suggests that these species rely on simple diffusion to supply oxygen to all the cells. Probably, the endoderm is the major site of oxygen consumption. The respiratory rate of the jellyfish *Cassiopeia* is oxygen dependent at low tensions (McCLENDON, 1917). Many examples of responses of other marine animals to variations in dissolved oxygen concentrations can be found in PROSSER and BROWN (1961; see also Chapter 3.31).

Acclimation. When allowing a period of acclimation to low levels of ambient oxygen tension, the response to declining oxygen pressure of the sea-star *Asterias forbesii* and the lamellibranch *Mytilus edulis* is such that oxygen consumption drops to an unusually low value, possibly due to neurosecretory factors (MALOEUF, 1937a). The best examples of the influence of acclimation to different oxygen concentrations on the metabolism of fishes come from work on freshwater species. SHEPARD (1955) used three size groups of *Salvelinus fontinalis* acclimated to a graded series of oxygen concentrations. His results demonstrate conclusively that —within the zone of respiratory dependence—fingerlings and fry acclimated to low oxygen levels remove oxygen at a greater rate than fish acclimated to high oxygen concentrations.

BEAMISH (1964) measured the standard metabolism of brook trout *Salvelinus fontinalis*, carp *Cyprinus carpio* and goldfish *Carassius auratus*, acclimated to different oxygen tensions. Down to a test tension of 80 mm Hg, the fishes respire at the same rate, regardless of the acclimation tensions. However, below 80 mm Hg, the standard Q_{O_2} values of the low-level oxygen-acclimated fishes are consistently lower. BEAMISH concluded that acclimation to low oxygen concentrations enhances the anaerobic fraction of standard metabolism. In *Carassius auratus* acclimated to low oxygen, the metabolism is affected in two ways (PROSSER and co-authors, 1957): (i) standard rate of oxygen consumption decreases; (ii) critical oxygen pressure shifts to lower values (from about $3 \cdot 1$ to $1 \cdot 5$ ml O_2/l). In low-oxygen acclimated *C. auratus*, brain and muscle tissue consume oxygen at a slower rate than in fish acclimated to high oxygen levels; liver tissue is not influenced.

In *Carcinus maenas*, decrease in the concentration of dissolved oxygen from 6 ml to 4 ml/l leads to the following responses: (i) oxygen consumption rate increases sharply, then drops significantly; (ii) ventilation volume varies (at $1 \cdot 5$ cm³ O_2/l, ventilation volume increases 4 to 5 times above normal; however, below 1 cm³ O_2/l water pumping stops after 1 hr); (iii) the amount of oxygen utilized increases greatly (ARUDPRAGASAM and NAYLOR, 1964). All these adaptive changes would help *C. maenas* to survive in isolated shallow pools.

In the European lobster *Homarus gammarus*, the dependency of oxygen consumption on oxygen concentration of the surrounding medium apparently does not vary over the temperature range 6° to 18°C (THOMAS, 1954); the same is true for *Calanus finmarchicus* from 5° to 15°C (MARSHALL and co-authors, 1935). On the basis of his experiments with *Asterias forbesii* and *Mytilus edulis*, MALOEUF (1937a) concluded that the critical oxygen pressure varies inversely with temperature. The puffer fish *Spheroides maculatus* removes about 45% of the dissolved oxygen from water passing over its respiratory surfaces, regardless of the ambient temperature (HALL, 1931). In contrast, *Branchiostoma lanceolatum* does not take up oxygen below 70% air saturation at temperatures from 4° to 10°C, while at 15°C some individuals utilize oxygen linearly down to less than 20% air saturation

(COURTNEY and NEWELL, 1965). GRAHAM (1949) demonstrated that the transition point at which the speckled trout *Salvelinus fontinalis* changes from oxygen-independent to oxygen-dependent metabolism varies with temperature, i.e., at 5°C the critical pressure is 100 mm Hg, while at 24·5°C it is 150 mm Hg.

Oxygen consumption of life-cycle stages other than the adult. Eggs of *Arbacia punctulata* respire at rates independent of the ambient oxygen concentration down to a low critical value. AMBERSON (1928) reported a critical value of 20 mm Hg, TANG and GERARD (1932) 50 mm Hg, for fertilized eggs. For unfertilized eggs, the critical value is 40 mm Hg (TANG, 1931). In eggs of the plaice *Pleuronectes platessa*, oxygen consumption in a closed vessel was claimed to decrease with time (DAKIN and CATHERINE, 1925); according to BURFIELD (1928), this decrease results from an accumulation of CO_2. For further information on effects of dissolved gases on egg development consult the section *Reproduction*.

Different life-cycle stages of one species may inhabit media with different oxygen regimes. Therefore, it is necessary to study the responses of all life-cycle stages in order to understand fully the role of respiratory gases in the ecology of marine animals. In marine trematodes, the free-living stages inhabit waters with widely fluctuating oxygen levels, while the parasitic stages live internally in their host, where the oxygen tension is rather constant but generally very low. The free-living cercaria of *Zoogonus laseus*, a tailless, sluggish form, consumes oxygen at a rate proportional to the external oxygen concentration when the tension decreases from 5% to 3%; but oxygen consumption becomes independent of the ambient tension when this decreases to 1·5% or 0·5%. The cercaria of *Himasthla quissetensis* is very active; its oxygen consumption decreases proportionately to the ambient oxygen tension from 5% to 0·5%. The cercaria becomes inactive when maintained at 0·5% oxygen for a prolonged period of time. In contrast, the larva of *Zoogonus laseus* remains normally active at this low oxygen level (HUNTER and VERNBERG, 1955; VERNBERG, 1963). The parasitic redia of *Himasthla quissetensis* is relatively independent of oxygen tension over a range from 5% to 1·5% oxygen, and even at 0·5% oxygen its metabolic rate is 61% of the rate in 5% oxygen (VERNBERG, 1963). In contrast, oxygen consumption of adult trematodes is, in general, oxygen dependent. Adults of *Gynaecotyla adunca*, for example, are oxygen dependent over the wide range of oxygen tensions from 1·5 to 21%; however, from 1·5% down to 0·5%, oxygen consumption rates remain the same; both 1·5% and 0·5% are values which would be found *in vivo* (VERNBERG, 1963).

Respiratory movements. A decrease in concentration of ambient dissolved oxygen causes increased respiratory movements (rate of pleopod beating) in certain amphipods and isopods (WALSHE-MAETZ, 1952, 1956). In the amphipod and isopod species studied by WALSHE-MAETZ, the marine forms move their pleopods twice as fast as their brackish-water counterparts; brackish-water species, in turn, have a rate twice that of their freshwater relatives. The semiterrestrial *Orchestia gammarella* and *Ligia italica* do not show any appreciable change in the rate of their respiratory movements with fluctuations in oxygen tension. While the oxygen consumption rates of *Gammarus pulex* from fresh water are higher than those of the marine *G. locusta* or *G. marinus*, the marine species exhibit higher rates of pleopod beating (Fox and SIMMONDS, 1933). In *Eriocheir sinensis*, the rate of scaphognathite beating increases when the oxygen content of the water decreases, and slows down when it increases (VAN HEERDT and KRIJGSMAN, 1939). A great variation in the response of respiratory movements to reduced oxygen and increased CO_2 has been demonstrated in numerous crustaceans by Fox and JOHNSON (1934), JOHNSON, M. L. (1936) and WOLVEKAMP and WATERMAN (1960).

Under 'normal' laboratory conditions, the air-breathing fish *Pseudapocryptes* surfaces every 2 to 10 mins to obtain air (DAS, 1934). If this fish is prevented from surfacing, it becomes very excited and swims upward at least every minute. Its opercular respiratory movements increase from a pre-experimental value of 6 to 80/min to at least 150/min. Complete asphyxiation and death results in 15 to 20 hrs.

When the ambient oxygen concentration is reduced, no inhalent stream is observed in *Branchiostoma lanceolatum*, probably due to the absence of ciliary activity (COURTNEY and NEWELL, 1965). In turn, the cessation of ciliary beating appears to be correlated with the low rate of oxygen uptake by the animal. During periods of high oxygen utilization, all frontal and lateral cilia and the gill bars are extremely active. In *Mytilus edulis* kept in anoxic conditions, the ciliary activity of the gill epithelium is greatly reduced. Upon return to oxygenated water, ciliary activity almost doubles, resulting in a rate about 25% higher than that of the controls (SCHLIEPER and KOWALSKI, 1958).

SOUTHWARD and CRISP (1965) report that cirral activity of various species of barnacles is little affected by an increased oxygen content of the water; but high oxygen concentrations appear eventually to reduce activity. If the ambient oxygen content is gradually reduced, an initial increase in cirral activity is followed by signs of anoxic distress and, eventually, by reduced activity.

Respiratory pigments. It has long been known that various environmental factors, including temperature, pH and external oxygen tension, influence the degree of oxygen saturation of respiratory pigments of marine animals (KROGH and LEITCH, 1919). Two values are of particular interest when discussing oxygen equilibrium curves: (i) the pressure necessary to half-saturate the pigment (P_{50} or Pu); (ii) the pressure required for saturation or loading (Pu or $P_{sat.}$). Interspecific differences in these values were reported for cod and plaice (scientific names not given). At 15°C, the Pu value is 70 mm Hg for cod and 40 mm for plaice, while the P_{50} value is 18 mm Hg for cod and 10 mm for plaice (KROGH and LEITCH, 1919). The review of MANWELL (1960), the work of REDMOND (1955), and the textbook of PROSSER and BROWN (1961) cite numerous examples, which illustrate the adaptive importance of oxygen-equilibrium values. Only a few recent examples need be included in the present chapter.

Half-saturation oxygen pressures equivalent to 20 to 26 mm Hg are similar for tropical and cold-water chitons, regardless of the great differences in habitat temperatures. REDMOND (1962b) suggested that these low values may be related to low oxygen demand, large gill area, ease of ventilation, and the occurrence of these molluscs in well-aerated sea water. The Bohr effect (shifting of the oxygen equilibrium curve to the right with increased CO_2 or acidity) shows no correlation between habitat and physiological significance in normal respiratory exchange in the four species studied: *Acanthopleura granulata, Chiton tuberculatus, Katherina tunicato* and *Mopalia muscora*. MANWELL (1963a) investigated the physiological importance of haemoglobin in some marine clams. The muscle haemoglobin of *Mercenaria mercenaria* has a very high oxygen affinity, which may be of significance in facilitating the diffusion of oxygen into that thick, non-vascular tissue. The pismo clam *Tivella stultorum*, which inhabits sandy open beaches, had the highest concentration of tissue haemoglobin; it is much more sensitive to oxygen lack than other clams. Tissue haemoglobin from the gill of the clam *Phacoides pectinatus* has a very low P_{50} of 0·19 mm Hg (READ, 1962).

In *Golfingia gouldii*, two types of haemerythrin were found by electrophoretic techniques (MANWELL, 1963b). Although there is a 37% difference in oxygen affinity in the two types, no apparent survival value was reported for either type based on population sampling.

The P_{50} value of 3·5 to 4 mm Hg for the land crab *Cardisoma guanhumi* is surprisingly low for a terrestrial animal (REDMOND, 1962b). Typically, respiratory pigments of animals from oxygen-rich habitats have a lower affinity for oxygen than animals from oxygen-poor habitats. The low P_{50} value of *C. guanhumi* may facilitate stable loading pressures even during thermal or osmotic stress.

In some species, such as *Limulus polyphemus* and *Busycon canaliculatum*, an increase in the concentration of ambient CO_2 increases the oxygen affinity of the respiratory pigment. This phenomenon is called 'Inverse Bohr Effect'. According to REDMOND (1955), this effect would have adaptive value to organisms living in regions of high CO_2 and low O_2 levels. However, unless the P_{O_2} in tissues is low, there would be considerable difficulty in allowing the pigment to release its oxygen.

In the dogfish shark *Squalus suckleyi*, the P_{50} value is 17 mm Hg at 11°C, and no Bohr effect is observed (LENFANT and JOHANSEN, 1966). These two responses may be beneficial to fairly active species like the dogfish for bottom-living in waters deficient of oxygen and high in CO_2.

Certain animals (references in FOX, 1955) exhibit an increase in blood pigment with a decrease in ambient oxygen concentration. However, Fox found a number of exceptions to this response. In particular, the haemoglobin level of the annelids *Arenicola marina* and *Scoloplos armiger* remains unchanged after prolonged exposure to poorly aerated water. If the haemoglobin of *Nereis diversicolor* is poisoned by CO, insufficient oxygen is taken up to maintain a high rate of oxygen consumption when the oxygen content of the medium drops below 61% (JÜRGENS, 1935).

Toadfish *Opsanus tau* often live in stagnant waters, where oxygen concentrations may be very low. Their haemoglobin has a high affinity for oxygen, allowing effective oxygen transport to the tissues even at low oxygen tensions (HALL and McCUTCHEON, 1938). In other fishes, the characteristics of haemoglobin appear to be more closely correlated with locomotory activity than with habitat differences in oxygen content. Mackerel *Scomber scombrus* and menhaden *Brevoortia tyrannus*, which are very actively swimming fishes, possess blood with high oxygen capacities and haemoglobin with a low affinity for oxygen, while the less active *Opsanus tau* and *Tautoga onitus* have blood with low oxygen capacities and haemoglobin with relatively high oxygen affinities. In *Brevoortia tyrannus* and *Spheroides maculatus* (the puffer fish), the concentrations of haemoglobin, phosphorus, iron, total nitrogen, and number of red blood cells increase during asphyxiation. The increase is proportional to the length of the asphyxiation period. Subsequent death of these fishes is probably due primarily to the increased acidity of blood and tissues, and to the retention of toxic substances (HALL and co-authors, 1926; HALL, 1928).

Oxygen storage. Oxygen may be stored temporarily in the body by respiratory pigments (see review by MANWELL, 1960). JONES and MARSHALL (1953) suggest that the swimbladder of fishes may store oxygen, since the proportion of oxygen in the swimbladder decreases when fish are kept in water of low oxygen content or are asphyxiated. The oxygen stored in the coelomic cavity appears to enable *Holothuria forskali* to survive periods of insufficient oxygen supply and move from regions of low to high oxygen tension (NEWELL and COURTNEY, 1965). COLLIP (1921) claims that the oxygen stored in the tissues of *Mya arenaria* is sufficient to maintain this clam when the external oxygen supply is temporarily cut off.

Heart rate. When *Mytilus edulis* is taken out of sea water, its heart rate is greatly reduced; it returns to about the pre-emersion rate when resubmersed (SCHLIEPER, 1955, 1957). Oxygen tensions below 1% (8 mm Hg) cause a gradual decrease in heart rate, but the rate returns to normal upon retransfer to sea water with 20% O_2. However, the rate of heart beat of *Mytilus edulis* greatly accelerates when this mussel is exposed to waters of reduced oxygen contents (11%, 6·4%, 6% and 3% O_2). 3% O_2 produced an initial increase in heart rate, but by 130 mins the heart stopped beating; at 6% O_2, the initial increase also occurred, but the accelerated rate was more or less maintained for 200 mins when the experiment was terminated (SCHLIEPER and KOWALSKI, 1957). In isolated heart ventricle preparations of the clam *Tivella stultorum*, the rate of beating was independent over a wide range of oxygen tension, i.e., from 3 to 160 mm Hg (BASKIN and ALLEN, 1963). The authors suggest that the control of heart beat does not reside in the heart.

Dorsal blood vessels. When the ambient oxygen tension decreases, the frequency of contraction of the dorsal blood vessel of *Nereis diversicolor* increases down to an oxygen content of 2·50 mg/l; at still lower oxygen contents, blood circulation decreases (JÜRGENS, 1935). LINDROTH (1938) obtained different information on *Nereis virens*. Oxygen tension differences between 50 mm and 250 mm Hg have no effect on pulsation rate of the dorsal vessel in 14 cm long individuals, but smaller ones have higher rates at the higher level of oxygen.

Blood sugar. According to STOTT (1932), the decapod crustaceans *Cancer pagurus, Carcinus maenas* and *Macropipus puber* respond to reduced concentrations of dissolved oxygen by increasing their blood sugar level.

Serum proteins. One of the principal features of polluted waters is low dissolved oxygen content. Frequently, hypoxic conditions occur in diurnal cycles, with the lowest values at night. When alternately exposed to water with a low oxygen content (3 ppm) for 8 hrs per day and with 'normal' oxygen content (8·3 ppm) for 16 hrs over a 9-day period, the serum protein pattern of bluegills *Lepomis macrochirus* and largemouth bass *Micropterus salmoides* changed significantly, but not that of the yellow bullhead *Ictalurus natalis* (BOUCK and BALL, 1965). This difference probably reflects the greater resistance of the bullhead to stress. Three possible explanations for the serum protein changes were suggested; (i) increased antibody formation; (ii) stress hormones may mobilize cellular proteins to provide additional energy; (iii) during hypoxia, insufficient energy may be produced to maintain cell membranes, resulting in leakage of cellular proteins. In addition to changes in serum proteins, hypoxia greatly accelerates the blood-clotting mechanisms and interferes with the digestive process by causing regurgitation.

Removal of oxygen from sea water. The amount of oxygen removed from sea water by organisms varies greatly, depending on many factors, including the type of feeding mechanisms, environmental factors and respiratory efficiencies. HAZEL-HOFF (1938) stated that particle-feeding marine animals remove 13% of the oxygen from sea water passing over respiratory surfaces, and that all other groups remove at least 53%. Two species of *Pecten* extract between 0·5 and 9% of the oxygen available (VAN DAM, 1954). At low oxygen tensions, the percent of oxygen removed from sea water by the tropical crabs *Calappa hepatica, Metopograpsus messor, Pseudosquilla ciliata, Pseudozius caystrus, Phymodius ungulatus, Platypodia grandulosa* and *Podophthalmus vigil* initially increases markedly, but later decreases (VAN WEEL and co-authors, 1954). The actual percent value varies with the species studied. To cite an example, *Podophthalmus vigil* removes 7·5% of the oxygen when the inflow level is 4·53 cm³ O_2/l, 25% when the inflow is 1·32 cm³ O_2/l, and 16·2% when the inflow drops to 1·17 cm³ O_2/l.

Octopus vulgaris removes 70% of the oxygen from sea water (WINTERSTEIN, 1925); the percent removal is not appreciably altered when the oxygen tension is low (HAZELHOFF, 1938). *Octopus dofleini* removes only 27% of the oxygen (JOHANSEN and LENFANT, 1966); this is quite different from the high value found for *O. vulgaris*; possibly, the difference is correlated with marked differences in habitat temperature.

The shark *Squalus suckleyi* extracts 25% of the oxygen from sea water (LENFANT and JOHANSEN, 1966). In the puffer *Spheroides maculatus*, the removal of oxygen remains at 46% over a wide range of ambient oxygen pressures (HALL, 1929). According to KEYS (1931), the higher oxygen demands of small specimens of *Fundulus parvipinnis* are not compensated for by any enhanced ability to extract oxygen from water low in oxygen content.

Behaviour. When the oxygen level of sea water drops, *Holothuria tubulosa* elevates its posterior end out of the water and presumably takes in air by its rectum (observed by TIEDEMANN, *in*: CARTER, 1931). *Holothuria forskali* moves toward a source of oxygen from an oxygen-poor medium (NEWELL and COURTNEY, 1965). Worms living in tubes built into mud soil apparently obtain oxygen by exposing a small portion of their tail end to the overlaying water. When the oxygen content of this water is experimentally decreased, more of the worm's tail is extended. If the worms are prevented from burrowing, they cluster with their tail ends protruding. The intensity of tail motion increases with decreased levels of oxygen in the water (FOX and TAYLOR, 1955). Tube-dwelling polychaetes show different responses in pumping when exposed to oxygen deficient water: *Nereis diversicolor* and *Sabella pavonina* respond with decreased pumping rates or complete cessation of pumping; *Chaetopterus variopedatus* pumps more violently; and *Arenicola marina* suspends external respiration, except for infrequent powerful bursts (WELLS, 1949; WELLS and DALES, 1951).

Reduction in the amount of oxygen dissolved in sea water influences differentially the rhythmic valve closure of various species of lamellibranchs (SALANKI, 1966). There is an immediate increase in the periodic closure rate in species of *Ostrea*, no detectable effect on *Cardium* and *Ensis*, and a delayed effect followed by an increased periodic closure by *Glycemeris*.

As the oxygen content is decreased, the ability of coho salmon *Oncorhynchus kisutch* and chinook salmon *Oncorhynchus tshawytscha* to maintain a sustained, maximum swimming speed is reduced. Lowering of the oxygen concentration to 3 mg/l, for example, causes a reduction in swimming speed of 30% (DAVIS and co-authors, 1963). Behavioural and structural adaptations of air-breathing fishes have been discussed by DAS (1934), and earlier in this chapter.

Under laboratory conditions, *Gammarus oceanicus* avoids regions of anoxia when observed in a preference chamber (COOK and BOYD, 1965). Studies of the behaviour of individuals at the aerated-anoxia water interface of the apparatus suggest that *G. oceanicus* possesses effective chemoreceptory mechanisms.

In the fish *Phoxinus phoxinus*, the basis for avoidance of water with undesirable low oxygen content seems to be random swimming and struggling that appears to be caused by dyspnoea (JONES, 1952). In contrast, BISHAI (1962) reported that 6 to 26 week-old fry of *Salmo salar* and *Salmo trutta* respond definitely and negatively to water of low oxygen concentrations. However, this behaviour varies with age, as alevins (1 to 4 weeks after hatching) show no response to abnormally low oxygen concentrations. Postlarval flounder *Paralichthys lethostigma* respond to low oxygen concentrations by swimming out of stress regions (DEUBLER and POSNER, 1963).

Carbon dioxide

Increased CO_2 tension accelerates the respiratory movements in certain crustaceans (WALSHE-MAETZ, 1956). Some crustaceans swim upward in relation to gravity and light when exposed to increased CO_2 pressures (VON BUDDENBROCK, 1952; UBRIG, 1952). ROAF (1912) found excess CO_2 to have no effect on respiratory movements of *Balanus balanoides*. In *Squilla mantis*, increased CO_2 accelerates respiratory movements of abdominal appendages (MATULA, 1912). The respiratory volume of an *Octopus* species increases up to 10 times the normal value when the CO_2 level is augmented, but there is little alteration in frequency of respiratory movements (WINTERSTEIN, 1925). Increased CO_2 levels reduce the frequency of heart beat in *Mytilus edulis* (SCHLIEPER, 1955).

Respiratory movements of the semiterrestrial isopod *Ligia oceania* are not influenced by excess CO_2; but the latter causes an accelerated pleopod rhythm in *Gammarus locusta* (FOX and JOHNSON, 1934). The rate of scaphognathite beating in *Eriocheir sinensis* increases when the CO_2 level is raised to 20 to 30 cm^3/l (VAN HEERDT and KRIJGSMAN, 1939).

Increased CO_2 levels reduce the rate of oxygen consumption of fishes maintained at a constant level of oxygen and at a steady rate of activity (BASU, 1959). The sensitivity of the fishes to CO_2 decreases with increasing acclimation temperature. POWERS and SHIPE (1928) and POWERS (1923) demonstrated that the metabolic rate of the herring *Clupea pallasii* is more sensitive to increased CO_2 tension than that of the salmon *Oncorhynchus kisutch* or the perch *Cymatogaster aggregatus*. Lowered pH results in similar responses in other marine fishes (PEREIRA, 1924).

Respiratory quotient

At low oxygen pressures (below 1 to 1·5% of atmosphere for *Uca pugnax* and *Uca pugilator*, and below 3·0 to 3·5% for *Sesarma cinereum*), the respiratory quotient (R.Q.) rises sharply above 1 (TEAL and CAREY, 1967); a value of 10 is reported for *U. pugnax*. TEAL and CAREY suggested that this increase in CO_2 release comes from carbonates of shell and tissues, as a result of the action of metabolic acids produced through fermentation at low oxygen pressures. According to BOSWORTH and co-authors (1936), it is necessary to cover the shell of crustaceans with collodion when making R.Q. measurements, in order to prevent respiratory CO_2 from reacting with carbonates. In various marine invertebrates and fishes, changes in oxygen tension have no effect upon the R.Q. values (BOSWORTH and co-authors, 1936). Studies involving *Asterias forbesii*, *Mytilus edulis* and a crayfish led MALOEUF (1937a) to conclude that, within limits, the R.Q. varies with the critical oxygen pressure in a fairly reciprocal manner.

The information presented above reveals a great diversity in regard to the effects of dissolved gases on metabolism and activity of aquatic animals. Our present knowledge does not yet allow us to formulate generalizations which can be applied to all species studied. Marine animals have evolved a large variety of physiological mechanisms which enable them to cope with a multiplicity of environmental conditions. New techniques, which permit simultaneous measurement of multiple physical and physiological parameters, promise to make possible a more thorough analysis of the mechanisms of metabolic adjustments to environmental stress.

(c) Reproduction

Little is known about the influence of dissolved gases on reproduction in marine animals when compared to the information at hand about the effects of other environmental factors, such as temperature (Chapter 3) and salinity (Chapter 4). This paucity of information may be, in part, the result of the technical difficulties in controlling the levels of dissolved gases in laboratory experiments. However, some phases of reproduction and embryonic development have received attention as far as the effects of deviated concentrations of dissolved ambient oxygen are concerned.

Gametes

In the absence of oxygen, *Arbacia punctulata* eggs become slightly reduced in volume, but endosmosis or exosmosis to water and ethylene glycol is not affected in either fertilized or unfertilized eggs (HUNTER, 1936). The unfertilized egg of *A. punctulata* remains normal for 8 hrs in the absence of oxygen, while the immobilization of the sperm, caused by oxygen lack, remains reversible if exposure periods do not exceed 3 to 4 hrs (HARVEY, 1930). Eggs and larvae of the chum salmon *Oncorhynchus keta* survive periods of low oxygenation by reducing their metabolic demands (WICKETT, 1954).

Other dissolved gases are also of importance. KELLEY (1946) related high mortalities in eggs of the Pacific herring *Clupea pallasii* to high concentrations of CO_2 or low pH values; however, the data presented by him are difficult to interpret. BURFIELD (1928) demonstrated that CO_2 profoundly decreases the rate of oxygen uptake by eggs of the plaice *Pleuronectes platessa*.

In the chum salmon *Oncorhynchus keta*, eggs are most sensitive to hypoxia between 100 and 200 centigrade degree days (ALDERDICE and co-authors, 1958). Reduced oxygen tensions cause a retardation in metabolic rate and in speed of embryonic development.

Fertilization

Spermatozoa of *Arbacia punctulata* lose their ability to fertilize eggs after 4 hrs exposure to sea water devoid of oxygen. If eggs are fertilized in the absence of oxygen, no fertilization membrane is formed (HARVEY, 1930). Oxygen lack may cause parthenogenetic development in *A. punctulata* (HARVEY, 1956).

Cleavage and early development

Oxygen lack arrests the rate of cleavage in the sea urchins *Paracentrotus lividus*, *Psammechinus microtuberculatus* and *Arbacia punctulata* and obliterates cleavage planes (HARVEY, 1927, 1956). When the ambient oxygen pressure sinks below 20 mm Hg, the rate of oxygen consumption of the eggs is reduced sharply, and cleavage becomes retarded (AMBERSON, 1928). TANG (1931) and TANG and GERARD (1932) reported the critical tension for fertilized and unfertilized *A. punctulata* eggs to be 50 and 40 mm Hg, respectively. In general, smaller fish eggs are not as resistant to anaerobiosis as larger eggs (SMITH, 1957). *Fundulus heteroclitus* eggs, when exposed to a complete vacuum for 4 days, will cleave for a short period of time before their development becomes arrested; when exposed to air, development continues. Eggs of the chum salmon *Oncorhynchus keta*, subjected to subnormal oxygen levels at early incubation stages, develop a number of abnormalities (ALDERDICE and co-authors, 1958). For additional metabolic responses of early life-cycle stages to variations in dissolved gases consult p. 1504.

Adult paradise fish *Macropodus opercularis* live in stagnant water and breathe atmospheric air which is stored temporarily in their suprabranchial respiratory organ. EBELING and ALPERT (1966) found that the growth rate of young *M. opercularis* fry is retarded when these are exposed to water with a subnormal oxygen level (2·5 ppm). However, the differentiation of their suprabranchial respiratory organ is not influenced markedly in any way to indicate a compensatory adaptation to low oxygen levels.

Combined effects of temperature, salinity and oxygen on egg mortality and embryonic development of the euryhaline fish *Cyprinodon macularius* have been examined by KINNE and KINNE (1962 a, b). Egg mortality is highest during the first few days following fertilization, with a second peak occurring shortly before and during hatching. At temperatures above 20°C, egg mortality is higher in water with 70% air saturation than in fully air-saturated water. In hypoxial water, egg mortality reaches 100% at 27·5°C, whereas in air-saturated water the 100% mortality level is attained at 36°C.

Reduced levels of dissolved oxygen retard the rate of embryonic development in *Cyprinodon macularius*. For example, in 100% air-saturated sea water (27·5°C, 35‰S) incubation time is 5·5 days, while in 70% air-saturated sea water it is 6·6 days (KINNE and KINNE, 1962b). When the oxygen saturation of the water is increased artificially to 300%, incubation time becomes significantly reduced in supranormal salinities and the ability to withstand high temperatures and high salinities increases. GARSIDE (1966) obtained similar results with freshwater-living brook trout *Salvelinus fontinalis* and rainbow trout *Salmo gairdneri*. It was pointed out by GARSIDE, as well as by SILVER and co-authors (1963), that their results do not agree with those predicted by ALDERDICE and co-authors (1958) on the basis of mathematical models.

Hatching

Eggs of salmon *Oncorhynchus keta* hatch 5 to 7 days faster when subjected to water through which hydrogen gas has been bubbled (TRIFONOVA, 1937). MILK-MAN (1954) reported that excess amounts of dissolved oxygen inhibit hatching of *Fundulus heteroclitus* eggs. He was able to control the time of hatching by bubbling oxygen through a flask containing eggs and sea water; the flask was then stoppered until several days after the expected hatching date. The eggs so treated hatch within minutes when washed in fresh sea water. This response, however, seems not entirely related to ambient oxygen levels, but may also be due to the presence of an inhibitory substance in the water which is heat sensitive.

When eggs of the chum salmon *Oncorhynchus keta* are subjected to subnormal dissolved-oxygen levels just prior to hatching, they hatch prematurely at a rate dependent upon the level of hypoxia; maximum premature hatching rate corresponds with the median lethal oxygen level (ALDERDICE and co-authors, 1958). In eggs of *Salmo salar*, the rate of oxygen consumption depends not only on the ambient oxygen tension, but also on the presence of the egg capsule (HAYES and co-authors, 1951). Embryos liberated from their capsule, consume oxygen at a higher rate. LINDROTH (1942) had been aware of the limitation of oxygen uptake by the egg membrane and emphasized that the most stressful period for developing eggs in respect of oxygen availability occurs just prior to hatching. The mean length of newly hatched alevins increases with increasing levels of ambient dissolved oxygen, even though the incubation time is reduced. This finding is in contrast to the typical relation observed between hatching size and speed of embryonic development. Higher temperatures, for example, accelerate embryonic development but cause, at the same time, a reduction in hatching length.

(d) Distribution

On the preceding pages of this chapter, many examples of the influence of dissolved gases on the ecology and distribution of marine animals have been presented. Marine animals have evolved a variety of physiological mechanisms which allow them to survive the 'normal' fluctuations in concentration of dissolved gases in their habitat. If abnormal fluctuations prevail, mass mortalities may result (BRONGERSMA-SANDERS, 1957; TULKKI, 1965). FRY (1957) writes:

> 'Any reduction of oxygen content below levels where active metabolic rate begins to be restricted is probably unfavorable to the species concerned. From the ecological point of view this incipient limiting level can be taken as the point where O_2 content begins to be unsuitable.'

The oxygen requirements of faunal assemblages have been used as evidence in support of theories on the origin of faunas occupying a given zoogeographical region. For example, a portion of the fauna in the Azov-Black Sea region is thought to have derived from that in the Caspian Sea since the post-Pleistocene. The Caspian fauna is not physiologically adjusted to life in habitats with low oxygen contents. Oxygen-poor habitats are colonized entirely by Mediterranean forms (CASPERS, 1957).

The oxygen requirements of the prawn *Penaeus indicus* change during their growth period in the estuary. Larger individuals show a greater respiratory dependency and higher mortality as the oxygen content of the water decreases, than do smaller prawns (SUBRAHMANYAN, 1962); it is possible to correlate these findings with the migration habit and seasonal distribution of this species. ROULE (1916) reported on the possible significance of differences in oxygen content of ponds and marine water for the migration of the mullet *Mugil*.

An abundant and varied fauna occupies the spaces between the sand grains of marine beaches. These interstitial animals are found at various substrate depths. According to GORDON (1960), animals living at depths greater than 5 to 10 cm are completely anaerobic when covered by the incoming tide. When the slope of the beach is slight, anaerobic conditions may persist even at low tide, if the capillary forces keep the sand saturated with water (see also Chapter 4.31). Little is known of the biochemical and physiological adaptations, which allow animals to endure such extreme conditions.

Rich collections of animals have been made in the oxygen-minimum layer of the sea, even though the oxygen content may be only $0 \cdot 1$ ml/l (MARSHALL, 1954; RICHARDS, 1957). However, little is known of the respiratory physiology of these forms, neither in terms of aerobic-anaerobic adaptions at the species level nor the total community respiratory balance sheet. In addition to residual biota, scattering layers have been recorded at the same depths as the oxygen-minimum layer by KANWISHER and EBELING (1957). These workers collected large numbers of the hatched fishes *Argyropelecus lychnus*, *A. affinis*, *Sternoptyx obscura*, *Vinciguerria lucetia* and *Ichthyococcus* sp. in the oxygen-minimum layer, all of which were bloated with large amounts of gas, which was mostly oxygen. This observation is different from that of JONES and MARSHALL (1953), who reported that, if fish are kept in low oxygen water, the proportion of oxygen in their swimbladder decreases. According to MOORE (1950), the vertical migration of the scattering layer over a fairly wide range of oxygen levels apparently does not hinder the organisms associated with it; although the scattering layer shifted from $274 \cdot 3$ to $731 \cdot 5$ m, the same species of animals were found throughout the entire scattering layer. The differences in oxygen content were at least $2 \cdot 0$ to $2 \cdot 5$ ml/l.

Approximately 54 to 90% of the gas in the swimbladder of deep-sea fishes is oxygen, while in surface-dwelling fishes, the oxygen content is slightly higher than in air (KANWISHER and EBELING, 1957). SCHOLANDER and co-authors (1956) established that the oxygen in the swimbladder comes from dissolved oxygen in the surrounding water. There are no significant differences in haemoglobin levels between deep-sea and surface fishes which could account for the increased oxygen (VAN DAM and SCHOLANDER, 1953). Based on calculations of the work required for fishes to migrate between different water depths, KANWISHER and EBELING (1957) stated that, if fishes with swimbladders make up a significant part of the deep scattering layer, it would be unrealistic to assume they attain neutral buoyancy at a given depth. Specific observations were made on the lantern fishes *Myctophum evermanni*, *M. spinosum* and *M. affine*. These fishes, while near the surface at night, have swimbladders with high percentages of oxygen, which is probably secreted at depth before their upward migration. Bathypelagic fishes without a gas-filled swimbladder have neutral buoyancy (DENTON and MARSHALL, 1958); their chief adaptations to life in greater depths are: poorly ossified skeletons, low protein content, and muscular modifications.

(3) Structural Responses

(a) Size

Exposure to dissolved oxygen levels of 21%, 10% and 4% air saturation for 35 days does not influence the size of growing individuals of *Arenicola marina*. However, newly hatched *Scoloplos armiger* exposed to 4% oxygen grow larger than individuals maintained at 21% oxygen, when observed after 35 days (Fox and TAYLOR, 1955). *Tubifex tubifex* from the mud of the River Thames (England), when maintained in one-fifth air-saturated water, grow about 3 times larger than those maintained in full-aerated water (Fox and TAYLOR, 1955). Although the cause and effect relation was not studied, KEYS (1931) reported that *Fundulus parvipinnis* with relatively long heads are more resistant to asphyxiation than individuals with normal-sized heads. This gain in resistance may be related to a possibly larger gill area in 'long heads'.

Although not strictly a subject of this review, it is justified to include a few references on morphological adaptations allowing most efficient use of gases for such important functions as buoyancy and diving (IRVING, 1939; DENTON and MARSHALL, 1958; DENTON, 1960, 1961; LANE, 1960; KUHN and co-authors, 1963; SCHOLANDER, 1964).

(b) External Structures

Newly hatched *Scoloplos armiger* develop 5 or 6 pairs of gills when maintained for 35 days in water with a dissolved oxygen content of 4%, while those kept at 21% have only 2 or 3 pairs; also the colour of the integument of marine poly-chaetes is influenced by the oxygen tension of the sea water (Fox and TAYLOR, 1955). After exposure to 4% oxygen for 35 days, the two ends of *Arenicola marina* become bright yellow; at 21% they become light reddish brown, and at 100%, dark brown. The presence of pure oxygen changes the pigment colour of the crown of *Sabella* from light to dark brown within 10 days. Based on field observations, FAUVEL (1958) suggested that the unusual pectinate gills, found in polychaetes of the family Ampharetidae in the Gulf of Guinea, are caused by warm, poorly oxygenated and polluted water. Additional modifications of external gill structures were observed when comparing these polychaetes with their counterparts from other regions along the west coast of Africa.

Embryos of the fish *Cyprinodon macularius* maintained in water of supranormal oxygen content (300% air saturation) tend to be more transparent than those developing at 100% saturation; their dorsal body surface tends to be slightly greenish in colour, and, in salinities above 35‰, their body length at hatching increases (KINNE and KINNE, 1962b).

(c) *Internal Structures*

No significant changes of internal structures (organs, tissues, cells) of marine animals due to different levels of dissolved gases in the surrounding water have come to the reviewer's attention.

Literature Cited (Chapter 9)

ALDERDICE, D. F., WICKETT, W. P. and BRETT, J. R. (1958). Some effects of temporary exposure to low dissolved oxygen levels on Pacific salmon eggs. *J. Fish. Res. Bd Can.*, **15**, 229–250.

ALLEN, M. B. (1963). Nitrogen fixing organisms in the sea. In C. H. Oppenheimer (Ed.), *Symposium on Marine Microbiology.* C. C. Thomas, Springfield, Ill. pp. 85–92.

AMBERSON, W. R. (1928). The influence of oxygen tension upon the respiration of unicellular organisms. *Biol. Bull. mar. biol. Lab., Woods Hole*, **55**, 79–92.

AMBERSON, W. R., MAYERSON, H. S. and SCOTT, W. J. (1924). The influence of oxygen tension upon metabolic rate in invertebrates. *J. gen. Physiol.*, **7**, 171–176.

ARUDPRAGASAM, K. D. and NAYLOR, E. (1964). Gill ventilation volumes, oxygen consumption and repiratory rhythms in *Carcinus maenas* (L.). *J. exp. Biol.*, **41**, 309–321.

BAAS-BECKING, L. G. M. (1925). Studies on the sulfur bacteria. *Ann. Bot.*, **39**, 613–650.

BALDWIN, E. (1957). *Biochemie.* (Transl. from Engl.: *Dynamic Aspects of Biochemistry.*) Verlag Chemie, Weinheim.

BANNISTER, W. H., BANNISTER, J. V. and MICALLEF, H. (1966). A biochemical factor in the zonation of marine molluscs. *Nature, Lond.*, **211**, 747.

BARKER, H. A. (1935). Photosynthesis in diatoms. *Arch. Mikrobiol.*, **6**, 141–156.

BARNES, H. (1959). *Apparatus and Methods of Oceanography*, Pt 1, Chemical. Allen and Unwin, London.

BASKIN, R. J. and ALLEN, K. (1963). Regulation of respiration in the molluscan (*Tivella stultorum*) heart. *Nature, Lond.*, **198**, 448–450.

BASU, S. P. (1959). Active respiration of fish in relation to ambient concentrations of oxygen and carbon dioxide. *J. Fish. Res. Bd Can.*, **16**, 175–212.

BAVENDAMM, W. (1924). *Die farblosen und roten Schwefelbakterien des Süß- und Salzwassers,* G. Fischer, Jena.

BAVENDAMM, W. (1932). Die mikrobiologische Kalkfällung in der tropischen See. *Arch. Mikrobiol.*, **3**, 201–276.

BEADLE, L. C. (1961). Adaptations of some aquatic animals to low oxygen levels and to anaerobic conditions. *Symp. Soc. exp. Biol.*, **15**, 121–131.

BEADLE, L. C. and BEADLE, S. F. (1949). Carbon dioxide narcosis. *Nature, Lond.*, **164**, 235.

BEAMISH, F. W. H. (1964). Respiration of fishes with special emphasis on standard oxygen consumption. III. Influence of oxygen. *Can. J. Zool.*, **42**, 355–366.

BERKELEY, C. (1921). Anaerobic respiration in some pelecypod mollusks. *J. biol. Chem.*, **46**, 579–598.

BISHAI, H. M. (1962). Reactions of larval and young salmonids to water of low oxygen concentration. *J. Cons. perm. int. Explor. Mer*, **27**, 167–180.

BLACK, E. C. (1951). Respiration in fishes. In W. S. Hoar, V. S. Black and E. C. Black (Eds), *Some Aspects of the Physiology of Fish*, Pt III. *Univ. Toronto Stud. biol. Ser.*, **59**, 91–111. (Publs Ont. Fish. Res. Lab., **71**.)

BLACKMAN, F. F. and SMITH, A. M. (1910). On assimilation in submerged water plants and its relation to the concentration of carbon dioxide and other factors. *Proc. R. Soc. (B)*, **83**, 389–412.

BORDEN, M. A. (1931). A study of the respiration and of the function of haemoglobin in *Planorbis corneus* and *Arenicola marina*. *J. mar. biol. Ass. U.K.*, **17**, 709–738.

BOSWORTH, M. W., O'BRIEN, H. and AMBERSON, W. R. (1936). Determination of the respiratory quotient in marine animals. *J. cell. comp. Physiol.*, **9**, 77–87.

BOUCK, G. R. and BALL, R. (1965). Influence of a diurnal oxygen pulse on fish serum proteins. *Trans. Am. Fish. Soc.*, **94**, 363–370.

BRAFIELD, A. E. and CHAPMAN, G. (1965). The oxygen consumption of *Pennatula rubra* ELLIS and some other anthozoans. *Z. vergl. Physiol.*, **50**, 363–370.

BRAND, T. F. VON (1927). Stoffbestand und Ernährung einiger Polychäten und anderer mariner Würmer. *Z. vergl. Physiol.*, **5**, 643–698.

BRAND, T. F. VON (1945). The anaerobic metabolism of invertebrates. *Biodynamica*, **5**, 165–295.

BREED, R. S., MURRAY, E. G. D. and SMITH, N. R. (1957). *Bergey's Manual of Determinative Bacteriology*, 7th ed., Williams and Wilkins, Baltimore.

BRIGGS, G. E. and WHITTINGHAM, C. P. (1952). Factors affecting the rate of photosynthesis of *Chlorella* at low concentrations of carbon dioxide and in high illumination. *New Phytol.*, **51**, 236–249.

BRITTAIN, E. G. (1957). Oxygen effects in photosynthesis. Thesis, University of Melbourne, Australia.

BROEKHUYSEN, G. J., JR. (1935). The extremes in percentages of dissolved oxygen to which the fauna of a *Zostera* field in the tidal zone at Nieuwdiep can be exposed. *Archs néerl. Zool.*, **1**, 339–346.

BRONGERSMA-SANDERS, M. (1957). Mass mortality in the sea. In J. W. Hedgpeth (Ed.), *Treatise on Marine Ecology and Paleoecology*, Vol. I. *Mem. geol. Soc. Am.*, **67**, 941–1010.

BROWN, A. C. (1964). Lethal effects of hydrogen sulphide on *Bullia* (Gastropoda). *Nature, Lond.*, **203**, 205–206.

BRUCE, J. R. (1928). Physical factors on the sandy beach. II. Chemical changes—carbon dioxide concentration and sulphides. *J. mar. biol. Ass. U.K.*, **15**, 553–565.

BUDDENBROCK, W. VON (1952). *Vergleichende Physiologie*, Vol. I, Sinnesphysiologie. Birkhaeuser, Basle.

BULL, H. O. (1931). Resistance of *Eurytemora hirundoides*, a brackish water copepod, to oxygen depletion. *Nature, Lond.*, **127**, 406–407.

BULLEY, N. R. (1969). Effects of light quality and intensity on photosynthesis and photorespiration in attached leaves. Thesis, Simon Fraser University, Burnaby, Canada.

BURFIELD, S. T. (1928). The absorption of oxygen by plaice eggs. *Br. J. exp. Biol.*, **5**, 177–182.

CARR, D. J. (1961). Chemical influences of the environment. *Handb. PflPhysiol.*, **16**, 737–794.

CARTER, G. A. (1931). Aquatic and aerial respiration. *Biol. Rev.*, **6**, 1–35.

CASPERS, H. (1957). Black Sea and Sea of Azov. In J. W. Hedgpeth (Ed.), *Treatise on Marine Ecology and Paleoecology*, Vol. I, Ecology. *Mem. geol. Soc. Am.*, **67**, 801–889.

CHAPMAN, R. E., COOK, W. R. I. and THOMPSON N. L. (1924). The effect of carbon dioxide on the tropic reactions of *Helianthus* stems. *New Phytol.*, **23**, 50–62.

COLEMAN, A. W. (1962). Sexuality. In R. Lewin (Ed.), *Physiology and Biochemistry of Algae*. Academic Press, New York. pp. 711–729.

COLLIP, J. B. (1921). A further study of the respiration processes in *Mya arenaria* and other marine molluscs. *J. biol. Chem.*, **49**, 297–310.

COOK, R. H. and BOYD, C. M. (1965). The avoidance of *Gammarus oceanicus* SEGERSTOLE (Amphipoda, Crustacea) of anoxic regions. *Can. J. Zool.*, **43**, 971–975.

COURTNEY, W. A. M. and NEWELL, R. C. (1965). Ciliary activity and oxygen uptake in *Branchiostoma lanceolatum* (PALLAS). *J. exp. Biol.*, **43**, 1–12.

DAKIN, W. J. and CATHERINE, M. G. (1925). The oxygen requirement of certain aquatic animals and its bearing upon the source of food supply. *Br. J. exp. Biol.*, **2**, 293–322.

DALES, R. P. (1958). Survival of anaerobic periods by two intertidal polychaetes, *Arenicola marina* (L.) and *Owenia fusiformis* DELLE CHIAJE. *J. mar. biol. Ass. U.K.*, **37**, 521–529.

DAM, L. VAN (1935). On the utilization of oxygen by *Mya arenaria*. *J. exp. Biol.*, **12**, 86–94.

DAM, L. VAN (1954). On the respiration in scallops (Lamellibranchiata). *Biol. Bull. mar. biol. Lab., Woods Hole*, **107**, 192–202.

DAM, L. VAN and SCHOLANDER, P. F. (1953). Concentration of hemoglobin in the blood of deep sea fishes. *J. cell. comp. Physiol.*, **41**, 522–524.

DAS, B. K. (1934). The habitats and structure of *Pseudapocryptes lanceolatus*, a fish in the first stages of structural adaptation to aerial respiration. *Proc. R. Soc. (B)*, **115**, 422–430.

DAVIS, G. E., FOSTER, J., WARREN, C. E. and DOUDOROFF, P. (1963). The influence of oxygen concentration on the swimming performance of juvenile Pacific salmon at various temperatures. *Trans. Am. Fish. Soc.*, **92**, 111–124.

DAVISON, R. C., BREESE, W. P., WARREN, C. E. and DOUDOROFF, P. (1959). Experiments on the dissolved oxygen requirements of cold-water fishes. *Sewage ind. Wastes*, **31**, 950–966.

DENTON, E. J. (1960). The buoyancy of marine animals. *Scient. Am.*, **203**, 118–128.

DENTON, E. J. (1961). The buoyancy of fish and cephalopods. *Prog. Biophys. biophys. Chem.*, **11**, 178–234.

DENTON, E. J. and MARSHALL, N. B. (1958). The buoyancy of bathypelagic fishes without a gas-filled swimbladder. *J. mar. biol. Ass. U.K.*, **37**, 753–767.

DEUBLER, E. E. and POSNER, G. S. (1963). Response of post-larval flounders, *Paralicthys lethostigma*, to water of low oxygen concentration. *Copeia*, **1963**, 312–317.

DIETRICH, G. and KALLE, K. (1963). *General Oceanography*, Wiley, New York.

DOUDOROFF, P. (1957). Water quality requirements of fishes and effects of toxic substances. In M. E. Brown (Ed.), *The Physiology of Fishes*, Vol. II. Academic Press, New York. pp. 403–430.

DOWNTON, W. J. S. and TREGUNNA, E. B. (1968). Carbon dioxide compensation—its relation to photosynthetic carboxylation reactions, systematics of the Gramineae, and leaf anatomy. *Can. J. Bot.*, **46**, 207–215.

DURNER, G., RÖMER, R. and SCHWARTZ, W. (1965). Untersuchungen über die Lebensgemeinschaften des Sulphuretums. *Z. allg. Mikrobiol.*, **5**, 206–221.

EBELING, A. W. and ALPERT, J. S. (1966). Retarded growth of the paradisefish, *Macropodus opercularis* in low environmental oxygen. *Copeia*, **1966**, 606–610.

ELIASSEN, E. (1955). The oxygen supply during ebb of *Arenicola marina* in the Danish Waddensea. *Univ. Bergen Årb.*, **12**, 3–9.

EMERSON, R. and ARNOLD, W. (1932). A separation of the reactions in photosynthesis by means of intermittent light. *J. gen. Physiol.*, **15**, 391–420.

EMERSON, R. and GREEN, L. (1934). Manometric measurements of photosynthesis in the marine alga *Gigartina*. *J. gen. Physiol.*, **17**, 817–842.

EMERSON, R. and GREEN, L. (1938). Effect of hydrogen-ion concentration on *Chlorella* photosynthesis. *Pl. Physiol.*, **13**, 157–168.

FAUVEL, P. (1958). Sur les Amphárétiens (Annélides Polychaètes) de la Côte Occidentale del l'Afrique. *Bull. Inst. océanogr. Monaco*, **1130**, 1–8.

FOGG, G. E. (1953). *The Metabolism of Algae*, Wiley, New York.

FOGG, G. E. (1962). Nitrogen fixation. In R. Lewin (Ed.), *Physiology and Biochemistry of Algae*. Academic Press, New York. pp. 161–170.

FOGG, G. E. (1966). *Algal Cultures and Phytoplankton Ecology*, University Wisconsin Press, Milwaukee, Wisc.

FORK, D. C. (1963). Action spectra for O_2 evolution by chloroplasts with and without added substrate for regeneration of O_2 evolving ability by far-red, and for O_2 uptake. *Pl. Physiol.*, **38**, 323–332.

K

FORK, D. C. and URBACH, W. (1965). Evidence for the localization of plastocyanin in the electron-transport chain of photosynthesis. *Proc. natn. Acad. Sci. U.S.A.*, **53**, 1307–1315.

FORRESTER, M. L., KROTKOV, G. and NELSON, C. D. (1966). Effect of oxygen on photosynthesis, photorespiration and respiration in detached leaves. I. Soybean. *Pl. Physiol.*, **41**, 422–427.

FOX, C. J. J. (1907). On the coefficients of absorption of the atmospheric gases in distilled water and sea water. Pt 1 Nitrogen and oxygen. *Publs Circonst. Cons. perm. int. Explor. Mer*, **41**, 1–23.

FOX, H. M. (1955). The effect of oxygen on concentration of haem in invertebrates. *Proc. R. Soc. (B)*, **143**, 203–214.

FOX, H. M. and JOHNSON, M. L. (1934). The control of respiratory movements in Crustacea by oxygen and carbon dioxide. *J. exp. Biol.*, **11**, 1–10.

FOX, H. M. and SIMMONDS, B. G. (1933). Metabolic rates of aquatic arthropods from different habitats. *J. exp. Biol.*, **10**, 67–74.

FOX, H. M. and TAYLOR, A. E. R. (1955). The tolerance of oxygen by aquatic invertebrates. *Proc. R. Soc. (B)*, **143**, 214–225.

FRY, F. E. J. (1957). The aquatic respiration of fish. In M. E. Brown (Ed.), *The Physiology of Fishes*, Vol. I. Academic Press, New York. pp. 1–63.

FRY, F. E. J. (1964). Animals in aquatic environments: fishes. In D. B. Dill, E. F. Adolph and C. G. Wilber (Eds), *Handbook of Physiology*, Sect. 4, Adaptation to the environment. American Physiological Society, Washington, D.C. pp. 715–728.

GAARDER, T. (1918). Über den Einfluss des Sauerstoffdruckes auf den Stoffwechsel. II. Nach Versuchen an Karpfen. *Biochem. Z.*, **89**, 94–125.

GAFFRON, H. (1940). Studies on the induction period of photosynthesis and light respiration in green algae. *Am. J. Bot.*, **27**, 204–216.

GAFFRON, H. and BISHOP, N. (1962). Photoreduction at λ705 mμ in adapted algae. *Biochem. biophys. Res. Commun.*, **8**, 471–476.

GALTSOFF, P. S., PRYTHERICH, U. F., SMITH, R. O. and KOEHRING, V. (1936). Effects of crude oil pollution on oysters in Louisiana waters. *Bull. Bur. Fish., Wash.*, **48**, 143–210.

GALTSOFF, P. S. and WHIPPLE, D. V. (1930). Oxygen consumption of normal and green oysters. *Bull. Bur. Fish., Wash.*, **46**, 489–508.

GARSIDE, E. T. (1966). Effects of oxygen in relation to temperature on the development of embryos of brook trout and rainbow trout. *J. Fish. Res. Bd Can.*, **23**, 1121–1134.

GESSNER, F. (1959). *Hydrobotanik*, Vol. II, Deutscher Verlag der Wissenschaften, Berlin.

GIETZEN, J. (1931). Untersuchungen über marine Thiorhodaceen. *Zentbl. Bakt. ParasitKde (Abt. 2)*, **83**, 183–218.

GORDON, M. S. (1960). Anaerobiosis in marine sandy beaches. *Science, N.Y.*, **132**, 616–617.

GRAHAM, J. M. (1949). Some effects of temperature and oxygen pressure on the metabolism and activity of the speckled trout, *Salvelinus fontinalis. Can. J. Res. (D)*, **27**, 270–288.

GUNKEL, W. (Ed.) (1967). Arbeitssitzung über Gewässerverölung, Ölbekämpfung und Ölabbau. *Helgoländer wiss. Meeresunters.*, **16**, 285–384.

HALL, F. G. (1928). Blood concentration in marine fishes. *J. biol. Chem.*, **76**, 623–631.

HALL, F. G. (1929). The influence of varying oxygen tensions upon the rate of oxygen consumption in marine fishes. *Am. J. Physiol.*, **88**, 212–218.

HALL, F. G. (1930). The ability of the common mackerel and certain other marine fishes to remove dissolved oxygen from sea water. *Am. J. Physiol.*, **93**, 412–421.

HALL, F. G. (1931). The respiration of puffer fish. *Biol. Bull. mar. biol. Lab., Woods Hole*, **61**, 457–467.

HALL, F. G., GRAY, I. E. and LEPKOVSKY, S. (1926). The influence of asphyxiation on the blood constituents of marine fishes. *J. biol. Chem.*, **67**, 549–554.

HALL, F. G. and McCUTCHEON, F. H. (1938). The affinity of hemoglobin for oxygen in marine fishes. *J. cell. comp. Physiol.*, **11**, 205–212.

HARDER, R. (1921). Kritische Versuche zu Blackman's Theorie der 'Begrenzenden Faktoren' bei Kohlensäure-Assimilation. *Jb. wiss. Bot.*, **60**, 531–571.

HARVEY, E. B. (1927). The effect of lack of oxygen on sea urchin eggs. *Biol. Bull. mar. biol. Lab., Woods Hole*, **53**, 147–160.

HARVEY, E. B. (1930). The effect of lack of oxygen on the sperm and unfertilized eggs of *Arbacia punctulata* and on fertilization. *Biol. Bull. mar. biol. Lab., Woods Hole*, **58**, 288–292.

HARVEY, E. B. (1956). *The American Arbacia and other Sea Urchins*. Princeton University Press, Princeton, N.J.

HARVEY, H. W. (1945). *Recent Advances in the Chemistry and Biology of Sea Water*, Cambridge University Press, London.

HAYES, F. R., WILMOT, I. R. and LIVINGSTONE, D. A. (1951). The oxygen consumption of the salmon egg in relation to development and activity. *J. exp. Zool.*, **116**, 377–395.

HAZELHOFF, E. H. (1938). Über die Ausnutzung des Sauerstoffs bei verschiedenen Wassertieren. *Z. vergl. Physiol.*, **26**, 306–327.

HECHT, F. (1932). Der chemische Einfluss organischer Zersetzungsstoffe auf das Benthos, dargelegt an Untersuchungen mit marinen Polychaeten, insbesondere *Arenicola marina* L. *Senckenbergiana*, **14**, 199–220.

HEERDT, P. F. VAN and KRIJGSMAN, B. J. (1939). Die Regulierung der Atmung bei *Eriocheir sinensis* MILNE-EDWARDS. *Z. vergl. Physiol.*, **27**, 29–40.

HENZE, M. (1910). Ueber den Einfluss des Sauerstoffdrucks auf den Gaswechsel einiger Meerestieren. *Biochem. Z.*, **26**, 255–278.

HERBERT, D. W. M. and SHURBEN, D. S. (1965). The susceptibility of salmonid fish to poisons under estuarine conditions. II. Ammonium chloride. *Int. J. Air Wat. Pollut.*, **9**, 89–91.

HIESTAND, W. (1940). Oxygen consumption of *Thyone briareus* (Holothuroidea) as a function of the oxygen tension and hydrogen ion concentration of the surrounding medium. *Trans. Wis. Acad. Sci. Arts Lett.*, **32**, 167–175.

HILL, R. and WHITTINGHAM, C. P. (1957). *Photosynthesis*, 2nd ed., Wiley, New York.

HONERT, T. H. VAN DEN (1930). Carbon dioxide assimilation and limiting factors. *Recl. Trav. bot. néerl.*, **27**, 149–286.

HOOD, D. W., DUKE, T. W. and STEVENSON, B. (1960). Measurement of toxicity of organic wastes to marine organisms. *J. Wat. Pollut. Control Fed.*, **32**, 982–993.

HUNTER, F. R. (1936). The effect of lack of oxygen on cell permeability. *J. cell. comp. Physiol.*, **9**, 15–27.

HUNTER, W. S. and VERNBERG, W. B. (1955). Studies on oxygen consumption in digenetic trematodes. II. Effects of two extremes in oxygen tension. *Expl Parasit.*, **4**, 427–434.

HUTCHINSON, G. E. (1957). *A Treatise on Limnology*, Wiley, New York.

HUTTON, W. E. and ZOBELL, C. E. (1949). The occurrence and characteristics of methane oxidizing bacteria in marine sediments. *J. Bact.*, **58**, 463–473.

HYMAN, L. H. (1929). The effect of oxygen tension on oxygen consumption in *Planaria* and some echinoderms. *Physiol. Zool.*, **2**, 505–534.

HYMAN, L. H. (1932). Relation of oxygen tension to oxygen consumption in *Nereis virens*. *J. exp. Zool.*, **61**, 209–221.

IRVING, L. (1939). Respiration in diving mammals. *Physiol. Rev.*, **19**, 112–134.

ISHIDA, S. (1935). On the oxygen consumption in the oyster, *Ostrea gigas* THURBERG under various conditions. *Sci. Rep. Tôhoku Univ.* (4), **10**, 619–638.

JACUBOWA, L. and MALM, E. (1931). Die Beziehungen einiger Benthos-Formen des Schwarzen Meeres zum Medium. *Biol. Zbl.*, **51**, 105–116.

JAMES, W. O. (1928). Experimental researches on vegetable assimilation and respiration. XIX: The effect of carbon dioxide supply upon the rate of assimilation of submerged water plants. *Proc. R. Soc.* (B), **103**, 1–42.

JOB, S. V. (1955). The oxygen consumption of *Salvelinus fontinalis*. *Univ. Toronto Stud. biol. Ser.*, **61**, 1–39. (Publs Ont. Fish. Res. Lab., **73**.)

JOB, S. V. (1957). The routine oxygen consumption of the milk fish. *Proc. Indian Acad. Sci.*, **45**, 302–313.

JOHANSEN, K. and LENFANT. C. (1966). Gas exchange in the cephalopod, *Octopus dofleini*. *Am. J. Physiol.*, **210**, 910–918.

JOHNSON, F. H. (1936). The oxygen uptake of marine bacteria. *J. Bact.*, **31**, 547–556.

JOHNSON, M. L. (1936). The control of respiratory movements in Crustacea by oxygen and carbon dioxide. II. *J. exp. Biol.*, **13**, 467–475.

JONES, F. R. H. and MARSHALL, N. B. (1953). The structure and functions of the teleostian swimbladder. *Biol. Rev.*, **28**, 16–83.

JONES, J. R. E. (1952). The reactions of fish to water of low oxygen concentration. *J. exp. Biol.*, **29**, 403–415.

JÜRGENS, O. (1935). Die Wechselbeziehungen von Blutkreislauf Atmung und Osmoregulation bei Polychäten (*Nereis diversicolor* O. F. MÜLL.). *Zool. Jb.* (*Abt. allg. Zool. Physiol. Tiere*), **55**, 1–46.

KALLE, K. (1945). *Der Stoffhaushalt des Meeres*, Akademische Verlagsanstalt, Leipzig.

KANWISHER, J. and EBELING, A. (1957). Composition of the swim-bladder gas in bathy-pelagic fishes. *Deep Sea Res.*, **4**, 211–217.

KELLEY, A. M. (1946). Effect of abnormal CO_2 tension on development of herring eggs. *J. Fish. Res. Bd Can.*, **6**, 435–440.

KEYS, A. B. (1930a). The measurement of the respiratory exchange of aquatic animals. *Biol. Bull. mar. biol. Lab., Woods Hole*, **59**, 187–198.

KEYS, A. B. (1930b). Influence of varying oxygen tension upon the rate of oxygen consumption of fishes. *Bull. Scripps Instn Oceanogr. tech. Ser.*, **2**, 307–317.

KEYS, A. B. (1931). A study of the selective action of decreased salinity and of asphyxiation on the Pacific killifish, *Fundulus parvipinnis*. *Bull. Scripps Instn Oceanogr. tech. Ser.*, **2**, 417–490.

KINNE, O. and AURICH, H. (Eds) (1968). Biological and hydrographical problems of water pollution in the North-Sea and adjacent waters. International Symposium, Helgoland, September 19–22, 1967. *Helgoländer wiss. Meeresunters.*, **17**, 1–530.

KINNE, O. and KINNE, E. M. (1962a). Effects of salinity and oxygen on developmental rates in a cyprinodont fish. *Nature, Lond.*, **193**, 1097–1098.

KINNE, O. and KINNE, E. M. (1962b). Rates of development in embryos of a cyprinodont fish exposed to different temperature-salinity-oxygen combinations. *Can. J. Zool.*, **40**, 231–253.

KRISS, A. E. (1961). *Meeresmikrobiologie-Tiefseeforschungen* (Transl. from Russ.), G. Fischer, Jena.

KRISS, A. E. and RUKINA, J. A. (1949). Microbiology of the Black Sea. (Russ.) *Mikrobiologiya*, **18**, 2.

KROGH, A. (1941). *The Comparative Physiology of Respiratory Mechanisms*, University of Pennsylvania Press, Pa.

KROGH, A. and LEITCH, I. (1919). The respiratory function of the blood in fishes. *J. Physiol., Lond.*, **52**, 288–300.

KUHN, W., RAMEL, A., KUHN, H. J. and MARTI, E. (1963). The filling mechanism of the swim-bladder. Generation of high gas pressures through hairpin countercurrent multiplication. *Experientia*, **19**, 497–511.

KUSNEZOW, S. J. (1959). *Die Rolle der Mikroorganismen im Stoffkreislauf der Seen*, Deutscher Verlag der Wissenschaften, Berlin.

LANE, C. E. (1960). The Portuguese man-of-war. *Scient. Am.*, **202**, 158–168.

LARIMER, J. L. and ASHBY, E. A. (1962). Float gases, gas secretion and tissue respiration in the Portuguese man-of-war, *Physalia*. *J. cell. comp. Physiol.*, **60**, 41–47.

LEDEBUHR, J. F. VON (1939). Der Sauerstoff als ökologischer Faktor. *Ergebn. Biol.*, **16**, 173–261.

LENFANT, C. and JOHANSEN, K. (1966). Respiratory function in the elasmobranch *Squalus suckleyi* G. *Resp. Physiol.*, **1**, 13–29.

LEWIN, R. A. (1956). Control of sexual activity in *Chlamydomonas* by light. *J. gen. Microbiol.*, **15**, 170–185.

 INDROTH, A. (1938). Studien über die respiratorischen Mechanismen von *Nereis virens* SARS. *Zool. Bidr. Upps.*, **17**, 367–503.

LINDROTH, A. (1941). Atmungsventilation der Polychaten. *Z. vergl. Physiol.*, **28**, 485–532.

LINDROTH, A. (1942). Sauerstoffverbrauch der Fische. II. Verschiedene Entwicklungs- und Altersstadien von Lachs und Hecht. *Z. vergl. Physiol.*, **29**, 583–594.

LLOYD, R. M. (1967). Oxygen-18 composition of oceanic sulfate. *Science, N.Y.*, **156**, 1228–1231.

LLOYD, R. M. and HERBERT, D. W. M. (1960). The influence of carbon dioxide on the toxicity of unionized ammonia to rainbow trout (*Salmo gairdnerii* RICHARDSON). *Ann. appl. Biol.*, **48**, 399–404.

LUE-KIM, H. (1969). Hydrostatic pressure in relation to the synchronous culture of algae in open and closed systems. Thesis, Simon Fraser University, Burnaby, Canada.

LUKINS, H. B. and FOSTER, J. W. (1963). Utilization of hydrocarbons and hydrogen by Mycobacteria. *Z. allg. Mikrobiol.*, **3**, 251–264.

LUTZ, B. R. (1930). The effect of low oxygen tensions on the pulsations of the isolated holothurian cloaca. *Biol. Bull. mar. biol. Lab., Woods Hole*, **58**, 74–84.

McCLENDON, J. F. (1917). The direct and indirect calorimetry of *Cassiopeia xamachana*: the effect of stretching on the rate of the nerve impulse. *J. biol. Chem.*, **32**, 275–296.

McFARLAND, W. N. and MOSS, S. A. (1967). Internal behavior in fish schools. *Science N.Y.*, **156**, 260–262.

McLEESE, D. W. (1956). Effects of temperature, salinity and oxygen on the survival of the American lobster. *J. Fish. Res. Bd Can.*, **13**, 247–272.

MALOEUF, N. S. R. (1937a). Studies on the respiration of animals. I. Aquatic animals without an oxygen transporter in their internal medium. *Z. vergl. Physiol.*, **25**, 1–28.

MALOEUF, N. S. R. (1937b). Studies on the respiration of animals. II. Aquatic animals with an oxygen transporter in their internal medium. *Z. vergl. Physiol.*, **25**, 29–42.

MANWELL, C. J. (1960). Comparative physiology: blood pigments. *A. Rev. Physiol.*, **22**, 191–244.

MANWELL, C. J. (1963a). The chemistry and biology of hemoglobin in some marine clams. I. Distribution of the pigment and properties of the oxygen equilibrium. *Comp. Biochem. Physiol.*, **8**, 209–218.

MANWELL, C. J. (1963b). Genetic control of hemerythrin specificity in a marine worm. *Science, N.Y.*, **139**, 755–758.

MANWELL, C. J., SOUTHWARD, E. C. and SOUTHWARD, A. J. (1966). Preliminary studies on hæmoglobin and other proteins of the Pogonophora. *J. mar. biol. Ass. U.K.*, **46**, 115–124.

MARSHALL, N. B. (1954). *Aspects of Deep Sea Biology*, Philosophical Library, New York.

MARSHALL, S. M., NICHOLLS, A. G. and ORR, A. P. (1935). On the biology of *Calanus finmarchicus*. VI. Oxygen consumption in relation to environmental conditions *J. mar. biol. Ass. U.K.*, **20**, 1–28.

MATULA, J. (1912). Die Regulation der Atemrhythmik bei *Squilla mantis. Pflügers Arch. ges. Physiol.*, **144**, 109–131.

MEYER, H. (1935). Die Atmung von *Asterias rubens* und ihre Abhängigkeit von verschiedenen Aussenfaktoren. *Zool. Jb. (Abt. allg. Zool. Physiol. Tiere)*, **55**, 349–398.

MILKMAN, R. (1954). Controlled observation of hatching in *Fundulus heteroclitus. Biol. Bull. mar. biol. Lab., Woods Hole*, **107**, 300.

MOLISCH, H. (1912). Neue farblose Schwefelbakterien. *Zentbl. Bakt. ParasitKde (Abt. 2)*, **33**, 55–62.

MOORE, B., EDIE, E. S., WHITLEY, E. and DAKIN, W. J. (1912). The nutrition and metabolism of marine animals in relationship to (a) dissolved organic matter and (b) particulate organic matter of sea water. *Biochem. J.*, **6**, 255–296.

MOORE, H. B. (1931). The muds of the Clyde Sea Area. III. Chemical and physical conditions; rate and nature of sedimentation; and fauna. *J. mar. biol. Ass. U.K.*, **17**, 325–358.

MOORE, H. B. (1950). The relation between the scattering layer and the Euphausiacea. *Biol. Bull. mar. biol. Lab., Woods Hole*, **90**, 181–212.

MORTON, J. E., BONEY, A. D. and CORNER, E. D. S. (1957). The adaptations of *Lasaea rubra* (MONTAGU), a small intertidal lamellibranch. *J. mar. biol. Ass. U.K.*, **36**, 383–405.

MYERS, J. and BURR, G. O. (1940). Some effects of high light intensity on *Chlorella. J. gen. Physiol.*, **24**, 43–67.

NEEB, O. (1962). *Hydrodictyon* als Objekt einer vergleichenden Untersuchung physiologischer Grössen. *Flora, Jena*, **139**, 39–95.

NEWELL, R. C. and COURTNEY, W. A. M. (1965). Respiratory movements in *Holothuria forskali* DELLE CHIAJE. *J. exp. Biol.*, **42**, 45–57.

NICOL, E. (1935). The ecology of a salt marsh. *J. mar. biol. Ass. U.K.*, **20**, 203–261.

NOMURA, S. (1926). The influence of oxygen tension on the rate of oxygen consumption in *Caudina*. *Sci. Rep. Tôhoku Univ. (Ser. 4)*, **2**, 133–138.

NOZAWA, A. (1929). The normal and abnormal respiration in the oyster, *Ostrea circumpicta* Pils. *Sci. Rep. Tôhoku Univ. (Ser. 4)*, **4**, 315–325.

OCHOA, S. (1948). Biosynthesis of tricarboxylic acids by carbon dioxide fixation. III. Enzymatic mechanisms. *J. biol. Chem.*, **174**, 133–157.

OLIVIER, L., BRAND, T. VON and MEHLMAN, B. (1953). The influence of lack of oxygen on *Schistosoma mansoni* cercariae and on infected *Australorbis glabratus*. *Expl Parasit.*, **2**, 258–270.

ORR, A. P. and MOORHOUSE, F. W. (1933). (a) Variations in some physical and chemical conditions on and near Low Isles Reef. (b) The temperature of the water in the Anchorage, Low Isles. (c) Physical and chemical conditions in mangrove swamps. *Scient. Rep. Gt Barrier Reef Exped.*, **2**, 87–110.

OSTERLIND, S. (1948a). Influence of low bicarbonate concentration on the growth of a green alga. *Nature, Lond.*, **161**, 319–320.

OSTERLIND, S. (1948b). The retarding effect of high concentrations of carbon dioxide and carbonate ions on the growth of a green alga. *Physiologia Pl.*, **1**, 170–175.

OSTERLIND, S. (1951). Inorganic carbon sources of green algae III. Measurements of photosynthesis in *Scenedesmus quadricauda* and *Chlorella pyrenoidosa*. *Physiologia Pl.*, **4**, 242–254.

PAASCHE, E. (1964). A tracer study of the inorganic carbon uptake during coccolith formation and photosynthesis in the coccolithophorid *Coccolithus huxleyi*. *Physiologia Pl.* (Suppl.) III, 1–82.

PAAUW, F. VAN DER (1932). The indirect action of external factors on photosynthesis. *Recl. Trav. bot. néerl.*, **29**, 497–620.

PACKARD, W. H. (1905). On resistance to lack of oxygen and on a method of increasing this resistance. *Am. J. Physiol.*, **15**, 30–41.

PATANE, L. (1946a). Sulla biologia de *Littorina punctata* (GM.). *Boll. Soc. ital. Biol. sper.*, **22**, 928–929.

PATANE, L. (1946b). Anaerobiosi in *Littorina neritoides* (L.). *Boll. Soc. ital. Biol. sper.*, **22**, 929–930.

PATEL, S. and SPENCER, C. P. (1963). The oxidation of sulphide by the haem compounds from the blood of *Arenicola marina*. *J. mar. biol. Ass. U.K.*, **43**, 167–175.

PEREIRA, J. R. (1924). On the influence of the hydrogen-ion concentration upon the oxygen consumption in seawater fishes. *Biochem. J.*, **18**, 1924–1926.

PFENNIG, N. (1965). Anreicherungskulturen für rote und grüne Schwefelbakterien. In H. G. Schlegel (Ed.), *Anreicherungskultur und Mutantenauslese*. G. Fischer, Stuttgart. pp. 179–189.

POWERS, E. B. (1920). The variation of the condition of sea-water, especially the hydrogen-ion concentration, and its relation to marine organisms. *Publs Puget Sound mar. biol. Stn*, **2**, 369–385.

POWERS, E. B. (1923). The absorption of oxygen by the herring as affected by the carbon dioxide tension of the sea water. *Ecology*, **4**, 307–312.

POWERS, E. B. and SHIPE, L. M. (1928). The rate of oxygen absorption by certain marine fishes as affected by the oxygen content and carbon dioxide tension of the sea-water. *Publs Puget Sound mar. biol. Stn*, **5**, 365–372.

PROSSER, C. L., BARR, L. M., PINC, R. D. and LAUER, C. Y. (1957). Acclimation of goldfish to low concentrations of oxygen. *Physiol. Zool.*, **30**, 137–141.

PROSSER, C. L. and BROWN, F. A., JR. (1961). *Comparative Animal Physiology*, 2nd ed., W. B. Saunders Co., Philadelphia.

PSHENIN, L. N. (1963). Distribution and ecology of *Azotobacter*. In C. H. Oppenheimer (Ed.), *Symposium on Marine Microbiology*. C. C. Thomas, Springfield, Ill. pp. 383–391.

RABINOWITCH, E. I. (1945). *Photosynthesis and Related Processes*, Vol. I, Interscience, N.Y.

RABINOWITCH, E. I. (1951, 1956). *Photosynthesis and Related Processes*, Vol. II, Pts 1 and 2, Interscience, N.Y.

RAWITCH-SCHTERBO, J. (1930). To the question of bacterial thin layer in the Black Sea according to the hypothesis of Professor Egunow. (Russ.; summary in Engl.) *Trudȳ sevastopol'. biol. Sta.*, **2**, 127–141.

READ, K. R. H. (1962). The hemoglobin of the bivalved mollusc, *Phacoides pectinatus* GMELIN. *Biol. Bull. mar. biol. Lab., Woods Hole*, **123**, 605–617.

REDMOND, J. R. (1955). The respiratory function of hemocyanin in Crustacea. *J. cell. comp. Physiol.*, **46**, 209–247.

REDMOND, J. R. (1962a). Oxygen-hemocyanin relationships in the land crab, *Cardisoma gaunhumi. Biol. Bull. mar. biol. Lab., Woods Hole*, **122**, 252–262.

REDMOND, J. R. (1962b). The respiratory characteristics of chiton hemocyanins. *Physiol. Zool.*, **35**, 304–313.

REISH, D. J. (1955). The relation of polychaetous annelids to harbor pollution. *Publ. Hlth Rep., Wash.*, **70**, 1168–1174.

REISH, D. J. and BARNARD, J. L. (1960). Field toxicity tests in marine waters utilizing the polychaetous annelid *Capitella capitata* (FABRICIUS). *Pacif. Nat.*, **1**, 1–8.

RHEINHEIMER, G. (1965). Mikrobiologische Untersuchungen in der Elbe zwischen Schnackenburg und Cuxhaven. *Arch. Hydrobiol.* (Suppl. Bd Elbe-Aestuar), **29** (2), 181–251.

RICHARDS, F. A. (1957). Oxygen in the ocean. In J. W. Hedgpeth (Ed.), *Treatise on Marine Ecology and Paleoecology. Mem. Geol. Soc. Am.*, **67**, 185–238.

RICHARDS, F. A. (1965). Dissolved gases other than carbon dioxide. In J. P. Riley and G. Skirrow (Eds), *Chemical Oceanography*, Vol. I. Academic Press, London, pp. 227–322.

RICKETTS, E. F. and CALVIN, J. (1948). *Between Pacific Tides*, Stanford University Press, Stanford, Calif.

RILEY, J. P. and SKIRROW, G. (Eds) (1965). *Chemical Oceanography*, Academic Press, New York.

ROAF, H. E. (1912). Contributions to the physiology of marine organisms. II. The influence of the carbon dioxide and oxygen tensions on rhythmical movements. *J. Physiol., Lond.*, **43**, 449–454.

ROULE, L. (1916). Observations comparitives sur la proportion d'oxygène dissous dans les laux d'un étang littoral (Étang de Thau) et dans les eaux marines littorales, et sur ses conséquences guant à la biologie des espèces migratrices des poissons. *C. r. Séanc. Soc. Biol.*, **79**, 434–436.

SALANKI, J. (1966). Comparative studies on the regulation of the periodic activity in marine lamellibranchs. *Comp. Biochem. Physiol.*, **18**, 829–843.

SALTON, M. R. J. (1964). *The Bacterial Cell Wall*, Elsevier, New York.

SAUNDERS, R. L. (1962). Respiration of the Atlantic cod. *J. Fish. Res. Bd Can.*, **20**, 373–386.

SCHLIEPER, C. (1955). Die Regulation des Herzschlages der Miesmuschel *Mytilus edulis* L. bei geöffneten und bei geschlossenen Schalen. *Kieler Meeresforsch.*, **11**, 139–148.

SCHLIEPER, C. (1957). Comparative study of *Asterias rubens* and *Mytilus edulis* from the North Sea (30 per 1000 S) and the Western Baltic Sea (15 per 1000 S). *Année biol.*, **33**, 117–127.

SCHLIEPER, C. L. and KOWALSKI, R. (1957). Weitere Beobachtungen zur ökologischen Physiologie der Miesmuschel *Mytilus edulis* L. *Kieler Meeresforsch.*, **13**, 3–10.

SCHLIEPER, C. L. and KOWALSKI, R. (1958). Ein zellulärer Regulationsmechanismus für erhöhte Kiemenventilation nach Anoxybiose bei *Mytilus edulis* L. *Kieler Meeresforsch.*, **14**, 42–47.

SCHÖBERL, P. and ENGEL, H. (1964). Das Verhalten der nitrifizierenden Bakterien gegenüber gelöstem Sauerstoff. *Arch. Mikrobiol.*, **48**, 393–400.

SCHOLANDER, P. F. (1964). Animals in aquatic environments: diving mammals and birds. In D. B. Dill, A. Adolph, C. G. Wilber (Eds), *Handbook of Physiology*. Sect. 4: Adaptation to the environment. American Physiological Society, Washington, D.C. pp. 729–740.

SCHOLANDER, P. F., DAM, L. VAN and ENNS, T. (1956). The source of oxygen secreted into the swimbladder of cod. *J. cell. comp. Physiol.*, **48**, 517–522.

SCHWARTZ, W. and SCHWARTZ, A. (1960). *Grundriß der allgemeinen Mikrobiologie*, Vol. I. De Gruyter, Berlin.

SHELFORD, V. E. (1918). The relation of marine fishes to acids with particular reference to the Miles Acid Process of Sewage Treatment. *Publs Puget Sound mar. biol. Stn*, **2**, 97–111.

SHELFORD, V. E. and POWERS, E. B. (1915). An experimental study of the movements of herring and other marine fishes. *Biol. Bull. mar. biol. Lab., Woods Hole*, **28**, 315–334.

SHEPARD, M. P. (1955). Resistance and tolerance of young speckled trout (*Salvelinus fontinalis*) to oxygen lack, with special reference to low oxygen acclimation. *J. Fish. Res. Bd Can.*, **12**, 387–446.

SILLEN, L. G. (1967). The ocean as a chemical system. *Science, N.Y.*, **156**, 1189–1196.

SILVER, S. J., WARREN, C. E. and DOUDOROFF, P. (1963). Dissolved oxygen requirements of developing steelhead trout and chinook salmon embryos at different water velocities. *Trans. Am. Fish. Soc.*, **92**, 327–343.

SINGH PRUTHI, H. (1927a). Preliminary observations on the relative importance of the various factors responsible for the death of fishes in polluted waters. *J. mar. biol. Ass. U.K.*, **14**, 729–739.

SINGH PRUTHI, H. (1927b). The ability of fishes to extract oxygen at different hydrogen ion concentrations of the medium. *J. mar. biol. Ass. U.K.*, **14**, 741–747.

SKIRROW, G. (1965). The dissolved gases—carbon dioxide. In J. P. Riley and G. Skirrow (Eds), *Chemical Oceanography*, Vol. I. Academic Press, London. pp. 227–322.

SMITH, E. L. (1937). Influence of light and carbon dioxide on photosynthesis. *J. gen. Physiol.*, **20**, 807–830.

SMITH, E. L. (1938). Limiting factors in photosynthesis: Light and carbon dioxide. *J. gen. Physiol.*, **22**, 21–35.

SMITH, S. (1957). Early development and hatching. In M. E. Brown (Ed.), *The Physiology of Fishes*, Vol. I. Academic Press, New York. pp. 323–359.

SOROKIN, C. (1962a). Inhibition of cell division by carbon dioxide. *Nature, Lond.*, **194**, 496–497.

SOROKIN, C. (1962b). Carbon dioxide and bicarbonate in cell division. *Arch. Mikrobiol.*, **44**, 219–227.

SOUTHWARD, A. J. and CRISP, D. J. (1965). Activity rhythms of barnacles in relation to respiration and feeding. *J. mar. biol. Ass. U.K.*, **45**, 161–185.

SPRUIT, C. J. P. (1962). Photoreduction and anaerobiosis. In R. Lewin (Ed.), *Physiology and Biochemistry of Algae*. Academic Press, New York. pp. 47–60.

STEEMANN NIELSEN, E. (1946). Carbon sources in the photosynthesis of aquatic plants. *Nature, Lond.*, **158**, 594–596.

STEEMANN NIELSEN, E. S. (1960). Uptake of CO_2 by the plant. *Handb. PflPhysiol.*, **5**, 70–84.

STOTT, F. C. (1932). Einige vorläufige Versuche über Veränderungen des Blutzuckers bei Dekapoden. *Biochem. Z.*, **248**, 55–64.

SUBRAHMANYAN, C. B. (1962). Oxygen consumption in relation to body weight and oxygen tension in the prawn *Penaeus indicus* (MILNE EDWARDS). *Proc. Indian Acad. Sci.*, **55**, 152–161.

SUNDNES, G. (1957). On the transport of live cod and coalfish. *J. Cons. perm. int. Explor. Mer*, **22**, 191–196.

SVERDRUP, H. U., JOHNSON, M. W. and FLEMING, R. H. (1942). *The Oceans*, Prentice-Hall, New York.

TAGATZ, M. E. (1961). Reduced oxygen tolerance and toxicity of petroleum products to juvenile American shad. *Chesapeake Sci.*, **2**, 65–71.

TAMIYA, H. and HUZISIGE, H. (1949). Effect of oxygen on the dark reaction of photosynthesis. *Stud. Tokugawa Inst.*, **6**, 83–104.

TAMIYA, H., IWAMURA, J., SHIBATA, K., HASE, E. and NIHEI, T. (1953). Correlation between photosynthesis and light independent metabolism in the growth of *Chlorella*. *Biochim. biophys. Acta*, **12**, 23–40.

TANG, P. S. (1931). The oxygen tension-oxygen consumption curve of unfertilized *Arbacia* eggs. *Biol. Bull. mar. biol. Lab., Woods Hole*, **60**, 242–244.

TANG, P. S. (1933). On the rate of oxygen consumption of tissues and lower organisms as a function of oxygen tension. *Q. Rev. Biol.*, **8**, 260–270.

TANG, P. S. and GERARD, R. W. (1932). The oxygen tension-oxygen consumption curve of fertilized *Arbacia* eggs. *J. cell. comp. Physiol.*, **1**, 503–513.

TEAL, J. M. (1959). Respiration of crabs in Georgia salt marshes and its relation to their ecology. *Physiol. Zool.*, **32**, 1–14.

TEAL, J. M. and CAREY, F. G. (1967). The metabolism of marsh crabs under conditions of reduced oxygen pressure. *Physiol. Zool.*, **40**, 83–91.

THOMAS, H. J. (1954). The oxygen uptake of the lobster (*Homarus vulgaris* EDW.). *J. exp. Biol.*, **31**, 228–251.

TOWNSEND, L. D. and CHEYNE, H. (1944). The influence of hydrogen ion concentration on the minimum dissolved oxygen toleration of silver salmon, *Onchorhynchus kisutch* (WALBAUM). *Ecology*, **25**, 461–466.

TOWNSEND, L. D. and EARNEST, D. (1940). The effects of low dissolved oxygen and other extreme conditions on salmonoid fishes. *Proc. Pacif. Sci. Congr.*, **3**, 345–351.

TRIFONOVA, A. N. (1937). La physiologie de la différenciation et de la croissance. I. L'équilibre Pasteur-Meyerhof dans le développement des poissons. *Acta Zool.*, **18**, 375–445.

TRUESDALE, G. A., DOWNING, A. L. and LOWDEN, G. F. (1955). Solubility of oxygen in water. *J. appl. Chem., Lond.*, **5**, 1–53.

TSENG, C. K. and SWEENEY, B. M. (1946). Physiological studies of *Gelidium cartilagineum*. I: Photosynthesis, with special reference to the carbon dioxide factor. *Am. J. Bot.*, **33**, 706–715.

TULKKI, P. (1965). Disappearance of the benthic fauna from the Basin of Bornholm (southern Baltic) due to oxygen deficiency. *Cah. Biol. mar.*, **6**, 455–463.

TURNER, J. S. and BRITTAIN, E. G. (1962). Oxygen as a factor in photosynthesis. *Biol. Rev.*, **37**, 130–170.

TURNER, J. S., TODD, M. and BRITTAIN, E. G. (1956). The inhibition of photosynthesis by oxygen. I. Comparative physiology of the effect. *Aust. J. biol. Sci.*, **9**, 494–510.

UBRIG, H. (1952). Der Einfluss von Sauerstoff und Kohlendioxyd auf die taktischen Bewegungen einiger Wassertiere. *Z. vergl. Physiol.*, **34**, 479–507.

UTERMÖHL, H. (1925). Limnologische Phytoplanktonstudien. *Arch. Hydrobiol.* (Suppl. Bd), **5**, 1–527.

VERNBERG, F. J. and VERNBERG, W. B. (1966). Interrelationship between parasites and their hosts. II. Comparative metabolic patterns of thermal acclimation of the trematode *Lintonium vibex* with its host *Spheroides maculatus*. *Expl Parasit.*, **18**, 244–250.

VERNBERG, W. (1963). Respiration of digenetic trematodes. *Ann. N.Y. Acad. Sci.*, **113**, 261–271.

VIDAVER, W. (1965). An oxygen requirement for photosynthesis. *Yb. Carnegie Instn Wash.*, **64**, 395–397.

WALDICHUK, M. and BOUSFIELD, E. L. (1962). Amphipods in low oxygen marine waters adjacent to a sulphite pulp mill. *J. Fish. Res. Bd Can.*, **19**, 1163–1165.

WALSHE-MAETZ, B. M. (1952). Environment and respiratory control in certain Crustacea. *Nature, Lond.*, **169**, 750–751.

WALSHE-MAETZ, B. M. (1956). Controle respiratoire et métabolisme chez les Crustacés. *Vie Milieu*, **7**, 523–543.

WARBURG, O. (1919). Über die Geschwindigkeit der photochemischen Kohlensäurezersetzung in lebenden Zellen. I. *Biochem. Z.*, **100**, 230–270.

WARBURG, O. (1920). Über die Geschwindigkeit der photochemischen Kohlensäurezersetzung in lebenden Zellen. II. *Biochem. Z.*, **103**, 188–217.

WARBURG, O. and KRIPPAHL, G. (1966). Über das einsteinsche Gesetz und die Funktion des roten Ferments bei der Photosynthese. *Biochem. Z.*, **344**, 103–120.

WARBURG, O., KRIPPAHL, G., GEWITZ, H. S. and VOLKER, W. (1959). Über den chemischen Mechanismus der Photosynthese. *Z. Naturf.*, **14**b, 712–724.

WASSINK, E. C., VERMEULEN, D., REMAN, G. H., and KATZ, E. (1938). On the relation between fluorescence and assimilation in photosynthesizing cells. *Enzymologia*, **5**, 100–118.

WEEL, P. B. VAN, RANDALL, J. E. and TAKATA, M. (1954). Observations on the oxygen consumption of certain marine Crustacea. *Pacif. Sci.*, **8**, 209–218.

WELLS, G. P. (1949). Respiratory movements of *Arenicola marina* L.: intermittent irrigation of the tube, and intermittent aerial respiration. *J. mar. biol. Ass. U.K.*, **28**, 447–464.

WELLS, G. P. and DALES, R. P. (1951). Spontaneous activity patterns in animal behaviour; the irrigation of the burrow in the polychaetes *Chaetopterus variopedatus* and *Nereis diversicolor*. *J. mar. biol. Ass. U.K.*, **29**, 661–680.

WHEDON, W. F. and SOMMER, H. (1937). Respiratory exchange of *Mytilus californianus*. *Z. vergl. Physiol.*, **25**, 523–528.

WICKETT, W. P. (1954). The oxygen supply to salmon eggs in spawning beds. *J. Fish. Res. Bd Can.*, **11**, 933–953.

WIESER, W. and KANWISHER, J. (1959). Respiration and anaerobic survival in some sea-weed-inhabiting invertebrates. *Biol. Bull. mar. biol. Lab.*, *Woods Hole*, **117**, 594–600.

WINTERSTEIN, H. (1925). Über die chemische Regulierung der Atmung bei den Cephalopoden. *Z. vergl. Physiol.*, **2**, 315–328.

WITTENBERG, J. B. (1960). The source of carbon monoxide in the float of the Portuguese man-of-war, *Physalia*. *J. exp. Biol.*, **37**, 698–705.

WOLVEKAMP, H. P. and WATERMAN, T. H. (1960). Respiration. In T. H. Waterman (Ed.), *The Physiology of Crustacea*, Vol. I. Academic Press, New York. pp. 35–100.

WOOD, E. J. F. (1965). *Marine Microbial Ecology*, Reinhold, New York.

WOOD, E. J. F. (1967). *Microbiology of Oceans and Estuaries*, Elsevier, Amsterdam.

WOOD, H. G. and STJERNHOLM, R. L. (1962). Assimilation of carbon dioxide by heterotrophic organisms. In I. C. Gunsalus and R. Y. Stanier (Eds), *The Bacteria: a Treatise in Structure and Function*, Vol. III. Academic Press, New York. pp. 41–117.

ZEUTHEN, E. (1955). Comparative physiology: respiration. *A. Rev. Physiol.*, **17**, 459–482.

ZOBELL, C. E. (1940). The effect of oxygen tension on the rate of oxidation of organic matter in sea water by bacteria. *J. mar. Res.*, **3**, 211–223.

ZOBELL, C. E. (1946). *Marine Microbiology*, Chronica Botanica, Waltham, Mass.

ZOBELL, C. E. (1962). The occurrence, effects, and fate of oil polluting the sea. In E. A. Pearson (Ed.), *Advances on Water Pollution Control*. Pergamon Press, Oxford. pp. 85–109.

10. ORGANIC SUBSTANCES

10.0 GENERAL INTRODUCTION

K. KALLE

(1) General Aspects of Dissolved and Suspended Organic Substances

For a long time, the opinion was widespread among marine ecologists that—apart from a few by-products of metabolism, biologically and chemically difficult to attack—substantial proportions of biologically utilizable organic material could not be present in dissolved form in sea water. This opinion was based on the assumption that the bacteria present everywhere in sea water would remove from it all biologically utilizable organic substances. Today, thanks to the refinement of micro-analytical methods, especially in the last decade, we know that the content of dissolved carbon present in organic combinations in the waters of the open oceans amounts to the order of milligrams C/l or 10^{-6} volume density.* Today, we assume, that bacteria can obtain organic nutrients from sea water only down to about this limiting value and not by direct uptake of the dissolved substance from the sea water itself, but via adsorption of organic substances at the surface of suspended particles, usually detritus particles of undetermined origin.

Acceptable measurements are now available from several sea areas on the concentrations of fatty acids, amino acids, and carbohydrates, as well as on a few other substances. The general rule has emerged that the quantitative relations between the concentrations of organic substances in sediments to those dissolved in sea water, to those suspended in sea water, are about 100 to 10 to 1. No doubt, we are only at the beginning; with further refinements of microchemical techniques, the presence of a large number of additional dissolved organic substances, possibly in very low concentrations, will be revealed. Such expectation is encouraged by recent results in the fields of sediment trace substance chemistry (VALLENTYNE, 1957) and atmosphere aerosol chemistry (JUNGE, 1963).

Since we know that some hormones, scent substances, sexually attractive substances, and others, produce organismic responses down to concentrations of 10^{-18} volume density (mg/km³), it seems appropriate to extend our studies on organic trace substances down to such extremely low concentration levels. Presently, our analytical techniques are sufficiently advanced to detect 10^{-9} volume density (mg/m³) by purely chemical methods, and 10^{-12} volume density by employing biological assay methods; however, a large concentration range is still beyond our reach. Future investigations may produce surprises in this modern field of marine ecology.

Estimations of the total dissolved carbon in sea water have been available for a relatively long time. Only in the last decade, however, have improved techniques led to an unequivocal elucidation of the relations involved (DUURSMA, 1963;

*The unit of volume density is taken to be 1 g/cm³.

MENZEL, 1964). We may summarize our present knowledge as follows: In the surface waters of the oceans, the amount of dissolved organic carbon undergoes small annual fluctuations; but it is practically constant—about 10^{-6} volume density (mg/l)—in the water masses beneath the euphotic zone.

In comparison to the inorganic trace elements, the oceans contain large reserve supplies of dissolved organic compounds, as is corroborated by the following calculation. If we assume that all organic carbon is derived from plant assimilation in near surface layers, and that roughly nine-tenths of the assimilation products undergo biological decomposition during their passage through the food chain, then we can put the annual addition, from which the residual carbon remaining in solution is supplemented, at 0·8 mg C/cm² per year (mean rate of primary production in the ocean = 8 mg C/cm² per year; KALLE, 1945). The mean amount of dissolved carbon stored in the sea, if one takes a value of 1 mg C/l and a depth of 3800 m, is 380 mg C/cm² of ocean surface. The average life-span of the carbon compounds is calculated accordingly as 380 : 0·8 = 475 years. This implies that the decomposition processes, at least in the deep sea, must be strongly retarded, since the organic substances concerned are far from being particularly stable and biologically unavailable; on the contrary, recent investigations have shown that they are basic biological substances: carbohydrates, fats and amino acids.

Experimental studies (BAADER and co-authors, 1960) have revealed that amino acids and carbohydrates dissolved in sea water are adsorbed to a large extent on the surfaces of suspended clay particles. It is not surprising, therefore, that their content in sediments is always considerably higher than in the free water. Recently, further interesting findings have been made which indicate the participation of dissolved organic matter in the cycling of nutrient elements in the uppermost water layers. Thus a considerable proportion of the dissolved organic products seems to exist in complex combination with phosphoric acid-containing compounds in the form of surface-active substances. Consequently, they accumulate first of all at the interface of microscopic air bubbles present everywhere in the surface waters of the oceans. With these bubbles they are carried to the sea surface and, subsequent to the bursting of the air bubbles, are transformed into a microscopically thin 'film'. Due to horizontal water currents, in which—as in the 'Langmuir circulation' —convection vortices with counter-current water displacements at the surface are formed, these surface films, enriched with organic matter, are pushed close together and concentrated (SUTCLIFFE and co-authors, 1963). Finally, in the convergence zones, the organic matter is transported via downwardly directed water currents to the depths again, where it precipitates into small fragile aggregate particles ('sea snow'), some thousandths of a millimetre up to some millimetres in size (RILEY, 1963). In this form, they represent a readily available food source for marine animals (BAYLOR and SUTCLIFFE, 1963).

(2) Measuring Organic Substances: Methods

Estimation of the content of organic matter in oceans and coastal waters had, for many years, to be restricted mainly to the total content of organic carbon because of lack of suitable methods. New sensitive methods—such as paper chromatography, thin layer chromatography and gas chromatography—which all depend on the phenomenon of selective adsorption of organic compounds, have greatly advanced our technical skills.

The advantage of the new chromatographic methods lies not so much in their increased sensitivity to organic substances occurring in extreme dilutions, but in their considerable capacity for separating the manifold mixtures of substances encountered.

Employing these new methods, it has, for example, been possible to determine quantitatively 22 amino acids in the small residue left on a glass plate by a thumb print. With the same methods, about 240 organic components were detected in the volatile aroma of coffee, of which 85 have so far been identified; in the aroma of white wine, 49 components have been identified, and 72 components in cocoa aroma. In marine ecology, a beginning has been made with the determination of different fatty acids, amino acids and carbohydrates. A few examples are mentioned below.

(3) Organic Substances in Oceans and Coastal Waters

The distribution of dissolved and particulate organic carbon in the Arabian Sea is listed in Table 10-1. This table exemplifies the fact, referred to above, that the amount of dissolved organic carbon is of the magnitude of 10^{-6} volume density and slowly but steadily decreases from the surface to the depth. Table 10-1 further indicates that the content of particulate carbon (mostly dead material) is roughly a tenth power lower. We would hardly be wrong if we assume the concentration of the organic carbon, fixed in living marine organisms, to be another order of magnitude lower, so that the above-mentioned general quantitative relation may be extended as follows: Amount of organic substances in sediments to those dissolved in sea water to those particulate suspended in sea water to those in living organisms $\approx 1000:100:10:1$.

A survey of the distribution of different fatty acids in the Gulf of Mexico is

Table 10-1

Dissolved and particulate organic carbon at different depths in the Arabian Sea (After MENZEL, 1964; modified)

Water depth (m)	Dissolved C (mg/l)	Particulate C (mg/l)	Water depth (m)	Dissolved C (mg/l)	Particulate C (mg/l)
1	1·00	0·157	400	0·64	0·017
12	0·92	0·155	600	0·60	0·054
19	0·96	0·204	800	0·20	0·021
50	0·74	0·040	1300	0·28	0·049
100	0·76	0·026	1600	0·36	0·026
200	0·78	0·018	2800	0·34	0·053
250	0·50	0·035	3200	0·36	0·030
Total C:	123 g C/m²			1275 g C/m²	

Table 10-2

Fatty acids (%) at different depths in the
Gulf of Mexico (After SLOWEY and co-authors,
1962; modified)

Fatty acids	Water depths			
	10 m	300 m	900 m	1900 m
C.10	—	—	7	—
C.10	0	0	6	6
C.12	6·5	12	42	94
C.14	20	11	15	0
C.14 =	4	7	0	0
C.16	35·5	35	22	0
C.16 =	16	14	0	0
C.18	8·5	9	0	0
C.18 =	1·5	5	0	0
C.18 = =	2	0	0	0
C.?	6	5	8	0
Total:	0·5	0·4	0·5	0·3 mg/l

presented in Table 10-2. Here also, the total concentration lies in the order of
magnitude of 10^{-6} volume density. Whilst fatty acids of higher molecular weights
(with 16 and 14 carbon atoms in the molecule) predominate at the surface, the
ratio shifts more and more to lower molecular weight fatty acids (with 12 C atoms
in the molecule) with increasing water depths. It appears as if, with increasing
age, the rate of degradation in the carbon chain slows down.

Analyses of sea water from the North Pacific Ocean, in which aliphatic acids
have also been determined, are presented in Table 10-3. Again, we find several
surprisingly high concentrations decreasing from glycollic acid with values of
10^{-6} volume density to acetic acid and formic acid down to lactic acid with 10^{-7}
to 10^{-8} volume density. Values which have been obtained in studies on carbohydrates are similar (DEGENS and co-authors, 1964). In contrast, the concentra-

Table 10-3

Organic acids (volume den-
sities*) in the North Pacific
Ocean (After KOYAMA and
THOMPSON, 1959; modified)

Glycollic acid	$8 - 140 \times 10^{-8}$
Acetic acid	$9 - 82$,,
Formic acid	$3 - 38$,,
Lactic acid	$3 - 8$,,

*Unit of volume density: 1 g/cm^3

tions of vitamins appear—as far as results are available (vitamins B_1, B_{12})—to lie several orders of magnitude lower at 10^{-11} to 10^{-12} volume density (VISHNIAC and RILEY, 1961).

True degradation products of a humus-like nature, which may be recognized in sea water by their yellow colour as well as their blue fluorescence in ultra-violet light, have been known for a long time. Exact data on their concentration in oceans and coastal waters have not yet been obtained.

10. ORGANIC SUBSTANCES

10.1 BACTERIA, FUNGI AND BLUE-GREEN ALGAE

W. GUNKEL

(1) Introduction

Organic substances are among the most important factors which control abundance, metabolism and distribution of micro-organisms in oceans and coastal waters. In most marine environments, the concentration of organic substances is low. This fact has already been elucidated in the *General Introduction* to this chapter.

Organic substances are especially important as nutrients for bacteria and fungi. While the numbers of these micro-organisms are generally very small in oceanic waters, their abundance increases significantly in the presence of organic substances produced by photosynthetic plankton organisms, seaweeds or animals, of organic pollutants from sewage outlets or ships, of fishery activities or of converging water bodies.

Most blue-green algae belong to the group of phototrophic micro-organisms which do not depend on organic substances. Some blue-green algae, however, are apochlorotic (non-chlorophyll containing organisms) and, hence, require organic substances as nutrients; others need specific 'micro-nutrients' for growth, such as vitamin B_{12}.

The term 'organic substances' comprises innumerable different compounds, which can be classified in different ways, for example, in regard to (i) their origin, (ii) their properties and importance for bacteria, fungi and blue-green algae, and (iii) their chemical composition.

(i) Classification of organic substances based on their origin in the marine environment:

  (a) Excretions of living organisms; for example, polysaccharides, organic acids, polypeptides and vitamins are excreted by algae (BROCK, 1966), glycollic acid by phytoplankton (Chapter 10.2)
  (b) Release from dead organisms of numerous different organic substances
  (c) Pollution of the sea due to human activities.

(ii) Classification of organic substances based on their properties and importance for bacteria, fungi and blue-green algae (some examples):

  (a) Nutrient
  (b) Chelating agent
  (c) Surface-active substance
  (d) Toxicant
  (e) Detoxicant.

(iii) Classification of organic substances based on their chemical composition (according to DUURSMA, 1965):

(a) Carbohydrates
(b) Proteins and their derivates
(c) Aliphatic carboxylic and hydroxycarboxylic acids
(d) Biologically active compounds
(e) Humic acids
(f) Phenolic compounds
(g) Hydrocarbons.

Our present knowledge of the influence of organic substances on bacteria, fungi and blue-green algae in oceans and coastal waters is quite insufficient. We do not know enough about the different kinds of organic substances present, their distribution in space and time and their importance for the species composition and ecological dynamics of assemblages of bacteria, fungi and blue-green algae. Considerable methodological and technological difficulties must be overcome before we can expect significant progress.

Most of the information available on functional and structural responses to organic substances of bacteria, fungi and blue-green algae is based on experimental evidence, produced under environmental circumstances which are not comparable to the situation met *in situ*.

A considerable amount of information has been produced on physiological aspects of the use of organic substances by selected micro-organisms, especially on metabolic pathways and biochemistry of decomposition. These aspects, and the importance of micro-organisms in the marine food chain, will be dealt with in Volumes II (Physiological Mechanisms), III (Cultivation) and IV (Dynamics) of this Treatise. Close relationships exist between the present chapter and Chapter 7.1.

The following books on marine microbiology contain information pertinent to the scope of this chapter: ZoBELL (1946), OPPENHEIMER (1963, 1968), WOOD (1965, 1967), DROOP and WOOD (1968); an important recent review on dissolved organic substances in sea water has been presented by DUURSMA (1965).

Readers interested in basic, general relationships between organic substances and micro-organisms are referred to the following textbooks: RIPPEL-BALDES (1955), THIMANN (1955), GUNSALUS and STANIER (1962), LAMANNA and MALLETTE (1965), PELCZAR and REID (1965), BROCK (1966), HAWKER and co-authors (1966) and SCHLEGEL (1969).

(2) Functional Responses

(a) *Tolerance*

Lower limits

In the open oceans, the concentration of organic substances is low, ususally in the order of a few mg/l (Chapter 10.0). For bacteria, the lower limits of population growth are quite close to this order of magnitude. Concentrations of organic substances below about 1 mg/l can no longer support bacterial growth. Maximum use of organic substances as energy source for bacteria is possible only in the presence of microparticles. Their surface areas function as sites of adsorption and concentration of organic substances, facilitating uptake and use by micro-organisms (Chapter 7.1).

The importance of solid surfaces as 'feeding and accumulation sites' of bacteria has been stressed in numerous papers. ZoBELL and ANDERSON (1936) and ZoBELL (1943, 1946) reported increases in bacterial numbers if the sea water used as culture medium was stored in glass containers. They pointed out that the yield in bacterial biomass depends on the ratio of glass container surface area to sea-water volume. In accordance with this finding, bacterial numbers per unit volume tend to be much higher in small containers than in large ones.

'Solid surfaces promote the activities of bacteria in dilute nutrient solutions primarily by absorbing organic matter thereby making it available to bacteria' (ZoBELL, 1946).

However, the increase in bacterial numbers is not parallel in all strains of the populations tested. Increase in total bacterial numbers leads to a decrease in the number of species present. This 'solid surface effect' interferes greatly with attempts to investigate quantitative relationships of organic substance concentrations and bacterial population dynamics under laboratory conditions. It is still not possible to design experiments which would allow strict comparisons between results obtained in the laboratory and in the sea. This basic difficulty is not restricted to studies on bacteria. Nearly all transformations of organic material are affected by adsorption at solid surfaces. In an attempt to overcome this difficulty, MCALLISTER and co-authors (1961) employed large-volume plastic spheres under *in situ* conditions in order to establish suitable surface to volume ratios.

KRISS (1961) contends that, besides increasing the nutrient level, chemical alterations take place at the water–glass interface, such that the nutrient material changes into a new form which is more readily available for the microbial metabolism.

JANNASCH (1954) studied the distribution of bacteria in the Gulf of Naples (Italy) and compared, by microscopical examination, the spatial arrangement of bacteria after filtration of sea-water samples through millipore filters. Where the water is poor in organic nutrients, the filters reveal the bacteria to be attached to particles. Where the water is rich in organic material, the bacteria exhibit an even distribution on the filters. This observation indicates that the bacteria do not depend on a surface in their natural environment.

FLOODGATE (1966) investigated the factors affecting the settlement of a marine bacterium. He comes to the conclusion that settlement does not depend on the organic material content of the solid surface, but is performed randomly. Subsequent to settling, growth and cell division are, of course, dependent on local concentrations of nutrients.

Numerous papers deal with nutrient concentrations limiting population growth in micro-organisms. HEUKELEKIAN and HELLER (1940) found that *Escherichia coli* fails to multiply when the concentration of the peptone medium sinks below 0·5 ppm, unless glass beads are added to increase the surface area. The glass beads exert beneficial effects on bacterial growth up to a peptone concentration of 25 ppm. JANNASCH (1963a, b, 1967) published papers of fundamental interest. He investigated bacterial population growth at low nutrient concentrations. The growth responses of several species of heterotrophic marine bacteria to limiting concentrations of lactate, glycerol and glucose in sea water were determined in a chemostat. In all cases, threshold concentrations of the limiting substrates were found, below which no growth could be detected. On the basis of their growth parameters, two types of bacteria could be distinguished; one type is adapted to life in the marine environment via its ability to grow at low nutrient concentrations, the other becomes inactivated in sea water (but survives). These investigations have been extended by VACCARO and JANNASCH (1966) and VACCARO (1969), and provide information about uptake patterns of heterotrophic marine micro-organisms. One of the most recent pertinent publications is the competent review by VAN UDEN (1969); it contains 60 references.

In most cases, it is not possible to state a definite threshold value for a given bacterium, fungus or blue-green alga, below which there is no growth. It seems particularly impossible to extrapolate threshold values obtained under laboratory conditions to the situation in oceans and coastal waters. Population growth is a dynamic response of numerous individuals, which depends to a large extent on the testing procedures; e.g., on whether growth is studied in batch cultures or in continuous chemostat cultures, on initial population densities and the final 'wash out' at a constant dilution rate; it depends further on the temperature, the nature of the substance and the species of micro-organisms tested. These and related matters will be dealt with in greater detail in the Volumes devoted to Physiological Mechanisms (II), Cultivation (III) and Dynamics (IV).

Most studies concerned with degradation of organic substances by bacteria focus on the question, whether the bacterium under consideration is able to degrade (use) a given substance. Little emphasis has been placed as yet on the problem of tolerances in regard to defined organic substances.

In regard to fungi and blue-green algae, there is no information available on lower threshold values under natural conditions. The determination of threshold values is complicated, due to multicellular growth, spore formation, sexual versus asexual reproduction and—in the case of blue-green algae—the process of photo-trophy.

Upper limits

Theoretically, all substances in oceans and coastal waters can become toxic if present at critically high concentrations. Organic substances, however, hardly ever attain concentrations in the marine environment which may approach the upper tolerance limit of bacteria, fungi and blue-green algae. In contrast, under laboratory conditions, marine micro-organisms are frequently exposed to critically high amounts of organic substances. Hence, much of our knowledge on tolerances of micro-organisms to critically high concentrations of organic substances pertains to laboratory studies.

The upper tolerance limit of a micro-organism to critically high amounts of organic sustances depends on its physiological condition. This well-known fact may be exemplified by referring to the resistance of bacteria to antimicrobial substances. A given concentration of an antibiotic, applied under defined conditions, may be lethal to active cells, but easily tolerated during the dormant state. Micro-organismic tolerances to organic substances may also be a function of other, simultaneously effective environmental factors such as light (Chapter 2), temperature (Chapter 3), salinity (Chapter 4) or water movement (Chapter 5).

In the following paragraphs, upper limits of tolerance to organic substances will be considered primarily in regard to bacteria, since little information is available on marine fungi, and blue-green algae do not normally experience critically high amounts of organic substances in their natural habitats.

Primary effects of high concentrations of organic substances. Under special circumstances, the concentration of organic substances may reach high *in situ* values; for example, (i) in near-shore sediments, especially at the mud–water interface, (ii) at the water–air interface during breakdown of a plankton bloom, (iii) in micro-environments (e.g., dead organisms), (iv) in tide pools, or (v) in beach areas with large amounts of dying or dead seaweeds piled up after heavy storms. Unusually high concentrations of organic substances may modify the species composition of the local microbial flora. Micro-organisms adapted to low nutrient concentrations may tend to decrease in number or to disappear, while forms which tolerate—or even prefer—extremely high concentrations of organic substances may increase rapidly and finally dominate the scene. However, there is no unequivocal evidence available to document such changes in interspecific relations. Taxonomic problems and methodological difficulties have, as yet, prevented detailed analyses. In contrast to tolerance experiments on higher plants or animals, where responses of individuals are studied, the bacteriologist must deal with populations, usually under quite specific environmental conditions. Even if more than 99% of the bacterial test population is not able to tolerate the conditions offered, one cell may eventually survive and, due to intensive multiplication, re-establish the population which, however, from now on, exhibits different response patterns in regard to the substance examined. The use of mixed populations results in an even more complicated situation, due to shiftings both at inter- and intraspecific levels. Most micro-organisms are very versatile and able to adjust to new environmental conditions very quickly. In fact, it is this very capacity which enables them to play their important role in the cycle of organic substances in the marine environment. Only some man-made substances (e.g. plastics) largely resist bacterial degradation.

Technically, the response of a single bacterial cell can be assessed by applying autoradiographical methods. However, such methods have, up to now, rarely been employed. The few cases that have come to the reviewer's attention deal with the ability of a given bacterium to use a defined substance, but not with lower or upper limits of tolerance.

Secondary effects of high concentrations of organic substances. Extensive degradation of organic substances requires large amounts of oxygen. Thus, complete bacterial degradation of 1 kg mineral oil will—according to calculations by ZoBell (1964)—use up all oxygen normally contained in about 400 m³ sea water (15°C, 35‰S). Such depletion of oxygen tends to bring about anaerobic conditions which, secondarily, reduce bacterial growth, since most bacteria are unable to multiply in the absence of oxygen. Furthermore, anaerobic conditions lead to the formation of hydrogen sulphide, which is toxic for most marine organisms, including bacteria.

The presence of hydrogen sulphide is easily recognized by the black colour, due to iron sulphides, of the sediments. A specialized group of bacteria is able to reduce sulphates to sulphides, thereby oxidizing organic substances; but this reduction proceeds quite slowly.

Anaerobic conditions and hydrogen sulphide formation are just one example illustrating a situation in which high amounts of organic substances tend to limit bacterial population growth due to secondary effects. Another, well-known example is the fact that yeasts, during fermentation processes, produce alcohols which—if allowed to accumulate—limit further growth and may finally cause death. Similar cases of inhibition and lethal effects due to substances formed during maximum rates of microbial metabolism have been reported in numerous studies conducted under laboratory conditions. Sugar fermentations often result in excretion of organic acids. These cause a reduction in pH, which can become a secondary limiting factor for the micro-organisms involved. A number of different substances excreted by marine bacteria have been shown to cause such limiting effects.

Antimicrobial substances. Marine algae, especially phytoplankters, are capable of producing antimicrobial organic substances. No definite quantitative data are available yet about the upper limits of micro-organismic tolerances to these substances. This lack of information is understandable, because the 'biologically active substances' are effective at extremely low concentrations. Hence, considerable technical difficulties exist in isolating them, making quantitative determinations and assessing their influences upon bacterial populations. The information available refers largely to qualitative aspects, at best to determinations of inhibition zones and inactivation rates in test strains, normally *Escherichia coli* and *Staphylococcus aureus*. The formation of antimicrobial substances in the marine environment represents one of the most important and exciting fields of marine microbiology. There is urgent need for detailed data on the upper limits of tolerances of bacteria, fungi and blue-green algae.

ZoBELL (1946) recommends the use of 'aged seawater' for preparing microbiological media. During ageing in the dark, toxic substances are decomposed. These can also be reduced or removed by application of heat, filtration through millipore filters and precipitation with aluminium and iron salts under slightly alkaline conditions. According to ZoBELL, the presence of organic bacteriostatic substances in sea water has definitely been established; while it is further known that the intensity of antimicrobial activities changes during the year, there are no data available yet on the extent to which antimicrobial substances may limit population growth in the sea.

During the last 25 years, a considerable amount of new pertinent information has been presented, especially in the reviews by SIEBURTH (1964, 1965a, b, 1968), who himself conducted fundamental studies on the formation of antimicrobial substances in ageing blooms of *Skeletonema* species and their influence upon bacteria. SIEBURTH (1959) discovered a strong antibacterial activity of acrylic acid arising presumably from dimethyl propriothetin. Enzymatic hydrolysis cleaves this substance to form dimethyl sulphide and acrylic acid. The antibacterial activity of acrylic acid was first observed in the Antarctic Ocean. Studies on bacteriologically sterile birds revealed gastro-intestinal antibiosis in penguins, due to their euphausiid diet; the euphausiids, in turn, had been grazing on blooms of the colonial phytoplankton form of *Phaeocystis pouchetii*. In addition to acrylic acid, terpenes, fatty acids, terpene fractions and chlorophyllides have been identified as antimicrobial substances. For further details the reader is referred to SIEBURTH's reviews.

BUCK and MEYERS (1965) reported a remarkable antiyeast activity in different plankton samples, sponges and alcyonarians. BAAM and co-authors (1966) isolated bacterial strains from sea-water samples with antagonistic effects against *Staphylococcus aureus* and *Salmonella typhosa*.

The upper tolerances of bacteria, fungi and blue-green algae to man-made forms of antibiotics, such as penicillin and streptomycin are of interest to marine ecologists concerned with cultivation of organisms and hatching experiments. They will be referred to in Volume III. The considerable amount of information on upper tolerance limits of bacteria and fungi causing diseases in man are beyond the scope of this Treatise. Contributions dealing with technical aspects of determining antimicrobial substances have been published by KERSEY and FINK (1957), KAVANAGH (1963), ZÄHNER (1965) and OBERZILL (1967).

Phenols and phenolic compounds. Phenols and phenolic compounds can exhibit limiting effects on marine micro-organisms. They originate from natural or man-made sources. Natural phenolic compounds exerting limiting effects have been reported from algae (SIEBURTH, 1968). The majority of phenolic compounds exhibiting limiting effects in the marine environment have been introduced by man; they affect a large variety of aquatic organisms from bacteria to fishes. This pollution problem is more acute in freshwater habitats than in oceans and coastal waters.

The upper tolerance limits of bacteria, fungi and blue-green algae to phenols and phenolic compounds vary significantly. Low concentrations cause no harm; there are even specialists among the bacteria which can degrade phenols and their derivates. Medium concentrations have bacteriostatic, high concentrations bactericidal effects. The importance of phenols and phenolic compounds as pollutants in the marine environment will be dealt with in Volume V.

The marine microbiologist employs phenolic compounds as a disinfectant to kill bacteria and fungi which contaminate his media, glassware or other laboratory equipment. Commercially available disinfectants containing phenol derivates usually also contain soaps, which form precipitates with the salts of the sea water (GUNKEL and RHEINHEIMER, 1968).

The bacteriostatic and bactericidal properties of organic substances used for disinfection and sterilization have been discussed in detail by WALLHÄUSER and SCHMIDT (1967). The properties of these substances have always been tested in pure cultures of a few micro-organisms only, e.g. *Staphylococcus aureus*, *Streptococcus pyogenes*, *Escherichia coli*, *Pseudomonas aeruginosa*, *Proteus vulgaris*, *Candia albicans*, *Penicillium chrysigenum*. Upper limits of tolerance depend, among other things, on temperature, number of micro-organisms present, duration of test, pH, and presence of other organic substances.

While bacteriostatic and bactericidal organic substances are of considerable importance in medicine, they are of fairly limited interest to the marine ecologist, except for methodological procedures requiring disinfection and the general inhibitory principle involved. Information on types and concentrations of disinfectants are available in numerous books dealing with microbiological methods.

Surface-active substances. The upper tolerance limits of micro-organisms to dissolved organic substances may be affected significantly by surface-active substances (see also Chapter 7.1). These are of great importance in aquatic habitats. At 25°C, pure water has a surface tension of 72 dynes/cm (LAMANNA and MALLETTE, 1965). The presence of organic substances (nutrients) in normal field concentrations can reduce the surface tension to 45 to 65 dynes/cm; this range seems to be tolerated well by most bacteria. According to LA REVIÈRE (1955a), 1% peptone decreases the surface tension of water by 18 dynes/cm, 1% yeast extract by 25 dynes/cm. Such a reduction in surface tension can suffice to disperse aggregates of bacteria.

In sea water, the surface tension—under otherwise identical conditions—is slightly higher than in pure water (SVERDRUP and co-authors, 1959). No information is available as to whether extreme intensities of surface tension may be lethal to micro-organisms in the open sea. Pollution of coastal waters due to surface-active substances has been reported in several cases. For details, the reader is referred to KRÜGER (1962, 1963), BOCK (1963), BOCK and WICKBOLD (1963), PRAT and GIRAUD (1964), LIEBMANN (1967), SILSBY (1968) and SMITH (1968).

The problem of 'overfeeding' during isolation and culturing. Proper assessment of the upper tolerance limits of micro-organisms to organic substances requires a critical evaluation of the concentrations of nutrients in media employed for isolation, cultivation and enumeration of marine bacteria (JANNASCH and JONES, 1959; GUNKEL, 1964; OPPENHEIMER, 1968). Several microbiologists assume that the high concentrations of peptones used in culture media allow only a limited number of bacteria cells to form colonies and that, in certain bacteria species, the concentrations surpass the upper limits of tolerance. For this reason, some authors recommended direct microscopical inspection of a concentrated water sample; however, most marine bacteria are so small that it is hardly possible to distinguish them from non-living organic or inorganic particles (e.g. LISTON *in*: OPPENHEIMER, 1968). It would be a mistake to attempt a microscopical count of the bacteria present, take the number obtained as 100, and then express plate counts as percentages. Interestingly, reduced concentrations of organic substances in culture media have, occasionally, led to higher numbers of 'micro-colonies', compared to those which can be counted by the naked eye.

Unfortunately, our present knowledge is based on a very limited number of samples and on plate counts of the colonies formed. We are still unable to offer a sufficient variety of culture conditions for all bacteria present to form colonies. Hence, the number of bacteria obtained must be assumed to be lower than that of the bacteria originally present in the sea-water sample. Furthermore, if bacterial aggregates are not carefully separated before cultivation, a group of bacteria may form one colony only. Worst for the marine ecologist: our present culture methods do not allow any conclusions regarding the physiological state of the bacteria under *in situ* conditions. With respect to the bacterial numbers obtained by plating procedures, the term 'total number of bacteria' is misleading; it should be replaced by the term 'viable number of bacteria'. A critical evaluation of the present status of our pertinent knowledge leaves no doubt: we need more information about the upper limits of tolerance to a large variety of organic substances before we can hope to achieve significant progress in the ecology of marine micro-organisms.

(b) Metabolism and Activity

Organic substances are required as sources of energy or carbon for metabolism and activity of bacteria, fungi and blue-green algae. In regard to bacteria and fungi, BROCK (1966) proposed distinguishing between macronutrients (required in large amounts as precursors or building blocks of cell structures) and micronutrients, such as vitamins, hormones and trace elements (required in minute amounts, difficult to detect analytically). This basic classification is not differentiated sufficiently to accommodate the different energy sources and nutritional requirements. A more detailed classification has been presented at the COLD SPRING HARBOR SYMPOSIUM (1946, p. 302; see also LAMANNA and MALLETTE, 1965, p. 511):

Classification and nomenclature based on energy sources
  A. Phototrophy
     Energy provided chiefly by photochemical reactions
     1. Photolithotrophy
        Growth depends on exogenous inorganic H-donators
     2. Photo-organotrophy
        Growth depends on exogenous organic H-donators
  B. Chemotrophy
     Energy provided entirely by dark chemical reactions
     1. Chemolithotrophy
        Growth depends on oxidation of exogenous inorganic substances
     2. Chemo-organotrophy
        Growth depends on oxidation or fermentation of exogenous organic substances
  C. Paratrophy
     1. Schizomycetotrophy
        Growth only in bacterial cells
     2. Phytotrophy
        Growth only in plant cells
     3. Zootrophy
        Growth only in animal cells

Classification and nomenclature based on ability to synthesize essential metabolites
  A. Autotrophy
     All essential organic metabolites synthesized
     1. Autotrophy *sensu stricto*
        Ability to reduce oxidized inorganic nutrients

2. Mesotrophy

Inability to reduce one or more oxidized inorganic nutrient(s); need for one or more reduced inorganic nutrient(s)

B. Heterotrophy

Not all essential organic metabolites synthesized; need for exogenous supply of one or more essential metabolite(s), growth factor(s) or vitamin(s)

C. Hypotrophy

Reproducing units (bacteriophages, viruses, genes, etc.) multiply by re-organization of complex structures of the host, as well as by uptake of materials from the external environment

The chemo-organotrophic micro-organisms only will be discussed here. Their metabolism and activity depend not only on the concentration but also on the nature of the organic substance concerned. Most sugars and proteins, for instance, are readily metabolized, i.e., decomposed. On the other hand, agar, cellulose and chitin are metabolized very slowly. In most cases, the metabolic pathways are well known. In fact, the degradation of organic substances and the chemical nature of the pertinent metabolic pathways make up the largest part of microbiological textbooks (several textbooks are listed in the introduction to this chapter). In contrast to heterotrophic micro-organisms, most phototrophic bacteria and blue-green algae do not require organic substances, but light (Chapter 2) as metabolic energy source.

Protein molecules, the principal nitrogenous constituents of plant and animal cells, are large and made up of about 20 different amino acids; they must be degraded before they can enter the bacterial cell. This process is called 'proteo-lysis'. It is accomplished by proteolytic, exocellular enzymes (proteinases) excreted by the bacteria. The proteinases degrade the protein into polypeptides, which, in turn, are converted by peptidases into amino acids. The amino acids can now be taken up by the bacterial cell, in which they may undergo further changes (de-amination. decarboxylation, release of indole, pyruvic acid, ammonia, acetic acid, butyric acid, carbon dioxide, acetate, pyruvate). The metabolic pathways vary considerably in different species and under different conditions. For example, glucose degradation can result in lactic acid, acetic acid, formic acid, ethyl alcohol, proprionic acid, formic acid, succinic acid, 2, 3 buthylene glycol, acetone or kojic acid. Most information on metabolic pathways of organic substances has been obtained on non-marine bacteria. However, it seems safe to assume that the situation is basically identical in marine forms.

Current research on micro-organisms is to a large extent concerned with molec-ular, biochemical and physiological aspects of metabolism. While such work is of utmost importance for the analysis of basic life processes, there is a deplorable paucity of ecological studies. Let us not forget that micro-organisms live in waters and soils—not in test tubes!

Of considerable importance in the marine environment are metabolic processes of micro-organisms which are able to degrade cellulose, chitin and agar-agar. Cellulose is the world's most plentiful, naturally occurring organic compound; it is present in all forms of the plant kingdom. Cellulose can be metabolized by micro-organisms which possess cellulase: this enzyme degrades cellulose to cellobiose, which is further transformed by cellobiase to glucose. Most heterotrophic bacteria are capable of metabolizing glucose. In nature, cellulose degradation makes up part of the carbon cycle. Unfortunately, our knowledge of cellulose-degrading bacteria is very limited.

In many cases, cellulose degradation produces effects which are undesirable for certain activities of man. Examples are the rotting of fishing nets and other man-made products. There are considerable difficulties in isolating and enumerating true cellulose-decomposing marine bacteria. Celluloses differ in their chemical composition, and chemically treated cellulose changes its characteristics. In fact, the term 'cellulose' appears to stand for a number of quite different organic substances. A practical handicap is also the very long incubation time needed for population growth of most cellulolytic micro-organisms.

Intensive studies about cellulose degradation and the bacteria involved have been published by OSTERTAG (1952) and KADOTA (1956). The latter paper includes descriptions of new species and summarizes information on previously described species; it also provides information on measures of disinfection against marine cellulose-decomposing bacteria, and on responses to pH and temperature. HALLI-WELL (1959) deals with cellulolytic activities, microbial and animal celluloses, enzymatic breakdown mechanisms and breakdown products.

Analytical studies on cellulolytic activity of fungi were lacking until MEYERS and REYNOLDS (1959) published their results on marine Ascomycetes. While indirect evidence has been available for enzymatic activity of fungi in regard to degrading cellulose, exact determinations are essential for appreciating the role of fungi in marine ecosystems. Using ashless cellulose powder, pine sulphite pulp, balsa wood and sodium carboxy-methyl cellulose, MEYERS and REYNOLDS established the presence of cellulolytic activity in *Lulworthia floridana*, *Torpedospora radiata*, *Ceriosporosis halima*, *Lignincola laevis*, *Peritrichospora integra*, *Areariomyces salinus*. Cell-free filtrates of these species acting on cellulose produce approximately 0·075 to 0·150 mg reducing sugars per ml. The cellulolytic activity of a given filtrate differs considerably with type and amount of substrate, and incubation conditions. The amount of effective cellulases produced by these fungi on wood in the sea is not determinable.

Most papers dealing with degradation of cellulases in the sea have been conducted from a practical viewpoint, without specific bacteriological investigations. Cordage, such as used for fishing nets, has been submersed in habitat water for different lengths of time and the resulting loss in strength determined. At the same time, different methods of preservation have been tested and physico-chemical parameters of the environment recorded (e.g. MESECK, 1929; MESECK and co-authors, 1934; VON BRANDT and KLUST, 1950; KLUST and MANN, 1954, 1955; VON BRANDT and co-authors, 1956).

In the marine environment, agar, a complex galactan, occurs in a large variety of seaweeds. Among marine micro-organisms, nearly 50 species of agar digesters have been described (HUMM, 1946). From 2 to 150 agar digesters were found in 1 ml of sea water, and up to 20×10^6 per g of coastal mud. Agar-digesting colonies are easily recognized during enumeration when employing the pour plate method. Bacterial colonies liquefy the agar or form holes.

Since the morphology of bacteria is monotonous if compared to that of higher organisms, metabolism and activity (especially the ability to metabolize certain sugars) provide the most important criteria for taxonomic classifications. The information available on bacterial responses to organic test substances easily fills several books (e.g. BREED and co-authors, 1957). However, this knowledge has been obtained under standardized culture conditions, which have little, if anything, in common with the ecological condition met in the natural habitat. Consequently, we know next to nothing of the capacities of a given bacteria species to degrade organic substances in its natural environment, where it may be confronted with an entirely different set of environmental conditions and where it has to face additional influences from co-existing organisms, including competitors. In view of the world-wide attempts to increase our knowledge on life in oceans and coastal waters, to make optimum use of living marine resources and to preserve a healthy marine environment for future generations of mankind, the degree of ignorance about the specific role of defined species of micro-organisms in the marine ecosystem is upsetting.

The dynamics of population growth as a metabolic response of bacteria to organic substances are reflected in the so called 'growth curve'. The growth pattern of a bacterial culture may be separated into a series of connected, consecutive phases, characterized by variations in growth rate (LAMANNA and MALLETTE, 1965). When plotted as a function of time, changes in the amount of bacterial protoplasm or in number of individuals result in a sigmoid curve. Transitions in growth patterns are indicated in such a curve as changes in slope. We may distinguish the following growth phases: (i) Lag phase (very long generation time); (ii) acceleration phase (decreasing generation time; phases (i) and (ii) are also known as adjustment phase); (iii) exponential phase (minimal, constant generation time); (iv) retardation phase (increasing generation time); (v) stationary phase (multiplication rate balanced by death rate); (vi) phase of decline (multiplication rate exceeded by death rate). Under culture conditions, types and sequence of bacterial growth phases remain the same, irrespective of whether the amount of bacterial protoplasm or the number of individuals is plotted. Nevertheless, increase in bacterial protoplasm does not always coincide completely with population increase or with multiplication rate. Maximum inconsistencies occur during the adjustment phase.

During exponential population growth and under optimum conditions, generation time can be as short as 20 mins or even shorter. Further growth may be increasingly limited due to (i) depletion of organic substances (nutrients), oxygen or inorganic salts, (ii) self-poisoning via accumulation of catabolic products.

Addition of defined organic substances to a multispecies culture tends to favour species with a greater ability to use these substances than others. Such selective growth promotion is the basic principle of the so-called enrichment cultures. In this way, the experimenter can reduce progressively the number of species present and finally obtain a nearly pure (i.e. monospecific) culture. Even small interspecific differences in generation time will allow use of this method.

We have, so far, considered largely 'promicrobial' effects of organic substances on metabolism and activity. Organic substances may, of course, also exert antimicrobial, i.e. inhibitory, influences on metabolic performance and activity of bacteria, fungi and blue-green algae. Critical, life-limiting antimicrobial effects have already been discussed under *Tolerance*. Antimicrobial organic substances may reduce significantly rates of metabolism and activity under habitat conditions, even to the extent of complete arrest of multiplication and to the formation of latent life stages such as cysts or spores.

Antimicrobial organic substances in the marine environment have received attention in the last few years only. There is still no detailed information available about the way in which they act on marine micro-organisms. However, we do have some knowledge of how they influence the metabolism of some bacteria, which endanger human health. Penicillin, for example, hinders the accumulation of some amino acids by *Micrococcus aureus* and stops the accumulation of glutamic acid. A bacterial strain exposed to penicillin can acquire resistance to such inhibitory effects. Apparently, reduction or loss of uptake capacity eventually causes an increase in the ability to synthesize the amino acids required. According to GALE (1949), penicillin blocks certain assimilation processes, while increase in resistance involves selection of mutants which depend less on assimilation than on synthesis for growth; if resistance is forced high enough, mutants are selected which can synthesize all amino acids required. Penicillin enters the bacterium via a lipoid component which is closely associated with the cell wall. Similarly, some bacteria are capable of counteracting metabolic effects of streptomycin; under continued influence of that substance, they may even become, at least temporarily, dependent upon streptomycin. Other antibiotics may affect the microbial metabolism quite differently. Gramicidin, for example, augments the rate of respiration and, at the same time, strongly inhibits uptake of phosphate. Polymyxin E exerts influences comparable to those of surface-active substances; in *Pseudomonas aeruginosa*, it causes leakages of pentose, phosphate and other substances.

In recent years, numerous intensive studies have been undertaken to elucidate the effects of antibiotic substances on the microbial metabolism. It is beyond the scope of this chapter to consider the vast amount of results obtained. It must suffice to point out that antibiotics can cause bacteriostatic and bactericidal effects. Penicillin acts solely upon growing cells and influences primarily the formation of the cell wall; presumably, it does not hinder the synthesis of muropolysaccharides but the interconnections of peptide chains (SCHLEGEL, 1969).

Surface-active substances may affect greatly metabolic processes of micro-organisms. All metabolic exchanges between micro-organisms and environment occur at the liquid phase of the cell boundary and hence depend strongly on the degree of surface tension. In view of the large number of papers dealing with surface-active substances and their influence on bacteria, fungi and blue-green algae, rigid selections have to be made. It must suffice to mention here a few immediately pertinent examples only. The importance of surface-active substances used to fight oil pollutions in the sea will be considered in Volume V.

MacLEOD and co-authors (1958) cultivated a marine bacterium of the genus *Flavobacterium*. In a chemically defined medium, this bacterium exhibits very slow population growth unless yeast extract is added. A similar growth-promoting effect is obtainable by adding a combination of a surface-active agent and three nucleotides. MacLEOD and co-authors concluded that the surface-active agent promotes growth because of its surface-active, rather than any other, properties. It seems surprising that a potent antibacterial agent such as the quaternary ammonium compound benzyl myristyl dimethyl ammonium chloride, in this case, promotes bacterial growth. This result is of great interest since it indicates that yeast extract—which is used as a component of many artificial media and whose growth-promoting properties are thought to originate mainly from its vitamin content—may act, at least in the present case, primarily via its surface-active properties. Possibly, organic substances, which are assumed to function as nutrients or vitamins, may act as surface-active substances also in other cases.

HALLMANN (1961) recommends the addition of a surface-active substance to culture media in order to suppress the motility (swarming) of the bacterium *Proteus vulgaris*, which is present in many putrifying materials and in faecal material; during cultivation procedures, it often spreads over the petri dish and overgrows other bacteria, thus preventing their colony formation. HALLMANN recommends the addition of 'Pril' (0·125%) or of 'Rei' (0·125—0·25%); those substances are widely used in households and have strong surface-active properties. Surface-active substances are also widely used to clean laboratory glassware, or are added to disinfectants in order to increase their effectiveness. In view of the information presented above, precautions must be taken to remove all traces of such substances to avoid undesired influences on experiments.

JONES and JANNASCH (1959) employed surface-active substances in an attempt to separate bacteria aggregates which occur under normal habitat conditions. The results obtained are hardly convincing, and BUCK and CLEVERDON (1961) failed to confirm their findings.

Interestingly, some bacteria are capable of producing surface-active substances (LA REVIÈRE, 1955a, b; OBERZILL, 1967). This ability is of importance in regard to oil pollution and degradation. With the help of the surface-active substances produced, bacteria may liberate oil from the minerals used to sink oil pollutions to the bottom of the sea. Hence the oil may ascend again and return to the surface.

(c) *Reproduction*

Only in some laboratory strains have bacteria been reported to reproduce sexually; it is not known whether sexual reproduction occurs in nature. Fungi exhibit a variety of modes of sexual reproduction under laboratory conditions; however, information on reproductive processes in the sea is quite limited. No sexual reproduction is known to occur in blue-green algae (ROUND, 1968); in regard to asexual reproduction, they do not depend on organic substances, but some species require vitamins.

In bacteria, asexual reproduction (multiplication) is identical to population growth and, hence, has been treated under *Metabolism and Activity*. Rate of multiplication depends directly on type and concentration of organic substances, as long as no other factors become limiting. Of the organic substances used as energy source, up to 40% of the bound organic carbon may be incorporated in multiplying cells.

(d) *Distribution*

Extensive information on the distribution of marine micro-organisms has been compiled by ZOBELL (1946). The general trends outlined in his book on the effects of organic substances on microbial distributions are still valid. Another important book has been published by KRISS (1961). KRISS presents data on the distribution of marine micro-organisms, which he and his associates have collected during world-wide cruises.

The distribution of heterotrophic bacteria and fungi are directly dependent on the presence of utilizable organic substances. In contrast to the extensive amount of data available on bacterial distributions, information about the influence of organic substances on distributions in marine fungi is sparse. Some pertinent data on yeast have been compiled by VAN UDEN and FELL (1968). Blue-green algae are, in general, restricted to coastal areas, and to water levels which provide sufficient light for photosynthesis. Organic substances play no important role as a factor governing their distribution pattern.

The basic dependence of many bacteria and fungi upon organic substances has produced a large number of papers. However, our pertinent knowledge has not advanced much beyond the fundamental facts outlined by ZoBELL (1946) and KRISS (1961).

Distributions in sediments

In a normal, unpolluted marine environment, the highest numbers of bacteria are found in the sediments; more precisely, in the surface layers of the sediments, which contain maximum amounts of organic substances. The upper centimetres of sediment cores, taken at different water depths, contained between 10 and 10^8 bacteria per g of wet mud. If counting is restricted to the uppermost millimetres, the bacterial numbers may be even higher.

Depending on the type of organic substances, their local concentration, water movement, temperature and salinity, the number of bacteria can vary considerably. In general, bacteria tend to be most abundant in shallow waters rich in a variety of organic substances. The mean abundancies range from 10^2 to 10^5 bacteria per g of wet sediment. Living bacteria have been collected from sediments down to water depths of 10,000 m and at all latitudes, from the Equator to the Poles. In fact, bacteria seem to be present in all life-supporting marine habitats. In sediments, their abundance usually decreases rapidly with increasing core depth, as does the content of organic substances.

In marine mud, the distribution is—according to REUSZER (1933)—directly correlated to the organic content; but the abundance of the bacterial population depends more on the degree of decomposition of the organic matter than on its total quantity.

Distributions in the free water

In the free water above the sediment, by far the highest numbers of bacteria and fungi are found in the euphotic zone. In general, the abundance of bacteria increases below the uppermost 5 to 10 m down to 25 or 50 m and then begins to decrease. Below 200 m, bacteria become rather rare, plate counts rarely revealing more than 10 bacteria per ml of sea water. In the ideal case, the vertical distribution curve of bacteria resembles that of the phytoplankton, which produces the organic material supporting bacterial growth. These facts, first presented by ZoBELL (1946), have been supported by the findings of numerous microbiologists for the various oceans. However, the picture may change completely due to intensive water movement or pollution. In shallow seas, like the North Sea, heavy storms stir up large amounts of bottom sediments. These are carried about for days or weeks in the water column and greatly affect bacterial distributions. Extensive vertical water mixing leads to an over-all increase in bacterial numbers and to the diminishing or disappearance of vertical distribution gradients.

In certain areas, the distribution of micro-organisms may be affected by water pollution (Volume V) due to organic substances. Modifying effects of water pollution typically lead to changes in horizontal distribution patterns, with abundance maxima near the place where polluted waters enter the sea. Examples in the North Sea have been presented by GUNKEL (1963). Seawards of the Elbe estuary, the numbers of bacteria decrease rapidly with increasing distance from the river mouth. In the estuary itself, pronounced vertical stratifications may build up. Waters with low salinity but high numbers of bacteria overlay high salinity waters with low bacterial numbers.

The general picture of microbial distributions is further complicated due to seasonal variations in phytoplankton abundance, production of organic anti-microbial substances, and changes in numbers of animals which feed on bacteria (e.g. predatory protozoans).

Local increases in bacterial numbers have frequently been recorded in the surface film of the water. A large number of different micro-organisms tend to accumulate near the water–air interface. The surface film serves as adsorption basis for micro-organisms, organic substances and miscellaneous particulate materials. Organisms associated with the surface film of water are referred to as neuston. SIEBURTH (1965) found considerably higher numbers of bacteria in the surface film than in the underlying water. Due to organic substances, the surface tension of the film was found to be lower than that of the water below.

KRISS (1961) suggested that bacteria may serve as direct indicators of the amount of organic substances present in different parts of the oceans. However, such simplification is not in keeping with many facts. ZoBELL (1946) writes:

'While the numbers of bacteria which may develop in seawater stored in the laboratory are directly proportional to content of utilizable organic matter, the bacterial population of seawater "in situ" is not necessarily indicative of the organic content of the water. A large bacterial population may more effectively utilize the organic content of seawater and reduce it to a lower level than a few bacteria, in which case there would be relatively little organic matter in the presence of large numbers of bacteria until the latter perish. The bacteria themselves contain very little organic matter, it requiring a bacterial population of about 10 Millions per ml to be equivalent to 1 mg of organic matter per litre. On the other hand, in certain regions there may be enough organic matter to provide for the rapid multiplication of bacteria, but the population may never exceed more than a few thousand per ml due to the activities of predators or other factors minimal to the prolonged survival of bacteria.'

(3) Structural Responses

Most micro-organisms have a fairly simple external morphology. Among the bacteria, predominant body shapes are rods, spheres and screws. Structural responses to increased amounts of utilizable organic substances consist largely of cell enlargements: short rods, hardly distinguishable from detritus, begin to increase in size; fission occurs and the new bacteria formed may remain in close contact, forming chains or other typical assemblages, for example, in species of *Sarcina* and *Streptococci*. Different species tend to exhibit different assemblage patterns.

Penicillin prevents the synthesis of cell-wall substance (but does not interfere with growth and reproduction), resulting in the formation of spheroblasts, i.e., cells without walls. Due to the lack of a rigid outer membrane, spheroblasts are round. Provided they maintain viability, they have been shown, under experimental conditions, to possess many of the physiological characteristics of their cell-walled counterparts. Spheroblasts without cell walls are also referred to as L-forms; most of these can change back to the original form upon removal of the inducing substance (penicillin). The morphological characteristics of the L-forms resemble largely those of the order Mycoplasmatales.

The formation of internal structures may also depend on the organic substances available. In general, number and size of cell components, such as granules, tend to increase in the presence of high amounts of utilizable organic substances. Volutin granules, also known as metachromatic granules, occur in many bacteria and fungi, as well as in algae and protozoans. The main constituent of these granules is polymetaphosphate. Lipid droplets, appearing as highly refractile globules, become more prominent in many bacteria as the cell grows in size and age. The lipids may be either neutral fats or granules of poly-hydroxybutyric acid. Polysaccharide granules, e.g. starch or glycogen, may also be found. It is assumed that granules represent stored food sources, but it is unlikely that this is their only function.

Under unfavourable environmental conditions, e.g. if organic substances become depleted, some bacteria transform into small ovals or spheres which exhibit low metabolic rates and high degrees of resistance to environmental stress. They are known as spores or endospores since they are formed intracellularly. If environmental conditions improve, the spore wall breaks, releasing a new vegetative cell. A typical spore features an exosporium, a spore coat, an outer membrane, a cortex and an inner membrane. All bacterial spores contain large amounts of depicolinic acid.

10. ORGANIC SUBSTANCES

10.2 PLANTS

G. E. Fogg

(1) Introduction

Marine phytoplankton and seaweeds are often considered as purely autotrophic organisms growing in a solution of mineral salts in water. This is certainly an over-simplification. As long ago as 1779 and 1781, PRIESTLEY noted that the growth of freshwater algae was promoted by certain organic substances and there is now abundant evidence that complete autotrophy is not an invariable concomitant of the ability to photosynthesize. In fact, most algae, marine species included, are affected in various ways by organic substances in the environment.

A great variety of organic substances, likely to be of biological significance, seems normally to be present in dissolved form in sea water (Chapters 1, 10.0). Some of these exogenous organic substances may supply an appreciable proportion of the carbon and energy requirements of phytoplankton; many plankton species require traces of particular organic substances if they are to grow at all. Other substances, present in minute concentrations in the water, may influence the course of development or onset of reproduction in the larger attached algae. In discussing these effects, it will be necessary to refer occasionally to freshwater species; there is no reason to suppose that there are any fundamental differences in reactions to organic substances between salt and freshwater species, and, for various reasons, investigations with the latter have been more extensive.

(2) Functional Responses

(a) Tolerance

It seems unlikely that the total concentration of dissolved organic substances in sea water can often be of much direct significance as a lethal factor for plants. The concentration in the open sea varies between narrow limits only (Chapter 10.0). In polluted inshore waters, secondary effects—such as competition from species promoted by organic substances, depletion of oxygen, production of hydrogen sulphide or of other toxic decomposition products—may affect the growth of marine species rather than the concentration of organic substances *per se*. However, marine algae in culture are particularly sensitive to organic substances; 100 to 300 mg/l of 'peptones' for example, drastically inhibit their growth (BOALCH, 1961; PROVASOLI and McLAUGHLIN, 1963).

It is to be expected that most responses of marine plants to individual organic substances will be basically of the usual type, in which the magnitude of the response is proportional to the concentration of the substance when this is low, but becomes independent of concentration when this is high. This situation is generally represented by an equation of the form:

$$\frac{k}{k_\infty} = \frac{C}{C + C_1}$$

in which k may be, for example, the relative growth factor of the organism at a concentration C, C_1 being a constant numerically equal to the concentration of the substance giving half the maximum relative growth constant, k_∞. There appears to have been no specific demonstration that this particular equation adequately describes the response of any marine alga to an organic substance, but that it does so has been assumed in studies on the heterotrophic assimilation of organic substrates by marine phytoplankton (PARSONS and STRICKLAND, 1962) and on the vitamin B_{12} requirement of the flagellate *Monochrysis lutheri* (DROOP, 1966a). It is, however, evident that the response of an organism to a given concentration of a substance depends both on the level of other simultaneously effective environmental factors and on the physiological state of the organism itself, that is to say, C_1 in the above expression is only constant for a particular set of circumstances.

By analogy with other organisms, one would expect marine algae to show non-genetic resistance adaptation to the presence of organic substances, but direct evidence of this seems to be lacking as yet. It has been shown that the photosynthesis of marine phytoplankton exhibits pronounced adaptation towards light intensity (RYTHER and MENZEL, 1959; STEEMANN NIELSEN and HANSEN, 1959) and presumably similar alterations in enzymic balance may be induced by organic substrates. As an indication of the kind and extent of such adaptation, two examples studied in freshwater algae may be mentioned. A strain of *Chlorella pyrenoidosa*, which normally contains no isocitrate lyase, forms this enzyme rapidly after a lag period of 40 mins when supplied with acetate as sole carbon source in the dark (SYRETT, 1966). By subculturing *Anacystis nidulans* in gradually increasing concentrations of antibiotics, resistant strains have been produced; for example, after 15 serial transfers a stable strain, resistant to 50,000 times the concentration of streptomycin as the original, was obtained (KUMAR, 1964).

(b) Metabolism and Activity

Effects of organic substances on the metabolism of marine plants have generally been assessed in terms of growth. However, growth is a vague term and it is necessary to state precisely what has been measured in each particular instance. In cultures of micro-algae in a limited volume of medium, various phases of growth may be distinguished (FOGG, 1965) and organic substances may affect these differently. Immediately following inoculation there may be a lag phase before cell division begins. This is followed by the **exponential phase**, in which the relative growth factor, defined as

$$k' = \frac{\log_{10} N - \log_{10} N_0}{t}$$

where N_0 and N are cell numbers at the beginning and end respectively of a period, t, remains nearly constant. Relative growth factors may also be obtained for cell volume, dry weight, chlorophyll or other parameters. Later the relative growth factor declines and the culture enters the **stationary phase**, in which growth may be measured as final yield or population density. Growth of macroscopic algae is measured in such terms as fresh or dry weight, length or area. Relative growth rates for these may be calculated but are rarely constant over any appreciable period.

In studying the effects of organic substances on a micro-organism or plant, it is essential for most purposes to have the organism in pure or axenic culture. If other organisms such as bacteria or fungi are present, their metabolism may obscure any direct effects of the added organic substance on the organism under study (Chapter 10.1). Bacteria and fungi may also produce growth factors, so that in their presence it is impossible to determine whether the organism studied has a requirement for these factors or not. Unless otherwise stated, all the work discussed in this chapter is said by its authors to have been performed with axenic cultures.

Organic substances as carbon and energy sources

Some marine organisms classifiable as algae undoubtedly depend on organic substances as energy sources. Colourless, **apochlorotic species** are frequent in the major groups of plankton algae—Chrysophyceae, Cryptophyceae and Dinophyceae (PRINGSHEIM, 1963). One such member of the Dinophyceae, *Oxyrrhis marina*, characteristic of brackish waters, is phagotrophic and requires various growth factors (p. 1559). However, its carbon requirements can be met simply with acetate or ethanol but not with any carbohydrate or amino acid (DROOP, 1959). Another colourless member of the same group, *Gyrodinium cohnii*, grows best on glucose, glycerol or acetate, a mixture of two or three of these being better than any one of them singly (PROVASOLI and GOLD, 1962). A few colourless diatoms are known. One is *Nitzschia putrida*, regularly found on *Fucus serratus*, more rarely on other *Fucus* species, but apparently on no other alga save *Pelvetia canaliculata*. Acetate, pyruvate and succinate are the best carbon sources; glucose and lactose also support its growth but not as well. Growth is greatest in the presence of peptone, yeast extract or beef extract (PRINGSHEIM, 1967). Two other *Nitzschia* species, *N. leucosigma* and *N. alba*, use lactate, succinate and glutamate as carbon sources, the last serving as a nitrogen source as well (LEWIN and LEWIN, 1967). From these examples, it will be seen that apochlorotic plants vary greatly in regard to the range of carbon sources which they are able to utilize but, generally speaking, most of them are 'acetate organisms', growing better on lower fatty acids than on any other substrate (PRINGSHEIM, 1963).

A few pigmented marine algae are capable of growing in the dark if suitable organic substrates are present, but records of failure to obtain growth in the dark are more numerous. Thus PINTNER and PROVASOLI (1963), in a study of five chrysomonads, failed to obtain growth in darkness with *Hymenomonas* sp. (which shows the most heterotrophic tendencies of the five) although a variety of carbon sources was tried. LEWIN (1963) tested isolates of marine littoral diatoms for their ability to grow in the dark on a medium containing tryptone, vitamins B_1 and B_{12}, and either glucose or lactate as a carbon source. Half of the 24 species of littoral pennate diatoms examined included strains capable of heterotrophic growth but only one centric diatom, *Cyclotella* sp., among fifteen species possesses this ability. As might be expected, the isolates capable of heterotrophic growth come from inshore habitats rich in organic matter, while planktonic forms from open waters show little tendency to heterotrophy. *Nitzschia marginata* multiplies as rapidly in the dark as in the light, but the value of k' for *Cyclotella* sp. in the dark is only half that in the light, and for *Nitzschia closterium* only one-sixth. KUENZLER (1965), in a study of glucose-6-phosphate utilization by sixteen species of unicellular algae representing Cyanophyceae, Chlorophyceae, Chrysophyceae, Bacillariophyceae, Cryptophyceae, Dinophyceae and Rhodophyceae, found that none can utilize this directly as a carbon source. The only one of them, *Cyclotella cryptica*, capable of growth on glucose in the dark may utilize the carbon of glucose-6-phosphate if phosphatase is available to split it, but this enzyme is not formed in the dark under the conditions used. DROOP and McGILL (1966) tested 39 strains of algae, mainly from supralittoral habitats, for their ability to grow in the dark with acetate or glycollate as carbon source. Of these, only twelve strains, including those of *Brachiomonas submarina*, *Chlamydomonas spreta*, *Haematococcus pluvialis*, *Stephanosphaera pluvialis* and *Oxyrrhis marina*, grow on acetate and none at all on glycollate. BOALCH (1961) failed to obtain growth of *Ectocarpus confervoides* in the dark on any of several carbon sources tested, even though these had no inhibitory effect in the light. FRIES (1963) was unable to obtain growth of the red alga *Goniotrichum elegans* in the dark on glucose, fructose or sucrose, although these sugars stimulate its growth in the light.

Nevertheless, there are various indications that heterotrophic growth of pigmented algae is important in the sea. Apparently autochthonous diatoms have been found in abyssal mud collected on the 'Galathea' expedition (WOOD, 1956); and healthy populations of dinoflagellates, coccolithophorids, and blue-green algae have been reported as occurring below the photic zone down to 4,000 m in various sea areas (BERNARD, 1963; KIMBALL and co-authors, 1963). There is no positive evidence that such populations grow actively, but the supposition that the organisms concerned are heterotrophic deserves critical investigation. It should be borne in mind that the ability to grow heterotrophically may depend on a particular metabolic balance so that, in addition to testing a wide variety of individual substrates, various combinations should be tried as well.

PARSONS and STRICKLAND (1962), using a method based on enzyme kinetics, demonstrated the heterotrophic uptake of glucose by natural populations of marine phytoplankton. Their results leave it an open question as to whether the uptake occurs by algae or by associated bacteria: further developments in their technique by HOBBIE and WRIGHT (1965) have yielded indications that, in fresh water at least, uptake by the algae is usually negligible compared with that of bacteria. However, natural populations of pelagic diatoms may normally be bacteria-free (DROOP and ELSON, 1966).

The assimilation of an organic carbon source in the light has little relation to the heterotrophic assimilation of the same substance in the dark. Studies with isotopically labelled substrates have shown that the pathway of oxidative assimilation (the dark process) and photo-assimilation of a given substance are quite distinct in, for example, *Chlorella* supplied with glucose (MARKER and WHITTINGHAM, 1966) or with acetate (GOULDING and MERRETT, 1966). It is probably generally true that, in the light, organic substrates are converted into cell material via the carbon reduction cycle at the expense of photochemically generated reducing power and adenosine triphosphate. The freshwater flagellate *Chlamydobotrys*, indeed, has become specialized for the photo-assimilation of acetate and is unable to grow either on this carbon source in the dark or by ordinary photosynthesis with carbon dioxide (WIESSNER and GAFFRON, 1964). This type of metabolism is of particular advantage at low light intensity, because more cell material can be synthesized with the limited assimilatory power available if the carbon source is already partly reduced, than if it is fully oxidized, as in carbon dioxide. The relative growth rate (k') of a planktonic freshwater species of *Chlorella* at low light intensity approximately doubles in the presence of 1 mg/l of glycollic acid, although this substance does not support the growth of this strain in the dark (Table 10-4).

Table 10-4

Effect of glycollic acid on a planktonic species of *Chlorella* (After SEN and FOGG, 1966; modified)

Illumination	k' without glycollate	k' with glycollate
Dark	0·000	0·000
500 lux	0.110 ± 0·0150	0·208 ± 0·0085
1600 lux	0·263 ± 0·0278	0·251 ± 0·0163

The rates of uptake of carbon by this *Chlorella* species from bicarbonate and glycollic acid are about equal when these two carbon sources are supplied together in equivalent concentrations (NALEWAJKO and co-authors, 1963). It will be seen from the figures just quoted, that supply of glycollate has no effect on k' when the light intensity approaches saturation. This illustrates a general rule: photo-assimilation of organic substances increases the growth rate only when light is limiting, and does not result in growth rates exceeding the maximum which can be achieved at saturating intensity with an inorganic carbon source. Exceptions to this rule are, however, to be found in organisms such as *Ochromonas malhamensis*, which has feeble photosynthetic powers and a tendency to phagotrophy (MYERS and GRAHAM, 1956).

These considerations are of first importance in assessing the effect of organic substances on the growth of algae in the light, but have rarely been taken into account. The observation, that supply of a particular organic substance increases final yield in an algal culture, reveals little about its photo-assimilation; the organic substance may exert this effect by acting as an inorganic carbon source after breakdown by respiration, as a growth factor, or as chelating agent (p. 1560). The fact that the substance has no effect on the relative growth rate of the alga when it is grown at saturating light intensity does not prove that the substance is not utilized as a carbon source; only experiments conducted under limiting light intensities can produce such evidence. Thus the nature of the stimulating effect on the growth of *Hymenomonas* sp., *Pavlova gyrans* and *Syracosphaera* sp. of lactate, pyruvate, acetate, glycerol, glucose and glycine, supplied to illuminated cultures (PINTNER and PROVASOLI, 1963) is not clear, because only final yields are given and the light intensity is not specified. In the experiments of DROOP and McGILL (1966), relative growth rates, determined at a light intensity which was probably limiting, provide evidence of photo-assimilation of acetate by various strains of *Chlamydomonas pulsatilla*, *Brachiomonas submarina* and *Haematococcus pluvialis*. It is unfortunate that, in parallel experiments on the effect of glycollate, the concentration employed was inhibitory, judging by the results of SEN and FOGG (1966). FRIES (1963) found that glucose, fructose or sucrose increase growth of the red alga *Goniotrichum elegans* by over 100% after 32 days.

It seems likely that photo-assimilation of organic carbon may be important in the sea in enabling greater productivity towards the bottom of the photic zone than would be possible were only inorganic carbon sources utilized. Such production would not be measured by the conventional radio-carbon technique for the determination of primary productivity, which is based on the assumption that only inorganic carbon sources are assimilated. The extent to which photo-assimilation of organic carbon occurs in the sea remains to be established.

Glycollic acid deserves special mention since, under certain conditions, it is liberated from the cells of phytoplankton as a major product of photosynthesis (FOGG and NALEWAJKO, 1964; WATT, 1966) and, as we have seen, may also serve as a carbon source. The planktonic *Chlorella* species seems unable to grow until a certain concentration of this substance has accumulated in the medium; the addition of 1 mg/l largely abolishes the lag phase exhibited by this strain in dilute culture at limiting light intensity (SEN and FOGG, 1966). A similar effect has been shown to occur with *Ditylum brightwelli* (PAREDES, *in*: FOGG, 1965) and with *Nitzschia closterium* (TOKUDA, 1966). Other substances, such as glucose and acetate, do not have this effect; hence the action of glycollic acid is probably specific (SEN and FOGG, 1966). But, since the lag phase shown by *Nannochloris oculata* in neutral medium lacking base is shortened by various weak acids, DROOP (1966b) considers this effect to be non-specific and connected with carbon dioxide absorption. Further investigation is called for and might throw light on the factors controlling the onset of phytoplankton growth in temperate waters in the spring. JOHNSTON (1963b) concluded from bio-assays of sea water that poor quality has been the general rule for sea waters before the spring bloom commences. Presumably, some modification of sea water must take place when light and stability become suitable, before the phytoplankton bloom is in full swing. The modifying factor might well be glycollic acid.

Organic substances as nitrogen sources

A variety of organic substances may serve as nitrogen sources for algae, but it is not always possible to predict from its molecular size or biochemical relations whether a given substance is utilized or not. In studies on the utilization of organic nitrogen, growth has usually been assessed in terms of final yield, which gives a valid basis for comparison provided that the nitrogen supply is limiting. PROVASOLI and GOLD (1962) reported that the colourless flagellate *Gyrodinium cohnii* uses various amino acids and amines as nitrogen sources and that combination of two or more sources produce better growth than single sources. Species of *Amphidinium* and *Gyrodinium* were found by PROVASOLI and McLAUGHLIN (1963) to utilize a variety of amino acids as nitrogen sources but complete amino acid mixtures, peptones or yeast hydrolysates do not promote growth and often are inhibitory. Five chrysomonads studied by PINTNER and PROVASOLI (1963) also utilize a wide variety of amino acids. GUILLARD (1963) compared the growth of 15 clones of centric diatoms, one pennate diatom, two estuarine flagellates, an oceanic green flagellate, and three coccolithophorids on various inorganic and organic nitrogen sources. Most organic sources sustained some growth above that in a medium free from combined nitrogen; but the α-amino acids were generally poor, except for bottom-dwelling diatoms, which grew as well on glutamate as on nitrate. In general, glutamine is a somewhat better source. Urea and uric acid are consistently better for most of the estuarine and neritic forms, but not for the oceanic ones. In some instances, the organic sources are toxic when the organism is growing on nitrate, for example: glycine to *Coccolithus huxleyi*, uric acid to *Cyclotella caspia* and one clone of *C. huxleyi*, and urea to *Chaetoceros pelagicus* and *Skeletonema costatum*.

Phaeodactylum tricornutum utilizes many organic sources, such as α-alanine, arginine, tryptophane, threonine and urea, as readily as nitrate-nitrogen (HAYWARD, 1965); other organic substances, such as asparagine, creatine, cysteine, glycine and lysine, support less growth. Pigment production is less when the alga is growing on amino acids than when nitrate is the nitrogen source. This does not seem to be due to chelation of trace elements by the amino acids. The growth of *Gymnodinium simplex*, *Nannochloris* sp., and *Chaetoceros gracilis* is supported to varying extents by glycine, glutamic acid, asparagine, urea and uric acid as nitrogen sources (THOMAS, 1966). Six species of red alga were found by FRIES (1961) to grow best on inorganic sources but amino and amide nitrogen can be utilized. Arginine gives excellent growth of *Rhodosorus marinus* but ornithine and glutamic acid are inhibitory to this species.

THOMAS (1966) concluded that labile amino nitrogen in sea water may sometimes be important for the growth of phytoplankton. The general impression is that the greater part of the dissolved organic nitrogen in sea water is unavailable to the algae but no critical studies appear to have been carried out to test this assumption. JONES (1967), using ^{15}N as a tracer, found that the nitrogenous extracellular products of the littoral blue-green alga *Calothrix scopulorum*, which include free amino acids and polypeptides, are taken up by various types of micro-organism, including *Chlorella marina*, and by seaweeds such as *Porphyra umbilicalis*, *Scytosiphon lomentarius* and *Fucus spiralis*. Cell fractionation experiments showed that some of the nitrogen is rapidly assimilated into protoplasmic components and some passively absorbed on to wall material. Nitrogen fixing algae, such as *C. scopulorum* evidently contribute appreciable quantities of combined nitrogen in the littoral zone, and JONES's results demonstrate that some of this nitrogen is directly available to other plants.

Organic substances as phosphorus sources

Nitzschia closterium f. *minutissima* (*Phaeodactylum tricornutum*) grows in the light with either inositol hexaphosphate (phytin) or glycerophosphate as phosphorus sources (CHU, 1946), and can also take up phosphorus from these substances in the dark (HARVEY, 1953). However, phosphorus of nucleic acid or lecithin is not used by this organism. According to FRIES (1963), glycerophosphate is a good source of phosphorus for various red algae. Five chrysomonads (PINTNER and PROVASOLI, 1963) and five dinoflagellates (PROVASOLI and McLAUGHLIN, 1963) utilize, besides inorganic phosphate, glycerophosphate, adenylic, guanylic and cytidylic acids. Organic compounds are hydrolyzed to give orthophosphate in the algal cultures (CHU, 1946). KUENZLER and PERRAS (1965) examined 27 clones of marine algae, representing the Chrysophyceae, Bacillariophyceae, Cryptophyceae, Cyanophyceae, Dinophyceae and Chlorophyceae; they found that phosphate-repressible alkaline phosphatases are produced, especially by members of the first two groups. These phosphatases, which appear to be situated near the cell surface, are synthesized when the alga become phosphorus deficient, synthesis ceasing if phosphate is restored to the medium. The activity of the enzymes is usually greatest above pH 9.0 but remains appreciable at the pH of ordinary sea water. The phosphatases enable the algae to split glucose-6-phosphate (the phosphate being assimilated and the glucose left in the medium), adenosine monophosphate and α-glycerophosphate.

These facts suggest that plankton algae may derive a substantial part of their phosphorus from organic compounds under natural conditions. This should be considered in relation to the finding of WATT and HAYES (1963) that the turnover time of dissolved organic phosphorus in inshore waters off Nova Scotia is 0.5 days as compared with 1.5 days for dissolved inorganic phosphorus and 2.0 days for particulate phosphorus.

Growth factors

A plant may be unable to synthesize a particular chemical grouping which is essential for carrying out its normal metabolism, and hence must obtain it from exogenous sources. The plant is then said to be **auxotrophic**. Since factors of this sort are usually involved in catalytic functions, they are required in trace amounts only and the amount of carbon assimilated in this form is negligible. Requirements of algae for growth factors have been the subject of several excellent reviews (LEWIN, 1961; DROOP, 1962; PROVASOLI, 1963; HUTNER and PROVASOLI, 1964) and need not be considered *in extenso* here. Experimental demonstration of these requirements necessitates, besides axenic cultures, meticulous precautions to avoid contamination with extraneous sources of growth factors. If the requirement is absolute, the final yield in a culture will be proportional to the amount of the growth factor available so long as it is limiting. DROOP (1961), for example, demonstrated a linear relationship between the final yield in cultures of *Monochrysis lutheri* and the concentration of vitamin B_{12} in the test medium. This relation is the basis for the use of algae for the bio-assay of this and other vitamins.

A sufficient number of species has now been examined to make it clear that requirements for one or more growth factors are common amongst both heterotrophic and photosynthetic marine algae. The ability to carry out photosynthesis does not imply that the plant is able to synthesize all the chemical structures necessary for life. The commonest requirements are for three factors which are also required as vitamins in mammalian nutrition. In order of frequency of the requirement these are vitamin B_{12} (cobalamin), B_1 (thiamine) and vitamin H (biotin or co-enzyme R). This is a surprisingly narrow and stereotyped range of requirements considering the variety of requirement found in other classes of micro-organisms and in higher animals.

The Cyanophyceae, Chlorophyceae and Bacillariophyceae seem to have a smaller proportion of auxotrophic members than the other groups, in which the majority of species examined have requirements for one or more vitamins. The groups with phagotrophic tendencies—Chrysophyceae, Cryptophyceae and Dinophyceae—have the highest incidence of auxotrophy. Auxotrophy is not, however, confined to unicellular plants. Several species of red alga are known to require vitamin B_{12} (PROVASOLI, 1963; IWASAKI, 1965).

Many different compounds having B_{12} activity exist, and the specificity of algae towards them varies greatly. Thus most members of the Chlorophyceae, Chrysophyceae and Dinophyceae resemble mammals in responding to B_{12} factor III and to certain synthetic analogues with nucleotide side chains such as benzimidazole as well as to cobalamin. Most Bacillariophyceae and Cryptophyceae, like *Escherichia coli*, have a wider specificity and, in addition to the compounds just mentioned, will also utilize pseudo-B_{12} and factors A and B (PROVASOLI, 1963). The specificity of the *Conchocelis* stage of *Porphyra tenera* is also of the *E. coli* pattern (IWASAKI, 1965).

Very few marine algae are known to have requirements for factors other than these three vitamins. A particularly interesting example is *Oxyrrhis marina*, which has already been mentioned as a phagotroph, for in addition to requiring vitamin B_{12}, thiamine and biotin, it needs a fat-soluble 'lemon factor' which is, apparently, ubiquinone. The amount needed is low, about 0·5 μg/million cells (DROOP and DOYLE, 1966). FRIES (1961) found that pyridoxamine, of the vitamin B_6 group, increases growth of the red seaweed *Nemalion multifidum* in the presence of vitamin B_{12}. In a different category, because the factor is needed for structural rather than catalytic purposes, are possible requirements for specific amino acids. One strain of the colourless flagellate *Gyrodinium cohnii* required histidine when first isolated, but can be trained to grow in its absence on ammonium salts as the only nitrogen source, although even then addition of histidine and betaine results in better growth (PROVASOLI and GOLD, 1962). Gibberellins sometimes have slight effects on the growth of phytoplankton (JOHNSTON, 1963c), but there is no evidence that these substances are essential for the growth of any algal species.

Without doubt many growth factors remain to be identified. Numerous common marine algae have defeated attempts to grow them in defined media; a possible reason for this, although certainly not the only one, is that unrecognized growth factors are required. After supplementing sea-water samples with mineral nutrients, chelating agents, vitamin B_{12} and thiamine, JOHNSTON (1963b) found that differences between them, in ability to support growth of *Skeletonema costatum* or *Peridinium trochoideum*, persist. He concluded that such differences are primarily a matter of content of labile, growth-promoting substances rather than of inhibitors, and that these substances have different effects on different species. Possibly, such substances are produced by algae themselves. KYLIN (1941) observed that the growth of *Ulva* and *Enteromorpha* is promoted by water taken from over the *Fucus–Ascophyllum* zone and concluded that this is due to substances other than mineral nutrients (see also p. 1562 and KINNE and BULNHEIM, 1970).

Inhibitors and antibiotics

KRAUSS (1962) reviewed the effects of inhibitors on algae and presented useful tables of inhibitory concentrations of various antibiotics and selective metabolic inhibitors; most of the information discussed by him relates to freshwater plants. JOHNSTON (1963a) studied the effects of various antimetabolites on marine phytoplankton in culture. The different species reveal different patterns of response, related species, or species occurring together in the sea at the same time, tending to show similar patterns. The effect of a given antimetabolite also depends on the concentrations of vitamins and metabolites in the medium. BONEY (1963) found that growth in cell number of sporelings of five red seaweed species is inhibited by 3-amino-1, 2, 4-triazole which is widely used as a herbicide and defoliant. The effects are more marked with intertidal species, such as *Antithamnion plumula*, than with the sublittoral species *Brongniartella byssoides*.

Probably of most ecological interest are inhibitory substances liberated by algae themselves. LEFÈVRE (1964) and his colleagues reported many examples of 'auto-antagonism' in freshwater algae, but information about the chemical nature of the postulated auto-inhibitors is, for the most part, lacking. Yellow ultra-violet absorbing inhibitory substances, liberated by various Phaeophyceae and by *Olisthodiscus* sp.—but not by representatives of the Rhodophyceae, Chlorophyceae, Bacillariophyceae or Cryptophyceae—were shown by CRAIGIE and MCLACHLAN (1964) to be flavonols or catechin-type tannins. MCLACHLAN and CRAIGIE (1964) reported that substances of this sort produced by *Fucus vesiculosus* are toxic to all species of unicellular marine algae tested. Concentrations completely inhibitory for growth vary from about 25 μg/ml for *Monochrysis lutheri* to more than 150 μg/ml for *Porphyridium* sp. The toxic fraction is ethanol-soluble and thermolabile and appears to act by causing cell lysis. Such polyphenolic materials may well be important in suppressing the growth of epiphytes (SIEBURTH and CONOVER, 1965). KHAILOV and LANSKAIA (1964) observed what appears to be a similar action of substances liberated by *Cystoseira barbata* on various plankton algae.

Chelation

Besides affecting algae through direct involvement in metabolism, organic substances may exert important indirect effects on their growth and activity as a result of chelating inorganic ions. Many different types of organic substance— hydroxy acids, amino acids, polypeptides and nucleotides—form chelates. Chela-tion may result in an inorganic ion remaining in solution under conditions which would otherwise result in precipitation and in its availability as a nutrient or toxicity being altered. Chelating agents, such as citric acid or ethylenediamine-tetra-acetic acid (versine or EDTA), are now normally included in artificial culture media to maintain trace elements in non-toxic available form. SPENCER (1958) has discussed the physical chemistry of metal-EDTA chelates in sea water. Many of the organic substances occurring in solution in natural sea waters are chelating agents and presumably exert important biological effects. DUURSMA and SEVEN-HUYSEN (1966), however, were not able to demonstrate that ions such as ferric occur naturally in sea water as chelates. JOHNSTON (1964) showed by bio-assay and mixed culture experiments that the supply of chelating agents is frequently the most crucial aspect of phytoplankton nutrition in sea water. Addition of a chelating agent alone to unfiltered sea water often gives better growth of phyto-plankton—presumably by bringing into solution trace metals, of which iron is, perhaps, the most important, originally present in colloidal or particulate form. However, some 'good' waters become 'poor' on addition of chelator. This might be due to an unduly high concentration of trace metals in the samples. The interrelationships of chelators, trace elements and algal growth are complex and call for further investigations.

(c) Reproduction

Without doubt, specific organic substances, some of which may be extracellular, are involved at crucial stages in reproduction such as induction of sporing and the release, chemotaxis and clumping of gametes. However, such information as is available regarding the nature and action of these substances mostly relates to freshwater algae (COLEMAN, 1962; LANG, 1965). One of the few studies on marine algae is that of COOK and ELVIDGE (1951) who demonstrated that the chemotactic attraction of sperms of *Fucus serratus* and *F. vesiculosus* to the ova can be simulated with dilute solutions of a number of simple organic substances—hydrocarbons, ethers and esters. It was concluded that the action is primarily physical in nature and that the volatile chemotactic agent produced by the ova is similar to but not identical with *n*-hexane.

(d) Distribution

In any of the various ways indicated above, organic substances may determine the distribution in space and time of marine algae. In this connexion, most attention has been focussed on vitamin B_{12}. Substances having B_{12} activity seem normally to be present in sea water; but whereas data obtained in the laboratory on the specificity of an alga towards various substances having vitamin activity can be used in interpreting the ecological situation, data on sensitivity cannot be so used since they are very much dependent on other factors (PROVASOLI, 1963). Thus there is uncertainty as to whether the concentration of vitamin B_{12} in sea water is ever so low as to exert a decisive influence on phytoplankton growth. While DROOP has maintained that the concentration can rarely be limiting, PROVASOLI has taken an opposite view (see discussion *in*: OPPENHEIMER, 1966; DROOP, 1970; KINNE and BULNHEIM, 1970).

The sensitivity of a plant to B_{12} differs according to the way in which the measurement is made, values obtained using final yield as the criterion being lower than those obtained using relative growth rate. Thus RILEY (*in*: OPPENHEIMER, 1966) stated that the requirement of B_{12} of *Skeletonema* species in culture of limited volume is 13 molecules per cell as compared with 100 to 150 molecules per cell in continuous culture in the chemostat. DROOP (1966a) found similar differences in the requirement for B_{12} of *Monochrysis lutheri* as determined in culture of limited volume and using the chemostat. However, internal inconsistencies in the chemostat's operation indicated that the situation is complicated by factors such as the release of protein-bound vitamin into the medium. Taking such complications into account, it does appear possible that the concentration of vitamin B_{12} may sometimes be a controlling factor in natural phytoplankton growth. One clear instance of this appears to have been established. MENZEL and SPAETH (1962), using the diatom *Cyclotella nana* as assay organism, found the concentration of B_{12} in the upper surface 50 m of the Sargasso Sea to vary from 0·03 mμg/l down to an undetectable amount from May to October. *Coccolithus huxleyii*, which does not require

B_{12}, was dominant in the phytoplankton under these conditions. However, in April, when the concentration of B_{12} rose to 0·06 to 0·1 mμg/l, evidently as a result of mixing in of deeper water, a bloom of diatoms, with *Rhizosolenia stolterfothii* and *Bacteriastrum delicatulum* predominating, occurred.

Evidence that other growth promoting substances or inhibitors affect plant successions are less definite. Following up the observation that brief blooms of *Skeletonema costatum* and *Olisthodiscus luteus* in Narragansett Bay, Rhode Island (USA), alternate, and that the two species are never abundant simultaneously, PRATT (1966) showed in culture experiments that each species somewhat inhibits its own growth and that while *S. costatum* does not inhibit *O. luteus*, *Skeletonema costatum* is inhibited by high concentrations but stimulated by low concentrations of filtrate from *Olisthodiscus luteus* cultures. The substance concerned is perhaps a tannin, similar to those substances studied by McLACHLAN and CRAIGIE (1964).

The results of bio-assays of North Sea and North Atlantic waters by JOHNSTON (1963b) suggest that growth promoters, but not inhibitors, may be important. The uptake of ^{14}C from ^{14}C bicarbonate in sea-water samples enriched with growth factors provides a rapid means of bio-assay (RYTHER and GUILLARD, 1959; CARLUCCI and SILBERNAGEL, 1966) which may assist in the further investigation of this possibility. MARGALEF (1958) observed a succession in the Gulf of Vigo (Spain), from species which can easily be grown in crude culture to a mixed community of forms which are extremely difficult to culture, and postulated that this is the result of the biochemical environment becoming more complex as the season advances so that it ultimately favours species which are best able to make use of organic matter. JOHNSTON (1963a) found a similar succession in waters off the northeast coast of Scotland and made the interesting observation that species at the beginning of the season, e.g. *Skeletonema* and *Thalassiosira*, are sensitive to antimetabolites such as sulphanilamide and benzimidazole, whereas forms occurring later, e.g. *Chaetoceros* spp. and *Rhizosolenia alata*, are most resistant. All this information suggests that further investigations of the dissolved organic substances of sea water and their effects on algae should throw much light on the spatial distribution and succession of phytoplankton.

(3) Structural Responses

(a) *Size*

Under this heading, we shall discuss the effects of growth substances, which LEWIN (1962) has defined as affecting the form or rate of growth without being essential for it (at least in exogenous supply), in contradistinction to growth factors (p. 1558) in the absence of an exogenous supply of which no growth occurs and which, if limiting, determine the final yield in cultures. The possibility that the growth substances of higher plants may play equally important roles in determining the growth of algae has been the subject of much investigation. Unfortunately, however, many of the results obtained are worthless. One reason for this is because it has not been realized that the growth substance, or the alcohol which may have been used to bring it into solution, may provide an appreciable source of carbon (STREET and co-authors, 1958), or, in the former case, of nitrogen, so that its action may not be basically different from those of other organic substances such as sugars or amino acids. Another point which has often not been realized is that an increase in mean cell size is frequently the result of inhibition of cell division rather than of positive promotion of cell extension.

It will be sufficient here to note that substances having auxin activity occur in sea water and in phytoplankton (BENTLEY, 1960) and that there is some acceptable evidence that such substances cause increases in size of algal cells (for further details, the reader is referred to CONRAD and SALTMAN, 1962, and LANG, 1965). THIMANN and BETH (1959) found that β-indolylacetic acid promotes elongation of the stem of *Acetabularia mediterranea* in cultures which are not bacteria-free, the optimal concentration being between 10^{-5} and 10^{-6} M. Gibberellic acid, β-indolylacetic acid, kinetin and adenine increase growth in the *Conchocelis* phase of *Porphyra tenera*, but combinations of these four substances are no more effective than they are separately (IWASAKI, 1965). The effect was measured in terms of dry weight, but filament length increased. Purines and pyrimidines were also found to increase plant growth.

Stimulation of cell division of sporelings of *Antithamnion plumula* and other red algae by low concentrations of various carcinogenic polycyclic aromatic hydrocarbons, applied as suspensions in sea water, has been reported by BONEY and CORNER (1962).

(b) External Structures

THIMANN and BETH (1959) found that, in addition to having the effect just mentioned, supply of β-indolylacetic acid also promotes formation of the cap of *Acetabularia mediterranea*. This effect seems to be independent of the other, since the optimal concentration is 100 times as great and since it can be produced by compounds, such as methyl indoleacetate, not active in promoting extension of the stem. Both effects are influenced by illumination and the composition of the nutrient medium; and both are still shown, although to a lesser extent, when *A. mediterranea* was enucleated.

Ulva lactuca grows poorly, forming a callus-like mass, in synthetic sea water. Nearly normal morphology, with a flat blade-like thallus, was obtained in a particular sample of natural sea water enriched with mineral nutrients, vitamins, adenine and kinetin. These results were repeatable while this sample of sea water lasted; other samples permitted only filamentous growth (PROVASOLI, 1958). Neither *Ulva lactuca* nor *Monostroma oxyspermum* could attain their normal morphology in axenic culture, but *M. oxyspermum* became normal if tannins from *Fucus* sp. were added to the medium (PROVASOLI, *in*: COSTLOW, 1969). Further investigation of these effects should elucidate the general problem of morphogenesis.

(c) Internal Structures

No information on the effects of organic substances on the differentiation of cells and tissues of marine algae appears to be available. Attention may, however, be drawn to work on unicellular algae in synchronous culture in which the effects of inhibitors and other organic substances on the course of the cell division cycle have been studied (TAMIYA, 1963).

10. ORGANIC SUBSTANCES

10.3 ANIMALS

C. G. WILBER

(1) Introduction

The important role of dissolved organic substances in the biology of marine animals has been stressed in the reviews of LUCAS (1947, 1949, 1955, 1961a,b). His view is that adaptation of organisms to metabolites released into the sea is a critical factor in ecology and evolution. It is held that the release of metabolites is important in mediating interrelationships within the community (LUCAS, 1961b).

The total amount of dissolved organic substances in oceans and coastal waters is remarkably large. In the oceans, the average amount is estimated to be about 5 mg/l. Since the sum of the volumes of all oceans is about 1.37×10^9 km³, the total amount of dissolved organic substances in all oceans is equal to about 6.75×10^{12} tons (PLUNKETT and RAKESTRAW, 1955).

Dissolved organic substances come from several sources: from decomposition of matter produced in the sea; from excretion of metabolic waste products by organisms; from diffusion of soluble body constituents (FOGG, 1959; Chapter 10.1); from sources outside the water. Loss of dissolved organic substances from the body of marine organisms may be of special significance. It is clearly unsound to expect that a marine organism be so constructed as to prevent absolutely the loss of materials from inside its body. Inevitably, there will be outward leakage of material. The degree of this leakage is unknown, but probably its magnitude is significant in the over-all production of dissolved organic material in oceans and coastal waters.

In discussing the biological effects of organic substances on marine animals, it is important to realize that the organic material formed during photosynthesis by phytoplankton in the ocean is the chief source of organic nutrients for heterotrophic organisms living in the sea. The rate of photosynthesis is apparently highest near the continents of South America, Antarctica and Africa. Moreover, organic substances are formed at higher rates in the Atlantic Ocean than in the Pacific or Indian Oceans.

An example of an important dissolved organic substance is vitamin B_{12} (Chapters 10.0, 10.1, 10.2). This vitamin is found in significant amounts, carried on particles of suspended material, in river water which enters estuaries. Along the coast of Georgia (USA), the so-called brown water contains very high concentrations of vitamin B_{12}; up to 6.4 $\mu g/g$ of solids has been reported (BURKHOLDER and BURKHOLDER, 1956). In sea-water samples, vitamin B_{12} concentrations may vary between 0.0027 and 0.130 $\mu g/l$. If recalculated in terms of dried solid material, one maximum concentration reported for B_{12} is 0.736 $\mu g/g$. The concentration of vitamin B_{12} varies extensively with time and point of water collection. The evidence at hand indicates that the major amount of vitamin B_{12} is associated with the organic fraction of the solids found in sea water. Suspended particles are important in vitamin nutrition, and bacteria are significant producers and carriers of vitamin B_{12} in the marine environment (Chapter 10.1).

In addition to vitamin B_{12}, the oceans carry in solution a large variety of organic substances. These include carbohydrates, proteins, lipids and organic acids. In the Atlantic Ocean, protein makes up about 1/3 of the total dissolved organic matter (about 5 mg/l). Fatty acids in sea water have been discussed by WILLIAMS (1965). Most of the dissolved organic substances are still unidentified. According to JØRGENSEN (1966), the few chemically identified organic substances in the sea make up only a small percentage of the total amount of dissolved organic compounds. An excellent review on the biochemistry and pharmacology of organic compounds of marine origin has been published (NIGRELLI, 1960).

Organic substances in the northeast Pacific Ocean include acetic acid (0·07 to 2·8 mg/l), formic acid (0·03 to 1·0 mg/l), lactic acid (0·00 to 0·13 mg/l) and glycollic acid (0·00 to 1·4 mg/l). Waters away from the coast contain less of these substances than inshore waters, deep inshore waters less than upper layers. The acids probably result from breakdown of high molecular weight organic compounds (KOYAMA and THOMPSON, 1964). Evidence is rapidly accumulating that organic materials play key roles in the biology and ecology of marine animals (VALLENTYNE, 1957).

Progress in knowledge on the ecological importance of dissolved organic substances in oceans and coastal waters is very slow, because of the extreme difficulties associated with isolating and identifying organic constituents from sea water. With the current improvement of pertinent methods, it is hoped that, in the near future, more knowledge on the ecological effects of these compounds will become available.

(2) Functional Responses

(a) Tolerance

Ectocrine substances in oceans and coastal waters are organic materials which exert potent biological effects, although they are present in small concentrations only. The term 'ectocrine' is used as distinct from the term 'endocrine', which is well known in physiology. Ectocrine substances may cause severe damage to marine animals and rapidly lead to death. The dinoflagellate *Gymnodinium veneficium*, for example, releases a toxin representing a large organic molecule, which acts specifically on the nervous system. Mass developments of *Gymnodinium* species may lead to the well-known phenomenon of 'red tide', with its catastrophic effects on marine animals, especially fishes. Species of *Gymnodinium* kill fish and warm-blooded animals, even if present in only relatively small concentrations (RAY and ALDRICH, 1965). Another organic substance produced by phytoplankton is the sugar rhamnoside which may occur, in the Gulf of Mexico, in concentrations as high as 50 mg/l. According to FOGG (1959), the feeding rate of oysters correlates with the concentration of rhamnoside in sea water.

There are, in natural sea water, organic chelating agents which may modify the toxicity of various substances such as copper, lead or mercury (Chapters 10.0,

10.1, 10.2). The numerous pharmacologically active substances in the sea have been pointed out by EMERSON and TAFT (1945).

The sea-cucumber *Actinopyga agassizi* manufactures and releases into the surrounding sea water a toxic substance called holothurin (NIGRELLI, 1952). Very small amounts of this substance kill the fish *Cyprinodon baconi* within a few minutes. Holothurin is a steroid saponin, the first such compound known of animal origin (NIGRELLI and co-authors, 1955).

There are not many well-documented reports of shellfish poisoning of higher animals. In the vicinity of the Gulf of Mexico, where blooms of the dinoflagellate *Gymnodinium breve* occur, experiments have been made in which adult oysters were exposed to laboratory cultures of *G. breve*. These oysters proved to be toxic when fed to chickens.

(b) Metabolism and Activity

Dissolved organic substances may stimulate or inhibit growth of marine animals. Such metabolic effects may play an important role in determining numbers and kinds of the different organisms involved in the nutrient cycle of the sea (FOGG, 1959).

Small amounts of dissolved carbohydrate influence the feeding rate of oysters, as indicated by variations in the rate of water pumping (COLLIER and co-authors, 1953). The identity of these carbohydrates is not known; they give a positive reaction with N-ethyl-carbazole. Pumping rates and shell gapes of several oysters indicate a parallel response to the concentration of carbohydrates in sea water. Augmentation of the carbohydrate concentration causes increasing pumping rates; the oysters exhibit lower threshold limits, below which they do not pump; temperature increase elevates these thresholds (COLLIER and co-authors, 1953). Maximum gape of oyster shells occurs only in carbohydrate concentrations above a threshold value: about 6 mg/l at 25°C, and 12 mg/l at 27°C. Oysters remove up to 50 mg of carbohydrate per hr from sea water, but the quantities vary (see also WINTER, 1969).

Addition of niacinamide to sea water causes the valves of the oyster *Crassostrea virginica* to gape and its pumping activity to decrease (COLLIER and co-authors, 1956). Apparently, the substance is not an irritant, otherwise the shells would close. Removal of the niacinamide from the water results in a rapid return to normal of pumping rate. Niacin has no comparable effect.

Substances with antimicrobial effects are known to be produced by several marine algae (OLSEN and co-authors, 1964; Chapters 10.1, 10.2). Such antimicrobial substances are complex and not always represented by a single chemical compound. In an investigation dealing with a small number of algae from a limited area, it was shown that at least six different antimicrobial substances occurred in these marine organisms. The effects of such substances on metabolism and activity of marine animals remain to be investigated.

Antibiotic substances are present in significant quantities in the oceans. They are produced by marine micro-organisms, plants or animals. Their chemical nature—in several cases not yet identified—involves, as a rule, fatty acids, terpenes, hydrocarbons and chlorophyllides (AUBERT and co-authors, 1966). Their action on marine animals is not known.

PÜTTER (1909) maintained that a large number of aquatic organisms, particularly the smaller forms, are able to use dissolved organic substances directly as food source. His monograph should be consulted for initial views on the role of dissolved organic material in the economy of marine animals. On the basis of the then available data, KROGH (1931) rejected PÜTTER's hypothesis. More recent investigations, however, clearly re-open the possibility that some organic substances, which are in solution in great amounts, may serve as nutriment for marine animals (e.g. MORRIS, 1955; Chapter 4.31).

As early as 1933, MAST and PACE demonstrated that the protozoan *Chilomonas paramecium* could thrive and reproduce, using dissolved substances only as nutrient source. In the process, it produced a growth influencing substance which diffused into the surrounding medium (MAST and PACE, 1938). This growth substance, reported to be heat labile, accelerated reproduction of the protozoan if present in small concentrations, but inhibited growth and was even lethal in high concentrations. The chemical nature of the substance was not identified, but tests indicated that the 'molecules of this substance are probably smaller than those of sucrose' (MAST and PACE, 1946). Other protozoans produce similar materials which diffuse into the surrounding water (MARBARGER, 1943).

In some instances, metazoans depend on protozoans to make use of organic nutrients in the water. *Artemia salina*, for example, shows little growth in a sea-water medium containing glucose and nitrate, in which marine bacteria are growing. However, in the same medium, in the presence of thriving populations of marine protozoans, *A. salina* grows well. Presumably, *A. salina* feeds on the protozoans, which, in turn, convert the dissolved organic nutrients (SEKI, 1965).

The fact that multicellular aquatic animals can be grown from egg to adult in an aseptic medium containing only soluble nutriment has been clearly demonstrated in experimental studies with *Artemia salina* (PROVASOLI and SHIRAISHI, 1959). In conclusion then, it seems reasonable to assume that, in nature, soluble organic compounds play a larger role in regard to metabolism and activity of marine animals than is usually thought. Dissolved organic substances *per se* do not seem to form a completely adequate nutritional source for multicellular animals, but they represent an important part of the nutritive cycle in the sea.

(c) *Reproduction*

Holothurin, which has been referred to previously in this chapter, produces developmental faults in cleaving eggs of sea-urchins. It also inhibits, to a degree, regeneration in planarians. NIGRELLI and JAKOWSKA (1960) conclude:

'The actions on the development of sea urchin eggs, regeneration of planaria, hemopoiesis in bone marrow of winterized frogs, and inhibition of tumors in mice suggest that holothurin may act as an antimetabolite.'

No unequivocal evidence is available yet on the possible effects of organic substances on rates and modes of reproduction in marine animals.

(d) Distribution

Aquatic organisms may condition their ambient medium via secretions and excretions. Such conditioning may significantly affect animal distributions. The nature and biological consequences of organismic metabolites released into the surrounding water represent important aspects of 'mass physiology' and ecology (ALLEE, 1931). In 1936, HARDY postulated a theory of animal exclusion to account for the occurrence of animal species in some, and their complete absence in other parts of the ocean (see also LUCAS, 1955).

Repeatedly, field studies have indicated an inverse quantitative relationship between maximum phytoplankton population densities and the abundance of marine animals. For example, the herring *Clupea harengus* tends to be excluded from regions of very dense phytoplankton. On a number of different occasions, it was possible to deduce whale distributions in Antarctic waters from measurements of phosphates contained in the upper water layers: areas of phosphate depletion indicate high phytoplankton production; dense phytoplankton populations tend to exclude euphausians, the food of the whales and hence whales themselves (HARDY, 1936, 1938).

The precise agents involved in the phenomenon of animal exclusion are not known. HARDY (1956) suggested the presence of an antibiotic effect. RYTHER (1954) proposed a chemical, produced by ageing phytoplankton.

Aquatic plants may produce organic substances which can be detected by fishes and serve as migratory guide-posts. Immature fish may be 'herded' into safe areas by 'attractants' produced by aquatic plants (HASLER, 1954). Salmon can detect stream odours and discriminate among specific odours of different streams. This capacity is of importance in migratory orientation (HASLER and WISBY, 1951). Morpholine is readily detected by fishes in concentrations as low as 1×10^{-6} ppm; it has been suggested as a test chemical to study the role of dissolved organic materials on orientation in migratory fish (HASLER, 1956). There can be hardly any doubt about the important role which soluble organic compounds can play as orientation clues for marine animals.

An interesting suggestion has been presented by HOOD (1955). Since the biochemical composition of organisms characteristic of the various oceanic environments differ, and since the materials excreted, or those resulting from decay and disintegration, depend on the donator organism, ecological differentiation of marine environments by chemical means may easily be accomplished via organic substances. Recent evidence seems to support the view expressed by HOOD. Hence, we may expect interesting new insights into the dynamics of animal distributions in the marine environment from future studies of organic substances present in the free water and in sediments.

(3) Structural Responses

A review of the available literature fails to reveal studies on the role of dissolved organic substances in oceans and coastal waters on size and external or internal structures of marine animals.

Literature Cited (Chapter 10)

ALLEE, W. C. (1931). *Animal Aggregations*, University of Chicago Press, Chicago.

AUBERT, M., AUBERT, J., GAUTHIER, M. and DANIEL, S. (1966). Origine et nature des substances antibiotiques présentes dans le milieu marin. Pts I to III. *Revue int. oceanogr. méd.*, **1**, 9–43.

BAADER, R. G., HOOD, D. W. and SMITH, J. B. (1960). Recovery of dissolved organic material in sea water and organic sorption by particulate material. *Geochim. cosmochim. Acta*, **19**, 297–308.

BAAM, R. B., GANDHI, N. M. and FREITAS, Y. M. (1966). Antibiotic activity of marine microorganisms. *Helgoländer wiss. Meeresunters.*, **13**, 181–187.

BAYLOR, E. R. and SUTCLIFFE, W. H., JR. (1963). Dissolved organic matter in sea water as source of organic food. *Limnol. Oceanogr.*, **8**, 369–371.

BENTLEY, J. A. (1960). Plant hormones in marine phytoplankton, zooplankton and sea water. *J. mar. biol. Ass. U.K.*, **39**, 433–444.

BERNARD, F. (1963). Density of flagellates and Myxophyceae in the heterotrophic layers related to environment. In C. H. Oppenheimer (Ed.), *Marine Microbiology*. C. C. Thomas, Springfield, Ill. pp. 215–228.

BOALCH, G. T. (1961). Studies on *Ectocarpus* in culture. II. Growth and nutrition of a bacteria-free culture. *J. mar. biol. Ass. U.K.*, **41**, 279–286.

BOCK, K. J. (1963). Biologische Untersuchungen an Waschrohstoffen. *Pro Aqua*, **1961**, 247–257.

BOCK, K. J. and WICKBOLD, R. (1963). Untersuchungen an biologisch abbaubaren Waschrohstoffen. *Seifen-Öle-Fette-Wachse*, **26**, 1–4.

BONEY, A. D. (1963). The effects of 3-amino-1, 2, 4-triazole on the growth of sporelings of marine red algae. *J. mar. biol. Ass. U.K.*, **43**, 643–652.

BONEY, A. D. and CORNER, E. D. S. (1962). On the effects of some carcinogenic hydrocarbons on the growth of sporelings of marine red algae. *J. mar. biol. Ass. U.K.*, **42**, 579–585.

BRANDT, A. VON and KLUST, G. (1950). Zelluloseabbau im Wasser. *Arch. Hydrobiol.*, **43**, 223–233.

BRANDT, A. VON, KLUST, G. and MANN, H. (1956). Zelluloseabbau in Fischteichen. *Arch. FischWiss.*, **7**, 31–40.

BREED, R. S., MURRAY, E. G. D. and SMITH, N. R. (1957). *Bergey's Manual of Determinative Bacteriology*, Baillière, Tindall and Cox, London.

BROCK, T. D. (1966). *Principles of Microbial Ecology*, Prentice-Hall, Englewood Cliffs, New Jersey.

BUCK, J. D. and CLEVERDON, R. C. (1961). The effect of Tween 80 on the enumeration of marine bacteria by the spread and pour plate method. *Limnol. Oceanogr.*, **6**, 42–44.

BUCK, J. D. and MEYERS, S. P. (1965). Antiyeast activity in the marine environment. I. Ecological considerations. *Limnol. Oceanogr.*, **10**, 385–391.

BURKHOLDER, P. R. and BURKHOLDER, L. M. (1956). Vitamin B^{12} in suspended solids and marsh muds collected along the coast of Georgia. *Limnol. Oceanogr.*, **1**, 202–208.

CARLUCCI, A. F. and SILBERNAGEL, S. B. (1966). Bioassay of sea-water. I. A ^{14}C uptake method for the determination of concentrations of vitamin B_{12} in sea-water. *Can. J. Microbiol.*, **13**, 175–183.

CHU, S. P. (1946). The utilization of organic phosphorus by phytoplankton. *J. mar. biol. Ass. U.K.*, **26**, 285–295.

COLEMAN, A. W. (1962). Sexuality. In R. A. Lewin (Ed.), *Physiology and Biochemistry of Algae*. Academic Press, New York. pp. 711–729.

COLLIER, A., RAY, S. M., MAGNITZKY, A. W. and BELL, J. O. (1953). Effect of dissolved organic substances on oysters. *Fishery Bull. Fish Wildl. Serv. U.S.*, **54**, (84), 166–185.

COLLIER, A., RAY, S. M. and WILSON, W. B. (1956). Some effects of specific organic compounds on marine organisms. *Science, N.Y.*, **124**, 220.

CONRAD, H. M. and SALTMAN, P. (1962). Growth substances. In R. A. Lewin (Ed.), *Physiology and Biochemistry of Algae*. Academic Press, New York. pp. 663–671.

COOK, A. H. and ELVIDGE, J. A. (1951). Fertilization in the Fucaceae: investigations on the nature of the chemotactic substance produced by eggs of *Fucus serratus* and *F. vesiculosus*. *Proc. R. Soc.* (*B*), **138**, 97–114.

COSTLOW, J. D., JR. (Ed.) (1969). *Marine Biology. Proceedings of the 5th International Interdisciplinary Conference*. New York Academy of Sciences, New York.

CRAIGIE, J. S. and MCLACHLAN, J. (1964). Excretion of coloured ultraviolet-absorbing substances by marine algae. *Can. J. Bot.*, **42**, 23–33.

DEGENS, E. T., REUTER, J. H. and SHAW, K. N. F. (1964). Biochemical compounds in offshore sediments and sea water. *Geochim. cosmochim. Acta*, **28**, 45–66.

DROOP, M. R. (1959). Water-soluble factors in the nutrition of *Oxyrrhis marina*. *J. mar. biol. Ass. U.K.*, **38**, 605–620.

DROOP, M. R. (1961). Vitamin B_{12} and marine ecology: The response of *Nonochrysis lutheri*. *J. mar. biol. Ass. U.K.*, **41**, 69–76.

DROOP, M. R. (1962). Organic micronutrients. In R. A. Lewin (Ed.), *Physiology and Biochemistry of Algae*. Academic Press, New York. pp. 141–159.

DROOP, M. R. (1966a). Vitamin B_{12} and marine ecology. III. Experiments with a chemostat. *J. mar. biol. Ass. U.K.*, **46**, 659–671.

DROOP, M. R. (1966b). Organic acids and bases and the lag phase in *Nannochloris oculata*. *J. mar. biol. Ass. U.K.*, **46**, 673–678.

DROOP, M. R. (1970). Vitamin B_{12} and marine ecology. V. Continuous culture as an approach to nutritional kinetics. *Helgoländer wiss. Meeresunters.*, **20**, 629–636.

DROOP, M. R. and DOYLE, J. (1966). Ubiquinone as a protozoan growth factor. *Nature, Lond.*, **212**, 1474–1475.

DROOP, M. R. and ELSON, K. G. R. (1966). Are pelagic diatoms free from bacteria? *Nature, Lond.*, **211**, 1096–1097.

DROOP, M. R. and MCGILL, S. (1966). The carbon nutrition of some algae: the inability to utilize glycollic acid for growth. *J. mar. biol. Ass. U.K.*, **46**, 679–684.

DROOP, M. R. and WOOD, E. J. F. (1968). *Advances in Microbiology of the Sea*, Vol. I. Academic Press, London.

DUURSMA, E. K. (1963). The production of dissolved organic matter in the sea as related to the primary gross production of organic matter. *Neth. J. Sea Res.*, **2**, 85–94.

DUURSMA, E. K. (1965). The dissolved organic constituents of sea water. In J. P. Riley and G. Skirrow (Eds), *Chemical Oceanography*, Vol. I. Academic Press, New York. pp. 433–475.

DUURSMA, E. K. and SEVENHUYSEN, W. (1966). Note on chelation and solubility of certain metals in sea water at different pH values. *Neth. J. Sea Res.*, **3**, 95–106.

EMERSON, G. A. and TAFT, C. H. (1945). Pharmacologically active agents from the sea. *Tex. Rep. Biol. Med.*, **3**, 302–338.

FLOODGATE, G. D. (1966). Factors affecting the settlement of a marine bacterium. *Veröff. Inst. Meeresforsch. Bremerh.* (Sonderbd), **2**, 265–269.

FOGG, G. E. (1959). Dissolved organic matter in oceans and lakes. *New Biol.*, **29**, 31–48.

FOGG, G. E. (1965). *Algal Cultures and Phytoplankton Ecology*, Athlone Press, London.

FOGG, G. E. and NALEWAJKO, C. (1964). Glycollic acid as an extracellular product of phytoplankton. *Verh. int. Verein. theor. angew. Limnol.*, **15**, 800–810.

FRIES, E. (1961). Vitamin requirements of *Nemalion multifidum*. *Experientia*, **17**, 75–76.

FRIES, E. (1963). On the cultivation of axenic red algae. *Physiologia Pl.*, **16**, 695–708.

GALE, E. F. (1949). The action of Penicillin on the assimilation and utilization of amino acids by gram-positive bacteria. *Symp. Soc. exp. Biol.*, **3**, 233–242.

GOULDING, K. H. and MERRETT, M. J. (1966). The photometabolism of acetate by *Chlorella pyrenoidosa*. *J. exp. Bot.*, **17**, 678–689.

GUILLARD, R. R. L. (1963). Organic sources of nitrogen for marine centric diatoms. In C. H. Oppenheimer (Ed.), *Marine Microbiology*. C. C. Thomas, Springfield, Ill. pp. 93–104.

GUNKEL, W. (1963). Daten zur Bakterienverteilung in der Nordsee. *Veröff. Inst. Meeresforsch. Bremerh.* (Sonderbd), **1**, 80–89.

GUNKEL, W. (1964). Bericht über das Helgoländer Arbeitstreffen der deutschen Gewässermikrobiologen. *Helgoländer wiss. Meeresunters.*, **11**, 327–340.

GUNKEL, W. and RHEINHEIMER, G. (1968). Bestandsaufnahme Bakterien. In C. Schlieper (Ed.), *Methoden der Meeresbiologischen Forschung*. G. Fischer, Jena. pp. 243–255.

GUNSALUS, I. C. and STANIER, R. Y. (1962). *The Bacteria*, Vol. IV, The Physiology of Growth. Academic Press, New York.

HALLIWELL, G. (1959). The enzymic decomposition of cellulose. *Nutr. Abstr. Rev.*, **29**, 747–760.

HALLMANN, L. (1961). *Bakteriologie und Serologie*, Georg Thieme Verlag, Stuttgart.

HARDY, A. C. (1936). Plankton ecology and the hypothesis of animal exclusion. *Proc. Linn. Soc. Lond.*, **148** (2), 64–70.

HARDY, A. C. (1938). Change and choice: a study in pelagic ecology. In G. R. DeBeer (Ed.), *Evolution—Essays presented to E. S. Goodrich*. University Press, Oxford. pp. 139–159.

HARDY, A. C. (1956). *The Open Sea*, Houghton Mifflin, Boston.

HARVEY, H. W. (1953). Note on the absorption of organic phsophorus compounds by *Nitzschia closterium* in the dark. *J. mar. biol. Ass. U.K.*, **31**, 475–487.

HASLER, A. D. (1954). Odour perception and orientation in fishes. *J. Fish. Res. Bd Can.*, **11**, 101–129.

HASLER, A. D. (1956). Perception of pathways by fishes in migration. *Q. Rev. Biol.*, **31**, 200–209.

HASLER, A. D. and WISBY, W. J. (1951). Discrimination of stream odors by fishes and its relation to present stream behavior. *Am. Nat.*, **85**, 223–238.

HAWKER, L. E., FOLKES, B. F., LINTON, A. H. and CARLILE, M. J. (1966). *Einführung in die Biologie der Mikroorganismen*. (Transl. from Engl.) G. THIEME, Stuttgart.

HAYWARD, J. (1965). Studies on the growth of *Phaeodactylum tricornutum* (BOHLIN). I. The effect of certain organic nitrogenous substances on growth. *Physiologia Pl.*, **18**, 201–207.

HEUKELEKIAN, H. and HELLER, A. (1940). Relation between food concentration and surface for bacterial growth. *J. Bact.*, **40**, 547–558.

HOBBIE, J. E. and WRIGHT, R. T. (1965). Competition between planktonic bacteria and algae for organic solutes. In C. R. Goldman (Ed.), *Primary Productivity in Aquatic Environments*. University of California Press, Berkeley. pp. 175–185. (*Mem. Ist. ital. Idrobiol.* (Suppl.), **18**.)

HOOD, D. W. (1955). *Chemical Oceanography*, Agriculture and Mechanical College, Texas. (Proj. 63, Ref. 55–11T.)

HUMM, H. J. (1946). Marine agar-digesting bacteria of the South-Atlantic coast. *Bull. Duke Univ. Mar. Stn*, **3**, 45–75.

HUTNER, S. H. and PROVASOLI, L. (1964). Nutrition of algae. *A. Rev. Pl. Physiol.*, **15**, 37–56.

IWASAKI, H. (1965). Nutritional studies of the edible seaweed *Porphyra tenera*. I. The influence of different B_{12} analogues, plant hormones, purines and pyrimidines on the growth of *Conchocelis*. *Pl. Cell. Physiol., Tokyo*, **6**, 325–336.

JANNASCH, H. W. (1954). Ökologische Untersuchungen der planktischen Bakterienflora im Golf von Neapel. *Naturwissenschaften*, **41**, 42.

JANNASCH, H. W. (1963a). Bakterielles Wachstum bei geringen Substratkonzentrationen. *Arch. Mikrobiol.*, **45**, 323–342.

JANNASCH, H. W. (1963b). Studies on the ecology of a marine *Spirillum* in the chemostat. In C. H. Oppenheimer (Ed.), *Marine Microbiology*. C. C. Thomas, Springfield, Ill. pp. 558–566.

JANNASCH, H. W. (1967). Growth of marine bacteria at limiting concentrations of organic carbon in seawater. *Limnol. Oceanogr.*, **12**, 264–271.

JANNASCH, H. W. and JONES, G. E. (1959). Bacterial populations in sea water as determined by different methods of enumeration. *Limnol. Oceanogr.*, **4**, 128–139.

JOHNSTON, R. (1963a). Antimetabolites as an aid to the study of phytoplankton nutrition. *J. mar. biol. Ass. U.K.*, **43**, 409–425.

JOHNSTON, R. (1963b). Sea water, the natural medium of phytoplankton. 1. General features. *J. mar. biol. Ass. U.K.*, **43**, 427–456.

JOHNSTON, R. (1963c). Effects of gibberellins on marine algae in mixed culture. *Limnol. Oceanogr.*, **8**, 270–275.

JOHNSTON, R. (1964). Sea water, the natural medium of phytoplankton. II. Trace metals and chelation, and general discussion. *J. mar. biol. Ass. U.K.*, **44**, 87–109.

JONES, G. E. and JANNASCH, H. W. (1959). Aggregates of bacteria in seawater as determined by treatment with surface active agents. *Limnol. Oceanogr.*, **4**, 269–276.

JONES, K. (1967). Studies in the physiology of marine blue-green algae. Ph.D. thesis, University of London.

JØRGENSEN, C. B. (1966). *Biology of Suspension Feeding*, Pergamon Press, Oxford.

JUNGE, C. E. (1963). *Air Chemistry and Radioactivity*, Academic Press, New York.

KADOTA, H. (1956). A study on the marine cellulose-decomposing bacteria. *Mem. Coll. Agric., Kyoto Univ. (Fish. Ser.* 6), **74**, 1–128.

KALLE, K. (1945). *Der Stoffhaushalt des Meeres*, Akademische Verlagsanstalt, Leipzig.

KAVANAGH, F. (Ed.) (1963). *Analytical Microbiology*, Academic Press, New York.

KERSEY, R. C. and FINK, F. C. (1957). Microbiological assay of antibiotics. In D. Glick (Ed.), *Methods of Biochemical Analysis*, Vol. I. Wiley, New York. pp. 53–79.

KHAILOV, K. M. and LANSKAIA, L. A. (1964). Some factors of the chemical action of *Cytoseira* on unicellular algae. (Russ.) *Trudȳ Sevastopol'. biol. Sta.*, **17**, 351–360.

KIMBALL, J. F., JR., COCORAN, E. F. and WOOD, E. J. F. (1963). Chlorophyll-containing microorganisms in the aphotic zone of the oceans. *Bull. mar. Sci. Gulf Caribb.*, **13**, 574–577.

KINNE, O. and BULNHEIM, H.-P. (Eds) (1970). Cultivation of marine organisms and its importance for marine biology. International Helgoland Symposium 1969. *Helgoländer wiss. Meeresunters* , **20**, 1–721.

KLUST, G. and MANN, H. (1954). Experimentelle Untersuchungen über den Zelluloseabbau im Wasser. *Vom Wass.*, **21**, 100–109.

KLUST, G. and MANN, H. (1955). Untersuchungen über den Einfluss chemischer Faktoren auf den Zelluloseabbau im Wasser. *Arch. Fisch Wiss.*, **5/6**, 249–275.

KOYAMA, T. and THOMPSON, T. G. (1959). Organic acids in sea water. *Int. oceanogr. Congr.*, **1**, 1–925.

KOYAMA, T. and THOMPSON, T. G. (1964). Identification and determination of organic acids in sea water by partition chromatography. *J. oceanogr. Soc. Japan*, **20**, 209–220.

KRAUSS, R. W. (1962). Inhibitors. In R. A. Lewin (Ed.), *Physiology and Biochemistry of the Algae*. Academic Press, New York. pp. 673–685.

KRISS, A. E. (1961). *Meeresmikrobiologie (Tiefseeforschungen)*. (Transl. from Russ.) G. Fischer, Jena.

KROGH, A. (1931). Dissolved substances as food of aquatic organisms. *Biol. Rev.*, **6**, 412–442.

KRÜGER, R. (1962). Die Behebung des Einflusses der Detergentien auf die Wasserwirtschaft. *Jb. text. ReinigGew.*, **1962**.

KRÜGER, R. (1963). Betrachtungen zum Detergentiengesetz. *Erdöl Kohle Erdgas Petrochem.*, **16**, 379–382.

KUENZLER, E. J. (1965). Glucose-6-phosphate utilization by marine algae. *J. Physiol., Lond.*, **1**, 156–164.

KUENZLER, E. J. and PERRAS, J. P. (1965). Phosphatases of marine algae. *Biol. Bull. mar. biol. Lab., Woods Hole*, **128**, 271–284.

KUMAR, H. (1964). Streptomycin- and penicillin-induced inhibition of growth and pigment production in blue-green algae and production of strains of *Anacystis nidulans* resistant to these antibiotics. *J. exp. Bot.*, **15**, 232–250.

KYLIN, H. (1941). Biologische Analyse des Meerwassers. *K. fysiogr. Sällsk. Lund Förh.*, **11** (21), 1–16.

LAMANNA, C. and MALLETTE, M. F. (1965). *Basic Bacteriology*, Williams and Wilkins, Baltimore.

LANG, A. (1965). Physiology of growth and development in algae. A synopsis. *Hdb. PflPhysiol.*, **15** (1), 680–715.

LEFÈVRE, M. (1964). Extracellular products of algae. In D. F. Jackson (Ed.), *Algae and Man*. Plenus Press, New York. pp. 337–367.

LEWIN, J. C. (1963). Heterotrophy in marine diatoms. In C. H. Oppenheimer (Ed.), *Marine Microbiology*. C. C. Thomas, Springfield, Ill. pp. 229–235.

LEWIN, J. and LEWIN, R. A. (1967). Culture and nutrition of some apochlorotic diatoms of the genus *Nitzschia. J. gen. Microbiol.*, **46**, 361–367.

LEWIN, R. A. (1961). Growth factors: metabolic factors limiting growth. Phytoflagellates and algae. *Hdb. PflPhysiol.*, **14**, 401–417.

LEWIN, R. A. (1962). Editorial note. In R. A. LEWIN (Ed.), *Physiology and Biochemistry of Algae*. Academic Press, New York. p. 669.

LIEBMANN, H. (1967). *Detergentien und Öle im Wasser und Abwasser*, R. Oldenbourg, München.

LISTON, J. (1968). On unresolved problems in marine microbiology. In C. H. Oppenheimer (Ed.), *Marine Biology, Proceedings of the 4th International Interdisciplinary Conference*. New York Academy of Sciences, New York. p. 180.

LUCAS, C. E. (1947). The ecological effects of external metabolites. *Biol. Rev.*, **22**, 270–295.

LUCAS, C. E. (1949). External metabolites and ecological adaptation. *Symp. Soc. expl Biol.*, **2**, 336–356.

LUCAS, C. E. (1955). External metabolites. *Deep Sea Res.* (Suppl.), **3**, 139–148.

LUCAS, C. E. (1961a). On the significance of external metabolites in ecology. *Symp. Soc. expl Biol.*, **15**, 190–206.

LUCAS, C. E. (1961b). Interrelationships between aquatic organisms mediated by external metabolites. In M. Sears (Ed.), *Oceanography*. A.A.A.S., Washington, D.C. pp. 499–517. (*Publs Am. Ass. Advmt Sci.*, **67**.)

McALLISTER, C. D., PARSONS, T. R., STEPHENS, K. and STRICKLAND, J. D. H. (1961). Measurements of primary production in coastal sea water using a large-volume plastic sphere. *Limnol. Oceanogr.*, **6**, 237–258.

McLACHLAN, J. and CRAIGIE, J. S. (1964). Algal inhibition by yellow ultraviolet-absorbing substances from *Fucus vesiculosus*. *Can. J. Bot.*, **42**, 287–292.

MacLEOD, R., HOGENKAMP, H. and ONOFREY, E. (1958). Nutrient and metabolism of marine bacteria. VII. Growth response of a marine Flavobacterium to surface active agents and nucleotides. *J. Bact.*, **75**, 460–466.

MARBARGER, J. P. (1943). The production of growth substance in *Colpidium striatum* (STOKES). *Physiol. Zool.*, **16**, 186–198.

MARGALEF, R. (1958). Temporal succession and spatial heterogeneity in phytoplankton. In A. A. Buzzati-Traverso (Ed.), *Perspectives in Marine Biology*. University of California Press, Berkeley. pp. 323–349.

MARKER, A. F. H. and WHITTINGHAM, C. P. (1966). The photoassimilation of glucose in *Chlorella* with reference to the role of glycollic acid. *Proc. R. Soc.* (*B*), **165**, 473–485.

MAST, S. O. and PACE, D. M. (1933). Synthesis from inorganic compounds of starch, fats, proteins and protoplasm in the colorless animal, *Chilomonas paramecium*. *Protoplasma*, **20**, 326–358.

MAST, S. O. and PACE, D. M. (1938). The effect of substance produced by *Chilomonas paramecium* on rate of reproduction. *Physiol. Zool.*, **11**, 360–382.

MAST, S. O. and PACE, D. M. (1946). The nature of the growth-substance produced by *Chilomonas paramecium*. *Physiol. Zool.*, **19**, 223–235.

MENZEL, D. W. (1964). The distribution of dissolved organic carbon in the Western Indian Ocean. *Deep Sea Res.*, **11**, 757–765.

MENZEL, D. W. and SPAETH, J. P. (1962). Occurrence of vitamin B_{12} in the Sargasso Sea. *Limnol. Oceanogr.*, **7**, 151–154.

MESECK, G. (1929). Untersuchungen über Netzimprägnierung und Netzbehandlung. *Z. Fisch.*, **27**, 295–335.

MESECK, G., MERTENS, H., SCHÖN, A. and RUMPHORST, H. (1934). Untersuchungen über den bakteriellen Abbau verschieden konservierter Netzzellulose in Küstengenwässern. *Z. Fisch.*, **32**, 399–458.

MEYERS, S. P. and REYNOLDS, E. S. (1959). Cellulolytic activity in lignicolous marine Ascomycetes. *Bull. mar. Sci. Gulf Caribb.*, **9**, 441–455.

MORRIS, R. W. (1955). Some considerations regarding the nutrition of marine fish larvae. *J. Cons. perm. int. Explor. Mer*, **20**, 255–265.

MYERS, J. and GRAHAM, J. R. (1956). The role of photosynthesis in the physiology of *Ochromonas*. *J. Cell. comp. Physiol.*, **47**, 397–414.

NALEWAJKO, C., CHOWDHURI, N. and FOGG, G. E. (1963). Excretion of glycollic acid and the growth of a planktonic *Chlorella*. In *Studies on Microalgae and Photosynthetic Bacteria*. Japanese Society of Plant Physiologists, Tokyo. pp. 171–183.

NIGRELLI, R. F. (1952). The effects of holothurin on fish, and mice with sarcoma 180. *Zoologica, N.Y.*, **37**, 89–90.

NIGRELLI, R. F. (Ed.) (1960). Biochemistry and pharmacology of compounds derived from marine organisms. *Ann. N.Y. Acad. Sci.*, **90** (3), 615–950.

NIGRELLI, R. F., CHANLEY, J. D., KOHN, S. K. and SOBOTKA, H. (1955). The chemical nature of holothurin, a toxic principle from the sea-cucumber (Echinodermata: Holothurioidea). *Zoologica, N.Y.*, **40**, 47–48.

NIGRELLI, R. F. and JAKOWSKA, S. (1960). Effects of holothurin, a steroid sapanin, from the Bahamian sea cucumber (*Actinopyga agassizi*) on various biological systems. *Ann. N.Y. Acad. Sci.*, **90** (3), 884–892.

OBERZILL, W. (1967). *Mikrobiologische Analytik*, Hans Carl, Nürnberg.

OLSEN, P. E., MARETZKI, A. and ALMODOVAR, L. A. (1964). An investigation of antimicrobial substances from marine algae. *Botanica mar.*, **6**, 224–232.

OPPENHEIMER, C. H. (Ed.) (1963). *Marine Microbiology*, C. C. Thomas, Springfield, Ill.

OPPENHEIMER, C. H. (Ed.) (1966). *Marine Biology, Proceedings of the 2nd International Interdisciplinary Conference*, New York Academy of Sciences, New York.

OPPENHEIMER, C. H. (Ed.) (1968). *Marine Biology, Proceedings of the 4th International Interdisciplinary Conference*, New York Academy of Sciences, New York.

OSTERTAG, H. (1952). Cellulose-Abbau durch Bakterien. *Ergebn. Hyg. Bakt.*, **27**, 149–322.

PARSONS, T. R. and STRICKLAND, J. D. H. (1962). On the production of particulate organic carbon by heterotrophic processes in sea water. *Deep Sea Res.*, **8**, 211–222.

PELCZAR, M. H. and REID, R. D. (1965). *Microbiology*, McGraw-Hill, St. Louis.

PINTNER, I. and PROVASOLI. L. (1963). Nutritional characteristics of some chrysomonads. In C. H. Oppenheimer (Ed.), *Marine Microbiology*. C. C. Thomas, Springfield, Ill. pp. 114–121.

PLUNKETT, M. A. and RAKESTRAW, N. W. (1955). Dissolved organic matter in the sea. *Deep Sea Res.* (Suppl.), **3**, 12–14.

PRAT, J. and GIRAUD, A. (1964). *The Pollution of Water by Detergents*. Organisation for Economic Co-operation and Development, Paris.

PRATT, D. M. (1966). Competition between *Skeletonema costatum* and *Olisthodiscus luteus* in Narragansett Bay and in culture. *Limnol. Oceanogr.*, **11**, 447–455.

PRIESTLEY, J. (1779, 1781). *Experiments and Observations Relating to Various Branches of Natural Philosophy; With a Continuation of the Observation on Air*, Vols I and II. J. Johnson, London.

PRINGSHEIM, E. G. (1963). *Farblose Algen*, G. Fischer, Stuttgart.

PRINGSHEIM, E. G. (1967). Zur Physiologie der farblosen Diatomee *Nitzschia putrida*. *Arch. Mikrobiol.*, **56**, 60–67.

PROVASOLI, L. (1958). The effect of plant hormones on *Ulva*. *Biol. Bull. mar. biol. Lab., Woods Hole*, **114**, 375–384.

PROVASOLI. L. (1963). Organic regulation of phytoplankton fertility. In M. N. Hill (Ed.), *The Sea*, Vol. II. Wiley, New York. pp. 165–219.

PROVASOLI, L. and GOLD, K. (1962). Nutrition of the American strain of *Gyrodinium cohnii*. *Arch. Mikrobiol.*, **42**, 196–203.

PROVASOLI, L. and McLAUGHLIN, J. J. A. (1963). Limited heterotrophy of some photosynthetic dinoflagellates. In C. H. Oppenheimer (Ed.), *Marine Microbiology*. C. C. Thomas, Springfield, Ill. pp. 105–113.

PROVASOLI, L. and SHIRAISHI. K. (1959). Axenic cultivation of the brine shrimp *Artemia salina*. *Biol. Bull. mar. biol. Lab., Woods Hole*, **117**, 347–355.

PÜTTER, A. (1909). *Die Ernährung der Wassertiere und der Stoffhaushalt der Gewässer*, G. Fischer, Jena.

RAY, S. M. and ALDRICH, D. V. (1965). *Gymnodinium breve*: Induction of shellfish poisoning in chicks. *Science, N.Y.*, **148**, 1748–1749.

REUSZER, H. W. (1933). Marine bacteria and their role in the cycle of life in the sea. III. The distribution of bacteria in the ocean waters and muds about Cape Cod. *Biol. Bull. mar. biol. Lab., Woods Hole*, **65**, 480–497.

REVIÈRE, J. W. M. LA (1955a). The production of surface active compounds by micro-organisms and its possible significance in oil recovery. I. Some general observations on the change of surface tension in microbial cultures. *Antonie van Leeuwenhoek*, **21**, 1–8.

REVIÈRE, J. W. M. LA (1955b). The production of surface active compounds by micro-organisms and its possible significance in oil recovery. II. On the release of oil from oil-sand mixtures with the aid of sulphate reducing bacteria. *Antonie van Leeuwenhoek*, **21**, 9–27.

RILEY, G. A. (1963). Organic aggregates in sea water and the dynamics of their formation and utilization. *Limnol. Oceanogr.*, **8**, 372–381.

RIPPEL-BALDES, A. (1955). *Grundriss der Mikrobiologie*, Springer, Berlin.

ROUND, F. E. (1968). *Biologie der Algen*, Georg Thieme Verlag, Stuttgart.

RYTHER, J. H. (1954). Inhibitory effects of phytoplankton upon the feeding of *Daphnia magna* with reference to growth, reproduction and survival. *Ecology*, **35**, 522–533.

RYTHER, J. H. and GUILLARD, R. R. L. (1959). Enrichment experiments as a means of studying nutrients limiting to phytoplankton production. *Deep Sea Res.*, **6**, 65–69.

RYTHER, J. H. and MENZEL, D. W. (1959). Light adaptation by marine phytoplankton. *Limnol. Oceanogr.*, **4**, 492–497.

SCHLEGEL, H. G. (1969). *Allgemeine Mikrobiologie*, Georg Thieme Verlag, Stuttgart.

SEKI, H. (1965). Studies on microbial participation to food cycle in the sea. II. Carbohydrate as the only organic source in the microcosm. *J. oceanogr. Soc. Japan*, **20**, 278–285.

SEN, N. and FOGG, G. E. (1966). Effects of glycollate on the growth of a planktonic *Chlorella*. *J. exp. Bot.*, **17**, 417–425.

SIERBUTH, J. McN. (1959). Gastrointestinal microflora of Antarctic birds. *J. Bact.*, **77**, 521–531.

SIEBURTH, J. McN. (1964). Antibacterial substances produced by marine algae. *Devs ind. Microbiol.*, **5**, 124–134.

SIEBURTH, J. McN. (1965a), Role of algae in controlling bacterial populations in estuarine waters. In *Pollutions Marines par les Microorganismes et les Produit Petroliers* (Symposium de Monaco, avril 1964). Commission Internationale pour l'Exploration Scientifique de la Mer Méditerranée, Paris. pp. 217–233.

SIEBURTH, J. McN. (1965b). Bacteriological samplers for air–water and water–sediment interfaces. In *Ocean Science and Ocean Engineering*. Transactions of Joint Conference of Marine Technology Society and the American Society of Limnology and Oceanography, Washington, D.C. pp. 1064–1068.

SIEBURTH, J. McN. (1968). The influence of algal antibiosis on the ecology of marine micro-organisms. In M. R. Droop and E. J. F. Wood (Eds), *Advances in Microbiology of the Sea*, Vol. I. Academic Press, London. pp. 63–94.

SIEBURTH, J. McN. and CONOVER, J. T. (1965). *Sargassum* tannin, an antibiotic which retards fouling. *Nature, Lond.*, **208**, 52–53.

SILSBY, G. C. (1968). The chemistry of detergents. In J. D. Carthy and D. R. Arthur (Eds), *The Biological Effects of Oil Pollution on Littoral Communities*. *Fid Stud.* (Suppl.) **2**, 7–14.

SLOWEY, J. F., JEFFREY, L. M. and HOOD, D. W. (1962). The fatty acid content of ocean water. *Geochim. cosmochim. Acta*, **26**, 607–616.

SMITH, J. E. (Ed.) (1968). '*Torrey Canyon*' *Pollution and Marine Life*, Cambridge University Press, Cambridge.

SPENCER, C. P. (1958). The chemistry of ethylenediamine tetra-acetic acid in sea water. *J. mar. biol. Ass. U.K.*, **37**, 127–144.

STEEMANN NIELSEN, E. and HANSEN, V. G. (1959). Light adaptation in marine phytoplankton populations and its interrelation with temperature. *Physiol. Pl.*, **12**, 353–370.

STREET, H. E., GRIFFITH, D. J., THRESHER, C. L. and OWENS, M. (1958). Ethanol as a carbon source for the growth of *Chlorella vulgaris*. *Nature, Lond.*, **182**, 1360–1361.

SUTCLIFFE, W. H., BAYLOR, E. R. and MENZEL, D. W. (1963). Sea surface chemistry and Langmuir circulation. *Deep Sea Res.*, **10**, 233–243.

SVERDRUP, H. U., JOHNSON, M. W. and FLEMING, R. H. (1959). *The Oceans, Their Physics, Chemistry and General Biology*, Prentice-Hall, Englewood Cliffs, N. J.

SYRETT, P. J. (1966). The kinetics of isocitrate lyase formation on *Chlorella:* Evidence for the promotion of enzyme synthesis by photophosphorylation. *J. exp. Bot.*, **17**, 641–654.

TAMIYA, H. (1963). Cell differentiation in *Chlorella. Symp. Soc. exp. Biol.*, **17**, 188–214.

THIMANN, K. V. (1955). *The Life of Bacteria*, Macmillan, New York.

THIMANN, K. V. and BETH, K. (1959). Action of auxins on *Acetabularia* and the effects of enucleation. *Nature, Lond.*, **183**, 946–948.

THOMAS, W. H. (1966). Surface nitrogenous nutrients and phytoplankton in the northeastern tropical Pacific Ocean. *Limnol. Oceanogr.*, **11**, 393–400.

TOKUDA, H. (1966). On the culture of a marine diatom *Nitzschia closterium.* Proceedings U.S.-Japan Conference on Cultures and Collections of Algae, Perspectives and Problems, Tokyo, 12–15 September, 1966. (Unpublished)

UDEN, N. VAN (1969). Kinetics of nutrient-limited growth. *A. Rev. Microbiol.*, **23**, 473–486.

UDEN, N. VAN and FELL, J. W. (1968). Marine yeasts. In M. R. Droop and E. J. F. Wood (Eds), *Advances in Microbiology of the Sea*, Vol. I. Academic Press, London. pp. 167–201.

VACCARO, R. F. (1969). The response of natural microbial populations in seawater to organic enrichment. *Limnol. Oceanogr.*, **14**, 726–735.

VACCARO, R. F. and JANNASCH, H. W. (1966). Studies on heterotrophic activity in seawater based on glucose assimilation. *Limnol. Oceanogr.*, **11**, 596–607.

VALLENTYNE, J. R. (1957). The molecular nature of organic matter in lakes and oceans, with lesser reference to sewage and terrestrial soils. *J. Fish. Res. Bd Can.*, **14**, 33–82.

VISHNIAC, H. S. and RILEY, G. A. (1961). Cobolamin and thiamin in Long Island Sound. *Limnol. Oceanogr.*, **6**, 36.

WALLHÄUSSER, K. H. and SCHMIDT, H. (1967). *Sterilisation, Desinfektion, Konservierung, Chemotherapie*, Georg Thieme Verlag, Stuttgart.

WATT, W. D. (1966). Release of dissolved organic material from the cells of phytoplankton populations. *Proc. R. Soc. (B)*, **164**, 521–551.

WATT, W. D. and HAYES, F. R. (1963). Tracer study of the phosphorus cycle in sea water. *Limnol. Oceanogr.*, **8**, 276–285.

WIESSNER, W. and GAFFRON, H. (1964). Role of photosynthesis in the light-induced assimilation of acetate by *Chlamydobotrys. Nature, Lond.*, **201**, 725–726.

WILLIAMS, P. M. (1965). Fatty acids derived from lipids of marine origin. *J. Fish. Res. Bd Can.*, **22**, 1107–1122.

WINTER, J. E. (1969). Über den Einfluss der Nahrungskonzentration und anderer Faktoren auf Filtrierleistung und Nahrungsausnutzung der Muscheln *Arctica islandica* und *Modiolus modiolus. Mar. Biol.*, **4**, 87–135.

WOOD, E. J. F. (1956). Diatoms in the ocean deeps. *Pacif. Sci.*, **10**, 377–381.

WOOD, E. J. F. (1965). *Marine Microbial Ecology*, Reinhold, New York.

WOOD, E. J. F. (1967). *Microbiology of Oceans and Estuaries*, Elsevier, Amsterdam.

ZÄHNER, H. (1965). *Biologie der Antibiotica*, Springer, Berlin.

ZOBELL, C. E. (1943). The effect of solid surfaces upon bacterial activity. *J. Bact.*, **46**, 39–56.

ZOBELL, C. E. (1946). *Marine Microbiology*, Chronica Botanica, Waltham, Mass.

ZOBELL, C. E. (1964). The occurrence, effects and fate of oil polluting the sea. In E. A. Pearson (Ed.), *Advances in Water Pollution Research*, Vol. III. Pergamon Press, Oxford. pp. 85–118.

ZOBELL, C. E. and ANDERSON, D. Q. (1936). Observations on the multiplication of bacteria in different volumes of stored sea water and the influence of oxygen tension and solid surfaces. *Biol. Bull. mar. biol. Lab.*, *Woods Hole*, **71**, 324–342.

11. IONIZING RADIATION

11.0 GENERAL INTRODUCTION

W. A. CHIPMAN

(1) General Aspects of Ionizing Radiation

One of the types of radiation to which all marine life is subjected is ionizing radiation; it constitutes high energy particulate and photon radiation from both natural and artificial nuclear reactions. The ionizations produced in the living substance from such radiation bring about changes in the atomic structure of materials within the cell and may result in observable biological effects. It is generally accepted that the effects of absorption of appreciable amounts of ionizing radiation are damaging and injurious to life processes.

(a) Physical Aspects

Ionizing radiation includes the electromagnetic radiation of γ and X rays and the fast-moving particles, including α particles, deuterons, neutrons, protons, mesons, electrons (β particles), positrons, and neutrinos. They are characteristics of nuclear transformations and interaction of energy with matter, and originate from cosmic sources or from unstable atoms of natural or artificial radio-active materials within the marine environment (sea water, bottom material) and within the organisms themselves.

The types of ionizing radiations differ greatly in their characteristics of mass, charge, etc., and, consequently, in their ability to penetrate and traverse living matter and to produce ionizing reactions. Both γ and X rays represent highly penetrating electromagnetic waves and can affect cells located deep within the organism. They are not ionizing themselves, but produce ionization by secondary interactions. Densely ionizing particles, such as α particles, dissipate their energy quickly and produce effects concentrated along the short path of the particle. The α particle, because of its high specific ionization and short path, may expend a large portion of its energy in a single cell, thus causing a high degree of cell damage. The β particle, or high-speed electron, will travel several hundred times farther through matter than an α particle of the same energy because of its smaller mass and smaller charge. It ionizes directly in its path and undergoes rapid deceleration, which causes the production of X radiation, an additional damaging factor. Within the organism, the emitted β particle may expend all its energy inside a body tissue, traversing many cells.

(b) Biological Aspects

One can consider biological responses to radiation as functional and structural changes of the total organism, whether it be unicellular or multicellular. In multicellular organisms, the interactions between cells are responsible for the well-being of the entire individual. Basically, radiation affects the complex molecules comprising functional and structural components of cells.

The primary effects of ionizing radiation are alterations of atomic structures or of charges within molecules of the living substance, which disrupt the binding energy and cause molecular breakage and fragmentation. Charged fragments react with adjacent atoms and molecules, new molecules are formed, and later, chemical interactions may occur involving the uncharged fragments of the original molecule.

If damage occurs in the molecule which absorbed the ionization energy (visualized as a particle hitting a target), the radiation injury may be termed a **direct effect**. If molecular damage is brought about by chemical reactions of free radicals formed as a result of the primary action of radiation, the radiation injury may be termed an **indirect effect**. Since the living material irradiated consists mostly of water, direct effects may be alterations of the bonds between hydrogen and oxygen atoms forming strongly reactive products, such as hydrogen dioxide (HO_2) and hydrogen peroxide (H_2O_2). These powerful products can break down the highly complex protein molecules and thus affect enzyme action, permeability of cell walls, etc. With damage to molecules composing cell walls, either via direct or indirect effects, cell products may pass readily between cells, and this interchange of material can result in temporary or permanent injury to the cell. Thus biological effects of radiation may be localized within the cells irradiated, or they may be transmitted to body cells remote from the point of injury.

(2) Measuring Ionizing Radiation: Methods

The unit of dose most frequently given for X or γ radiation is the **roentgen (r)**. It is defined as the quantity of X and γ radiation which will produce 1 electrostatic unit of charge, either negative or positive, in 1 cm³ of air at standard temperature and pressure (0°C and 760 mm Hg). One electrostatic unit of charge is equivalent to $2 \cdot 083 \times 10^9$ ion pairs; the roentgen, therefore, is also equivalent to the production of $2 \cdot 083 \times 10^9$ ion pairs per cm³ of air. Converted to ion pairs per g, this figure becomes $1 \cdot 61 \times 10^{12}$ ion pairs per g, which represents an energy of 83 ergs. Since soft body tissues can absorb 10 ergs more than air, 1 roentgen equals 93 ergs per g soft tissue, or 1 rep (roentgen equivalent, physical). 1000 roentgen are also referred to as 1 Kr.

Another unit of dose is the **rad** (radiation absorbed dose) which is a measure of the absorbed energy of any type of ionizing radiation in any medium. One rad represents an energy of absorption of 100 ergs/g. One rad, with energy absorption of 100 ergs/g of matter, is only slightly larger than the more familiar roentgen unit; it is approximately the energy deposited by $1 \cdot 1$ roentgen of γ rays. The only qualifying statement necessary in regard to the rad is one of type and energy of radiation. 1000 rad are also referred to as 1 Krad.

Exposure to internal radiation may be measured in rads, but it is frequently expressed in micrograms (10^{-6} g) of the material, or in microcuries ($2 \cdot 2 \times 10^6$ disintegrations/min).

The determination of exposure to internal emitters depends on the nature of the radio-active material and the amount and site of the deposition in the organism.

The amount present is very difficult to measure accurately. The quantities may be estimated from excretion rates and biological half-lives of the radionuclides (the times required for biological loss of half of the radio-active material). Radiation damage from internal emitters is often similar in appearance to that caused by comparable amounts of external radiation.

(3) Radiation Sources and Levels of Exposure in Oceans and Coastal Waters

Marine life has been subjected to small doses of ionizing radiation from natural sources over geological periods of time. In recent years, man has added slightly to the radio-activity of oceans and coastal waters and their biota as a result of his development of nuclear weapons and the use of nuclear power for peaceful purposes.

Cosmic radiation is an important contributor to the ionizing radiation received by marine organisms, but only at the sea surface or close to it. The ionizing component of these rays decreases markedly with water depth; it is less than one-third the surface value at 10 m, and only slightly more than one-hundreth of it at 100 m. For this reason, marine plants (Chapter 11.2) limited to the lighted surface layer receive greater doses than most marine animals (Chapter 11.3) which are more apt to spend their time at depths below the surface. At sea level the average absolute intensity has been given by FOLSOM and HARLEY (1957) as 35 millirads per year.

Considerable radiation dosage can be received by organisms living within, or closely associated with, the ocean sediments of the sea bottom. These sediments contain appreciable quantities of the intrinsic isotopic components of the earth's crust, such as radio-isotopes of the uranium and thorium series, and particularly those of radium. A complete listing of the quantities of naturally occurring radio-nuclides in marine sediments, as well as in sea water, has been presented by MAUCHLINE and TEMPLETON (1964). FOLSOM and HARLEY (1957) state that micro-organisms buried in true deep-sea sediments must endure exceptionally high exposure to radium; they receive 40 to 620 millirads per year, depending upon the type of sediment.

External ionizing radiation comes to marine organisms also from the radio-isotopes of the sea water in which they live. Sea water has a characteristic natural radio-activity, due to many different primordial radio-isotopes arising from erosion of continental areas and those entering the sea surface from nuclear reactions taking place in the atmosphere from cosmic-ray bombardment of elements. It now also contains small amounts of artificially produced fission products and neutron-induced radionuclides from man's atomic energy activities, mostly those from world-wide fallout of radio-active materials which were injected into the strato-sphere by thermo-nuclear explosions. Sources of these—other than from world-wide fallout—such as close-in fallout from nuclear weapons tests, from chemical reprocessing plants and other nuclear energy installations, contribute for relatively short periods of time only, or add but little to the radio-activity in the sea, and are restricted to rather limited ocean areas. Fairly complete discussions of sources and degree of contamination of the marine environment from artificial radio-active products have been presented by MAUCHLINE and TEMPLETON (1964) and CHIPMAN (1966).

M

Tables 11-1 and 11-2 give the concentrations of some of the more important natural radio-isotopes of sea water and those of several artificially produced isotopes of long half-life added to the sea due to man's nuclear activities. Other radio-isotopes of short half-life, including both fission products and neutron-induced nuclides, may be present in quantities of little importance except for unusual situations. The values given for the long-lived radionuclides added to oceans and coastal waters from contamination may have increased slightly after the observations included in the tables were made, due to the testing of nuclear devices since 1960, but this would be of little significance. It can be seen from Table 11-1 that the main contributor to the radio-activity in the sea is the natural radio-isotope ^{40}K which accounts for more than 90% of the natural radio-activity of ocean water.

Table 11-1

Concentrations of some primordial radio-active isotopes in oceans (After CHIPMAN, 1966; modified)

Isotope	Half-life (years)	Estimated average concentration in ocean water ($\mu\mu$Ci/l)
Thorium-232	$1\cdot42 \times 10^{10}$	<$0\cdot0022$
Thorium-230	8×10^{4}	<$0\cdot0549$
Uranium-235	$7\cdot13 \times 10^{8}$	$0\cdot045$
Uranium-238	$4\cdot5 \times 10^{9}$	$1\cdot0$
Radium-226	$1\cdot622 \times 10^{3}$	$0\cdot098$
Rubidium-87	$4\cdot7 \times 10^{10}$	$2\cdot62$
Potassium-40	$1\cdot25 \times 10^{9}$	277

Table 11-2

Concentrations of some naturally (cosmic-ray) and artificially produced isotopes in oceans (After CHIPMAN, 1966; modified)

Isotopes	Half-life (years)	Estimated concentration in surface ocean water ($\mu\mu$Ci/l)
Natural cosmic-ray produced		
Carbon-14	5570	$0\cdot092–0\cdot138$
Tritium	$12\cdot26$	$0\cdot685–4\cdot89$
Artificially produced		
Carbon-14	5570	$\sim0\cdot0014$
Promethium-147	$2\cdot8$	$0\cdot002–0\cdot028$
Cerium-144	$0\cdot78$	$0\cdot003–0\cdot08$
Cesium-137	28	$0\cdot049–0\cdot12$
Strontium-90	28	$0\cdot086–1\cdot0$
Tritium	$12\cdot26$	$0\cdot98–9\cdot8$

The tissues of the marine biota contain many of the same radio-active isotopes that are characteristically present in sea water. However, in tissues the isotopes are generally present in much greater quantities since biological concentration processes of elements are natural phenomena. The tendency of living matter to collect sources emitting ionizing radiations leads to internal radiation exposure of cellular components which, in many instances, yields much higher radiation dosages than the surrounding sea water. Of the natural radio-active materials present in marine organisms, ^{40}K is, again, the most abundant one, as shown in the tabulation prepared by MAUCHLINE and TEMPLETON (1964). In this tabulation, the ^{40}K values for the living substance range from 1 to 15 $\mu\mu$Ci/g in algae, from 0·1 to 4·0 $\mu\mu$Ci/g in invertebrates, and amount to less than 4·0 $\mu\mu$Ci/g in fishes. The estimations of ^{40}K included in the tabulation are based on chemical determinations of tissue potassium and the known isotopic ratio of the ^{40}K in natural potassium.

Numerous investigators have measured the concentrations of many of the different radionuclides in the tissues of marine plants and animals. There are, of course, tremendous variations in the amounts of radio-activity present since these vary with the species and the environmental situation. Concentrations of fission products and neutron-induced radio-activities are generally quite low and likely to add only very little to the dose of radiation coming from natural sources. The ecological importance of such feeble doses of ionizing radiation received by a marine organism in the natural environment is extremely difficult to assess.

The relative contributions to the total dose of radiation received by a marine organism from external or internal radio-isotope sources can differ with its body size (FOLSOM and HARLEY, 1957). Considering the β particle emission of ^{40}K, the short range of these particles limits the external ^{40}K dose received by a large organism, but may result in a relatively large dose received by a very small organism. Also, due to the fact that a ^{40}K β particle originating inside a small organism will deposit most of its energy outside the organism's body, external ^{40}K may be the only important source even if this radio-isotope is more concentrated within the small organism, a situation very different from that in the large organism where the β particle absorption allows the dose from the internal source to be more important than that from the external ^{40}K.

Table 11-3 gives the relative doses received by large marine fishes and small micro-organisms from the cosmic radiation, from the radio-activity of the ambient sea or sediment and from the internal radio-activity at different locations in their habitat. It emphasizes (i) the attenuation of the ionizing component of cosmic radiation with water depth, (ii) the importance of body size as regards dosage proportions from external versus internal radio-activity (from ^{40}K) and (iii) the intensive external radiation in the bottom clays (from radium). It is obvious that, in the marine environment, the doses of ionizing-radiation received by marine organisms (given in millirads per year) are extremely small.

Table 11-3

Radiation dosage (millirads per year) received by two types of marine organisms from natural sources (After FOLSOM and HARLEY, 1957; modified)

Type of organism and its location in the sea	Type of radiation			Approx. total dose
	Cosmic	Sea	Internal	
Large fishes				
Near sea surface	35	0·9	28	64
At 100 m depth	0·5	0·9	28	30
Micro-organisms (Mean radius 0·01 mm or less)				
Near sea surface	35	3·6	— *	39
At 100 m depth	0·5	3·6	— *	<5
Buried in deep-sea sediment	0·000	40–620**	— *	40–620

* insignificant dose
** dose from radium in clay

11. IONIZING RADIATION

11.1 BACTERIA, FUNGI AND BLUE-GREEN ALGAE

W. A. CHIPMAN

(1) Introduction

The fact that ionizing radiation affects the life processes of micro-organisms is, of course, well known. Numerous studies have been and are being made on growth-inhibiting and lethal effects of ionizing radiation on bacteria and fungi; such studies have a direct bearing on the use of irradiation techniques in the preservation of foods and other agricultural products and in the sterilization of medicinals and medical supplies. A few investigations have also been made on the possible use of ionizing radiations for controlling objectionable growth of blue-green algae in domestic water supplies.

As pointed out and discussed by ZoBELL (1946) and WOOD (1965, 1967), bacteria and fungi play an important ecological role in estuaries and marine environments. In the past, very little attention has been paid to their physiology and ecology. Almost nothing is known of their responses to ionizing radiation. We must resort, therefore, to studies made on bacteria and fungi of non-marine environments. Such an approach appears justified since there is, according to WOOD (1967, p. 41), considerable difficulty in defining the difference between marine and non-marine micro-organisms; they may have the same morphology and enzyme systems. MacLEOD (1965), after intensive study, has not found any characteristics which could be used to separate 'marine bacteria' from bacteria found in other environments.

A review of the literature published prior to 1952 on the effects of ionizing radiation on bacteria has been written by ZELLE and HOLLAENDER (1955), a more recent account on the effects of ionizing radiation on micro-organisms by POLLARD (1966). Reviews on particular aspects of the effects of ionizing radiations on bacteria have been published by BONÉT-MAURY (1963), BRIDGES (1963), DAVYDOFF-ALIBERT (1963), DUPUY (1963), LEY (1963) and BRIDGES and MUNSON (1968). POMPER and ATWOOD (1955) and POMPER (1965) reviewed the effects of ionizing radiations on fungi. The very small amount of literature available on the effects of ionizing radiation on blue-green algae has not yet warranted a review.

(2) Functional Responses

(a) Tolerance

Lethality and inactivation. Relative tolerances of micro-organisms

In micro-organisms, lethal effects of ionizing radiation are commonly assessed on the basis of their ability to multiply in an appropriate growth medium following exposure, relative to unirradiated control series. In bacteria, for instance, plating of known dilutions of a cell suspension on agar containing the desired nutrient medium is followed by relative colony counts made of irradiated and unirradiated samples. The relative colony count is then plotted against dose administered and thus a 'survival curve' obtained. In many instances, plotting of the data, with the dose plotted linearly and the per cent survival logarithmically, results in a straight line, i.e. in an exponential survival curve. In other cases, however, a sigmoidal type of curve may be obtained. Often the dose-survival curve may have a 'shoulder', which can vary greatly in magnitude, and is characterized by a lack of sensitivity for low radiation doses. A typical dose-survival curve for cell cultures exposed to ionizing radiation is presented in Fig. 11-1. The curve is linear, or exponential, except for a small shoulder evidenced at the lower dose levels. The radiosensitivity, based on mitotic death, is given by the slope, D_0, calculated for the 37% survival along the exponential portion of the curve. The greater the D_0

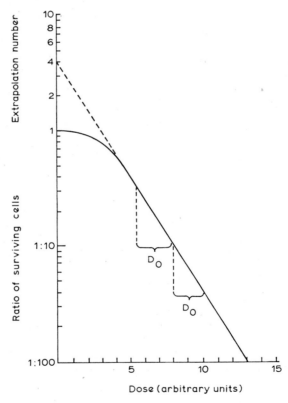

Fig. 11-1: Typical dose-survival curve of irradiated cells in culture. The ratio of surviving cells is plotted logarithmically against arbitrary linear dose units. Cell sensitivity, D_0, is shown as 37% survival (63% mitotic death) on the exponential portion of the curve; in this representation D_0 is 2·2. The extrapolated curve gives 'n', the extrapolation number, expressing the breadth of shoulder of the dose-survival curve. (After ALEXANDER, 1966; modified.)

value, the greater is the tolerance of the cells, and the less steep is the survival curve. The extrapolation number 'n' gives the breadth of the shoulder ($n = 1$: no shoulder).

Further examination of the irradiated micro-organism may be required in order to obtain a better understanding of lethal effects. Sometimes, alternative methods of measuring lethality to that of simple colony counts are necessary. Possible errors in estimating lethal doses of ionizing radiation may result from failure to consider 'delayed death' and from basing dose-survival curves on spore germination as an indicator of eventual colony formation on agar substrates (SCHWINGHAMER, 1958). Working with rust fungi (obligate parasites), SCHWINGHAMER found that the dose required for complete inhibition of germination was far in excess of the limiting dose for 'ultimate death' as determined on the host plant.

Bacteria. THORNLEY (1963a) discusses the radiation tolerance among bacteria. She states that the doses required for 90% inactivation of different bacteria vary by a factor of more than fifty. *Micrococcus radiodurans* and some of the clostridial spores (particularly those of *Clostridium botulinum* type A) are the most resistant bacteria. Spores of some *Bacillus* species are moderately resistant. Among vegetative bacteria, the most tolerant strains (micrococci and streptococci) are gram positive, while the most sensitive ones (pseudomonads and Flavobacteria) are gram-negative. Many genera of both groups occupy an intermediate position. According to DUPUY (1963), the less sensitive gram-negative bacteria have effective doses (LD_{99}) ranging from 25 Krads to 250 Krads. Among the particularly sensitive species are *Pseudomonas aeruginosa*, *Escherichia coli*, and *Serratia marcescens*. Of the more tolerant species (including those of the genus *Bacillus*), *Diplococcus pneumoniae* is perhaps the most tolerant one with a sterilizing dose that may reach 500 Krads. Very tolerant are species of *Micrococcus* and spores of the genera *Bacillus* and *Clostridium*.

A very high level of tolerance to ionizing radiation has been reported by FOWLER and co-authors (*in:* THATCHER, 1963) for a pseudomonad (a group which is normally radiosensitive). This bacterium was found in the coolant water circulating in proximity to fuel rods of a U.S. nuclear reactor where the radiation was in excess of 10^6 to 10^7 rep (roentgen equivalent physical units; Chapter 11.0). In another instance, a species of trichomic bacteria was observed in the secondary circuit of the nuclear reactor Budapest-Csilleberc (Hungary) where the intensity of radiation was approximately 10r/hr (HORTOBAGYI and VIGASSY, 1967).

In a recent paper, BRUCE and co-authors (1969) report the radiation responses of twelve strains, representing four families and eight genera of bacteria. Their data are particularly valuable for comparative purposes since the observations were made under comparable conditions. The bacteria exhibited a range of LD_{90} from 4.5 to 450 Kr with differently shaped survival curves. From their data, Table 11-4 has been prepared, showing the bacteria studied, the LD_{90} dose, and the D_0 dose.

Spores of the two genera *Bacillus* and *Clostridium*, although the first is aerobic and the second anaerobic, show the same degree of tolerance. PEPPER and co-authors (1956) demonstrated that 24 species of the 25 studied require more than 1000 Krad, and 2 species more than 2,000 Krad, for their destruction. This high tolerance is related to their resistance as spores, since KOH and co-authors (1956)

Table 11-4

Relative tolerance of bacteria to ionizing radiation (Based on data by BRUCE and co-authors, 1969)

Bacteria	LD_{90} (Kr)	D_0 (Kr)
Micrococcus radiodurans	450	70
,, ,, wild	735	91
,, ,, UV-38	205	45
,, ,, UV-17	105	13
Bacillus cereus var. *mycoides*	83	36
Sarcina lutea	60	9·0
Escherichia coli B/r (ORNL)	37	15
,, ,, B/r (CSH)	27	13
Staphylococcus aureus var. *albus*	27	7·8
Achromobacter/Alcaligenes sp.	16	5·2
Escherichia coli B	8·0	4·5
Proteus mirabilis	7·7	3·0
Paracolobactrum intermedium	7·2	3·0
Escherichia coli B_{s-1}	4·5	1·9
Bacillus subtilis	21	3·3

found that the vegetative cells of various species of *Bacillus* are as sensitive to radiation as those of other gram-positive bacteria.

As might be expected, since vegetative cells are much less tolerant to ionizing radiation than spores, germinating spores become more sensitive. The use of substances inducing germination resulted in a marked increase in sensitivity of germinating spores of *Clostridium botulinum* PA3679 (KAN and co-authors, 1958), and of *Bacillus cereus* spores (WOESE, 1958).

BRIGGS (1966) measured the tolerance of the spores of six species of *Bacillus* to γ irradiation. *B. cereus*, *B. stearothermophilus* and *B. megatherium* exhibited a pronounced shoulder on their dose-survival curves extending to about 200, 500, and 600 Krads, respectively; the remaining portion of the curve was linear. The other species showed linear relations over the whole dose range. *B. stearothermophilus*, *B. licheniformis* and *B. subtilis* were resistant to radiation, all three having a D_0 value of 220 Krads.

Fungi. BRIDGES and co-authors (1956) compared the relative tolerance of different fungi to the lethal effects of cathode rays. Yeasts were more resistant than moulds. The lethal dose for the *Aspergillus* species investigated was approximately 300,000 reps (roentgen equivalent physical units; Chapter 11.0), for the *Penicillium* species approximately 200,000 reps. The yeasts and moulds investigated were, in general, considered moderately resistant when compared with the tolerances of vegetative cells of non spore-forming bacteria and bacterial spores reported by other investigators.

Numerous studies have been conducted on the tolerance to ionizing radiation of fungi species important in the spoilage of different food products. In most instances, the doses of γ irradiation necessary to give complete sterilization or virtually complete inactivation were investigated. CHOWDHURY and co-authors (1966) report complete sterilization of rice at ^{60}Co γ irradiation with doses ranging from 192 to 225 Kr. Doses of 170 to 185 Kr brought about inactivation of the fungi *Mucor fragilis*, *Aspergillus ochraccous*, *A. ustus*, *A. tamarii*, *A. versicolor*, and *Rhizopus oryzae*. BARKAI-GOLAN and co-authors (1966) report that *Rhizopus nigricans* was quite resistant to γ irradiation. Doses of 100 and 200 Krads did not alter the developmental rate of the fungus in culture as compared to that of unirradiated controls. While a dose of 300 Krads slightly reduced the multiplication rate, the formation of sporangia remained unaffected at this dose. Irradiation of *Botrytis cinerea* with 100 Krads did not affect the developmental rate in culture; but increasing the dose to 200 Krads resulted in an inhibition of the growth of the inoculum in culture. Irradiation of *B. nigricans* with 100 Krads somewhat reduced the growth rate. BARKAI-GOLAN and co-authors (1968) found that a 40 Krad dose of γ irradiation was lethal for the spores of *Trichothecium roseum*. This fungus is extremely radiosensitive. A 100 Krad dose was lethal to *Penicillium cyclopium* and *P. viriclicatum* while it reduced somewhat the growth rate of *Fusarium* sp. In the fungus *Alternaria tenuis*, a 300 Krad dosage was sublethal.

SCHWINGHAMER (1958) made radiation dose-survival evaluations for urediospores of the rust fungi *Melampsora lini*, *Puccinia graminis* f.sp. *tritica*, *P. graminis* f. sp. *avenae*, and *P. coronata* var. *avenae*, exposed to X rays, γ rays, fast neutrons, thermal neutrons, and ultra-violet light. Sigmoidal survival curves were obtained for all species for the five types of radiation used. Urediospores of *Melampsora lini* were notably more radiosensitive than spores of the other three species. An X-ray dose of 100 Kr to *M. lini* had little effect on spore germination, but resulted in almost 100% lethality later, as measured by its inability to infect the host plant.

As in the bacteria, vegetative fungi stages are much more radio-sensitive than spores. The X-ray sensitivity of hyphae (vegetative stage) of the rust fungus *Melampsora lini* in host infections (irradiation of cotyledons of uniformly inoculated flax seedlings) was approximately 20 times that of urediospores (resting stage) irradiated at 52% relative humidity and 10 times that of urediospores irradiated at the 98% level (SCHWINGHAMER, 1958).

Blue-green algae. POLIKARPOV (1966) quotes NIKITIN as stating that a dose of external ionizing radiation of 8 Kr was tolerated by the blue-green alga (Cyanophyta) *Oscillaria limosa*. BONHAM and PALUMBO (1951) found that a dose of 8 Kr of hard X rays (200 KV, 20 MA) reduced the culture population of *Chroococcus* sp. by 50% within 4 to 12 days. A dose of soft X rays (55 KV, 15 MA) averaging about 28 Kr also resulted in 50% lethality. However, BONHAM and PALUMBO were not able to reduce the population of cultures of *Synechococcus* sp. to a 50% level even with doses as high as 100 Kr.

GODWARD (1962) determined the effects of accelerated β rays, i.e. high energy electrons (10^6 e.v.), on different algae. She reports doses of 100 to 200 Kr as allowing ultimate survival of cultures of the blue-green alga *Anabaena* sp.; most of the other algae examined were considerably less tolerant. MARTIN and DERSE (1968) studied the effects of different exposures to ^{60}Co γ irradiation on the population densities of cultures of different species of blue-green algae. Although doses of 50 or 100 Krads reduced the population density somewhat 10 to 15 days after irradiation, doses from 100 to 200 Krads were necessary for significant reductions in populations of *Microcystis aeruginosa* and *Anabaena circinalis*. *Aphanizomenon flos-aquae* was slightly less radioresistant to the γ irradiation; the population was very markedly reduced 7 and 8 days after irradiation with 100 Krads.

Blue-green algae have been shown to grow abundantly when subjected to chronic irradiation in natural streams and other bodies of water contaminated with a mixture of fission and activation products from atomic reactors. In such contaminated waters and bottom sediments, they are relatively high in radio-activity due to their pronounced capacity for concentrating many radionuclides present in such water. In the impounded waters of the Par Pond, which receives the reactor effluent cooling water of the U.S. Atomic Energy Commission Savannah River Plant, blue-green algae contain high concentrations of radionuclides (HARVEY, 1964). Relatively high concentrations, as compared to the low concentrations in the river water, are found in indigenous algae collected routinely near and below the Savannah River Plant, including the abundantly growing blue-green algae *Phormidium* sp. (SAVANNAH RIVER LABORATORY, 1966). HARVEY (1969) reports both the adsorption and absorption of radionuclides in laboratory experiments with *Plectonema boryanum* cultured from waters in the area of the Savannah River Plant. The biota living in the radio-active contaminated White Oak Lake, which receives chronically radio-active pollution from the U.S. Atomic Energy Commission installations at Oak Ridge, Tennessee, is being studied continuously by AUERBACH (1966). Included in these studies are measurements of radionuclide concentrations in the periphyton growing on different substrates. Dominant forms of the algal vegetation, particularly in the deeper less-lighted regions, are species of the blue-green algae genera *Microcystis* and *Oscillatoria*.

A noteworthy instance of blue-green algae tolerating chronic high-level exposures to radiation has been reported by HORTOBAGYI and VIGASSY (1967). These authors found a strain of the blue-green alga *Romeria gracilis* in the primary circuit of the reactor coolant of the 2·5 MW Budapest-Csilleberc research reactor. During operation, very high levels of radiation are attained and, as a consequence of the neutron radiation of the water and its nitrogen and dissolved salts, induced radio-activity is also present. The intensity of the neutron radiation was approximately 10^{13} n/cm^2 sec, and of the γ radiation approximately 2×10^8 r/hr. The specific activity of the primary water was approximately $2·5 \times 10^{-7}$ Ci/l, and, 1·5 days after removal from the reactor, a 50-cm^3 sample gave a dose rate of about 0·5 mr/hr. *R. gracilis* was also present in the water and sediments of the spent fuel holding tanks which received approximately 10^4 r/hr irradiation. In the secondary circuit, besides *R. gracilis*, there were a trichomic bacterium, the green alga *Ankistrodesmus falcatus* var. *spirilliformis* (*Raphidium contortum*), and the blue-green alga *Synechocystis minuscula*. In the secondary circuit, the intensity of radiation was approximately 10 r/hr.

Delayed expression of lethal response

DICKSON (1932) observed that X rays exert little effect on the per cent germination of the ascospores of the fungi *Chaetomium cochliodes* at doses at which the subsequent growth of the spores is affected. With increasing doses, more and more of the spores develop only a small mycelium. In X-ray, γ-ray, and neutron treatments of the urediospores of rust fungi, SCHWINGHAMER (1958) observed that inhibition of spore germination occurs only at dose levels far greater than those required for inhibition of host infection. The inability of urediospores to infest the host plant after germination represents a phenomenon of 'delayed death' similar to that described for many micro-organisms and cells of higher plants (LEA, 1955; KIMBALL, 1957).

DEERING (1968) reported studies on the effects of γ irradiation on populations of the water mould *Blastocladiella emersonii*. One of the responses observed is the development of a giant sporangium of about 200 μ in diameter (normal sporangia are about 60 μ in diameter) sometimes giving off zoospores after a lag. Another type of response is the development of a slower-growing germling with the usual rhizoids, followed by an abrupt cessation of growth at a diameter of about 30 μ, with no zoospore emission.

Tolerance of different life-cycle stages

STAPLETON (1952) showed that the form of the inactivation curve obtained after X irradiation of *Escherichia coli* B/r cells depends on the life-cycle stage. *Micrococcus durans* exhibits a phasic response to X irradiation during its growth cycle (SERIANNI and BRUCE, 1968); the cultures progress from a relatively sensitive state during the exponential portion of the growth cycle to a highly resistant state during the stationary phase. The shift in tolerance is characterized by a marked extension of the shoulder of the dose-survival curve and a threefold increase in the LD_{99} from 250 Kr to over 700 Kr.

Survival of colony-forming germlings of the water mould *Blastocladiella emersonii* irradiated at various stages of their nuclear division up to the four nuclear stage was measured by DEERING (1968). They exhibit the lowest degree of tolerance just after nuclear division, and the highest just before. Of several hypotheses which attempt to explain this variation, differential repair of damage seems to be the most plausible.

ZHESTYANIKOV (1966) studied cytologically the time course of repair and development of radiation injury in cells of *Escherichia coli* B. The cells were exposed to X rays for 1 min at a dose rate of 6 and 15 Krads/min and then incubated for 24 hrs. In one series of tests, the bacteria were kept at 45°C for 1, 2 or 3 hrs and then at 37°C for the remaining time. In another series they were kept first at 37°C and then at 45°C. It was found that the development of irradiation injury is confined to the time of first cell division and that post-irradiation recovery is accomplished during the period of first cell division.

Studies on the tolerance of synchronous cultures of *Escherichia coli* to X irradiation by CLARK (1968) revealed that the tolerance is correlated with the replication cycle of deoxyribonucleic acid (DNA). The sensitivity to X irradiation in the wild type is attributed to the presence of nuclear targets plus DNA repair mechanisms. The effects of nuclear targets were observed in the recombination-deficient (rec-) mutant B/r, but the sensitivity reflected by changes in the slope of inactivation curves was absent. A study of different growth conditions indicated that maximal tolerance to X rays occurs toward the middle of the division cycle.

TRGOVCEVIĆ and KUĆAN (1969) compared the variations of sensitivity to γ irradiation during the progression of *Escherichia coli* B and strain B_{s-2} through various phases of growth with the ability of the same cells to break down their DNA during post-irradiation incubation. Both survival and DNA degradation vary during cell progression in a very similar manner. Among several strains examined, those with higher radioresistance show lower abilities to break down their DNA; moreover, radiosensitive growth phases display more intensive degradation of DNA. A general correlation demonstrating direct proportionality of the percentage degradation and the negative logarithm of survival is obtained and it is proposed that the degradation of DNA in *E. coli* does not constitute a part of the repair process.

Causes of cellular inactivation and radiation injury

The direct effect theory (Chapter 11.0) claims that a sensitive (nuclear) site is affected by an ionizing particle, while the indirect effect theory states that ionization produces chemical changes, such as the formation of free radicals, which account for the radiobiological responses observed. According to POMPER (1965), the most reasonable view of the mechanism of ionizing radiation injury in micro-organisms retains the feature of the target theory of a sensitive site and combines this with the indirect theory of radiation-induced chemical events which, in turn, affect sensitive sites. The modified direct-action model suggests that the primary chemical change occurs directly at the sensitive site.

An interpretation of microbial inactivation and recovery phenomena has been presented by HAYNES (1966). According to this interpretation, reproductive death in bacteria exposed to ionizing radiation arises primarily from the formation of structural defects in DNA which tend to block normal DNA replication. In many cells, viability is enhanced through the action of certain enzymic processes that repair these defects before the onset of replication. The analysis of dose-response curves suggests that survival is determined jointly by the nature and distribution of initial DNA defects and the probability of their subsequent repair. The apparent radiotolerance of cells can be modified by altering either the physico-chemical responses involved in the formation of the defects or by interfering with the bio-chemical repair process.

Tolerance as a function of repair capacities

ZHESTYANIKOV (1967) considers various factors which may account for radio-tolerance of bacterial cells. Of special importance are DNA content, presence of intracellular radioprotective substances, and the repair capacity of the cell. Reasons are presented for the elimination of the first two factors as potent variables, and an hypothesis put forward which attempts to explain variation in cell radiotolerance as a function of repair capacities.

HILL and SIMSON (1961) report that relative differences in radiotolerance between *Escherichia coli* strain B and the mutant strains B/r (more resistant than

B) and two types of B_s (more sensitive) were studied with respect to inactivation by employing ultra-violet radiation, X rays and decay of incorporated ^{32}P. The differences recorded could not be ascribed to variations in number of nuclei or in content of DNA or RNA. Recent evidence, however, indicates quite clearly that certain radiation-induced defects in DNA are susceptible to repair in *Escherichia coli* B/r and that differences in sensitivities can be ascribed to differences in repair capacities. BRIDGES and MUNSON (1968) present evidence which confirms the hypothesis that the basic unreplicated genome is the target for γ radiation damage in *E. coli* and which attributes strain differences in γ-ray sensitivity to differences in repair activities.

It seems plausible that differences in tolerance of different species of bacteria and other micro-organisms to ionizing radiations are probably due primarily to differences in their capacities to repair radiation induced cellular damage. CALKINS (1967) discusses the similarities in the radiation responses of *Escherichia coli* and the protozoan *Tetrahymena pyriformis* and points out that the responses of both organisms indicate the possession of an activated or induced radiation repair system counteracting radiation-induced lesions; he also reports results of studies on the mechanisms for repair of radiation injuries. HAYNES (1966) reviews work, which clearly demonstrates that differences in bacterial cell tolerance to radiation are closely related to differences in their capacities to repair induced damage. Referring to the classical target theory of lethality, he points out that the theory erred in associating lethality so closely with the initial absorption events, and that the theoretical equations of this theory used to describe survival curves should be modified to allow for the possibility of repair. HAYNES presents new mathematical models to explain modifications in the target theory as expressed by different types of dose-inactivation curves. Tables published by ALEXANDER (1966) reveal great variations in tolerance of different bacteria species to ionizing radiation. ALEXANDER reviews evidence which indicates quite clearly that the differences in tolerance are probably due to differences in capabilities to repair radiation-induced damage to cellular DNA. BRUCE and co-authors (1969) compare the radiotolerance of twelve bacteria strains, representing four families and eight genera (Table 11-4) and the binding potential of p-hydroxymercuribenzoate. They report a linear relationship between LD_{90} and p-hydroxymercuribenzoate binding. Their data suggest that the degree of radiotolerance is determined primarily by the presence of a cellular repair system. The radiotolerance of the repair system is, in turn, a function of the sulphydryl content of the cell.

Factors modifying quantitative aspects of tolerance

Type of ionizing radiation. Ionizing radiations may differ greatly in characteristics of mass, charge, etc., and consequently in their ability to penetrate living matter and to produce ionization effects within the cell (Chapter 11.0). The two major classes of radiation, fast electrons and heavy particles (such as protons, deuterons, alpha particles, and others), differ considerably in the amount of energy released and in regard to energy distribution within the irradiated cell. Fast-moving electrons, produced by photons X or γ, or by accelerators, easily become scattered within the irradiated substance and, as a result, the tracks are not straight. From this scattering, and from the secondary ionizations associated with each primary ionization, a pattern of rather randomly produced ionizations results, which is distributed throughout the cell volume. In contrast, the much more slowly moving heavy particles release significantly more energy, mostly along a short straight track.

Ionization along a short straight track, induced by irradiation with heavy particles, can be used advantageously as a line probe to obtain information on the passage through important parts of a micro-organism (POLLARD, 1966). In addition to this, fast-moving particles may be used in certain instances as line probes of varying efficiency. Separation of ionizations along the track can be altered by varying the energy of the fast-moving particle. This variable energy along the track is called linear energy transfer (LET) or ionization density.

Using different types of radiation, an absorbed dose of the same number of rads will cause different biological consequences. To compensate for these differences, one can apply a coefficient of relative biological efficiency (RBE). The RBE is measured in terms of the reported doses producing the same biological effect for the radiation under consideration and for a radiation of reference, usually that for X rays.

Survival rates of *Escherichia coli* and of spores of *Bacillus mesentericus* following β irradiation and exposure to α particles have been determined by LEA and co-authors (1936), to γ rays by LEA and co-authors (1937), and to neutrons and X rays by LEA and co-authors (1941). Using the same bacterial strains employed by LEA and his associates, SPEAR (1944) studied neutron-irradiation lethality; he presented a curve based on his data and all data produced by LEA and his associates; this curve illustrates a systematic decrease in the mean lethal dose (MLD) ratio of *B. mesentericus* spores to *E. coli* cells as the ionization density increases. SAVANT and co-authors (1964) investigated the RBE of protons of 152 MeV based on the dose-survival curve of *Salmonella typhi* and report that the RBE of protons of 152 MeV, and of ^{60}Co γ rays, is about equal to 1. WAMBERSIE (1967), using survival rates of *Escherichia coli* as criterion, measured the RBE of X rays of classical energy and of low energy and found no significant differences. DEERING (1963) observed the lethal effects of heavy ions on a strain of *E. coli*. The radiation sources used included 140-MeV oxygen ions, 40-MeV α particles, 1-MeV electrons, ^{60}Co γ rays, and 250-Kvp X rays. All dose-survival curves obtained were exponential. Within the ranges of experimental errors, all radiations, except oxygen ions, gave the same survival curves as a function of dose. The RBE for oxygen ions (aerobic irradiation), as compared to the other radiations, was about 0·45. RBE's of less than 1 for the heavy ions have also been observed at very high LET's with many biological systems.

DEWEY and HAYNES (1967), using *Micrococcus radiodurans*, obtained dose-survival curves for a number of 10 MeV/amu* heavy-ion beams. All curves, even for the highest LET available, were found to have a broad shoulder of approximately 1 Mrad. For LET values up to about 1000 MeV-cm²g (i.e. helium, lithium, and boron beams) the apparent limiting slope of the curves increases by a factor of about 3, but there is little change in the magnitude of the shoulder over this range of LET. For LET values greater than 1000 MeV-cm²/2 (i.e. carbon, oxygen, neon,

* amu = atomic mass unit

and argon beams) the shoulder increases and the slope decreases, indicating a general decrease in the RBE over this range of LET.

POWERS and co-authors (1968) obtained survival curves for dry spores of *Bacillus megatherium* following irradiation with stripped atoms accelerated in a heavy-ion linear accelerator. Nine different ions were used, each at 8·3 MeV/ nucleon, with specific energy losses (LET) varying from 5 keV/μ (total) for the ionized deuteron up to 500 keV/μ for the Ne^{10+} ion. For the class of damages seen only when oxygen was present, constant values are obtained with increasing LET values up to 190 keV/μ; beyond this level, the values decrease to zero at approximately 500 keV/μ. Other conditions of exposure gave somewhat different results. According to POWERS and co-authors, the cross-sectional values correspond with actual geometric dimensions of the spores; a high LET particle is lethal in any part of the protoplast; lower cross-sectional values, smaller than morphologically distinct parts of the spores, are yielded by lower LET particles.

Experimental evidence indicates that fungi do not respond to changes in the LET of radiation exposure as do the bacteria. Exposure of fungi to radiations of heavy particles produces higher lethality than exposure to X rays or γ rays. Densely ionizing radiations (neutrons and α particles) are more effective in regard to their lethal action on spores of the fungus *Aspergillus terreus* than γ and X rays (STAPLETON and MARTIN, 1949). They are, however, less effective in inducing mutations. Densely ionizing radiations are, similarly, more effective in producing lethality in spores of *Aspergillus niger* than are more disperse radiations (ZIRKLE, 1940). Survival studies on heterokaryotic conidia of *Neurospora crassa* following exposure to X rays, helium ions, and carbon ions (using a heavy-ion linear accelerator) yielded exponential curves in each instance (DE SERRES and co-authors, 1967). The slopes of the two heavy-ion curves were significantly different from each other and from the X-ray curve. Employing comparable doses, much lower survival levels were obtained with both heavy ions than with X rays. The estimate of RBE's for inactivation of the heterokaryotic conidia amount to 5·3 for carbon ions and to 1·7 for helium ions.

Dose-rate effects. From higher plants and animals it is well known that an acute, quick exposure to ionizing radiation is more effective than the same dose applied over a longer period of time. Almost without exception, the few investigations conducted on dose-rate effects on inactivation of bacteria show no such effect. This is surprising since the reduction in lethality effects, brought about by the lengthening of exposure time, is generally considered to be the result of prompt repair activities initiated almost immediately after radiation injury. A review of the scanty literature on the effects of dose rate on the inactivation of bacteria has been presented by LEY (1963). The following discussion is based on that review.

Variation in the intensity of α particle flux has no effect on the percentages of organisms surviving a given dose (LEA and co-authors, 1936). This fact has been confirmed in experiments during which a given dose was administered in a number of fractions: no effect on survival was observed. Lack of an intensity effect of α particles and X rays was reported by LEA and co-authors (1941). Survival rates of *Escherichia coli* B/r following irradiation over a wide dose range (8000 fold) revealed no dose-rate dependence (BELLAMY and LAWTON, 1954). In mixed suspensions of both vegetative and spore-forming bacteria irradiated with ^{60}Co γ rays at dose rates ranging from 13,000 to 116,000 rep/hr (variation of the distance between sample and radiation source), only the total irradiation dose determined the extent of bacterial inactivation (TARPLEY and co-authors, 1953). Varying the beam current of a Van de Graaff accelerator, GOLDBLITH and co-authors (1953) found no dependence on dose rates of 2000 rep/min or of 3×10^6 rep/min in *Escherichia coli*. Irradiation of *Bacillus subtilis* spores, also accomplished by employing a Van de Graaff accelerator, revealed no rate dependence (EDWARDS and co-authors, 1954).

In contrast to the findings reported above, a dose-rate effect was found in X-ray inactivation of dry spores of *Bacillus megatherium* (POWERS and co-authors, 1958). As the dose rates changed from 300 r/min to 65,000 r/min, the slope of the exponential portion of the inactivation curve increased, indicating that the inactivation efficiency increases with dose rate.

TITANI and co-authors (1958) and PERSHINA and co-authors (1966) studied dose-rate effects by employing ^{60}Co γ sources. Both groups of investigators report the unexpected finding that protracted exposure to the source had a greater bactericidal effects than a comparable dose applied during a brief period. There is some question as to the effects of other environmental factors during the long periods required for the application of low dose rates. It can be calculated from the γ dose rate administered in the experiments of PERSHINA and co-authors that the low dose of 100 Kr would require an exposure of 15·5 hrs in the case of *Escherichia coli*, and a dose of 800 Kr an exposure of about 76 hrs in *Bacillus anthracis*. TITANI and co-authors suspended their test bacteria in distilled water during the long exposure periods.

In fungi, dose-rate effects on survival rates of species of Dematiaceae (Fungi Imperfecti) were observed by ZHDANOVA and P'YANKOV (1967) following ^{60}Co γ irradiation at doses of 300, 400 and 500 Krad, applied in intervals of 50 to 2540 rad/sec.

Role of oxygen. The so-called 'oxygen effect', which refers to the decrease in radiotolerance with increasing oxygen content in the surrounding medium at the time of irradiation, is a well-known, general phenomenon in the responses of all types of organisms to radiation stress. In 1951, HOLLAENDER and co-authors studied the inactivation of *Escherichia coli* as a function of oxygen tension and saturation of the buffer solution in which the cells were suspended. Lowering of the oxygen tension with different gases changed the slope of the survival curve, demonstrating an increase in tolerance to X irradiation. Other authors have verified this finding, notably HOWARD-FLANDERS and ALPER (1957) who irradiated different bacterial strains in the presence of oxygen or nitrogen under rigorously controlled conditions.

The role of ambient oxygen in affecting radiation tolerance has received, and is still receiving, much attention, along with the closely allied problem of chemical protection against radiation from many substances added to the irradiation medium. The numerous pertinent investigations have been discussed in reviews by ZELLE and HOLLAENDER (1955) and by DAVYDOFF-ALIBERT (1963).

From the information available so far, it seems most likely that reduction in oxygen tension or addition of a protective compound influences the indirect effect of radiation, although modification of the direct effect on metabolically important molecules is also conceivable. In the presence of oxygen, irradiation leads to the formation of the radical HO_2, which, in turn, forms peroxide (H_2O_2) or organic peroxides. A reduction in oxygen tension would reduce the concentrations of peroxide and HO_2 radicals formed during irradiation. The protective compounds could exert their effects, either by reducing the oxygen concentration within the cell or its immediate surrounding, or by competing for the active products formed due to radiation.

Importance of water content of the cell. In bacteria, the water content of the cell affects the tolerance to ionizing radiation. *Escherichia coli*, irradiated while suspended in concentrated sucrose, which causes dehydration, exhibits increased tolerance to X rays, indicating that a rather appreciable part of the tolerance capacity is somehow related to the water content in the cell (POLLARD, 1966). BHATTACHARJEE (1961) compared the X-ray tolerance of *E. coli* cells in the wet conditions with that in the dry condition following vacuum desiccation of the cells; the values obtained for 37% survival doses were 3·6 and 12·4 Kr in wet and 24-hr dried samples, respectively. TETTEH and CORMACK (1968) investigated the variation of X-ray tolerance of *Serratia marcescens* with changes in relative humidity under aerobic and anaerobic conditions using filter and aerosol techniques for drying. The resulting radiotolerance of the filter-mounted cells ranged from 0·61 Krad^{-1} at 98% relative humidity to a maximum of 0·31 Krad^{-1} at 80%; it decreased with decreasing relative humidity to 0·28 Krad^{-1} at about 55%, and then increased again to 0·15 Krad^{-1} at 40% and below.

Increased tolerance of bacterial cells to ionizing radiation, when irradiated in a frozen state, also stresses the effect of cell water on the radiation-induced damage. Irradiation of *Staphylococcus aureus* in a frozen state with X rays was reported by BELLAMY and LAWTON (1954) to increase the 37% survival dose from 5000 to 6000 rep for cell suspensions at room temperature to about 25,000 rep for cells irradiated at −78°C. Five strains of bacteria, one of *Pseudomonas*, one of *Alcaligenes*, two of *Streptococcus*, and *Escherichia coli* B/r, were used by MATSUYAMA and co-authors (1964) to study the radiotolerance of vegetative cells to ^{60}Co γ rays at temperatures of 10° to 13°C and −79°C. From the dose-survival curves, the increased radiotolerance in the frozen state, relative to that obtained at room temperature, was lowest for the *Pseudomonas* strain with the highest dose ratio of 8·5.

In fungi, water content likewise affects resistance to irradiation. STAPLETON and HOLLAENDER (1952) found that drying and dehydration reduce the tolerance to X-ray irradiation of *Aspergillus terreus* spores, and that there is an interaction between water content and anoxia. Spores with 25% water content (desiccated) have an LD_{99} of 126 Kr, those with 42% water content (normal) an LD_{99} of 74 Kr, and those with 80% water content (in water suspension) an LD_{99} of 52 Kr. At a dosage of 60 Kr, anoxia permits about threefold more survivors at 42% water content, but about tenfold more survivors at 80%. The tolerance of urediospores of rust fungi to X-ray, γ-ray, and ultra-violet light treatment is markedly affected by spore water content, as controlled by the atmospheric humidity maintained during irradiation (SCHWINGHAMER, 1958). The tolerance remains relatively constant at humidity levels of 52% (about 45% water content) or less, but decreases sharply at higher levels; a twofold decrease occurs at 98% humidity (about 70% water content).

Chemical modifiers. Various chemical agents are known to modify the effects of ionizing radiation on bacterial and fungal cells. Information obtained from the use of such chemical modifiers, which decrease or increase radiation tolerance, provides a better understanding of the mechanisms of the biological responses to radiation in these micro-organisms.

Besides considering alkali halides as agents enhancing the lethal effect of ionizing radiation on micro-organisms, MATSUYAMA and co-authors (1967) briefly discuss other chemical modifiers which increase the bactericidal action of ionizing rays, and present a lengthy list of such agents. Many substances act as protective agents against injurious effects of ionizing radiation; among those most intensively studied are cysteine, cysteamine, BAL, and chloramphenicol. The last named is of particular importance since it inhibits protein synthesis while allowing nucleic acids, mainly RNA, to accumulate.

THORNLEY (1963b) discusses bacterial inactivation as related to metabolic conditions before and after irradiation. She considers pre-irradiation and post-irradiation treatments and points out that post-irradiation treatments can be regarded in three main groups: (i) treatments concerned with energy supply; (ii) treatments with metabolic inhibitors (other than respiratory ones) and with inhibiting conditions in general; and (iii) treatments in which complex media decrease the damage. The latter does not apply to mutation studies, but, for example, to inactivation work with *Escherichia coli* B/r; inactivation appears to be due possibly to the necessity for adaptive enzyme synthesis to repair radiation damage.

(b) Metabolism and Activity

In suspensions of bacterial cells, there is a progressive reduction of vital cell functions with increasing doses of ionizing radiations. The more obvious reductions include the loss of the ability of the cells to divide (evidenced by differences in turbidity of liquid cultures or the number of visible bacteria colonies on solid media following irradiation, as compared to normal unirradiated suspensions), respiratory failure, and loss of motility. This sequence of reductions of vital cell functions is followed, finally, by death and dissolution of the cells.

There are great differences in the doses required to suppress multiplication and those necessary to affect respiration or motility. For example, for *Escherichia coli* in broth medium, it takes 200 Kr to inactivate a culture containing 10 cells per cm³, but several hundred thousands of Kr are required to stop respiration entirely. In *Salmonella typhi*, motility is retained completely up to 800 Kr (BONÉT-MAURY, 1963). A number of authors have pointed out that bacterial cells, sufficiently injured by radiation to inhibit cell division, still retain a number of vital functions (LURIA, 1939; BONÉT-MAURY and co-authors, 1943; BILLEN and co-authors, 1952;

and others). These cellular activities retained are affected by higher doses of ionizing radiation and will be considered below.

Respiration

Studies by BILLEN and co-authors (1952) showed that cells of *Escherichia coli* B/r continue some of their synthetic properties following exposure to X rays of 60 Kr. Manometric measurements of cellular respiration revealed only slight changes, even with the highest dose used (90 Kr). At this maximum dose, respiration was normal for the first 40 mins, thereafter it decreased below that of the control cells.

Bioluminescence

X irradiation of the luminous *Bacterium issatchenkoi* depresses the light-emitting enzymatic luciferin-luciferase system (ZOTIKOV, 1960). Protracted irradiation reduces bioluminescence about twice as strongly as transient irradiation. Doses of the order of $1 \cdot 5 \times 10^6$ r depress bioluminescence intensity by 80 to 85% of the original luminescence value of the bacteria when measured directly after irradiation. A partial and temporary recovery of the light-emitting enzymatic system occurs after irradiation with doses of the order of 100 to 200 Kr.

In *Photobacterium* (*Pseudomonas*) *fisheri*, which is symbiotic on fish, bioluminescence is suppressed rapidly at first, then more slowly within 1 sec of exposure to X rays at dose rates between 2 and 7 Krad per min (JACOBSON, 1966). The effect is essentially the same both *in vivo* and *in vitro*; it appears to be related to a fleeting increase of cellular oxygen uptake, which vanishes after exposure. The increase is a metabolic phenomenon linked to cytochrome respiration, and probably the result of radio-induced interference with oxidative phosphorylation.

Cell permeability

A number of investigators have observed leakage of cell components from irradiated *Escherichia coli* cells into the surrounding medium. BILLEN (1957) reports a leakage of adenosine triphosphate (ATP) following X-ray exposure. The leakage of β galactosidase was reported by POLLARD and VOGLER (1961) to occur after γ irradiation. ACHEY and POLLARD (1967) found leakage of thymine from the irradiated cells. POLLARD and GRADY (1967) obtained indirect evidence that γ irradiation of *E. coli* cells at a dose of 27 Kr is followed by a period during which portions of both DNA and RNA are lost. This was demonstrated by differences in banding in a CsCl density gradient between irradiated and normal cells; one band was at the normal position in the gradient while the other two bands shifted to a lighter region. However, from different checks with labelled cells an examination of the nature of the leakage products made at a later time (POLLARD and WELLER, 1968) indicated that the great majority of the degradation and ejection is caused by degradation of DNA and no good evidence could be given for the concomitant degradation and ejection of messenger RNA.

POLLARD and WELLER (1968) conducted extensive studies on the character and time dependence of this radiation-induced loss by the cell. Their investigations were made, in part, to answer the question as to whether the cell membrane is sensitive to the initial action of ionizing radiation; and, if so, whether the phenomenon of permeability is really causally related to the entire syndrome of radiation damage. They found that leakage of components associated with DNA appear in the medium with no lag. The absence of lag was observed even for cells in which the tracer label was far from the polymerase, indicating that the enzyme causing DNA degradation is not polymerase. Leakage in all cells occurred long after irradiation and the authors suggest that this is due to a developing imbalance in enzyme content due to radiation effects on transcription.

Other cellular responses

A variety of metabolic processes are inhibited by doses considerably greater than those necessary for suppression of the capacity of cells for multiplication (SHEKHTMAN and ZANIN, 1966). The authors describe studies on the effect of ionizing radiation on the absorption of oxygen, assimilation of glucose, increase in dry mass, synthesis of ATP, and the activity of the cytochrome system of intestinal bacilli.

Besides discussing the effects of ionizing radiation on respiration and cell permeability, POLLARD (1966) presents brief reviews on investigations concerned with irradiation effects on other cellular functions, including enzyme formation, DNA formation, uptake of phosphate and sulphate, formation of lipid, incorporation of amino acids, and effects on cellular enzymes themselves. Summarizing these effects of irradiation, POLLARD points out that immediate effects on enzymes and all aspects of the cell, including damage to the membrane, occur at doses as high as 10^6 rad. Immediate effects on ribosomes occur at doses of 10^5 rad, as evidenced by a fall in amino-acid incorporation. Most effects, however, are delayed. Very probably, delayed effects result from a variety of causes, including an increase in free DNA-ase and RNA-ase, incorporation of radiation created analogues in DNA, breaks and crosslinks in DNA, and production of faulty synthesizing enzymes.

Low doses of ionizing radiation applied to bacteria or fungi can initiate biochemical changes, often evidenced promptly after the start of the irradiation. Such a response was shown in the observations of JACOBSON (1966) on the X-ray effects on bioluminescence and respiration in the luminescent *Pseudomonas fisheri*. When applied at low dose rate, X rays increase the catalase activity of conidia of *Aspergillus niger*; a dosage of 5400 rads increases the catalase activity to a maximum (TAI, 1962). FORSSBERG (1943) observed a radiation response of the sporangiophora of *Phycomyces blakesleanus* to very low doses of X rays and γ rays. This was evidenced by rapid growth-rate changes. That these reversible, fluctuating growth-rate changes in the X-irradiated sporangiophores of *Phycomyces* (species not stated) are associated with biochemical cell reactions was demonstrated by FORSSBERG and NOVAK (1960). The initial decrease in growth rate reaches a minimum value about 3 mins after a 1·2 r dose given in 5 secs; the decrease amounts to about 60 to 70% of the normal value. At this stage, or even earlier, acid-labile organic phosphates have increased to about 30% above normal. Growth rate increases, and acid-labile phosphorus decreases in the following 5 to 10 mins, whereupon the steady-state level is eventually reached some 15 to 20 mins after irradiation. A chain of biochemical events, proceeding at a rapid rate, accompanies these growth reactions, indicated by changes in lactic-acid concentration.

Further investigations on radiation responses of *Phycomyces* (species not stated) in regard to growth rate and biochemical changes were made by FORSSBERG and co-authors (1960). They report an interaction between visible light and ionizing rays. Cultures grown under illumination differ from those grown in darkness in respect of some biochemical responses (lactic-acid concentration changes). Since the effects of ionizing radiation on growth-rate changes were found to be the same in cultures grown in light and in dark, some reactions at the biochemical level do not seem to be directly connected with an action on the growth-rate mechanism.

Indication of radiation-induced metabolic changes during the development of a bacterial spore has been reported by GOULD and ORDAL (1968). They found that *Bacillus cereus* spores (strain PX), exposed to γ radiation, become progressively more activated with increasing dose, i.e. they germinate more rapidly in the presence of germinants than unirradiated spores.

(c) Reproduction

Mutation induction

Genetic changes and mutations resulting from exposure to ionizing radiation have been studied extensively in bacteria and fungi. A few investigations refer also to mutation induction by ionizing radiation in blue-green algae. A rather comprehensive review of radiation-induced mutations in bacteria and fungi has been published by ZELLE and HOLLAENDER (1955), some information on mutation induction in fungi by POMPER and ATWOOD (1955). Mutation induction in *Escherichia coli* has been discussed quite fully, including recent investigations, by BRIDGES and MUNSON (1968).

Bacteria. In the (generally asexually reproducing) bacteria, there is a strong indication that the fundamental unit of inheritance is the gene, and that gene mutations are responsible for the heritable variation observed (ZELLE and HOLLAENDER, 1955). Observations of particular importance include mutations exhibiting increased resistance to bactericidal agents, including radiation, and reverse cases of biochemical mutations which lack the ability to synthesize certain nutrients. Strains of these mutants or variants, which have formed spontaneously or were induced by mutagenic agents, have been isolated and maintained in culture. An example is the development of strains of *Escherichia coli* B with different radiotolerances from a given parent strain (WITKIN, 1947).

It has long been known that ionizing radiation increases the number of mutants and variants observed in irradiated bacterial suspensions. Many estimations of such changes in mutation rate and their relation to the dose and type of radiation employed have been made. Previous findings are fully covered in the review by ZELLE and HOLLAENDER (1955).

In a recent study, DEERING (1963) reports differences in mutagenesis caused by high LET particles. Using 140-MeV oxygen ions, 40-MeV α particles, 1-MeV electrons, ^{60}Co γ rays, and 250-kvp X rays, cell survival and reversion to tryptophan independence have been investigated in cultures of *Escherichia coli*, strain WP-2 (tryptophan requiring). Doses giving common survival fractions induce about one-half as many reversions for the oxygen ions as do the other radiation sources tested. This fact indicates that heavy ions have a greater tendency to kill the bacterial cell than to cause partial damage manifested as a mutation. Complete curves were obtained for survival and reversion as a function of radiation dose for the different radiation sources employed.

Fungi. Early work demonstrated the induction of morphological mutations from exposures to X rays. For example, NADSON and PHILIPPOV (1925, 1928) report frequent heterogamic conjugations in cultures of normally isogamic *Mucor genevensis*; they were able to isolate two strains of this mould, one showing an increased amount of zygote formation with a decrease in sporangia, the other exhibiting the reverse condition. Exponential curves relating mutation rate to dose have been obtained in studies on mutation frequency in a number of different fungi (HOLLAENDER and ZIMMER, 1945; SANSOME and co-authors, 1945; STAPLETON and MARTIN, 1949; GILES, 1951).

Frequencies of X-ray induced mutations and the genetic structure in *Neurospora crassa* have been studied by DE SERRES and his co-workers (DE SERRES and OSTERBIND, 1962; DE SERRES, 1964; WEBBER and DE SERRES, 1965). They have developed a specific-locus technique in a genetically marked two-component heterokaryon which permits the recovery of mutations resulting from gene inactivation from either intragenic or extragenic alterations. A comparison was made by DE SERRES and co-authors (1967) on the effects of 250-kvp X rays, 40-MeV helium ions, and 108-MeV carbon ions on mutation induction with a heterokaryon of *Neurospora crassa*. The RBE's for mutation induction at two separate loci in the *ad*-3 region are 1·8 for helium ions, and 5·0 for carbon ions. The relative efficiency of carbon ions for mutation induction is higher in *N. crassa* than in other micro-organisms. This difference was attributed to the recovery of a class of mutations (extragenic alterations) which has not been studied with other test organisms.

Blue-green algae. Techniques have been developed which made it possible to elucidate the genetic system (UNIVERSITY OF ROCHESTER, N.Y., 1964). A number of culture strains were obtained in experiments employing various chemical algicides and radiations, including the β particles from ^{32}P incorporated in the medium.

KUMAR (1964) repeatedly treated the unicellular blue-green alga *Anacystis nidulans* with X rays and ^{32}P during successive subcultures. By comparison with the untreated control strain, he observed that the strain treated with ^{32}P was relatively more resistant to streptomycin. The X-rayed strain was relatively more sensitive to isoniazid than the control strain. In old cultures, the cells of the X-rayed strain were significantly smaller than those of the untreated control strain. KUMAR also studied cell and heterocyst dimensions of the nitrogen-fixing alga *Chlorogoea fritschii* after irradiation. In the irradiated material, the cell diameter and heterocyst breadth proved to be somewhat greater than in the unirradiated material.

In *Anacystis nidulans*, an obligate photo-autotrophic blue-green alga, no pigment mutagenic effect was obtained after irradiation with X rays; however, pigment mutations were produced after treatment with nitrosomethylurea (STOLETOV and co-authors, 1965).

(d) *Distribution*

The doses of ionizing radiation in oceans and coastal waters are extremely small (Chapter 11.0). There is no evidence available which would suggest significant influences of *in-situ* amounts of ionizing radiation on the distributions of bacteria, fungi and blue-green algae. (For light effects on the distribution of micro-organisms consult Chapter 2.1.)

(3) Structural Responses

Ionizing radiation can cause delay or inhibition of cell division; with growth and other cellular activities continuing, structurally abnormal individuals may be formed. Some strains of *Escherichia coli* are known to have a strong tendency to form filaments. These strains are also unusually sensitive to ionizing radiation. LURIA (1939), who studied inactivation curves following irradiation of *E. coli* with α particles and X rays, observed (by microscopic examination) that some of the cells were killed immediately, while others continued to grow without dividing and ultimately developed into long filamentous forms. These filamentous forms either divided a few times and died, or recovered and proceeded to develop normal colonies. Exposure to both α and X rays caused filamentous forms to appear, but the proportion of filamentous forms was higher with X rays.

BAZILL (1967) describes two ways in which filamentous forms may be induced: (i) by treatments which stop DNA replication but not RNA and protein synthesis, such as radiation or thymine starvation; (ii) by agents which do not directly affect DNA synthesis. Weakness or porosity of the outer wall and subsequent inhibition of cell division could be caused by any agent (for example penicillin) which selectively inhibits outer wall synthesis. BAZILL also presents evidence that unbalanced lethal growth in bacteria, following damage by ionizing radiation or thymine starvation, is the result of loss of the ability to synthesize an outer cell wall that is strong enough for its role in cell division.

Water mould *Blastocladiella emersonii*, exposed to γ irradiation, developed a giant sporangium of about 200 μ in diameter instead of the normal sporangium of 60 μ (DEERING, 1968).

(4) Conclusions

While it is true that marine bacteria, fungi, and blue-green algae are subjected to ionizing radiation in their natural environments, both from external sources and from decay of internally deposited radionuclides, nothing is known of ionizing radiation as an environmental factor affecting their life processes. In order to understand how these micro-organisms respond to ionizing radiation, it is necessary to consider the responses of similar types of micro-organism of non-marine environments. We have every reason to assume that the basic responses to ionizing radiation are similar in marine and non-marine representatives.

Bacteria, fungi and blue-green algae are much more tolerant to ionizing radiation than are higher plants (Chapter 11.2) and animals (Chapter 11.3). Doses giving virtually complete inactivation are generally in the hundreds of thousands of rads, and in some instances may reach a million rads or more. Fungi, in general, are more resistant to radiation killing than are vegetative bacteria; maximum radio-tolerance appears to be exhibited by certain bacterial spores.

Different species and strains may differ greatly in their ability to tolerate radiation. Their tolerance may, in addition, vary at different periods of their life cycle and at different stages of cell division.

The best evidence available on the mechanisms of ionizing radiation injury in bacterial cells indicates that the primary site for radiation injury is the DNA, and that radiation produces its biological consequences either by direct effects or by indirect effects through the intermediary of enzymes. Repair of the injury induced is considered an important factor modifying the effects of radiation. Comparable mechanisms exist, presumably, in cells of other micro-organisms.

The literature reviewed indicates quite convincingly that the variations observed in the radiotolerance of different species and strains are attributable to inherent differences in the capacity to repair radiation-induced damage. Differences in the capability to repair injury to vitally important macro-molecules of the irradiated cell are postulated to be responsible also for the development of mutant strains of Escherichia coli which exhibit marked differences in radiotolerance.

A number of physical and chemical factors and the type of radiation may alter the degree of tolerance to ionizing radiation as assessed by dose-inactivation curves. In bacteria, the relative biological efficiency (RBE) of different radiation sources, relative to X-ray effects, is close to 1; there is a decrease in efficiency (RBE's less than 1) when more highly ionizing particles are employed; this is especially evident from the more recent work with high LET particles. In fungi, however, exposure to radiations with heavy particles produces higher lethality rates than exposure to X or γ rays.

Surprisingly, the majority of studies made on dose-rate effects show no greater killing effect in bacteria for a dose given quickly than for a comparable dose applied over a longer period of time. The general fact, established on other organisms, that there is a decrease in radiotolerance in the presence of oxygen holds also for the micro-organisms considered here. There is a marked increase in radiotolerance when bacteria and fungi are irradiated in a dry state. Chemical agents, which are known to alter the radiotolerance of mammalian cells in culture, also increase or decrease the radiotolerance of bacterial cells when added to the medium in which these are suspended.

Ionizing radiation may affect the metabolism of bacteria and fungi (and, presumably, also of blue-green algae) in ways other than that expressed as a change in cell multiplication. It may produce alterations of respiratory rates, cell permeability, biochemical syntheses and enzyme action. The radiation exposures necessary to induce significant changes in these processes are much higher than those delaying or inhibiting cell multiplication or causing reproductive death.

Among the non-lethal effects of ionizing radiation on bacteria, fungi and blue-green algae, perhaps the most intensively studied phenomenon is the production of viable heritable mutations. Mutation rates increase markedly with the radiation dose administered and are related to the type of radiation employed. Changes in mutation induction rate are a function of the ionization density; notable differences are obtained when heavy ions are used.

11. IONIZING RADIATION

11.2 PLANTS

W. A. Chipman

(1) Introduction

Our considerations of the responses to ionizing radiation of marine plants, other than those discussed in the preceding chapter, are limited to algae. A number of the higher plants, such as the salt-marsh grasses *Spartina alterniflora* and *Juncus roemerianus* and others, play an important role in the ecology of an estuary. They are known to accumulate radio-active materials from stratospheric fallout (SCHELSKE and co-authors, 1965) and to contain relatively high concentrations of radionuclides added to experimental ponds (DUKE and co-authors, 1966). Such accumulations can contribute to the passage of radio-activity to the estuarine fauna. Nothing is known, however, of the responses of these marine plants themselves to such accumulations or to external ionizing radiation.

The large marine algae are concentrators of different fallout radionuclides. Many species of seaweeds are of considerable importance in different areas of the world (BONEY, 1965); because of their use as (or in) food, feed and fertilizer, the accumulating ability for radio-active materials has been the subject of many investigations. The uptake of fallout radio-activity has been studied in the edible seaweed *Porphyra tenera* and several other littoral seaweeds of Japan (TSURUGA, 1962). The importance of seaweeds in radio-active waste disposal was demonstrated at the Windscale Atomic Works in England when the radio-activity, accumulated by the edible *Porphyra umbilicalis*, limited the permissible rate of radio-active waste release into the sea (DUNSTER, 1958). Studies of the concentrating ability of marine algae have been going on at Scripps Institution (La Jolla, California, USA) for several years and revealed that the kelp *Macrocystis pyrifera* is a notable radio-activity concentrator (FOLSOM, 1959a,b). FOLSOM presents calculations of approximate dosages from radionuclides contained in this seaweed. The fixation of large amounts of fallout radio-activity by *Sargassum fluitans* and *S. natans* has been reported by ANGINO and co-authors (1965). A number of investigators, including SCOTT (1954), POLIKARPOV (1961, 1966), BARINOV (1964) and GUTKNECHT (1965), have studied the accumulation of specific radionuclides by different species of larger algae.

Many of the smaller species of marine algae are known to be effective concentrators of radionuclides. Following the testing of nuclear devices in the Pacific Ocean, phytoplankton was found to contain extremely high levels of radio-activity (HARLEY, 1956; SEYMOUR and co-authors, 1957). Numerous investigators report uptake and accumulation levels for specific radio-active materials by different species of algae (literature *in*: WALLAUSCHEK and LÜTZEN, 1964).

Although many algae have been grown in laboratory cultures and employed in numerous experimental irradiation studies, there is a great paucity of information on the responses of marine species to radiation. It seems advisable, therefore, to include in this chapter the responses of all classes of algae (except the Cyanophyta which have characteristics very different from the others and have already received attention in the preceding chapter), even though many of the species referred to live in freshwater environments.

GODWARD (1962) summarized the earlier literature on the effects of ionizing radiation on algae. Some studies of viability and growth of several species of green algae following irradiation have been reviewed by HOWARD and HORSLEY (1960).

(2) Functional Responses

(a) *Tolerance*

Lethality and survival in algae

Algae are known to be tolerant to relatively high intensities of ionizing radiation. It is of interest that no great killing of the larger marine algae was evident as a result of the atomic bomb explosions in the test area at Bikini in the Pacific Ocean (BLINKS, 1952). It can be assumed that the algae were exposed to continual and, at first, fairly intense mixed β and γ radiation, which decreased exponentially, depending on the half-life of the fission products. When examined 1 year after the detonations, no marked alteration in the many normal life functions tested was found. The algae did not show an accumulation of radio-activity in their tissues. They were, however, receiving and absorbing—as indicated by the shielding effect of the algae to the survey meter—ionizing radiation from the substrate. At the time of survey, 1 year after the detonations, the substrate measured 2 to 3 times background in the milder radio-active regions, and 50 times background in the 'hotter' coral heads nearest to the blast site.

Experimental irradiation of algae in laboratory cultures has demonstrated that the tolerance of algae to ionizing radiation is generally high. Algae are able to survive exposures many times greater than those inducing severe damage and death in higher plants and animals (Chapter 11.3), and even continue to grow vigorously.

The sensitivity of an irradiated experimental population is commonly expressed in terms of a dosage to which 50% of the exposed individuals succumb, the median lethal dose (LD_{50}). Such values may not have comparable meaning unless the conditions under which they were obtained are qualified. The criteria used as death point by investigators often differ. Not all algal cells die at the same time after irradiation, and the effects may be scored after greatly different time intervals, depending on the characteristics of the species under study and the techniques employed. With populations of proliferating algal cells, investigators of radiation-induced lethality most frequently consider viability on the basis of loss of reproductive integrity, such as failure to produce a chain of a predetermined number of cells, or the failure to produce viable colonies. The tolerance of algae, expressed in LD_{50} values for X and γ radiation, is shown in Table 11-5.

The irradiation dose administered to algal cultures in lethality tests has often caused only apparent extinction of the cultures; a few survivors repopulated the culture if given sufficient time. This repopulation of 'extinct' cultures has been reported by BONHAM and PALUMBO (1951), WICHTERMAN (1957) and GODWARD

Table 11-5

Relative tolerance of algae to X or γ radiation (Based on data from various authors)

LD$_{50}$ (in roentgens)	Alga	Author
2,000*	*Micrasterias truncata*	MARČENKO (1965)
2,500–5,000*	*Brachiomonas submarina*	DUCOFF and co-authors (1964)
3,000–3,400**	*Oedogonium cardiacum* (zoospores)	HORSLEY and FUČIKOVSKY (1961)
4,500	*Chlamydomonas reinhardi*	JACOBSON (1957)
5,000	*Chlamydomonas reinhardi*	NYBOM (1953)
5,000	*Chlamydomonas eugametos*	NYBOM (1953)
5,000	*Chlamydomonas moewusi*	NYBOM (1953)
9,000	*Chlorella ellipsoidea*	SHEVCHENKO (1965)
10,000	*Chlamydomonas* sp.	POSNER and SPARROW (1964)
11,000	*Chlorella vulgaris*	SHEVCHENKO (1965)
11,000	*Ankistrodesmus* sp.	BONHAM and PALUMBO (1951)
12,000	*Dunaliella salina*	RALSTON (1939)
14,500	*Chlorella vulgaris*	ZAKHAROV and TUGARINOV (1964)
16,000	*Mesotaenium caldariorum*	LANGENDORFF and co-authors (1933)
18,000	*Chlorella* sp.	BONHAM and PALUMBO (1951)
23,000	*Chlorella pyrenoidosa*	POSNER and SPARROW (1964)
32,000	*Euglena gracilis*	WICHTERMAN (1957)
40,000†	*Chlorella* sp.	BONHAM and PALUMBO (1951)
60,000†	*Euglena gracilis* (colourless strain)	KASINOVA (1964)
120,000†	*Euglena gracilis* (green strain)	KASINOVA (1964)

* LD$_{37}$; ** rads; † soft X-ray

(1960). For a number of species, the radiation exposures beyond which no cells remain viable have been determined (GODWARD, 1960). In a later publication, GODWARD (1962) uses the term 'ultimate survival' in considering the tolerance of algal species to radiation. 'Ultimate survival' is the capacity of a culture to maintain at least 1 cell which finally proceeds to multiply at a normal rate after the culture has received a nearly lethal dose and then been allowed to recover for several weeks under conditions favourable for growth.

Some of the findings of GODWARD (1962) and her associates on the β-radiation doses allowing 'ultimate survival' of cultures for a number of different algae (highest dose allowing survival, in rads) are listed below:

1,000,000	*Chlorella pyrenoidosa*
100,000	*Chaetomorpha melagonium* (zoospores)
40,000	*Spirogyra subechinata*
20,000–50,000	*Eudorina elegans*
20,000–50,000	*Mougeotia* sp.
20,000–50,000	*Zygnema cylindricum*
20,000–50,000	*Cosmarium subtumidum*
15,000	*Spirogyra crassa*
10,000	*Chaetomorpha melagonium*

Other investigators have reported the following highest doses of X rays (in r) allowing 'ultimate survival' of cultures:

600,000	*Chara vulgaris*	MOUTSCHEN (1957)
100,000	*Chlorella* sp.	BONHAM and PALUMBO (1951)
55,000	*Euglena gracilis*	WICHTERMAN (1957)
15,000	*Spirogyra crassa*	GODWARD (1954)

Environmental aspects affecting lethality in algae

Lethality may be a function of the type of ionizing radiation used. BONHAM and PALUMBO (1951) found hard X rays to be approximately twice as efficient in *Chlorella* sp. as soft X rays. GILET and co-authors (1963) observed in *Chlorella pyrenoidosa* that, for the same amount of energy absorbed per cell, soft X rays appear to be twice as efficient as X rays of medium energy after acute exposures. However, hard X rays have been employed almost exclusively for X-irradiation lethality experiments with algae and the results of the different studies are quite generally comparable.

Recovery processes following radiation injury are known to start promptly. They may explain, at least in part, the fact that high radiation intensities administered during short exposures are more damaging to living cells than lower intensities of comparable dosage given over a longer period of time. The effect of various doses of γ radiation, from radium delivered at various intensities, on lethality and colony formation of *Eudorina elegans* was studied by HALBERSTAEDTER and LUNTZ (1929). FORSSBERG (1933) investigated the deleterious effects of fixed X-ray doses delivered at different intensities on the per cent mortality and the retardation of cell division in three species of unicellular algae (*Scenedesmus brasiliensis, Chlorella vulgaris*, and *Mesotaenium calderiorum*). Both mortality and retardation of cell division increased rapidly with increasing intensity of X-ray exposure.

Although chronic exposure to radiation for considerable periods has less effect than a single acute exposure of the same magnitude for the flagellate *Chlamydomonas* sp., POSNER and SPARROW (1964) found *Chlorella pyrenoidosa* to be more tolerant to acute doses of γ radiation than to doses of chronic irradiation. This unexpected result is supported by findings of POSNER (1965) and KÖSSLER (1965). KÖSSLER suggests that the high sensitivity of *Chlorella pyrenoidosa* to chronic exposures may be related to different reaction mechanisms for high and low doses, and related to the metabolic state. ARNAUD (1966) irradiated *Chlorella* sp. cultures continuously during the cellular cycle with γ radiation and observed a sensitization of the cells from continuous irradiation. There was an indication that the total dose received by the cell line plays an important role, which led ARNAUD to the supposition that an accumulation of non-lethal damages was responsible for this radiation sensitization. DAVIES and THORBURN (1968) compared the survival of cells of three strains of *Chlamydomonas reinhardi* exposed at a high or low dose rate to ^{60}Co γ radiation. The results can be accounted for in terms of the repair of sublethal damage at low dose rates. Strains known to differ in genetically controlled repair activities show marked differences in response to dose rate. In the presence of oxygen, a dose-rate effect was found in all strains, but in hypoxia, the mitigating effect of a protracted irradiation was observed only in strains having a certain type of repair activity.

Recovery from radiation injury is evidenced also by differences in the effectiveness of fractionated doses and single acute exposures of similar magnitude. In experiments with *Chlamydomonas reinhardi*, JACOBSON (1957, 1962) noted that when the exposure to the X-ray beam was interrupted, survival rates after a given dose increased. Two irradiations at 4500 r each, separated by a 20-min interval, resulted in higher survival rates than uninterrupted irradiation at 9000 r, the effect being more like that from 8000 r than from 9000 r. The recovery process is temperature dependent; little or no increase in survival from fractionated doses took place at temperatures below 10°C. Further evidence that this recovery may be related to cellular metabolism was revealed when recovery seemed to be somewhat inhibited by such antimetabolites as dinitrophenol and chloramphenicol.

From numerous studies using dose fractionation experiments with cells of a rather wide variety of organisms, it is now well established that repair of sublethal cellular damage from ionizing radiation takes place rapidly and that, given sufficient time for complete recovery, the recovered cells respond to subsequent irradiation in the same way as cells that have not been irradiated. This would mean that the survival curve of recovered cells receiving subsequent like doses of irradiation would be characterized by the same parameters (D_0 and the extrapolation number n) as the survival curve of normal unirradiated cells. DAVIES (1966) studied the patterns and rates of recovery in diploid spores of *Chlamydomonas reinhardi* from γ irradiation from a ^{60}Co source during synchronous growth and cell division. The criterion of survival was the production of a 100-cell colony of haploids from a single diploid cell. Recovery rates and patterns from the irradiation at different stages of the cell cycle, determined by obtaining full survival curves after dose fractionation treatment, showed that D_0 values of irradiated cells are similar to those of unirradiated cells at the same stage of development. The rate of normalization of the curve shoulder varies slightly at different times in the cycle and, furthermore, is more rapid immediately after irradiation than at later times. HILLOVÁ (1967) studied the recovery from X-ray induced damage in *Chlamydomonas reinhardi* with the technique of dose fractionation. After the same conditioning dose, D_0 of recovered cells increased with longer fractionation intervals reaching the maximum value for the interval during which the repair was completed. For longer intervals it remained constant. Complete recovery from sublethal damage was found 2 hrs after the conditioning dose. However, in consequence of the increase in D_0 of recovered cells, the net survival after a given dose of X rays was greater than might have been expected on the basis of theoretical assumptions.

HILLOVÁ and DRÁŠIL (1967) report that the values of D_0 after the acute and fractionated dose (fractionation interval 200 mins and sufficiently long for complete recovery) are 2550 and 7090 rads, respectively. With the use of iodoacetamide, which sensitizes *Chlamydomonas reinhardi* to X rays when present in 1×10^{-4} M, the effect was found to be about three times higher in the recovered cells after the second exposure given in the presence of the chemical. The resulting D_0 had the same value (1600 rads) as after the acute dose given in the presence of iodoacetamide. Incubation of irradiated cells with iodoacetamide during the fractionation interval, sufficiently long for complete recovery, causes inhibition of the recovery mechanism. The fact that, in normal as well as recovered cells, iodoacetamide reduces D_0 to the same value is also considered to be a result of inhibition of the recovery mechanism. BRYANT (1968) reports his findings on the survival and recovery of *Chlamydomonas reinhardi* following separated doses from a linear accelerator electronic beam at a dose rate of 25,000 r/min. Doses were separated by intervals of 10 mins to 4 hrs. There was a rapid increase in recovery during the first half hr between doses, after which the recovery rate decreased, and the curve flattened to a plateau between 2 and 4 hrs. The results indicate that oxygen is required for repair of sublethal damage, and that this repair occurs even if oxygen is excluded during irradiation. Similarly, HOWARD (1968), using split X-ray exposures with *Oedogonium cardiacum* and testing for loss of clonogenic ability, showed that part of the X-ray damage is repaired in a 2-hr interval. The ability of the cells to repair is independent of the presence of oxygen during irradiation. When less than 0·01 μM/l oxygen is present in the water surrounding the cells during the recovery interval, little or no recovery occurs. When 0·12 μM/l oxygen is present, recovery is as complete as in air-saturated water. Recovery, apparently, is a metabolic process which requires an energy source in the cell.

DAVIES and co-authors (1969) studied the survival curves of haploid and diploid cell stages of *Chlamydomonas reinhardi* following acute and fractionated doses of γ rays and heavy ionizing particles having LET (linear energy transfer) values up to 1298 MeVg^{-1} cm^2. The shoulder region of the acute survival curves is the smaller the higher the LET, the reduction being greater for the diploid than for the haploid. In no instance does the response become exponential. Recovery on dose fractionation occurs at all LET's, and the magnitude of the recovery is correlated with the size of the shoulder.

Post-irradiation exposure to light appears to influence the survival of algae subjected to injurious exposures of ionizing radiation; however, present evidence is not fully convincing. JACOBSON (1957) found that illumination applied between fractionated doses of X rays slightly increases survival in *Chlamydomonas reinhardi*. During exposure of *C. reinhardi* to X radiation, illumination decreases survival (NYBOM, 1953; and others). According to GODWARD (1962), visible light, if administered sufficiently soon after irradiation, has some restorative effect on X-irradiated algae. However, immediate exposure of X-irradiated *C. reinhardi* cells to photoreactivating light did not result in photoreactivation (RYZNAR and DRÁŠIL, 1967). The curves for survival in the light and dark were the same.

Post-irradiation dark exposure of cultures for 24 or 48 hrs and subsequent return to alternating light and dark periods had no effect on increasing the survival of the unicellular desmid *Micrasterias truncata*, but a period of 3 to 4 days in the dark following irradiation resulted in a distinct increase in survival when returned to the normal light and dark periods (MARČENKO, 1965).

'Immediate' and 'delayed' death in algae

Doses of ionizing radiation that are not lethal in the sense of bringing about immediate visual evidence of protein coagulation, autolysis, etc., temporarily suppress division in most types of cells. The cells do not enter into recognizable stages of division, although synthesis in preparation for the next division may have been completed. If irradiated at the time that the division process is underway, the cells usually complete that division. Division delay following irradiation has been quite fully discussed by LEA (1955). Death of the cells before completing the first post-irradiation division is often termed 'interphase death'. During division delay, however, many metabolic activities continue, including uptake of nutrients, liberation of energy, and RNA and protein synthesis. Continued growth with somewhat greater cell size in irradiated algal cultures may give evidence of damage leading to ultimate cellular death (PORTER and KNAUSS, 1954; PORTER and WATSON, 1954; JACOBSON, 1957, 1962; POSNER, 1965; and others).

Interphase death, or death without division, was reported by KASINOVA (1964) as the only post-irradiation death seen in cultures of the unicellular green flagellate *Euglena gracilis* following exposure to soft X rays. HELLWIG (1963) observed the effects of X rays on single-cell division of the diatom *Nitzschia liniaris* and reports that with low doses, several post-irradiation divisions took place before the mechanisms of cell division came to a halt, but with high doses, only one further division was possible.

In the colonial green alga *Pandorina morum*, radiation-induced division delay apparently does not occur. HALBERSTAEDTER and BACK (1942) found that, although large doses (300,000 r) of X ray are needed to bring about loss of motility and cytolysis within 45 mins (immediate death); doses over 4000 r result in death of the colony at the time it would normally have liberated its daughters. Replication, the simultaneous successive division of each member of the colony, is not affected by X-ray doses sufficient to cause cytolysis and death-at-division in most of the colonies.

In the Chlamydomonidae (Volvocales) in which each cell bursts to liberate four or more motile spores, death may not be preceded by division; but JACOBSON (1962) noted, as did NYBOM (1953), that most cells killed by X irradiation divide before dying. JACOBSON (1957) studied the effect of irradiation on subsequent colony formation in *Chlamydomonas reinhardi* with X rays at a high dose rate. He found that death rarely occurs until one to five cell generations have elapsed. According to JACOBSON (1962), replication is not affected and death usually occurs at a time when 4 or 8 daughter cells have been produced. In the normal division cycle of *Chlamydomonas reinhardi*, 4 or 8 nuclei are produced by successive mitoses before cytoplasmic division begins, at which time a multinucleate cell is rapidly reduced to a colony containing a corresponding number of uninucleate cells. Prior irradiation does not interrupt this series of division events, but prevents the unicellular progeny from undergoing further mitosis. In some cases death is not preceded by division ('early' death, as contrasted with 'late' death, evidenced by failure to produce viable daughter cells). The relative frequencies of 'early' and 'late' death depend on the dose given. DUCOFF and co-authors (1965) established that replication following exposure to X rays occurs in *Brachiomonas submarina* even in individuals that cannot survive to form colonies.

The observations by PORTER and KNAUSS (1954) regarding radiation effects on the growth of cultures of *Chlorella pyrenoidosa*, following exposure to several radio-isotopes in their growth medium, show that time may be required for the development of radiation injury. Thus they found in one instance, where the estimated dose rate from ^3H_2O was 13,000 rep/day, no difference in population numbers at

30 hrs. However, dose-dependent reduction in cell numbers occurred at a later time; this reduction was apparent at 48 hrs after dose rates of only 1600 rep/day.

HORSLEY and FUČIKOVSKY (1963) report variations in radiosensitivity during the cell cycle of *Oedogonium cardiacum*. Measuring survival following X irradiation of synchronously developing zoospores, they found a 6-fold difference in sensitivity throughout the cell cycle. The least sensitive stage was the early interphase, and the most sensitive one the late interphase or pre-prophase. These observations parallel the changes in sensitivity to the induction of chromosome aberrations found in higher plants. HORSLEY and FUČIKOVSKY's single-cell irradiations have also shown that the number of times which non-survivors divide depends upon the stage of the cell at the time of irradiation.

An analysis of lethal responses in *Oedogonium cardiacum* following X irradiation at different cell stages was made by HORSLEY and co-authors (1967). They studied the non-surviving progeny from *O. cardiacum* sporelings following increasing doses of X radiation up to 16,000 r given at different stages during the first cell cycle. Radiotolerance was measured by three expressions of radiation damage, viz. formation of giant cells, nuclear fragmentation and anucleate cells, and is compared with the radiosensitivity of these cells as measured by loss of reproductive ability. HORSLEY and co-authors report a correlation between loss of cell reproductive ability and radiation damage leading to giant cell formation. No correlation was observed between loss of cell reproductive ability and nuclear fragmentation or anucleate cell formation. Loss of cell reproductive ability includes all forms of lethal radiation damage, whereas radiotolerance measured by any single criterion reveals only a part of the total damage.

Relative importance of cytoplasmic and nuclear lesions

The relative importance of cytoplasmic and nuclear lesions in irradiation-induced cell death can be ascertained in different ways. Radiations of limited penetration power may be employed, and nucleus removal, or transplantation of nuclear and cytoplasmic material between irradiated and non-irradiated cells be used.

HOLWECK and LACASSAGNE (1931a, b) employed the low penetrating α particles from a polonium source to investigate the irradiation-produced lesions and the organelles affected in the colourless chlamydomonad *Polytoma uvella*. The lesions produced were classified into four types (A to D); the dose required to produce the observed effects increases from A to D:

(A) This most frequent type is characterized by normal growth and more or less normal divisions leading to sterile daughter cells and subsequent cytolysis; it results from α particles striking the nucleus, not affecting growth but altering the chromatin.

(B) Growth without division, leading to abnormal size, followed by cytolysis; it results from injury to the centrosome, preventing division, but not growth or motility.

(C) Loss of motility, with growth and cytolysis, without division.

(D) This rarest type shows loss of motility and cytolysis without growth; it results, as in C, from damage to the kinetic centres.

In a similar manner, PETROVÁ (1942) compared the effects of α particles from a polonium source on the nucleus and cytoplasm of filaments of the alga *Zygnema* sp. With rays of moderate intensity, and long enough to reach the centrally situated algal nucleus, mortality was proportional to dosage; the stricken cells became abnormal, failed to divide, and ultimately died. When, through the use of cellulose nitrate and mica filters, the penetrating power of the particles was cut down to less than the distance to the cell nucleus, the mortality became very much less and no longer proportional to dosage. In other experiments the dosage was made great enough to cause cell death in the first 24 hrs. Such death was not dependent on penetration of the α particles to the nucleus; consequently it was called 'plasma-death' as opposed to 'nuclear-death'. The dosage required for 'plasma-death' was about 700 times that required for 'nuclear death'. In a more recent study on *Zygnema* sp., PETROVÁ (1963) concludes that the direct effect of ionizing radiation manifests itself by long-lasting damage which can be caused only by a 'hit' of the nucleus. The indirect effect appears as a temporary effect caused by irradiation of the cytoplasm.

Some of the biological and biochemical effects of radiation can be investigated on parts of cells isolated prior to exposure. The cells of the marine alga *Acetabularia mediterranea* are easily enucleated. SIX (1958) studied the effects of X rays on the nucleated cell part of this alga. The irradiation led to a reduction of the regenerative capacity, to a decrease of cyst formation of the regenerated cells, and to a lowering of the viability of the cysts. After a dose of 400,000 r, the regenerative capacity was almost completely destroyed. The capacity for the formation of reproductive gametes was lost after 40,000 r. It was concluded that the observed functions of the nucleated cell part are essentially determined by the nucleus. BACQ and co-authors (1957) report, also from observations on *A. mediterranea*, that the survival capacity of anucleate fragments is less than that of nucleate fragments following X irradiation; the mortality rate of cell and fragments was not proportional to the dose administered; dosages of 3000 r resulted in more rapid death than dosages of 10,000 r and 100,000 r; stimulation of growth and morphogenesis from irradiation was observed, which was not dependent upon the presence of the nucleus. MOUTSCHEN (1957) and GILLET (1963c) report cell elongation from cytoplasmic effects of X irradiation in the internodal cells of *Chara vulgaris* in which the nucleus does not divide. ERRERA and co-authors (1958) conducted studies on the nucleo-cytoplasmic relations both in isolated and nucleus-containing cytoplasm of *A. mediterranea* after X irradiation. They found that X radiation affects survival of nucleated and anuclear fragments. In regard to the apparent greater sensitivity of the nucleus to radiation, they express the view that the nucleus is not more sensitive to ionizing radiation than the cytoplasm, but, inasmuch as it controls essential and often very specific processes, this sensitivity has a large probability of manifesting itself, particularly when phenomena of cell division, survival, and mutation are involved. In the resting cell, the cytoplasmic phenomena may be of great importance, and determine the changes in cellular behaviour after more or less pronounced delays.

Nature of radiation-induced nuclear lesions

It seems evident that ionizing radiation produces reproductive death in algae primarily through direct effects on the nucleus. Its effects on the cytoplasm as a

N

direct target, or its role in the development of nuclear functions, cannot be fully ignored, however. The action of ionizing radiation on the nucleus appears to represent a direct damage to the genetic material. Some evidence of such action has been obtained from investigations on algae.

The DNA content of the nucleus plays an important part in radiation-induced lethality in many types of cells (DAVIES and EVANS, 1966). BERGER (1967), however, in his studies on the ability of synchronized unicellular Volvocales to divide after X irradiation, found no correlation of X-ray tolerance in representatives of different species with their DNA content, but could demonstrate a correlation between X-ray tolerance and division factors in *Haematococcus pluvialis, Chlamydomonas chlamydomonas, C. moewusii, C. eugametos* and *C. reinhardi*.

As there is considerable interspecific variation in the number of nuclei per cell in algae (in some cases, also intraspecific differences during the life cycle), variations in radiotolerance may be related to differences in nuclear organization (HOWARD and HORSLEY, 1960). While differences in nuclear volume or DNA content are not considered, HOWARD and HORSLEY present a brief review of the literature on radiation effects on some species of green algae and conclude that some of the differences in radiotolerance can be attributed to multinuclearity.

Differences in ploidy may alter the sensitivity of cells to radiation. GILLET (1962a, 1963c) studied the effects of apical cell X-ray exposure on survival, growth and morphogenesis of telomes in two chromosome races ($n = 14$ and $n = 28$) of the alga *Chara vulgaris*. He found that the rate of growth is higher and ontogenetic development less affected in the strain with 14 chromosomes; thus the euploid strain appears to be more radiotolerant.

SPARROW and co-authors (1967) published a very informative paper in which cellular radiosensitivity (D_0) was correlated with the chromosome volume of 79 organisms ranging from viruses to higher plants and animals. A plot of D_0 versus interphase chromosome volume results in a series of 8 regression lines whose slopes do not differ significantly from -1. For organisms within each regression group (radiotaxon), D_0 is inversely related to average chromosome volume. The constants indicated by each of the -1 slopes are expressed in terms of amounts of energy (eV) absorbed at D_0 by the average chromosome of each species, and this amount of energy is nearly constant for organisms in each radiotaxon. Organisms in radiotaxon I absorb approximately 200 eV per chromosome at D_0, while those in radiotaxon VIII absorb more than 16,000 times as much. The organisms in these radiotaxa do not show any consistent relation with respect to classical taxonomy, genetic complexity, or level of ploidy. These radiotaxa are probably a reflection of other chromosomal parameters, both structural and functional, the sums of which result in a specific level of energy absorption at D_0. It is concluded that the size of the chromosome is a highly significant index of cellular radiotolerance but that the survival response is also influenced by other variables.

Nuclear lesions can appear as structural changes in chromosomes, and the loss of parts of chromosomes, as a consequence of post-irradiation cell division, may result in death of some or all daughter cells. Structural changes in chromosomes of irradiated algae, including fragmentation and bridge-formation, have been studied by GODWARD (1962) and associates. Chromosome fragmentation was observed in *Eudorina elegans* and *Chaetomorpha melagonium* following irradiation. Fragmentation in *E. elegans* was seen in all divisions of cells that had received a sublethal dose; however, in the cultures which ultimately survived, the chromosomes appeared normal. In *C. melagonium*, nuclei with more than one or two fragmentations were non-viable. Because of peculiarities in the chromosome morphology of some algae, loss of radiation-induced fragments does not occur subsequent to division. Without loss of chromosome fragments upon cell division, continued growth is possible. GODWARD (1954) reported continued growth despite X-ray-induced chromosome fragmentation in the alga *Spirogyra crassa*. Almost all fragments passed to the poles during mitosis. DODGE and GODWARD (1963) studied nuclear effects following X-ray exposures in the dinoflagellate *Prorocentrum micans*. X irradiation caused anaphase bridges to appear; after low doses, these apparently consisted of entire strands; after doses of over 5000 rads, the bridges were composed of large numbers of fragments which, nevertheless, separated into daughter nuclei.

(b) Metabolism and Activity

Metabolism

Photosynthesis and respiration. Compared to cell division and growth, photosynthesis in algae appears to be relatively insensitive to ionizing radiation. Using γ radiation from ^{60}Co on *Chlorella* sp. cells, SPIKES and co-authors (1958) found that dosages of approximately 1,500,000 r were required for complete inhibition of photosynthesis. Dosages in this range partially destroy the chlorophyll in the cell.

Photosynthesis and respiration of a thermophilic strain of *Chlorella pyrenoidosa* were measured by KÖSSLER (1964, 1965) when subjected to either acute or chronic doses of γ radiation. Chronic exposures to 40 r/hr from a ^{137}Cs source caused a depression of respiration after 5 to 6 days in constant darkness. This depression amounted to 30 to 50% of the initial value. With intermittent light, such radiation exposures influenced neither endogenous respiration nor photosynthesis. Decreased respiration caused by γ irradiation could be restored by exposures to light. This photorestoration, after damaging the system by γ rays, was also attained if photosynthesis was blocked by complete removal of CO_2. The effect can be explained as photo-reactivation after incidence of ionizing radiation. When the system was irradiated with acute doses of very high dose rate (^{60}Co, 1000 r/min), in contrast to the results obtained from chronic exposure, high radiation resistance was observed. No depression of respiration was to be seen, and doses above 100,000 r gave an increased CO_2 release. The rate of photosynthesis was not changed even by doses as high as 960,000 r.

ZILL and TOLBERT (1958) report a decreased CO_2 fixation after γ irradiation of *Chlorella pyrenoidosa* cells, which was greater immediately after the irradiation. Normal CO_2 fixation was found after a 5-hr post-irradiation exposure to light. However, continued normal oxygen release, despite decreased CO_2 fixation, suggests independent action of γ radiation on the carbon cycle and the photolytic system.

Permeability of cellular membranes. The large cells of the green alga *Nitella flexilis* are well known as one of the groups of excitable plant cells. Besides maintaining a resting potential across the cellular membrane due to selective permeability of ions, they exhibit the phenomenon of an action potential, or propagated potential, when stimulated by light, heat, pressure, chemicals, or electricity. Rather extensive discussions of resting potentials and action potentials of cells are given by GIESE (1962). Ionizing radiation has now also been shown to be a stimulating agent for action potentials in plant cells. Leaf movement in the sensitive plant *Mimosa pudica*, stimulated by moderate doses of 50 kv X rays (4000 to 5000 r)—the so-called 'radionastic movement', observed by HUG and MILTENBURGER (1962)— is due to the development of an action potential followed by turgor (HUG and co-authors, 1964).

In *Nitella flexilis*, ionizing radiation changes the resting potential of the cell: it alters the selective permeability of the cell membrane. BERGSTRÖM (1962) irradiated young internodal cells of this alga with 5·3 Mev ^{210}Po α particles and measured the resting potential of the cells over a period of 48 hrs after the irradiation. A significant decrease in the potential could be obtained with doses of more than 20,000 rep. With a dose of 50,000 rep, the decrease was 50% of the initial value in about 24 hrs. Electron microscope studies showed that, in all cases in which there was a drop in the potential due to the irradiation, there occurred a separation of the protoplasm from the cell wall.

ESCH and co-authors (1964) studied the effects of soft X rays (50 kv, 18 ma) on the resting potential of cells of *Nitella flexilis*; with doses between 1000 and 40,000 r, the resting potential starts to decrease immediately after the beginning of the irradiation. The decline reaches a maximum several minutes later; then the potential returns to the original level. ESCH (1966), using ^{60}Co γ rays on the internodal cells of this alga, describes the relation of dose to changes in the resting potential. The drop in resting potential, which begins immediately, increases up to approximately 20 mv by 50,000 r. With doses from 50,000 r to 200,000 r, there is no further drop. With doses from 200,000 r up to 800,000 r, the potential decreases faster and faster, and drops suddenly to zero. The potential change at a given dose is greater in a potassium-free bath solution. This result supports the hypothesis that the observed changes in resting potential are caused by a radiation-induced decrease in the mobility of potassium in the cell wall, which is the same as a decreased permeability. Doses higher than 100,000 r apparently cause an increase in over-all permeability of the cell wall, and physiological death of the cell indicated by a sudden potential drop to zero. As pointed out by ESCH, the above findings on radiation-induced changes in resting potential (permeability changes) with dose are in agreement with the observations of GILLET (1964b) on permeability of the cell wall of *Nitella flexilis* measured by the penetration time of NaOH solution into cells following exposure to 50 kv X rays (soft). The permeability increases with increasing dose, and shows an abrupt change to infinite values after doses higher than 750,000 r.

Activity

Cyclosis. GILLET (1962b) reported that X irradiation of the internodal cells of *Nitella flexilis*, at doses of 50,000 or 500,000 r, causes an immediate, but momentary, arrest of cyclosis (cytoplasmic streaming). After irradiation, cyclosis increases, at first following an exponential law, and then slows down so that after an hour a constant value is reached. The different modalities in the re-establishment of the cytoplasmic streaming depend upon the dose received. Irradiation of an internodal cell induces a transitory acceleration of cyclosis in the adjacent internodal cell (GILLET, 1963d). For at least 8 hrs, the streaming speed was higher than in control internodal cells. GILLET suggests that this phenomenon is probably due to the transfer of certain substances from the irradiated cell. Cyclosis in *Nitella flexilis* cells, however, appears to be particularly radioresistant (GILLET, 1964a). Large X-ray doses were necessary to reduce the rate of streaming permanently, and 800,000 r were required to stop it completely.

GILLET (1964b) investigated the relation between the X-ray induced changes in cyclosis and those similarly produced in the cellular permeability of *Nitella flexilis*. It had been observed earlier (GILLET, 1962b) that the resumption of cyclosis following X-ray induced arrest does not allow return to the rate of streaming observed before irradiation; the re-established rate is dose dependent. In this study, following irradiation with various doses of soft X ray, cell permeability was measured from the rate of penetration of a NaOH solution and the rate of cyclosis ascertained. Immediately after irradiation, except for doses of 500,000 r or higher, there is a relation between the changes induced by X rays in the speed of streaming and those observed in cell permeability. After strong irradiation, the structural lesions of the cytoplasm represent the most important features. After a few hours, permeability and cyclosis are correlated, to a certain extent, with ionic exchanges. Cell permeability increases with increasing doses of X ray, but an abrupt change occurs to infinite values after doses higher than 750,000 r.

Restoration of cyclosis, slowed by X irradiation, was reported by GILLET (1965) as following two exponential processes; one corresponds to alteration in proto-plasmic viscosity and the other, inactivation of the motive force. The period of both processes increases similarly with dosage, but it is the second which especially determines the radiosensitivity of cyclosis of the *Nitella flexilis* internodal cell. Other reactions of unknown nature appeared, particularly at doses of 500,000 and 750,000 r, which temporarily reduced the speed of recovery.

GILLET and KLERCKX (1965) employed ^{45}Ca incorporated into the cells of *Nitella flexilis* to study the relation of X-ray induced changes in calcium exchange and those in the rate of cytoplasmic streaming. Exposures of 250,000 r and 500,000 r slow cyclosis to about 75 and 50%, respectively. However, a radio-assay for ^{45}Ca immediately after and during 30 hrs post-irradiation shows no difference in the rate of loss of the radionuclide from the irradiated and the unirradiated control cells.

Flagellate movement. Irradiation with ionizing rays or particles affects the swimming activity of flagellates at doses other than those bringing about complete immobilization and immediate death. Exposure of *Euglena gracilis* to β-particle radiation causes temporary loss of motility (GODWARD, 1962). *Euglena gracilis* cells, irradiated with X rays at doses of 16,500 r and higher, have a greatly altered swimming behaviour (WICHTERMAN, 1955). The irradiated individuals, while still active like the controls, swim in an erratic manner, often in circles. With higher doses, the euglenoid movement slows down, and later the cells become immobilized. In *Polytoma uvella*, irradiated with α particles from a polonium source, with higher radiation exposures the flagella become affected; in some instances, the basal granules of one of the paired flagella was leading to injured cells capable only of rotation (HOLWECK and LACASSAGNE, 1931a, b).

(c) Reproduction

Much of the experimental work discussed previously in this chapter deals with effects of ionizing radiation on asexual processes of reproduction, including the formation of specialized motile zoospores and several types of non-motile spores. Sexual reproduction involving cell and nuclear unions, occurs in most types of algae. Unfortunately, there is almost no information available on responses of gamete production, gamete release or embryonic development of algae to ionizing radiation.

MOUTSCHEN and DAHMEN (1956) studied the effects of various doses of X rays on spermatogenesis following irradiation of antheridia of *Chara vulgaris*. In examining the nuclear changes of pre-mitosis, mitotic, and telophase stages and completed spermatozoids, they observed that most of the aberrations (fragmentation of chromosomes, fusion of chromocentres, amitosis, etc.) induced by irradiation are merely excess frequencies of the normally occurring phenomena. Spermatogenesis can proceed to completion after doses up to 10,000 r, but it becomes more promptly damaged at higher irradiation levels.

LAWRENCE (1965) made observations on the action of a non-lethal dose of γ radiation from a ^{60}Co source on recombination in the algal flagellate *Chlamydomonas reinhardi*. Different samples of zygotes, germinating on membrane filters, were treated with a dose of 1890 rads at 3780 rads/min at 30-min intervals from 3 to 7·5 hrs after the start of germination. 5·5 hrs after the start of germination, recombination was depressed to less than 85% of the control value; after 6·5 hrs, the recombination value increased nearly 10%; irradiation with this dosage at other times after the start of germination did not affect recombination.

(d) Distribution

The doses of ionizing radiation which have been measured in oceans and coastal waters are very low. It seems unlikely, therefore, that marine plants may be significantly affected in their general distribution patterns by ionizing radiation. The information at hand is very limited. (For light effects on plant distribution consult Chapter 2.2.)

(3) Structural Responses

GILLET (1963a) irradiated *Chara vulgaris* with X rays at doses from 1560 r to 100,000 r under screening, in order to direct the irradiation only to selected portions of the growing points. He examined the tissues 6 weeks after the irradiation and classified the morphogenetic and cytological abnormalities found. These abnormalities can be explained by damage to specific cells. The type of abnormality produced depends upon the nature and extent of cellular damage in the irradiated portion of the growing point. All abnormalities observed represent structural simplifications. There is no evidence of genetic changes, and development tends to return to normal patterns in cells derived from damaged cells. GILLET (1963b) discusses the extent of the damage as related to dose and recovery from different exposures. He reports the following cytological damages: supernumery and vesiculated nucleoli, 'stickiness' of chromosomes, failure of the mitotic apparatus to perform normally, clumping, and apparent amitosis.

Frequently, some irradiated cells enlarge without division and grow to extreme sizes; they are referred to as 'giant cells.' X irradiation of cultures of the alga *Mesotaenium caldariorum*, which normally multiplies by a single nuclear division followed by cell division, results in the development of multinucleate giant cells (FORSSBERG, 1934). The effects of X irradiation on the mitotical growth of the vegetative spores of *Oedogonium cardiacum* manifest themselves in easily recognizable giant-cell formations—usually at the distal end of the chain—granulation, and increased density of the cytoplasm (HORSLEY and FUČIKOVSKY, 1961). Giant cells, in which the nucleus was abnormally large (probably polyploid, and apparently unable to divide further) were also formed in cultures of *Chlorella pyrenoidosa* and *Mougeotia* sp. after β irradiation (GODWARD, 1962); all such cells died after a few weeks. Growth to abnormal size, without division, followed by cytolysis was observed for some cells after α particle irradiation of the colourless flagellate *Polytoma uvella* (HOLWECK and LACASSAGNE, 1931a, b).

(4) Conclusions

In view of the scarcity of information on responses of marine plants to ionizing radiation, this chapter draws heavily on knowledge obtained on non-marine species.

The tolerances of algae to ionizing radiation are expressed in terms of dosages that either reduce test population size to the 50% level or prevent 'ultimate survival' of irradiated populations. Algae are more resistant to ionizing radiation than higher plants or animals (Chapter 11.3). Radiotolerance of algae depends on factors which are well-known for affecting resistance to irradiation also in other groups of organisms, and on repair processes following radiation-induced injury.

In cultures of proliferating alga cells, death commonly occurs after the first cell division following irradiation. Such 'delayed death' varies considerably, both in regard to the time period elapsed and the external appearance of the test material. The degree of this variation is due to the different cellular and nuclear organizations and modes of cell multiplication in the different algae species tested. The most apparent effect of ionizing radiation on algae is failure of a population to increase in number as a result of an arrest of cell division which is followed by cellular death. In regard to the cytoplasm of algae cells, both direct and indirect effects of ionizing particles and rays have been observed; such radiation-induced injuries can be important in cell viability. However, the nucleus appears to be less radiotolerant than the cytoplasm. Low radiation exposures of nuclei result in cellular inactivation and ultimate cell death. The critical injury is damage of the genetic material. Chromosome aberrations leading to failure of mitosis are the likely cause of reproductive death.

Many functions of algae are considerably more tolerant to ionizing radiation than cell multiplication. Photosynthesis is especially radioresistant. High doses of radiation are necessary to alter cell growth, cyclosis, and selective permeability of the cell membrane.

Structural changes induced by irradiation with ionizing particles and rays are, of course, of a cellular nature in algae. Continued growth without cell division results in the formation of abnormally large cells ('giant cells'). Increased granulation and changes in cytoplasmic viscosity have been reported as well as abnormal chromosomes, chromosome fragmentations, and bridge-formations.

11. IONIZING RADIATION

11.3 ANIMALS

W. A. Chipman

(1) Introduction

This chapter presents an account of our present knowledge of the responses of marine animals to ionizing radiation. There is no convincing evidence that marine animals respond in any way to ionizing radiation of any type at radiation levels present in their natural environment. It is necessary, therefore, to turn to information produced by experimental irradiation of different types of animals in order to obtain an understanding of the tolerances and the functional and structural responses of marine animals to this type of radiation.

Under the extremely low intensities of ionizing radiation which exist in oceans and coastal waters, it appears unlikely that any damaging effects on individuals or populations would be apparent; this assumption has, so far, proved to be correct. Even under the locally abnormal environmental situation of extremely high radiation and release of radio-active products from the testing of nuclear weapons in the Bikini-Eniwetok area of the Pacific Ocean, there were no gross effects attributable to the radiation or radio-active materials added to the sea water. Also there have been no recognizable mutations in the thousands of fishes that have been examined (DONALDSON, 1964). There was evidence from histological examination of reef fishes from an area close to a test site, that some fishes had damaged thyroids; this could be attributed to relatively large amounts of iodine radio-isotopes taken up following the great abundance of these radio-isotopes in the environment immediately after the detonation of the nuclear device (GORBMAN and JAMES, 1963). No effects on marine organisms have been found which could be attributed to routine discharge into the Irish Sea of thousands of curies of radio-active materials from the Windscale atomic processing plant; surveys have revealed stability of the bottom configuration of mud and sand and the fauna associated with them (MORGAN, 1960, 1962).

However, radio-active materials are present in oceans and coastal waters, and the amounts added from man's activities are increasing, particularly in estuaries and closed-in coastal areas. It is in these areas that extreme environmental stresses may be exerted on the fauna due to sudden and drastic changes in temperature and salinity, and that the complex responses to these interacting factors may be augmented by the addition of radiation stress, thus causing ultimately adverse effects on populations. Ionizing radiation must, therefore, be considered a potentially effective environmental factor in marine environments.

While we find a tremendous number of studies on the uptake and accumulation of radio-active materials by a wide variety of marine animals, both in the natural environment and under laboratory conditions (WALLAUSCHEK and LÜTZEN, 1964; POLIKARPOV, 1966), there is almost no information on the effects of ionizing radiations from such accumulations, or from the external environment, on these organisms. Indeed, only a few studies have been performed on experimental irradiation of marine animals. Adequate treatment of the information available necessitates the inclusion of some observations made on non-marine forms. Many marine animals are closely related to their freshwater-living counterparts, and we may assume that the general nature of their responses to radiation is similar. However, lack of knowledge regarding marine animals should not lead to uncritical extrapolation of results obtained on freshwater-living relatives.

(2) Functional Responses

(a) Tolerance

Tolerance of different animal groups

The tolerance of animals to ionizing radiation is expressed in terms of that radiation dose which causes death of the individuals of a test population. Since not all individuals have the same sensitivity, the tolerance of the test population is expressed in terms of the dose to which half of the irradiated individuals succumb within a given time, the median lethal dose or LD_{50}. For comparative purposes, the period of time ordinarily selected for mammals is 30 days. While this period may be satisfactory for animals with long life spans, it obviously must be shorter for many lower animals with short life spans. For these, periods of weeks or days have been used. For protozoans, the LD_{50} dose is generally expressed in terms of hours after irradiation (e.g., 24, 48 or 72 hrs).

Although many evaluations of tolerance to ionizing radiation of invertebrates and fishes have been made on the basis of LD_{50}–30 day values, such comparisons may not give a true indication of their relative sensitivity in some instances. Injuries following radiation exposure and leading to death may differ with radiation dose. In fishes, there exist several distinct ranges of X-ray doses in which the causes of death are different (EGAMI and co-authors, 1963b). In certain species of marine invertebrates LD_{50}–30 day values were obtained which indicate similar tolerances to radiation, while their LD_{50}–40 day values show one species to be twice as resistant as the other (WHITE and ANGELOVIC, 1966a). These authors point out that median lethal dose–time curves describe the relative tolerances better than the frequently used LD_{50}–30 day values. They recommend determining the LD_{50} doses for a number of different time intervals following irradiation in order to obtain a better understanding of the relative tolerance of different populations or species.

The results obtained by different investigators on the radiotolerance of animal species are not strictly comparable for a number of reasons, including differences in radiation energy administered, dose rate, and other physical and biological conditions under which the observations were made. Comparison of LD_{50} results, however, serves a useful purpose in revealing relative tolerances of different groups of organisms.

Discussing the relative tolerance of aquatic organisms to ionizing radiation, many authors (e.g. DONALDSON, 1964; MAUCHLINE and TEMPLETON, 1964; TEMPLETON, 1965) refer to a table by DONALDSON and FOSTER (1957) showing tolerance ranges for different groups of aquatic organisms; these data are presented

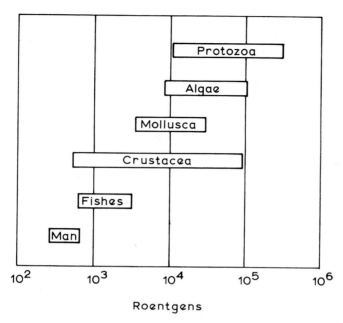

Roentgens

Fig. 11-2: Relative tolerances of different groups of aquatic
organisms to radiation in terms of dosages of X or γ
radiation required to kill 50% of the exposed individuals
in a given period of time. (After DONALDSON, 1964.)

in Fig. 11-2; they indicate that 'primitive' forms are more tolerant to the ionizing
radiations of X and γ rays than those with a more complex functional and
structural organization.

Table 11-6 has been prepared in order to demonstrate the relative tolerance of
some marine invertebrates and fishes to X or γ radiation. It is evident from this
table that marine invertebrates are more radiotolerant than marine fishes and that
some of the higher marine invertebrates approach the sensitivity of the fishes
more closely than do the lower forms of invertebrates.

The greater tolerance of the lower phyla of marine invertebrates to ionizing
radiation is also clearly demonstrated in results obtained on coelenterates and
sponges, not shown in Table 11-6. In coelenterates, BACQ and HERVE (*in*: BACQ
and ALEXANDER, 1961) found a lethal dose of about 300,000 r for *Anemonia
sulcata*; test individuals survived a dose of 100,000 r. In the colonial hydrozoans
Clava multicornis and *Coryne loveni*, KOROTKOVA and TOKIN (1965) found 450,000
rads of β radiation to be lethal, but no adverse effects were noted at 100,000 rads;
Laomedea flexuosa survived doses of 450,000 rads. The same authors found two
species of calcareous sponges, *Leucosolenia complicata* and *Leucosolenia variabilis*,
to be quite radioresistant; the lethal doses were probably greater than 100,000 rads
of β radiation. The experimental conditions did not allow the administration of
higher doses of this radiation to the sponges.

As can be seen from Table 11-6, significant differences exist in radiotolerance
of different species belonging to the same major group. For instance, among the

Table 11-6

Relative tolerance of some marine animals to ionizing radiation (Original)

Group and species	LD$_{50}$ (30 days) in roentgens	Authors
Fishes (juvenile or postlarval)		
Micropogon undulatus	1,050	WHITE and ANGELOVIC (1966a)
Fundulus heteroclitus	1,120	,, ,, ,, ,,
Mugil cephalus	1,450	,, ,, ,, ,,
Lagodon rhomboides	3,000	,, ,, ,, ,,
Eucinostomus sp.	3,500	,, ,, ,, ,,
Paralichthys lethostigma	5,550	,, ,, ,, ,,
Crustaceans		
Palaemonetes pugio	1,500	REES (1962)
Gammarus salinus	1,700	HOPPENHEIT (1969)
Gammarus zaddachi	1,700	,, ,,
Gammarus duebeni (females)	3,500	,, ,,
,, ,, (males)	3,900	,, ,,
Uca pugnax	8,000	REES (1962)
Callinectes sapidus	56,600	ENGEL (1967)
Calliopius laeviusculus	10,000*	BONHAM and PALUMBO (1951)
Allorchestis angustus	10,000*	,, ,, ,, ,,
Artemia salina	14,450**	WHITE and co-authors (1966)
Echinoderms		
Arbacia punctulata	38,900	WHITE and ANGELOVIC (1966a)
Molluscs		
Thais lamellosa	20,000†	BONHAM and PALUMBO (1951)
Urosalpinx cinerea	38,000†	HARGIS and co-authors (1957)
Nassarius obsoletus	37,500	WHITE and ANGELOVIC (1966a)
Crassostrea virginica	99,000	PRICE (1965)
Mercenaria mercenaria	109,000	,, ,,

* LD$_{50}$–5 weeks;   ** LD$_{50}$–25 days for mixed adult population (13,000 rads);   † LD$_{50}$–30 days calculated by WHITE and ANGELOVIC (1966a).

marine fishes tested, the southern flounder *Paralichthys lethostigma* has an LD$_{50}$ over five times greater than that of the most sensitive fish listed, the Atlantic croaker *Micropogon undulatus*. Radiosensitivity is often greatly different even in species of one and the same genus. Comparing typical sigmoid X-ray dose-survival curves obtained for 24-hr periods, WICHTERMAN (1957) found, in seven species of *Paramecium*, LD$_{50}$ values varying from about 160,000 r to nearly 400,000 r.

The enormous differences in tolerance of different animals to ionizing radiation, ranging from hundreds of thousands of r in lower invertebrates to about a thousand r or less in vertebrates, cannot, at present, be explained. As pointed out by BACQ and ALEXANDER (1961), we do not understand at all how some invertebrates achieve resistance to ionizing radiation at doses known to be destructive to proteins as well as other cellular components (and likely to induce 'molecular death').

Effects of fractionated doses and dose rates

Besides the total dose of ionizing radiation administered, there are a number of variables which may affect the tolerance of the test population. There is, first of all, a greater tolerance to fractionated doses of radiation than to a single acute exposure of similar magnitude. Repair processes in cells and tissues, following repeated exposure to ionizing radiations, may proceed in the intervals between successive treatment.

WHITE (1965a) observed that embryos of the euryhaline fish *Fundulus heteroclitus* show a remarkable ability to survive comparatively high doses of X-ray irradiation if these doses are fractionated and given in daily exposures. EGAMI and ETOH (1966) made a similar observation in the fish *Oryzias latipes*: separated doses of radiation cause less damage than a single dose of similar magnitude; they also attempted to measure the rate of recovery from fractionated radiation and found that, if the interval between the two divided doses at 23°C is 3 days or longer, the lethal effect of the irradiation decreases. However, dose fractionation does not reduce detrimental effects if the interval is 1 day or shorter, or if the temperature at which the fish are held subsequent to exposure is 11°C or lower. IWASAKI (1963) reduced metabolic rates of encysted dry eggs (arrested embryonic development) of brine shrimp *Artemia salina* considerably by storage at dry ice temperature during rest periods between exposures to fractionated doses and found no remarkable difference in hatchability in comparison to eggs which had received single exposure treatment.

The nature of the repair processes taking place following ionizing radiation injury in cells and whole organisms has received considerable attention in recent years. It is now well established from investigations on cells from many different biological systems that actively metabolizing cells repair sublethal radiation damage rapidly and that the systems responsible for recovery are not attenuated by repeated exposures. After sufficient time for complete recovery from damage due to acute sublethal exposure to ionizing radiation, the cells respond to another similar exposure exactly as normal un-irradiated cells. The steps involved in the repair processes following whole-body irradiation in multicellular animals do not progress simply with the lapse of time after irradiation; several hypotheses have been formulated to interpret the recovery patterns in a number of organisms.

EGAMI (1969) gives an analysis of the recovery processes in the fish *Oryzias latipes* after examining mortality of the individuals exposed to two-fraction X irradiations at different time intervals under different temperature conditions. Taking into consideration histological and autoradiographical findings (ETOH, 1968; HYODO-TAGUCHI, 1968; RASTOGI and co-authors, 1968) that intestinal epithelium and haematopoietic tissue of goldfish *Carassius auratus* irradiated with 1000 to 2000 r increase mitotic activity and the cell population in these organs after a temporary depression, EGAMI describes and interprets the phases of recovery from radiation injury in *Oryzias latipes* as follows: (i) A decrease in original injury takes place within 3 hrs after irradiation. This corresponds to the rapid recovery reported by other authors and probably reflects intracellular repair processes in critical organs. The duration of this stage is temperature dependent. (ii) The second phase, starting about 3 hrs and lasting up to 48 hrs after irradiation at 24°C, is subject to considerable fluctuations. These might be attributed to differences in radiosensitivity among the cells involved. Such differences could result if, at the time of the first irradiation, the cells in critical tissues were hetero-geneous with respect to the stage of the cell cycle and became synchronized so that they exhibited a difference in susceptibility to the second irradiation at different times after the first. (iii) In the third phase, recovery from radiation injury is evident 3 days or more after the first irradiation at 24°C. In the critical organs damaged, cells may be replaced by newly divided cells from undamaged cell clusters during this period. Recovery processes during this phase can be interpreted on the basis of cell population kinetics in critical organs. If the fish were kept at a low temperature (4°C) during the interval between two fractionations, neither recovery nor repopulation of cells took place in the organs.

Irradiation affects plants (Chapter 11.2) and mammals usually more intensively if delivered quickly at a high dose rate; inactivation of micro-organisms by irradiation, however, is considered to be independent of dose rates (LEA, 1955; LEY, 1963; Chapter 11.1). In marine invertebrates and fishes, it has been demon-strated that a long-term low-level radiation exposure is less dangerous than an acute dose of similar magnitude. Again, this difference is related to repair processes compensating for radiation-induced damage.

Employing single acute exposures to ^{60}Co γ rays, WHITE and ANGELOVIC (1966b) report LD_{50} values for post-larvae of the summer flounder *Paralichthys lethostigma* obtained at different time intervals following irradiation. Dosages of 3075 r killed 50% of the fish after 40 days, 1925 r after 50 days. A cumulative dose of 2245 r, delivered at a dose rate of 12·8 r/hr over a period of 184 hrs, was found to produce a mortality of only 28% during the 120-day observation period (WHITE, 1965b).

ENGEL (1967) observed that the blue crab *Callinectes sapidus* is quite tolerant to continuous ^{60}Co γ-ray exposure. A dose rate of 29·0 rads/hr resulted in significant mortalities only after 30 days of irradiation. Survival rates of crabs exposed to 3·3 or 7·3 rads/hr did not differ from those of the controls during 70 days of exposure. The cumulative exposure of the two lower doses during the experiment amounted to 5105 and 11,502, respectively.

At exposure rates of a very low level—although significantly above the natural background level of a few thousandths of a rad per year—continuous exposure produces no observable effects on marine animals. It is likely that responses to continuous irradiation may be masked due to recovery processes. At very low dose rates, there is a competition between injury and restoration and, if restoration keeps pace with injury, the two processes may be in equilibrium and hence irradiation has no obvious final effect (BACQ and ALEXANDER, 1961).

DONALDSON and BONHAM (1964) subjected chinook (*Oncorhynchus tshawytscha*) and coho (*O. kisutch*) salmon to about 0·5 r/day from a ^{60}Co source of γ rays for a period beginning shortly after fertilization of the eggs and ending when the alevins began to swim; this procedure resulted in a total dose of 33 to 37 r for the chinooks and of 40 r for the cohos. Up to the time of release of the smolts for their seaward migration, little, if any, differences in survival, growth, vertebral numbers and structural anomalies, opercular defects, and sex ratios were observed in experimentals and controls.

BLAYLOCK (1969) compared the fecundity of the common mosquito fish *Gambusia affinis affinis*, that had inhabited a small lake contaminated with radio-active waste for over 100 generations, with that of a control population. The gamma dose rate from the bottom sediments of the contaminated lake to the *G. affinis affinis* population was calculated to be 10·9 rads/day. Other sources of ionizing radiation to the fish were considered to be negligible. In the irradiated population there occurred a significantly larger brood size than in the non-irradiated population, although considerably more dead embryos and abnormalities were observed in the irradiated brood. It is suggested that an increased fecundity may be a means by which a natural population of organisms, having a short life cycle and producing a large number of progeny, can adjust rapidly to an increased mortality caused by radiation.

A number of investigators have maintained species of marine invertebrates and fishes in sea water containing radio-active materials for long periods of time, and demonstrated that these organisms appear to be quite resistant to irradiation during such long exposure. MORGAN (1960) kept plaice and lobsters in radio-active sea water for periods of up to 6 years without observable effects on these animals. RICE (1965) reports that cultures of the marine copepod *Tigriopus californicus*, fed on an algal diet and maintained in sea water containing 45 μCi/l of ^{137}Cs, revealed no deleterious effects and that the population was at about its original level even after 3 years.

Tolerance as a function of metabolic state and age

It has long been established that the radiotolerance of cells is closely correlated with their metabolic state; actively dividing cells, or tissues with cells in a rapid rate of division, are more sensitive to ionizing radiation than resting cells or tissues in which cell multiplication is less active. The dry and encapsuled 'eggs' of the brine shrimp *Artemia salina* (actually the early embryonic stage in arrested development) has a very low metabolic rate; a number of investigators (BONHAM and PALUMBO, 1951; RUGH and CLUGSTON, 1955; and others) have demonstrated that embryos in the dried stage are much more radiotolerant than those resuming development after water uptake.

In many marine invertebrates and vertebrates, radiotolerance increases with age; this increase is particularly pronounced during embryogenic and early post-embryonic life stages. NAKANISHI and co-authors (1964) report a marked increase of radiotolerance with increasing age in the embryo of *Artemia salina* following γ irradiation from ^{60}Co, administered at different time intervals up to 15 hrs after initiation of the development. A pronounced difference in radiotolerance of nauplii and adults of *Artemia salina* was shown by WHITE and co-authors (1966) who found LD_{50}–25 day values amounting to 450 rads in nauplii, but to 13,000 rads in a mixed male and female adult population. Further evidence of increasing tolerance to ionizing radiation with increasing age has been obtained in developing embryos of the teleost *Fundulus heteroclitus* (SOLBERG, 1938). The dose of 800 r of X radiation produces more deleterious effects on unfertilized eggs of the Atlantic silverside *Menidia menidia* than on fertilized eggs (ENGEL and co-authors, 1965a). Acute doses (300 r, the lowest dose administered) of X radiation given to the 2–16 cell stage of *Fundulus heteroclitus* resulted in 59% mortality among the developing embryos, and a dose of 900 r resulted in 94% mortality (WHITE, 1965a). The

LD_{50}–30 days for juveniles of this species (Table 11-6) was, however, 1120 r (WHITE and ANGELOVIC, 1966a).

Tolerance under multivariable conditions

Since the degree of damaging effects of ionizing radiation in marine animals is closely related to their metabolic state and rate of cellular activity, time of death due to radiation is likely to change whenever environmental factors alter the rates of metabolism. In general, the median lethal dose can be expected to decrease with increasing temperature, since metabolic rate tends to increase with temperature. Conversely, the median lethal dose will increase with decreasing temperature which tends to reduce the rate of metabolism and to delay the development of processes leading to radiation damage.

ETOH and EGAMI (1965) subjected the ricefish *Oryzias latipes* to various doses of X rays and maintained the irradiated fish at different water temperatures. They found that the post-irradiation survival time was shortened at higher water temperatures. On the other hand, at a low temperature of 6°C, the majority of the irradiated fish survived more than 120 days even if irradiated with high doses of 8000 r or more. If fish were transferred to water of 23°C after having been kept at low temperatures for a period of 21 days, their survival time was approximately the same as in irradiated fish subjected continuously to 23°C after irradiation. HYODO (1964) observed in the goldfish *Carassius auratus* that 92·4% of the test individuals, which had been exposed to whole body irradiation with 8000 r of X rays, survive more than 100 days if kept at 4°C, while most similarly irradiated individuals die within 10 days if kept continuously at 22°C. Low temperatures inhibit damage of the intestinal epithelium. If fish kept at 4°C for 40 or 100 days were transferred to water of 22°C, they died within about 10 days and exhibited characteristic damage in their intestinal epithelium. In another paper, HYODO (1965) demonstrated that irradiated goldfish kept at low temperatures show neither acute radiation death nor histological changes in their intestinal epithelium over a period of at least 100 days; however, after a much longer period of time, lethal damage and histological changes in the intestinal epithelium do occur; thus radiation injury is retarded rather than avoided at the low test temperatures. Similar observations have been made in regard to the development of damage to haematopoietic tissues of the goldfish *Carassius auratus* by AOKI (1964) and to changes in the median lethal dose of the crucian carp *Carassius carassius* by GROS and co-authors (1958).

ANGELOVIC and co-authors (1966) found that the fish *Fundulus heteroclitus* and the grass shrimp *Palaemonetes pugio*, both typical estuarine species, were more tolerant to γ rays from ^{60}Co at the low salinities characteristic of their normal brackish-water environment; the magnitude of variations in LD_{50} values at different salinities varied as a function of water temperature.

Estuarine animals are often subjected to sudden and severe changes in salinity and temperature, and their survival depends on their ability to adjust to such changes within their environment (KINNE 1966, 1967). It is important to realize that these environmental factors rarely act independently; one factor can greatly affect the tolerance of an organism to another (Chapter 12). The tolerance to ionizing radiation may be greatly influenced by environmental stresses due to other factors. A striking instance, in which extreme environmental changes in salinity and temperature resulted in severe stress and caused high rates of mortality to fish that had previously been subjected to radiation, was described by ENGEL and co-authors (1965a). They subjected juvenile striped mullet *Mugil cephalus* to different sublethal doses of X rays and then maintained the fish in running sea water in the laboratory. Sudden and extensive mortalities occurred in all groups of irradiated fish following a severe storm, drastic reduction in salinity and lowering of temperature. These stresses did not result in death of the unirradiated controls; hence the added stress from ionizing radiation appears to have been the cause of the sudden mortality.

A careful study of the responses to combined effects of temperature (Chapter 3), salinity (Chapter 4), and ionizing radiation on the mortality of the euryhaline fish *Fundulus heteroclitus* was made in a series of experiments by ANGELOVIC and co-authors (1969). Multiple regression analyses were used to derive equations describing the effects of all factors which had a significant influence on the mortality of the fish, at three different time intervals after irradiation. The effects of temperature were significant at all three intervals, and the effects of salinity, radiation dose, and the temperature–salinity interaction were each significant at one or more of the different time intervals after irradiation. Various combinations of different levels of temperature and salinity yielded different LD_{50} values. At the upper end of their temperature range, the fish tolerated more radiation in low salinities; at the lower end of their temperature range, the tolerance was reversed. (For further information on factor combinations consult Chapter 12.)

(b) *Metabolism and Activity*

Metabolism

Rate of growth. A general effect of ionizing radiation on aquatic animals is the retardation of growth, although growth stimulation from very low exposures has been observed in a few instances. Continuous irradiation of the blue crab *Callinectes sapidus* with the γ rays of ^{60}Co retarded growth at the higher dose rates (ENGEL, 1967). ENGEL and DAVIS (1965b) observed that growth rates of the brine shrimp *Artemia salina* were reduced when they were maintained in sea water containing ^{137}Cs (0·2 to 1·0 mCi/l) or ^{32}P (0·6 and 1·23 mc/l). WELANDER and co-authors (1949) found a marked growth retardation in young trout *Salmo gairdnerii*, during their period of fastest growth following X radiation, and FOSTER and co-authors (1949) report that growth rates of young *S. gairdnerii*, during the first year of life, were directly affected by the amount of radiation received by the parent fish prior to spawning. Reduction in growth rate of *S. gairdnerii* fed ^{32}P daily (0·06 μCi/g body weight) became apparent within 17 weeks (WATSON and co-authors, 1959).

Various investigators have noted a tendency of radiation to retard rates of embryonic development of marine invertebrates and fishes. A delay in the hatching of *Artemia salina* following irradiation of the encysted 'egg' has been observed by ENGEL and FLUKE (1962) and NAKAZAWA and YASUMASU (1964). LYNCH (1958) reports slow growth and impeded development of the larvae of the salt-water bryozoan *Bugula turrita* into zoids following X radiation. RUGH and CLUGSTON

(1955) frequently observed developmental retardation due to irradiation in *Fundulus heteroclitus*; the same phenomenon was reported by WHITE (1964) following various doses of X rays administered during the 2 to 16 cell stages.

There is evidence that retardation in developmental rate and organ differentiation resulting from X or γ radiation, as observed in the embryos of marine invertebrates and fishes, begins at the cellular level by a suppression of cell division or an increase in the interval between subsequent divisions. In irradiated embryos of *Fundulus heteroclitus*, in which the rate of growth was retarded, cell division in the interval of rapid mitotic activity is either suppressed completely or retarded for a given period of time (WHITE, 1964). BONHAM and WELANDER (1963) have shown the most radiation-sensitive period of egg development in the silver salmon *Oncorhynchus kisutch* to be a particular stage during the mitosis of the single cell. The interval between fertilization and first cleavage of the egg of the sea-urchin *Arbacia punctulata* is prolonged by X-ray irradiation (HENSHAW, 1932). CATHER (1959) determined the radiosensitivity of the various mitotic stages of the snail *Ilyanassa obsoleta* to X rays by measuring the cleavage delay produced in the eggs. The cleavage delay increases with dose; a dosage of 4000 r in the telophase causes the greatest delay in the next cleavage. HENLEY and COSTELLO (1957) observed some retardation in cleavage of the fertilized eggs of the marine annelid *Chaetopterus pugamentaceus* following exposure to relatively low doses of X rays.

A number of investigators have added radio-isotopes to the medium in order to follow the effects of ionizing radiation on growth and development of fertilized eggs of marine invertebrates and fishes, but the dosages employed have not been defined. It may be assumed they were extremely high in comparison to dosages of ionizing radiation to which eggs of marine animals are exposed in their natural environment. As calculated by FEDOROV (1965), a large marine egg (0·52 to 0·6 cm in diameter) would receive 200 to 300 times the dose rate of natural ^{40}K β radiation from a concentration of 50 to 500 $\mu\mu$ Ci ^{90}Sr + ^{90}Y per litre; or 180 to 250 times that of natural ^{40}K β radiation from identical concentrations of ^{137}Cs.

COSTELLO and co-authors (1952) have shown that sea water containing 32 μCi/ml of ^{32}P affects the mitotic division of the eggs of the marine annelid worm *Chaetopterus pugamentaceus*; and GREEN and ROTH (1955) found that rather high levels of this and other radionuclides affect cleavage rates in eggs of the sea-urchin *Arbacia punctulata*. POLIKARPOV and IVANOV (1961, 1962) and FEDOROV and PODIMAKHIN (1962) report effects on the egg development of several marine fishes following the addition of ^{90}Sr + ^{90}Y to the experimental sea water; both mortality and number of abnormal offspring increased when the concentration surpassed 200 $\mu\mu$Ci ^{90}Sr/l. However, BROWN and TEMPLETON (1964) found no significant mortality increase or augmentation of abnormal development in fertilized eggs of plaice *Pleuronectes platessa* at concentrations less than 100 μCi/l sea water.

In a paper presented at the Oceanographic Congress in Moscow in 1966, IVANOV (1967) reported that when the eggs of a number of Black Sea fishes were incubated in solutions of ^{90}Sr – ^{90}Y with an activity of 10^{-7} to 10^{-9} Ci/l or in a ^{144}Ce solution with an activity of 10^{-6} Ci/l, their mitotic activity decreases and the number of chromosome aberrations increases as a function of ambient radio-activity. In the freshwater fish *Tinca tinca*, incubation of fertilized roe in aqueous solutions of

^{90}Sr – ^{90}Y with a concentration from 10^{-10} to 10^{-5} Ci/l has no appreciable effect either upon the rate of development of the embryos or upon the quantitative yield of normal and abnormal prelarvae from this roe (KULIKOV and co-authors, 1968). The survival rate of the prelarvae for 10 days after hatching from roe incubated in radio-active water does not differ from the control. Prelarvae hatched from roe incubated in radio-active solutions are just as resistant to supplementary external γ irradiation (dose 800 r) as the controls.

At fairly low doses of ionizing radiation, an interesting effect has been observed both in plants and animals, an effect that is not yet understood. Although higher doses of radiation induce a depression of growth, low doses may, in some instances, produce an apparent stimulation resulting in increased growth and vigour. ENGEL (1967) subjected young blue crabs *Callinectes sapidus* to different dosages of continuous irradiation from a ^{60}Co γ radiation source and observed at a radiation level of 3·2 rads/hr a distinct increase in growth rates. The mean percentage increase in carapace width of crabs exposed to this dose rate was greater (1% confidence level) than the means of the other two irradiated groups subjected to higher dose rates; it was also greater than the mean obtained for the controls (5% level). WHITE and co-authors (1966) studied the growth rates of 2-day old brine shrimp *Artemia salina* for 21 days after ^{60}Co γ ray irradiation of different dosages. Exposures of 500 and 2500 rads stimulated growth; after 10 days, the test animals were larger than the controls, and on the 21st day those receiving 500 rads were significantly larger than the controls. Brine shrimps which had received 500 rads reached sexual maturity sooner than did the controls or other irradiated test series. Following a dosage of 2500 rads, the brine shrimps exhibited sexual activities on the 18th day; in the control series, however, no mating pairs were observed up to the end of the experiment on the 21st day.

WELANDER (1968) reports a stimulating effect on growth in length and weight of fry following single X-ray doses to different embryological stages of salmonoids (*Oncorhynchus tshawytscha* and *Salmo gairdnerii*). There appeared to be an abrupt growth stimulating effect at the lowest dose administered (21 r) at each stage. The largest lengths and weights attributed to a radiation stimulation (as well as the lowest mortalities) were found in the ambient temperature groups at low dose. In 1969, WHITE (1969) published the results of his investigations on the interactions of chronic ^{60}Co γ irradiation (0·83 and 1·28 rads/hr over 15, 30 or 45 days) with temperature and salinity on the growth of postlarval pinfish *Lagodon rhomboides*. From analysis of variance for each of 9 body characteristics, WHITE found that temperature appears to control the growth of irradiated and unirradiated fish more than salinity or radiation. The interactions among the 3 variables of radiation, temperature and salinity also cause significant changes in the measured characteristics. Fish exposed to low levels of radiation are slightly longer and have greater body depths than unirradiated fish or those exposed to higher levels of radiation, suggesting a stimulation of growth by the γ ray exposure. The stimulating effect of low radiation doses on marine animals requires further investigation.

Moulting. In crustaceans, moulting (shedding of the exoskeleton) is associated with growth. Ionizing radiation has been shown to affect the moulting process in a number of marine species. Along with other factors, irradiation affects the moulting frequency of newly hatched grass shrimp *Palaemonetes pugio* (WHITE and co-authors, 1966). REES (1962) made some observations on the frequency of moulting in various irradiated groups of the fiddler crab *Uca pugnax*, following exposures to the γ rays of ⁶⁰Co. He found that acute doses of 9750 r and higher completely inhibit moulting. At doses of 4875 r and less, moulting does occur, but the intermoult period is prolonged. Individuals exposed to 975 r have a mean intermoult period of 106 days; those receiving 4875 r, of 212 days; the intermoult period for the unirradiated controls is 95 days. Somewhat different results were obtained by ENGEL (1967) for the blue crab *Callinectes sapidus*. ENGEL exposed young crabs to continuous ⁶⁰Co γ rays at rates of 3·2, 7·3, and 29·0 rads/hr; the total radiation doses received over 70 days by the three groups of crabs were 5105, 11,502 and 45,693 rads respectively. In the first 48 days, irradiated crabs of all groups, as well as the controls, moulted at about the same rates. From the 48th day until the end of the experiment after 70 days, crabs receiving 29·0 rads/hr did not moult. Moulting at the two lower levels—3·2 and 7·9 rads/hr—was not different from that of the controls throughout the experiment. The intermoult times were alike for irradiated and control crabs; however, differences in the width of the moulted carapaces did occur in the different groups.

HOPPENHEIT (1969) observed the effects of single acute exposures of X radiation on moulting of the euryhaline amphipod *Gammarus duebeni*. Few or no moults occur following the second day after irradiation with 20,000 r, and after the 30th day in individuals irradiated with 2500 to 10,000 r; moulting is delayed after exposure to 5000 to 10,000 r.

Strains of *Artemia salina* were exposed to single X-ray doses up to 100,000 rads by BALLARDIN and METALLI (1966) in order to assess the effects of irradiation on oogenesis and the moulting rhythm. It was concluded that *A. salina* exhibits two types of death after irradiation; at doses up to 60,000 rads, death is not preceded by any alteration in the moulting rhythm; at higher doses, death is associated with the blocking of the moulting rhythm and with acute modification of other physiological functions.

Regeneration. Many marine animals show remarkable powers of regeneration and are able to rebuild completely lost or mutilated body structures, even to the extent of replacing a large portion of the body itself. The capability of regenerating injured or lost body parts is particularly pronounced in several species of echinoderms. Unfortunately, the effects of irradiation on regeneration processes of marine animals have received little attention so far.

Evidence from investigations on different types of animals shows clearly that radiation tends to prevent or to slow down regeneration by injuring regenerative cells. Radiation does not appear to affect the nature of the body part regenerated, that is to say, it does not interfere with the so-called 'organizers'.

LIU (1948) investigated the effects of X rays on the restitution of dissociated cells of the sponge *Microciona* sp. At the higher doses, some aggregation of the dissociated cells takes place, but regeneration proceeds to an ill-defined reticular stage. Even with survival of some groups of rounded-up cells for as long as three weeks, no differentiation occurs. At lower doses, regeneration beyond the aggregation stage is abnormal and delayed. Flagellated chambers form very late. Irradiation apparently damages the archeocytes responsible for the reformation of choanocytes.

Regeneration has been studied in the two calcareous sponges *Leucosolenia complicata* and *L. variabilis* following exposure to β radiation from a $^{90}Sr + ^{90}Y$ source by KOROTKOVA and TOKIN (1965). A small portion of the body wall of the ocular tube was removed in a large number of sponges of each species immediately after irradiation with doses of 25,000, 50,000 and 100,000 rads. Exposure to the lowest dose does not influence appreciably the rate of regeneration; but in sponges receiving the two higher doses, restoration of injured body walls is completed, on an average, 2 to 3 days later than in the non-irradiated controls. No morphological peculiarities were observed in the regenerates of irradiated sponges, and no differences were apparent in the regenerated body walls of the two species of sponges tested.

KOROTKOVA and TOKIN (1965) studied also the effects of β irradiation on regeneration in the colonial hydroids *Clava multicornis* and *Laomedea flexuosa*. Dosages of β radiation of 2000, 25,000, and 100,000 rads were applied either before or after the removal of hydranths or before sectioning hydranths or stolons. At the two higher doses, regeneration of hydranths removed from their stolons is suppressed, and often stolon-like 'braids' are formed at the site of the removed hydranth. However, at 2000 rads, rates of hydranth regeneration increase. Subsequent to removal of the apical hydranth portion, irradiation at 100,000 rads delays regeneration slightly; the regenerated part does not reach the same size as that of the controls. Sectioning of the entire hydranth longitudinally all the way to the stolon is followed by wound healing in both control and irradiated individuals, but no regeneration of the missing portion takes place in either case. Histological examinations revealed a somewhat more complete healing of the wound in the control specimens than in the irradiated ones. Sectioning of hydranth bases resulted in the formation of a smooth oval-shaped body in controls and in experimentals irradiated with a dose of 25,000 rads. However, experimentals irradiated with 100,000 rads produced an irregularly shaped body with imprints of asymmetrically arranged tentacles on the surfaces.

In annelid worms, regeneration is accomplished by totipotent mesodermal cells, neoblasts, which constitute a persistent cell stock and are capable of migrating to the site of tissue injury. Some neoblasts replace damaged cells and others give rise to ectoderm, mesoderm, and endoderm of regenerating segments. It has been reported (O'BRIEN, 1946) that inhibition of regeneration from irradiation in the oligochaete *Nais paraguayensis* results from a selective inhibitory effect on the neoblasts. These cells are more susceptible to radiation than the endodermal epithelium and sufficient exposure causes their eventual disintegration. Both anterior and posterior regeneration in this naid is inhibited due to irradiation effects on the migration and proliferation of neoblasts. STÉPHEN-DUBOIS (1956) observed that irradiation of the entire worm prevents all regeneration of the amputated posterior end of the polychaete *Nereis diversicolor*. When only a few segments anterior to the wound were irradiated, some regeneration was possible, apparently due to immigration of uninjured mesodermal cells from segments not exposed to radiation. Similar observations were reported by BOILLY (1962) on the inhibiting action of X rays on regeneration in the polychaete *Syllis amica*. Whole body irradiation or regional irradiation inhibited caudal regeneration from the cut anterior portion of the worm, suggesting that the regenerative cells are located in the segment preceding the section.

Oxygen consumption. It might be expected that exposure of marine animals to ionizing radiation results in metabolic changes which would be reflected in altered rates of oxygen consumption. Some evidence obtained on laboratory mammals indicates that a slight increase in oxygen consumption occurs during irradiation; this increase may be followed by a decrease immediately after the irradiation. The post-irradiation decrease, however, may be of a temporary nature. Although there is no unanimity among experimentors, most investigations have shown that radiation does not alter the oxygen consumption of small invertebrates or their embryos. This conclusion was also reached by BOELL (1952) on the basis of his own studies and a review on radiation effects regarding respiratory metabolism.

Numerous investigators have studied the oxygen consumption of gametes and fertilized eggs of species of the sea-urchin *Arbacia* and of the annelid *Chaetopterus*; both are favorite biological materials for studying X-ray effects on early development, embryonic growth and differentiation. In spite of the great alteration in rates of cleavage and later embryogenesis due to irradiation, no changes in oxygen consumption, attributable directly to radiating eggs or developing zygotes, have been observed (CHESLEY, 1934; EVANS, 1940), although there appeared to be a reduced respiratory rate after irradiation of *Arbacia punctulata* sperm (BARRON and co-authors, 1949). Irradiation of eggs or sperm of *Arbacia* species does not suppress the rapid increase in respiratory rate which normally occurs when they join during fertilization (CHESLEY, 1934; EVANS, 1940).

Regarding the more recent investigations of radiation effects on oxygen consumption of invertebrates, SENEGAR (1964) did not find any change in oxygen consumption of the planarian *Dugesia dorotocephala* after X-ray exposure to dosages well above the LD_{50} value. However, ENGEL and co-authors (1966b) reported some changes in the rate of oxygen consumption of both nauplii and adults of *Artemia salina* immediately following exposures to relatively high doses of ^{60}Co γ radiation. They found an increase in the rate of oxygen consumption in nauplii, a dose-dependent decrease in adult males, and little change in adult females.

Differences in the observed effects of ionizing radiation on oxygen consumption may be explained in part by interactions of radiation and environmental conditions. In the ciliate protozoan *Tetrahymena pyriformis* GL strain, VAN DE VIJVER (1967) observed that when the protozoans were suspended in tap water, oxygen consumption decreased with increasing radiation dose, reaching 70% at 500,000 r. Cells irradiated in a 0·017 M phosphate buffer of pH 7·2 show a markedly smaller decrease in respiration rate, even at 500,000 r (30% inhibition). Respiration inhibition increases with post-irradiation time, even when the protozoans are introduced into the phosphate buffer after exposure. On the other hand, if the ciliates are transferred immediately after irradiation to fresh nutritive medium, 20 hrs after the irradiation they show enhancement of oxygen uptake. Salinity may affect animal responses to ionizing radiation. Subnormal or supranormal salinities can result in increased or decreased respiratory rates (Chapter 4.3). Combined effects of ionizing radiation and salinity may be responsible for the changes in metabolic rate observed. Effects of radiation and salinity on respiratory rates of the brine shrimp *Artemia salina* nauplii have recently been reported by ANGELOVIC and ENGEL (1968). Irradiated nauplii (^{60}Co γ radiation) were maintained in salinities ranging from 5‰ to 200‰. At 5‰ and 50‰ S, irradiated nauplii respire at rates which are about 5 to 15% below the control level at all doses tested (10,000 to 80,000 rads). Respiration of brine shrimp nauplii maintained at 100‰ and 150‰ S, and irradiated with doses up to 40,000 rads, is accelerated above that of the controls; but nauplii which have received a dose of 80,000 rads respire at about the same rate as the controls. At 200‰ S, respiration of irradiated nauplii increases at a dose of 10,000 rads, and then decreases again with further increasing doses of γ radiation.

Activity

It seems that many marine animals are capable of sensing X or γ rays; often even extremely low doses of radiation elicit increased locomotory activity and behavioural responses. The mechanisms involved in perception of ionizing radiation are not known. In numerous instances the responses resemble those caused by light (Chapter 2.3).

When protozoans are subjected to definitely sublethal doses of X or γ rays, they respond with an 'avoiding reaction' (WICHTERMAN, 1957). Ciliates react to lower doses of radiation with a temporary increase in swimming activity; the higher the dose, within this lower range, the stronger is the beat of their cilia. Higher, damaging doses cause a reduction of the ciliary activity and, finally, complete loss of swimming ability. Looking for biochemical changes involved in such responses, YAMAGUCHI (1962) measured intracellular contents of Na, K and Ca and found that they increase parallel to the dose-dependent changes in locomotory activity during the early responses following X-ray irradiation of *Paramecium caudatum*, but decrease again parallel to the decrease in activity. Flagellates also exhibit a change in their swimming activity at lower sublethal doses of X radiation. Many irradiated individuals swim actively, but in a less forwardly directed path; some swim in circles or change directions erratically over short distances. Amoebae, when irradiated with low X-ray doses, likewise show a transitory increase in locomotion rate. At higher doses they withdraw their pseudopods and become spherical (CHATTERJEE, 1968).

BROWN (1963) observed that the common planarian *Dugesia dorotocephala* displays a significant orientational response to increased ^{137}Cs γ radiation, even if the increase is no greater than six times the natural background intensity. The planarians are able to sense the direction of the weak gamma source and turn away from it. Marine snails respond similarly to extremely low exposure levels from the same source; their responses vary with the time of day and the snail's geographic orientation (BROWN and co-authors, 1962).

HUG (1958) observed a reflex-like retraction of the tentacles of different species of snails when exposed to weak doses of X or γ rays, and BORN (1960) described a reflex contraction of the mantle in pulmonate snails under similar exposure conditions. The tentacles remain retracted for the duration of exposure; they are extended again as soon as the irradiation ceases. The threshold for the response was found to vary from 1·5 to 5 r/sec depending upon the species. The time between start of irradiation at threshold rate and initiation of response varied from 5 to 15 secs. HUG (1960) demonstrated such instantaneous reactions in various marine invertebrates, namely, the snail *Paludina* sp., the mussel *Mytilus edulis*, the anthozoans *Actinia equina* and *Metridium senile*, and the sea-urchin *Echinus miliaris*; the sea-urchin was found to be particularly sensitive.

Also vertebrates may perceive external sources of ionizing radiation at very low intensities; this has been demonstrated by employing conditioned reflexes as criteria. GARCIA and BUCHWALD (1963), using X rays as unconditioned and shock as conditioned stimulus, demonstrated that rats respond to doses as low as 0·050 r/sec. The response occurred in less than 1 sec after the onset of irradiation. The development of a conditioned reflex, using γ radiation and electric shock, made it clear that also fish (carp) can detect radiation (TSYPIN and KHOLODOV, 1964); radiation source was a ^{60}Co γ ray delivered at the low rate of 0·5 to 0·1 r/sec over a period of 5 to 10 secs. Irradiation of the fish before the development of the conditioned reflex caused no locomotory responses. However, after 4 to 13 combined irradiation and electric stimuli, most of the fish began to move when irradiated only. Further evidence that the carp *Cyprinus carpio* can detect the presence of ^{60}Co and ^{137}Cs γ radiation was reported by PRAVDINA (1966). Food consumption was much less when the radiation sources were installed in an access labyrinth than in an area reached by a different path.

The cladoceran *Daphnia magna* is able to perceive X rays (BAYLER and SMITH, 1958). If a red light beam is allowed to pass through an opening in a lead shield, the daphnids swim to and fro in the beam. However, if sufficient X rays enter the water through the opening, daphnids passing under the opening swim toward the bottom; they come up again after reaching the water area protected by the lead shield. When the X-ray beam was turned on suddenly, those daphnids which were exposed to it swam downward and away from the red light source, and when the beam was turned off, those exposed swam up and toward the light.

Functional aspects of visual mechanisms are known to be altered by irradiation, for example, changes in brightness discrimination and changes in thresholds of light and dark adaptation (see also Chapter 2.32). These changes generally follow irradiation at higher doses; histological modifications in visual structures often offer explanations for the changed behavioural response. A negation of squid *Loligo pealli* embryo tropisms, positive heliotropism and negative geotropism, was observed following exposure to X rays by RUGH (1950). Histological changes (retinal disorganization and lens involvement) could be established only 3 to 4 days after exposure to 800 r or more. RUGH postulated that the retina becomes functionally damaged some time before there is histological evidence.

ENGEL and co-authors (1965a) observed an interesting delay of a behavioural response in post-larvae of the marine teleost *Eucinostomus lefroyi* following exposure to X rays and return to normal sea water. In contrast to the unirradiated control fish, experimentals exhibited, 26 days after irradiation, a typical vertical stratification pattern, related to the radiation dose received, which persisted over the next 17 days. While the controls maintained a position near the bottom of their aquarium, experimentals which received the highest dose (3200 r) remained near the surface of their aquarium. Experimentals which had received intermediate doses stratified at intermediate levels. Shifting the position of the aquariums in relation to the source of light did not change this pattern of behaviour. Stratification was not present at night, when all fish sank to the bottom. If lights were turned on suddenly at night, there was no immediate change in the position of the fish; however, 30 mins later, the stratification was apparent again.

Behaviour of marine animals viewed as part of the 'radiation sickness syndrome', which may follow lethal or near lethal exposures, has been described by a few investigators. Responses and movements slow down and feeding ceases. If the irradiation dose received does not result in death, the test animals slowly recover; they resume feeding and regain their normal activity. In decapod crabs which normally show a very pugnacious behaviour, the temporary loss of this attitude is very striking. Loss of the characteristic antagonism between individuals, a sluggish locomotory behaviour and refusal to accept food, has been observed following heavier irradiation of the fiddler crab *Uca pugnax* (REES, 1962) and the blue crab *Callinectes sapidus* (ENGEL, 1967). ENGEL also described the response of individuals subsequent to exposure to the high dose of 64,000 rads: *C. sapidus* loses the ability to maintain its positional equilibrium and exhibits catatonic seizures with the chelae open and all appendages rigid. Except for the pumping action of gill appendages and slight movements of the antennae, the crabs appear to be dead; they show only slight recovery before death.

(c) *Reproduction*

Unlike many other environmental factors, ionizing radiation has not been shown to affect the reproductive behaviour of marine animals. Radiation can, however, affect the actual reproductive processes, both sexual and asexual. The effects produced are the result of cellular damage.

In some of the lower invertebrates, exposure to X or γ rays inhibits or greatly modifies the rate of reproduction. In many instances, irradiation results in the production of non-viable offspring. Protozoa (usual method of reproduction: binary fission) respond with inhibition or delay of cell division. Individuals of the ciliate *Didinium nasutum*, exposed to a dose of 100,000 r, require a 20% longer period than the controls to reach the 4th division; higher doses progressively delay division until it is completely inhibited (SHEPARD and co-authors, 1956). In the freshwater-living cnidarian *Hydra littoralis*, PARK (1958) observed sensitivity of tissues to X rays and noted that the initiation of new buds is almost completely suppressed at an X-ray dose which only slightly arrests tissue differentiation and has no effect on 10-day survival. In *Hydra fusca*, similar exposures to X rays destroy interstitial cells (BRIEN and VAN DEN EECKHOUDT, 1953). These authors reported that, for 10 days after exposure, appearance, budding and regeneration are normal, but the young hydras produced are not viable. FREEMAN (1964) studied the asexual reproduction by budding in the tunicate *Perophora veridis*; as γ-ray irradiation from a ^{137}Cs source increases, budding is inhibited; complete inhibition of asexual reproduction occurs at a dose of 5000 r.

None of the females of the grass shrimp *Palaemonetes pugio* that had been irradiated with [60]Co γ rays at doses ranging from 975 to 39,000 r produced eggs at any time during the observations, whereas all the females of the unirradiated control group had become ovigerous after 40 days (REES, 1962). Three females of a group receiving 4875 r were ovigerous at the time of irradiation; within 4 days after exposure, all had released their eggs unhatched.

X-ray exposures up to 60,000 r of virgin females inhibited the first egg laying in the brine shrimp *Artemia salina* (BALLARDIN and METALLI, 1966). No apparent effects were observed by GROSCH (1962) after X irradiation of adult *A. salina*, but later there was a decline and even extinction of the culture populations. BALLARDIN and METALLI (1968) assessed the effects of X radiation (500 and 1000 rads) on juveniles of several generations of *A. salina* in terms of number of eggs produced, hatchability and percentage of larvae that survived to the adult stage. The irradiated lines did not show a change in performance compared to the control line until the 7th generation. In the 8th generation there was a drop in most of the performances studied. The average relative number of descendants that survived to reproductive age decreased to about half after 3500 rads and about a quarter after 7000 rads total accumulated dose.

The effects of radiation on eggs and spermatozoa of a number of marine invertebrates and fishes have been studied. Most investigators produced evidence that eggs are more sensitive to X-ray irradiation than sperms. This has been reported for the clam *Spisula solidissima* (RUGH, 1953); the ascidian *Molgula manhattensis* (GROSCH and SMITH, 1957) and for the fishes *Fundulus heteroclitus* (RUGH and CLUGSTON, 1955) and *Menidia menidia* (ENGEL and co-authors, 1965a). However, SOLBERG (1938) found *Fundulus heteroclitus* sperm to be relatively radio-sensitive to doses of 2000 r, and ENGEL and co-authors (1965a) observed considerable mortality in developing eggs of *Menidia menidia* produced from sperm irradiated with doses of 800 r. The main mortality appeared during late cleavage stages or the embryonic shield stage. In contrast, maximum mortality of developing eggs irradiated before fertilization with unirradiated sperm occurred 11 to 12 days after heart and eyes had developed.

It is well known that various higher vertebrates become temporarily sterilized after irradiation with moderate X-ray doses; irradiation of their gonads with larger doses is followed by permanent sterility. Examples of the sterilizing effects of X rays on fishes have been published by EGAMI and HYODO (1965) and EGAMI and KONNO (1966). Irradiation of either male or female ricefish *Oryzias latipes* with moderate doses of X rays affects their reproductive capacity. Irradiation of the female temporarily reduces the number of eggs laid; irradiation of the male reduces the percentage of fertilized eggs. Higher doses of X rays result in permanent sterility in both males and females.

(d) Distribution

The doses of ionizing radiation received by marine animals in oceans and coastal waters are extremely low. It is unlikely, therefore, that the general patterns of animal distribution are significantly affected by present-day ionizing radiation. The pertinent information is still very scanty. (In regard to light effects on animal distribution consult Chapter 2.3.)

(3) Structural Responses

In oceans and coastal waters, no structural changes in marine animals are known which can be attributed to ionizing radiation. Experimental irradiation may induce changes due to damage of cells and tissues. Some structural responses become apparent only after considerable periods of time (e.g. irradiated embryos may reveal structural changes only during subsequent ontogenetic development); other structural responses, such as cell or tissue injuries in adult individuals, may appear quite promptly after treatment. Apparently, modifications arising as genetically induced anomalies have not yet been reported for marine species, but investigations along this line are under way; they concentrate on the study of long-term exposure of developing eggs and larvae of salmon (*Oncorhynchus tshawytscha* and *O. kisutch*) to low doses of γ radiation, and on the offspring of such irradiated smolt and their return from seaward migrations to the spawning areas (DONALDSON and BONHAM, 1964).

In the preceding section on functional responses of marine animals to ionizing radiation, it was necessary to touch upon certain structural aspects; more complete and more detailed studies dealing with radiation effects on external and internal structures will now be considered. These studies are limited to fishes, most of them to freshwater fishes. However, the responses to ionizing radiation appear to be similar in freshwater and marine fishes.

(a) Size

Ionizing radiation may presumably affect final body size. The information at hand is, however, insufficient for a critical assessment.

(b) External Structures

Exposure to X rays during early development of fishes results in a stunting of growth and marked abnormalities in the development of fins. Reduction in the number of dorsal and anal fin rays in the trout *Salmo gairdnerii* was observed by WELANDER (1954); the extent of the reduction in anal fin rays following irradiation at different stages of development of the ricefish *Oryzias latipes* was studied by EGAMI (1964). Exposure to X rays of post-larval fish resulted in abnormal development of dorsal and anal fins in the marine teleost *Eucinostomus lefroyi* (ENGEL and co-authors, 1965a). These authors describe a striking effect in the appearance of the spiny portion of the dorsal fin. While the soft rays of the dorsal fin are not affected, the state of the spiny rays ranges from normal to complete absence. ENGEL and co-authors reported also stunting of body growth, a dose-dependent reduction of the eye diameter, and differences in the development of pigmentation on body and fins. X rays applied daily during the first 43 days of embryogenesis and larval development cause a significant reduction in the number of dorsal fin rays in *Fundulus heteroclitus* but no significant effect on its number of anal fin rays (WHITE and ANGELOVIC, 1966c).

It is well known that adult goldfish *Carassius auratus* respond to X- or γ-ray exposure by augmenting external pigmentation. SHECHMEISTER and co-authors

(1962) described the transitory nature of this phenomenon. The way in which ^{60}Co γ rays influence melanophore eruption was studied in some detail by ETOH (1962), EGAMI and co-authors (1963a), and ETOH and EGAMI (1964).

In mammals, low doses of ionizing radiation cause reddening of the skin and limited cell damage; high doses may result in severe skin burns with extensive cellular injuries. The basal tissue, sweat glands and hair follicles are particularly radiosensitive. In fishes, the general nature of radiation injuries to the skin is similar: damage to basal epithelial cells, scale sac cells (homologous to hair follicles), and mucous-secreting cells (homologous to sweat glands). In two fresh-water fishes, the loach *Misgurunus anguillicaudatus* and the glass catfish *Kryptopterus biciruhrs*, whole body irradiation with hard X rays at doses between 5000 and 10,000 r severely injured the skin, and death followed in a few days (ETOH and NAKAO, 1962). In some of the irradiated individuals, a light-coloured, severe inflammation appeared on the skin soon after irradiation; in heavily irradiated skin, mucous cells were exuviated, the naked dermis was exposed and the scale sac filled with a lymph-like fluid. These symptoms resemble those of fourth degree X-ray burns in mammalian skin. *K. biciruhrs*, irradiated in the posterior half of the body with 12,000 r of soft X rays, lost their caudal fins within 4 days. However, whole body irradiation with hard X rays at doses from 5000 to about 10,000 r applied to two species of saltwater fish, the goby *Chasmichthys dolichognatus* and the puffer *Fugu rubripes*, resulted in no microscopically observable skin changes during a period of 7 to 17 days after irradiation. These findings indicate that fresh-water fishes are more radiosensitive than saltwater fishes. However, ENGEL and DAVIS (1965a) noted that individuals of the saltwater pinfish *Lagodon rhomboides* died within a week after whole body irradiation with γ rays of a ^{60}Co source at a dosage of 5000 r; there were conspicuous subcutaneous haemorrhages, and most of the caudal and pectoral fins were sloughed off.

(c) Internal Structures

Ionizing radiations administered to the whole bodies of fishes affect the clotting time and the cellular components of the peripheral blood. In two species of salt-water fishes, *Lagodon rhomboides* and *Fundulus heteroclitus*, sublethal and relatively low doses of X rays produced a transitory increase in blood clotting time (ENGEL and co-authors, 1965b). ENGEL and co-authors (1966a) studied the effects of a moderate dose of γ rays from ^{60}Co on the cellular components of the blood of *Lagodon rhomboides*, sampled at intervals over a total period of 34 days. A dose of 2000 r reduced the number of thrombocytes markedly during the first 7 days. The number then slowly increased, and by the end of the experiment it had returned to the control level. The number of leucocytes increased during the first 24 hrs, and then decreased to reach its lowest level on the 3rd day after irradiation. From the 3rd to the 21st day, the number steadily increased until it was above the control level (overcompensation) and subsequently decreased. This radiation exposure had almost no effect on erythrocyte numbers, haematocrit values, or haemoglobin levels.

Gross pathological effects on internal organs of fishes, resulting from ionizing radiation, have been described by a number of investigators. Inflammation and haemorrhages in different parts of the body, due to a breakdown of the vascular system, give evidence of tissue damage induced by radiation. After applying low to moderate dosages of X rays to the entire body of goldfish *Carassius auratus*, SHECHMEISTER and co-authors (1962) observed an inflammation of the outer surface of intestines, kidney, and gonads. WELANDER and co-authors (1949) describe the quantitative relation of haemorrhages to dose, in whole body X-ray irradiated adult rainbow trout *Salmo gairdnerii*. They report gonadal haemorrhage at 500 r, formation of haemorrhagic areas in the peritoneum at 750 r, muscular haemorrhages at 1000 r or more, and mass haemorrhage in individuals which had received 1500 and 2500 r. *Salmo gairdnerii* fed daily with the radionuclide ^{32}P, giving β radiation to surrounding tissues, responded with a reduction in growth and haemorrhages in liver and muscles at moderate doses; large daily amounts of ^{32}P caused damage to the gastro-intestinal tract, anterior kidney, and a reduction in leucocyte numbers (WATSON and co-authors, 1959). In the marine pinfish *Lagodon rhomboides*, subcutaneous haemorrhages were noted, following exposure to 5000 r of ^{60}Co γ radiation, as well as changes in the cellular components of its blood (ENGEL and DAVIS, 1965a). Post-larvae of the winter flounder *Paralichthys lethostigma*, subjected to continuous irradiation for 184 hrs at a dose rate of 63·4 r/hr from a ^{60}Co gamma source, responded with damage in the intestinal mucosa, especially destruction of the villi (WHITE, 1965b).

Marked effects on internal organs and tissues following X-ray exposure during early embryonic stages of eggs of chinook salmon *Oncorhynchus tshawytscha* were ascertained by histological examinations of the organs of embryos and larvae (WELANDER and co-authors, 1948). The extent of the injuries was proportional to the X-ray doses administered. The gonads proved to be the most sensitive organs; they responded with a distinct reduction in the number of primordial germ cells at doses as low as 250 r, delivered to the developing eggs. There was a reduction in the number of cells and a temporary retardation in the development of haematopoietic tissue of the anterior portion of the kidney; this demonstrates the high sensitivity of the haematopoietic tissue to radiation. In similar experiments, low and moderate doses of X ray delivered to the early embryonic stages of the ricefish *Oryzias latipes* resulted in abnormal testis formation and often in the development of testes lacking germ cells (EGAMI and co-authors, 1966).

More detailed studies have been made on organs and tissues known to be particularly sensitive and expected to play an important role in causing radiation death in fishes. Histopathological effects on the intestinal mucosa, induced by whole body irradiation, of goldfish *Carassius auratus* were studied by HYODO and AOKI (1963). They classified the effects according to the degree of damage into: (i) no observable effects; (ii) hypertrophy of epithelial cells and their nuclei; (iii) more pronounced hypertrophy with a considerable decrease in the cell population of the mucosal layer and irregular arrangement of epithelial cells; (iv) collapse or denuding of the epithelial layer. These effects were found to be proportional to the X-ray dose administered. In later studies, factors affecting the development of intestinal radiation damages, and acute intestinal radiation death, were investigated (HYODO, 1964, 1965, 1966). Cell proliferation and generation cycle in the intestinal epithelium following lethal and sublethal X-ray exposures of *Carassius auratus* were studied with autoradiographic techniques by HYODO-TAGUCHI (1968), who observed that the lethal dose (8000 r) results in complete inhibition of DNA synthesis and cell mitosis soon after irradiation, whereas with sublethal doses (1000 and 2000 r), recovery from damage proceeds after an initial fall.

AOKI (1964) conducted detailed histopathological studies of the anterior portion of the kidney, the chief haematopoietic organ, following acute total body irradiations of adult *Carassius auratus*. Within 3 days, all doses administered (1000, 2000, 4000 and 8000 r) decreased the weight of this organ and reduced the number of haematopoietic cells, especially in fish receiving the three higher doses, which turned out to be lethal after 10 to 14 days. After exposure to 1000 r, the number of haematopoietic cells returned to normal within 21 days. Detailed studies of the early repair of X-ray induced damage in the cells of the haematopoietic tissues of *C. auratus* were made by ETOH (1968) using autoradiographic techniques to follow the incorporation of ^{3}H-thymidine injected at varying time intervals after irradiation. Recovery in mitotic and DNA synthetic activities was observed 72 hrs after the irradiation with exposures of 1000 or 2000 r. There was no indication of recovery after exposures to 8000 r. Similar haematopoietic damages and adjustments (repair processes) were found in haematopoietic kidney tissues of the marine teleost *Lagodon rhomboides*, after administering sublethal doses of ^{60}Co γ irradiation and using ^{59}Fe as tracer (ENGEL, 1969). The damaging radiation effects on the haematopoietic function of the kidney, suffered after exposure to 2000 rads, were apparent 1 day after irradiation and were followed by a recovery which was quite complete 22 days after irradiation.

An interesting study on the nature of histological effects in the ovary of the loach *Misgurunus anguillicaudatus* was made by EGAMI and AOKI (1966). In an earlier study on *Oryzias latipes*, EGAMI and HYODO (1965) had observed that, although high X-ray doses destroy oocytes of all stages of development, low doses reduce only the number of large oocytes. Irradiation of the whole body or a part of it (shielding parts of the ovaries) revealed that the low doses (2000 r) of whole body irradiation affect the ovary largely via a reduction of gonadotropin secretion by the pituitary gland influenced by radiation, while high doses (16,000 r) damage ovaries directly, and severely injure oocytes of all stages. After X-ray irradiation at 100 to 2000 r of *Oryzias latipes* males, EGAMI and KONNO (1966) observed a temporary decrease in testicular weight, and their histological examinations revealed that large numbers of spermatogonia and spermatocytes had been affected, but that the spermatids and spermatazoa were more radioresistant. Irradiation of males at 8000 r resulted in a complete degeneration of their testes.

(4) Conclusions

So far as is known, the levels of natural ionizing radiation present in oceans and coastal waters elicit no specific responses by marine animals; they cause no observable damages or injuries. Functional and structural responses to ionizing radiation are, therefore, documented here on the basis of laboratory experiments conducted under artificially increased radiation intensities.

The radiotolerance of marine animals differs greatly, irrespective of their taxonomic or ecological relationships. However, lower invertebrates exhibit considerably higher tolerances to radiation than do members of 'higher' phyla and marine fishes; they are hundreds of times more resistant. The reasons for such an enormous decrease in radiation tolerance with increased complexity of animal organization are still unknown.

Radiation tolerance varies as a function of dose-rate, time patterns of exposure, and rate of metabolism. These parameters affect the quantitative relations between radiation injury and repair processes. Other environmental factors, such as temperature (Chapter 3), may accelerate, retard or qualitatively modify these processes. Low temperatures delay the onset, but cannot prevent, radiation damage. At extremely low doses of radiation, the equilibrium between the rate of development of radiation damage and recovery processes, may give the impression that no radiation injury has occurred. This fact may explain, in part, the failure to observe responses of marine animals to radiation at *in situ* exposure levels.

Ionizing radiation affects marine animals primarily at the cellular level; metabolically highly active cells are more heavily disturbed or damaged than less active or resting ones. Irradiation effects are, therefore, most evident during embryonic development; detrimental effects—generally a delay or an arrest of cell division—often modify the developmental pattern during tissue and organ differentiation. The basic responses of marine animals do not differ essentially from those known for freshwater or terrestial animals.

Regeneration of body parts, a phenomenon commonly observed among marine animals, is delayed or inhibited by radiation, but the nature of the regenerate part is not affected.

The reproductive potential of marine animals subjected to ionizing radiation is primarily disturbed at the cellular level. Asexual reproduction by budding, common in many coelenterates, is either delayed or inhibited; the new individual produced is often not viable. Sexual reproduction is primarily affected via gametogony or ripe gametes.

Some marine animals appear to be able to perceive extremely low intensities of ionizing radiation, both X rays and γ rays; they respond by performing avoiding reactions. Functional responses of visual mechanisms (light receptors) due to ionizing radiation occur in marine animals, just as they do in freshwater and terrestial ones. The perception mechanisms involved are not known.

Structural responses to radiation involve, for example, the pattern of development of fins in fishes due to cellular injury during early ontogeny. Structural components which are most responsive to radiation are (as in mammals) highly specialized cells, such as those producing the cellular components of the blood, the blood cells themselves, gamete-producing and other specialized gonadial cells, the haematopoietic cells of the kidney, cells of the skin and of the alimentary canal.

Literature Cited (Chapter 11)

ACHEY, P. M. and POLLARD, E. C. (1967). Studies on the radiation-induced breakdown of deoxyribonucleic acid in *Escherichia coli* 15 T⁻L⁻. *Radiat. Res.*, **31**, 47–62.

ALEXANDER, P. (1966). Verschiedene Empfindlichkeit von Zellen gegenüber ionisierenden Strahlen. *Nova Acta Leopoldina* (*N.F.*), **31**, 131–151.

ANGELOVIC, J. W. and ENGEL, D. W. (1968). Interaction of gamma radiation and salinity on respiration of brine shrimp (*Artemia salina*) nauplii. *Radiat. Res.*, **35**, 102–108.

ANGELOVIC, J. W., WHITE, J. C., JR. and DAVIS, E. M. (1969). Interactions of ionizing radiation, salinity and temperature on the estuarine fish, *Fundulus heteroclitus*. In D. J. Nelson and F. C. Evans (Eds), *Symposium on Radioecology*. Proceedings of the Second National Symposium held at Ann Arbor, Michigan, May 15–17, 1967. CONF-670503, U.S. Atomic Energy Commission Doc. TID-4500. pp. 131–141.

ANGELOVIC, J. W., WHITE, J. C., JR. and ENGEL, D. W. (1966). Influence of salinity on the response of estuarine animals to ionizing radiation. *Ass. SEast. Biol. Bull.*, **23**, 29.

ANGINO, E. E., SIMEK, J. E. and DAVIS, J. A. (1965). Fixing of fallout material by floating marine organisms, *Sargassum fluitans* and *S. natans*. *Publs Inst. mar. Sci. Univ. Tex.*, **10**, 173–178.

AOKI, K., 1964. The effect of temperature on the development of histological damage of hematopoietic tissue following whole-body X-irradiation in the goldfish, *Carassius auratus*. In *Annual Report 1963, National Institute of Radiological Sciences*, Science & Technology Agency, Japan, NIRS-**3**, 35–36.

ARNAUD, M. (1966). Irradiations continues aux rayons X de cultures de Chlorelles. Thesis, Grenoble University (France), Faculté des Sciences. (U.S. Atomic Energy Commission Doc. No. NP-17125).

AUERBACH, S. I. (1966). Progress in terrestrial and freshwater ecology. In Health Physics Division, Annual Progress Report for period ending July 31, 1966. *U.S. Atomic Energy Commission Report* ORNL-4168, 61–119.

BACQ, Z. M. and ALEXANDER, P. (1961). *Fundamentals of Radiobiology*, 2nd ed., Pergamon Press, Oxford.

BACQ, Z. M., VANDERHAEGHI, F., DAMBLOM, J., ERRERA, M. and HERVE, A. (1957). Effets des rayons-X sur *Acetabularia mediterranea*. *Expl Cell Res.*, **12**, 639–648.

BALLARDIN, E. and METALLI, P. (1966). Effects of X-rays on oogenesis and rhythm of molting in *Artemia salina* LEACH. (Ital.) *Boll. Zool.*, **33**, 195–196. (RT/BIO (67) 26)

BALLARDIN, E. and METALLI, P. (1968). Estimates of some components of fitness in diploid parthenogenetic *Artemia salina* irradiated over several generations. (Ital.) *Atti Ass. genet. ital.*, **13**, 341–345. (Conference on Genetics, Naples, Italy, CONF-671057 Vol. 13.)

BARINOV, G. V. (1964). Radioactive isotopes and algae. (Russ.). *Priroda, Mosk.*, **7**, 82–83.

BARKAI-GOLAN, R., KAHAN, R. S. and TEMKEN-GORODEISKI, N. (1968). Sensitivity of stored melon fruit fungi to gamma irradiation. *Int. J. appl. Radiat. Isotopes*, **19**, 579–583.

BARKAI-GOLAN, R., TEMKEN-GORODEISKI, N. and KAHAN, S. R. (1966). The effect of gamma irradiation on the development of the fungi *Botrytis cinerea* and *Rhizopus nigricans* that cause rot in strawberry fruits. National and University Institute of Agriculture, Rehovot, Israel. Volcani Institute of Agriculture Research.

BARRON, E. S. G., GASVODA, B. and FLOOD, V. (1949). Studies on the mechanism of action of ionizing radiations. Effect of X-ray irradiation on the respiration of sea urchin sperm. *Biol. Bull. mar. biol. Lab., Woods Hole*, **97**, 44–50.

BAYLER, E. R. and SMITH, F. E. (1958). Animal perception of X-rays. *Radiat. Res.*, **8**, 466–474.

BAZILL, G. W. (1967). Lethal unbalanced growth in bacteria. *Nature, Lond.*, **216**, 346–349.

BELLAMY, W. D. and LAWTON, E. J. (1954). Problems in using high voltage electrons for sterilization. *Nucleonics*, **12**, 54–57.

BERGER, M. (1967). Die Teilungsfähigkeit einzelliger synchronisierter Volvocales nach Röntgenbestrahlung. *Int. J. Radiat. Biol.*, **12**, 477–486.

BERGSTRÖM, R. M. (1962). The effect of 5·3 Mev Po²¹⁰ alpha particles on the resting membrane potential and ultrastructure of *Nitella flexilis*. *Annls Med. exp. Biol. Fenn.*, **40**, 45–53.

BHATTACHARJEE, S. B. (1961). Action of X-irradiation on *E. coli*. *Radiat. Res.*, **14**, 50–55.

BILLEN, D. (1957). Modification of the release of cellular constituents by irradiated *Escherichia coli*. *Archs Biochem. Biophys.*, **67**, 333–340.

BILLEN, D., STAPLETON, G. E. and HOLLAENDER, A. (1952). The effect of X radiation on the respiration of *Escherichia coli*. *J. Bact.*, **65**, 131–135.

BLAYLOCK, B. G. (1969). The fecundy of a *Gambusia affinis affinis* population exposed to chronic environmental radiation. *Radiat. Res.*, **37**, 108–117.

BLINKS, L. R. (1952). Effects of radiation on marine algae. *J. cell. comp. Physiol.*, **39** (Suppl. 2), 11–18.

BOELL, E. J. (1952). Effects of radiations on respiratory metabolism. *J. cell. comp. Physiol.*, **39** (Suppl. 2), 19–42.

BOILLY, B. (1962). Inhibition de la régénération caudale par irridation X chez *Syllis amica* QUATREFAGES (Annélide Polychète). *C.r. hebd. Séanc. Acad. Sci.*, Paris, **255**, 1414–1416.

BONÉT-MAURY, P. (1963). Influence de la nature du rayonnement sur l'inactivation des micro-organismes. *Int. J. appl. Radiat. Isotopes*, **14**, 36–38.

BONÉT-MAURY, P., PERAULT, R. and ERICHSEN, M. C. (1943). Mise en évidence, par la respirométrie, de l'action bactériostatique des radiations ionisantes. *Annls Inst. Pasteur Paris*, **69**, 189–192.

BONEY, A. D. (1965). Aspects of the biology of the seaweeds of economic importance. *Adv. mar. Biol.*, **3**, 105–253.

BONHAM, K. and PALUMBO, R. (1951). Effects of X-rays on snails, Crustacea, and algae. *Growth*, **15**, 155–188.

BONHAM, K. and WELANDER, A. D. (1963). Increase in radioresistance of fish to lethal doses with advancing embryonic development. In V. Schultz and A. W. Klement, Jr. (Eds), *Radioecology*. Reinhold, New York. pp. 353–358.

BORN, W. (1960). Zur Auslösung von Reflexen bei Schnecken durch Röntgen- und Alpha-strahlen. *Strahlentherapie*, **112**, 634–636.

BRIDGES, A. E., OLIVO, J. P. and CHANDLER, V. L. (1956). Relative resistance of micro-organisms to cathode rays. II. Yeasts and molds. *Appl. Microbiol.*, **4**, 147–149.

BRIDGES, B. A. (1963). Conditions after irradiation affecting survival and recovery of micro-organisms. *Int. J. appl. Radiat. Isotopes*, **14**, 48–51.

BRIDGES, B. A. and MUNSON, R. J. (1968). Genetic radiation damage and its repair in *Escherichia coli*. In M. Ebert and A. Howard (Eds), *Current Topics in Radiation Research*, Vol. IV. Wiley, New York. pp. 95–188.

BRIEN, P. and EECKHOUDT, J. P. VAN DEN (1953). Bourgeonnement et régénération chez les Hydres irradies par les rayons X. *C.r. Séanc. Soc. Biol.*, **237**, 756–758.

BRIGGS, A. (1966). The resistance of spores of the genus *Bacillus* to phenol, heat, and radiation. *J. appl. Bact.*, **29**, 490–504.

BROWN, F. A., JR. (1963). An orientation response to weak gamma radiation. *Biol. Bull. mar. biol. Lab.*, Woods Hole, **125**, 206–225.

BROWN, F. A., JR., WEBB, H. M. and JOHNSON, L. G. (1962). Orientational responses in organisms effected by very small alterations in gamma (Cs-137) radiation. *Biol. Bull. mar. biol. Lab.*, Woods Hole, **123**, 488–489.

BROWN, V. M. and TEMPLETON, W. L. (1964). Resistance of fish embryos to chronic irradiation. *Nature, Lond.*, **203**, 1257–1259.

BRUCE, A. K., SANSONE, P. A. and MacVITTIE, T. J. (1969). Radioresistance of bacteria as a function of p-hydroxymercuribenzoate binding. *Radiat. Res.*, **38**, 95–108.

BRYANT, P. E. (1968). Survival after fractionated doses of radiation: Modification of anoxia of the response of *Chlamydomonas*. *Nature, Lond.*, **219**, 75–77.

CALKINS, J. (1967). Similarities in the radiation response of *Escherichia coli* and *Tetrahymena pyriformis*. *Int. J. Radiat. Biol.*, **13**, 283–288.

CATHER, J. N. (1959). The effects of X-radiation on the early cleavage stages of the snail, *Ilyanassa obsoleta*. *Radiat. Res.*, **11**, 720–731.

CHATTERJEE, S. (1968). X-ray induced changes in the cell body of amoeba. *Z. Biol.*, **116**, 68–80.

CHESLEY, L. (1934). The effect of radiation upon cell respiration. *Biol. Bull. mar. biol. Lab.*, Woods Hole, **67**, 259–272.

CHIPMAN, W. A. (1966). Food chains in the sea. In R. Scott Russell (Ed.), *Radioactivity and Human Diet*. Pergamon Press, Oxford. pp. 421–453.

CHOWDHURY, S. U., HOSSAIN, M. M. and YUSUF, Q. M. (1966). Survey and radio-sterilization doses of fungi causing spoilage of rice in storage. Atomic Energy Centre, Dacca (Pakistan), Radiobiology Division, Report AECD-RR-2.

CLARK, D. J. (1968). Effects of ionizing radiation on synchronous cultures of *Escherichia coli* B/r. *J. Bact.*, **96**, 1150–1158.

COSTELLO, D. P., HENLEY, C. and KENT, D. E. (1952). Effects of P^{32} on mitosis in *Chaetopterus* eggs. *Biol. Bull. mar. biol. Lab., Woods Hole*, **103**, 298–299.

DAVIES, D. R. (1966). Patterns and rates of recovery in synchronous populations of algal cells exposed to gamma radiation. *Radiat. Res.*, **29**, 222–235.

DAVIES, D. R. and EVANS, H. J. (1966). The role of genetic damage in radiation-induced cell lethality. *Adv. Radiat. Biol.*, **2**, 243–353.

DAVIES, D. R., HOLT, P. D. and PAPWORTH, D. G. (1969). The survival curves of haploid and diploid *Chlamydomonas reinhardtii* exposed to radiations of different LET. *Int. J. Radiat. Biol.*, **15**, 75–87.

DAVIES, D. R. and THORBURN, M. A. P. (1968). The effect of dose rate in relation to hypoxia, and the role of repair systems in the green alga *Chlamydomonas reinhardi*. *Radiat. Res.*, **35**, 401–409.

DAVYDOFF-ALIBERT, S. (1963). Influence de la nature du milieu ambiant pendant l'irradiation sur la survie bactérienne. *Int. J. appl. Radiat. Isotopes*, **14**, 41–44.

DEERING, R. A. (1963). Mutation and killing of *Escherichia coli* WP-2 by accelerated heavy ions and other radiations. *Radiat. Res.*, **19**, 169–178.

DEERING, R. A. (1968). Radiation studies of *Blastocladiella emersonii*. *Radiat. Res.*, **34**, 87–100.

DEWEY, D. L. and HAYNES, R. H. (1967). Sensitivity of *Micrococcus radiodurans* to densely ionizing radiations. In J. H. Lawrence (Ed.), *Biology and Medicine Semiannual Report, Spring 1967*. California University, Berkeley. pp. 138–144. (U.S. Atomic Energy Commission Report UCRL-17481.)

DICKSON, H. (1932). The effects of X-rays, ultraviolet light, and heat in producing saltants in *Chaetomium cochliodes* and other fungi. *Ann. Bot.*, **46**, 389–405.

DODGE, J. D. and GODWARD, M. B. E. (1963). Some effects of X-rays on the nucleus of a dinoflagellate. *Radiat. Bot.*, **3**, 99–104.

DONALDSON, L. R. (1964). Evaluation of radioactivity in the marine environment of the Pacific Proving Ground. In S. H. Small (Ed.), *Nuclear Detonations and Marine Radioactivity*. Norwegian Defense Research Establishment, Kjiller, Norway. pp. 73–83.

DONALDSON, L. R. and BONHAM, K. (1964). Effects of low-level chronic irradiation of chinook and coho salmon eggs and alevins. *Trans. Am. Fish. Soc.*, **93**, 333–341.

DONALDSON, L. R. and FOSTER, R. F. (1957). Effects of radiation on aquatic organisms. In *The Effects of Atomic Radiation on Oceanography and Fisheries*. *Publs natn. Res. Coun., Wash.*, **551**, 96–102.

DUCOFF, H. S., BUTLER, B. D. and GEFFON, E. J. (1964). X-ray survival studies on the alga *Brachiomonas submarina* BOHLIN. *Radiat. Res.*, **23**, 446–453.

DUCOFF, H. S., BUTLER, B. D. and GEFFON, E. J. (1965). The effect of radiation on replication in *Brachiomonas submarina* BOHLIN. *Radiat. Res.*, **24**, 563–571.

DUKE, T. W., WILLIS, J. N. and PRICE, T. J. (1966). Cycling of trace elements in the estuarine environment. I. Movement and distribution of zinc-65 and stable zinc in experimental ponds. *Chesapeake Sci.*, **7**, 1–10.

DUNSTER, H. J. (1958). The disposal of radioactive liquid wastes into coastal waters. In *Proceedings of the 2nd United Nations Conference on the Peaceful Uses of Atomic Energy, held in Geneva, 1–13 Sept. 1958*, **18** (Waste treatment and environmental aspects of atomic energy), 1–10.

DUPUY, P. (1963). Micro-organismes résistant aux radiations ionisantes. *Int. J. appl. Radiat. Isotopes*, **14**, 29–35.

EDWARDS, R. B., PETERSON, L. J. and CUMMINGS, D. G. (1954). The effect of cathode rays on bacteria. *Fd Technol. Champaign*, **8**, 284–290.

EGAMI, N. (1964). Effect of irradiation on embryonic development of skeletal structure in medaka, *Oryzias latipes*. 1. Effect of X-irradiation on number of anal fin rays. In *Annual Report 1963, National Institute of Radiological Sciences*, Science & Technology Agency, Japan, NIRS-**3**, 38–39.

EGAMI, N. (1969). Kinetics of recovery from injury after whole-body X-irradiation of the fish, *Oryzias latipes* at different temperatures. *Radiat. Res.*, **37**, 192–201.

EGAMI, N. and AOKI, K. (1966). Effects of X-irradiation applied to a part of the body on the ovary of the loach, *Misgurunus anguillicaudatus*. In *Annual Report 1965, National Institute of Radiological Sciences*, Science & Technology Agency, Japan, NIRS-**5**, 28.

EGAMI, N. and ETOH, H. (1966). Effect of temperature on the rate of recovery from radiation-induced damage in the fish, *Oryzias latipes*. In *Annual Report 1965, National Institute of Radiological Sciences*, Science & Technology Agency, Japan, NIRS-**5**, 28–30.

EGAMI, N., ETOH, H., TACHI, C., AOKI, K. and ARAI, R. (1963a). Role of the pituitary gland in melanization of the skin of the goldfish, *Carassius auratus*, following X-irradiation. In *Annual Report 1962, National Institute of Radiological Sciences*, Science & Technology Agency, Japan, NIRS-**2**, 38–39.

EGAMI, N., ETOH, H., TACHI, C. and HYODO, Y. (1963b). Relationship between survival time after exposure to X-rays and dose of the irradiation in the fishes, *Carassius auratus* and *Oryzias latipes*. In *Annual Report 1962, National Institute of Radiological Sciences*, Science & Technology Agency, Japan, NIRS-**2**, 33–34.

EGAMI, N. and HYODO, Y. (1965). Effects of X-irradiation on oviposition of the fish, *Oryzias latipes*. In *Annual Report of Radiological Sciences*, Science & Technology Agency, Japan, NIRS-**4**, 44–45.

EGAMI, N., HYODO, Y. and ITO, Y. (1966). Inhibitory effect of X-irradiation of embryos on testis formation in *Oryzias latipes*. In *Annual Report 1965, National Institute of Radiological Sciences*, Science & Technology Agency, Japan, NIRS-**5**, 31–32.

EGAMI, N. and KONNO, K. (1966). Effect of X-irradiation on fertility of the male of the fish, *Oryzias latipes*. In *Annual Report 1965, National Institute of Radiological Sciences*, Science & Technology Agency, Japan, NIRS-**5**, 30–31.

ENGEL, D. W. (1967). Effects of single and continuous exposures to γ-radiation on the survival and growth of the blue crab, *Callinectes sapidus*. *Radiat. Res.*, **32**, 685–691.

ENGEL, D. W. (1969). Effect of sublethal gamma irradiation on the iron metabolism of the pinfish, *Lagodon rhomboides*. In D. G. Nelson and F. C. Evans (Eds), *Symposium on Radioecology*. Proceedings of the Second National Symposium held at Ann Arbor, Michigan, May 15–17, 1967. CPNF-670503, U.S. Atomic Energy Commission Doc. TID-4500. pp. 152–156.

ENGEL, D. W., ANGELOVIC, J. W. and DAVIS, E. M. (1966a). Effects of acute gamma irradiation on the blood constituents of pinfish, *Lagodon rhomboides*. *Chesapeake Sci.*, **7**, 90–94.

ENGEL, D. W., ANGELOVIC, J. W. and DAVIS, E. M. (1966b). Effects of cobalt-60 gamma rays on the respiratory metabolism of *Artemia salina* nauplii and adults. In *Annual Report, Radiobiological Laboratory, U.S. Fish & Wildlife Service* (=Circular 244) **1965**, 38–39.

ENGEL, D. W. and DAVIS, E. M. (1965a). Radiation effects program. Physiological effects: Blood studies. In *Annual Report, Radiobiological Laboratory, U.S. Fish & Wildlife Service* (=Circular 217) **1964**, 26–27.

ENGEL, D. W. and DAVIS, E. M. (1965b). Effects of radiation on brine shrimp. In *Annual Report, Radiobiological Laboratory, U.S. Fish & Wildlife Service* (=Circular 217) **1964**, 28 only.

ENGEL, D. W. and FLUKE, D. J. (1962). The effect of water content and post-irradiation storage on radiation sensitivity of brine shrimp cysts (eggs). *Radiat. Res.*, **16**, 173–181.

ENGEL, D. W., WHITE, J. C., JR. and DAVIS, E. M. (1965a). Effect of radiation on marine organisms. Effects of X-radiation on the early life stages of fish. In *Annual Report, Radiobiological Laboratory, U.S. Fish & Wildlife Service* (=Circular 204) **1963**, 32–38.

ENGEL, D. W., WHITE, J. C., JR. and DAVIS, E. M. (1965b). Effect of radiation on marine organisms. Blood characteristics of irradiated and unirradiated fish. In *Annual Report, Radiobiological Laboratory, U.S. Fish & Wildlife Service* (=Circular 204) **1963**, 29–30.

ERRERA, M., FICQ, A., LOGAN, R., SKREB, Y. and VANDERHAEGHI, F. (1958). Relations nucleocytoplasmiques au cours de l'effet du rayonnement sur les cellules vivantes. In *Proceedings of the 2nd United Nations Conference on the Peaceful Uses of Atomic Energy, held in Geneva, 1–13 Sept. 1958.* **22** (Biological effects of radiation), 475–478.

ESCH, H. E. (1966). Dose dependent gamma irradiation effects on the resting potential of *Nitella flexilis. Radiat. Res.*, **27**, 355–362.

ESCH, H., MILTENBURGER, H. and HUG, O. (1964). Die Beeinflussung elektrischer Potentiale von Algenzellen durch Roentgenstrahlen. *Biophysik*, **1**, 380–387.

ETOH, H. (1962). Radiation effect on the regulating mechanism of fish melanophores. In *Annual Report 1961, National Institute of Radiological Sciences*, Science & Technology Agency, Japan, NIRS-1, 53–54.

ETOH, H. (1968). Changes in incorporation of ^{3}H-thymidine into hematopoietic cells of goldfish during recovery period from radiation injury. In *Annual Report 1967, National Institute of Radiological Sciences*, Science & Technology Agency, Japan, NIRS-7, 38.

ETOH, H. and EGAMI, N. (1964). Responses to Na$^+$, K$^+$, atropine and adrenaline of melanophores induced by X-irradiation in the fin of the goldfish, *Carassius auratus*. In *Annual Report 1963, National Institute of Radiological Sciences*, Science & Technology Agency, Japan, NIRS-3, 37–38

ETOH, H. and EGAMI, N. (1965). Effect of temperature on survival time of the teleost, *Oryzias latipes*, after irradiation with various doses of X-rays. In *Annual Report 1964, National Institute of Radiological Sciences*, Science & Technology Agency, Japan, NIRS-4, 42–44.

ETOH, H. and NAKAO, Y. (1962). Histological studies on irradiated fish skin. In *Annual Report 1961, National Institute of Radiological Sciences*, Science & Technology Agency, Japan, NIRS-1, 56–57.

EVANS, T. C. (1940). Oxygen consumption of *Arbacia* eggs following exposure to Röntgen radiation. *Biol. Bull. mar. biol. Lab., Woods Hole*, **79** (Suppl.), 361 only.

FEDOROV, A. F. (1965). Radiation doses for some types of marine biota under present day conditions. *Bull. Inst. océanogr. Monaco*, **64** (1334), 1–28.

FEDOROV, A. F. and PODIMAKHIN, V. N. (1962). The world ocean should be protected from radioactive contamination. *Priroda, Mosk.*, **11**, 47–50.

FOLSOM, T. R. (1959a). Approximate dosages close to submerged radioactive layers of biological interest. *Proc. Ninth Pacif. Sci. Congr.*, **16**, 170–175.

FOLSOM, T. R. (1959b). Comparison of radioactive dosage from potassium with estimated dosages from uranium and radium in marine biospheres. *Proc. Ninth Pacif. Sci. Congr.*, **16**, 176–178.

FOLSOM, T. R. and HARLEY, J. H. (1957). Comparison of some natural radiations received by selected organisms. In *The Effects of Atomic Radiation on Oceanography and Fisheries*. U.S. National Academy of Sciences. (*Publs natn. Res. Coun., Wash.*, **551**, 28–33.)

FORSSBERG, A. (1933). Der Zeitfaktor in der biologischen Wirkung von Röntgenstrahlen. II. Untersuchungen an Algen und Drosophilapuppen. *Acta radiol.*, **14**, 399–407.

FORSSBERG, A. (1934). Über die Einwirkung der Röntgenbestrahlung auf die Entwicklung von *Mesotaenium caldariorum. Acta radiol.*, **15**, 603–607.

FORSSBERG, A. (1943). Studien über einige biologischen Wirkungen der Röntgen- und gamma-Strahlen insbesondere an *Phycomyces blakesleanus. Acta radiol.* (Suppl.), **49**, 1–143.

FORSSBERG, A. G. and NOVAK, R. (1960). Growth rate and biochemical changes in *Phycomyces* at low dose-levels of X-rays. *Int. J. Radiat. Biol.* (Suppl. Immediate Low Level Effects Ionizing Radiations, Proc. Symp., Venice, 1959), 205–215. (*Nucl. Sci. Abstr.*, **15**, 3939).

FORSSBERG, A. G., NOVAK, R., DREYFUS, G. and PEHAP, A. (1960). The radiation sensitivity of *Phycomyces*, interaction of visible light and ionizing radiation. *Radiat. Res.*, **13**, 661–668.

FOSTER, R. F., DONALDSON, L. R., WELANDER, A. D., BONHAM, K. and SEYMOUR, A. H. (1949). Some effects on embryo and young rainbow trout (*Salmo gairdnerii* RICHARDSON) from exposing the parent fish to X-rays. *Growth*, **13**, 119–142.

FREEMAN, G. (1964). The role of blood cells in the process of asexual reproduction in the tunicate *Perophora veridis. J. exp. Zool.*, **156**, 157–183.

GARCIA, J. and BUCHWALD, N. A. (1963). Perception of ionizing radiation. A study of behavioral and electrical responses to very low doses of X-ray. *Boln Inst. Estud. méd. biol. Univ. nac. Méx.*, **21**, 391–405.

GIESE, A. C. (1962). *Cell Physiology*, 2nd ed., W. B. Saunders, Philadelphia.

GILES, N. H., JR. (1951). Studies on the mechanism of reversion in biochemical mutants of *Neurospora crassa*. *Cold Spring Harb. Symp. quant. Biol.*, **16**, 283–313.

GILET, R., SANTIER, S. and VILLEROUX, A. (1963). Effet comparé des rayons X de moyenne énergie et des rayons X mous sur le taux de survie de la Chlorophycée *Chlorella pyrenoidosa* CHICK. *C.r. hebd. Séanc. Acad. Sci.*, Paris, **257**, 1985–1988.

GILLET, C. (1962a). Actions de rayons X sur *Chara vulgaris* L. III. Rôle de l'euploidie dans la reponse de *Chara* à l'irradiation. *Radiat. Bot.*, **2**, 195–204.

GILLET, C. (1962b). Effets immédiates du rayonnement X sur le courant cytoplasmique de *Nitella flexilis* AG. *Bull. Acad. r. Belg. Cl. Sci.*, **48**, 1161–1168.

GILLET, C. (1963a). Actions des rayons X sur *Chara vulgaris* L. I. La morphogenèse anormale induite par l'irradiation. *Revue Cytol. Biol. vég.*, **26**, 59–76.

GILLET, C. (1963b). Actions des rayons X sur *Chara vulgaris* L. II. Modalités des effets de l'irradiation sur le développement et la cytologie nucléaire de l'appareil végétatif. *Revue Cytol. Biol. vég.*, **26**, 77–96.

GILLET, C. (1963c). Actions des rayons X sur *Chara vulgaris* L. IV. L'elongation, après irradiation, des cellules internodales de deux races chromosomique. *Radiat. Bot.*, **3**, 155–161.

GILLET, C. (1963d). Modification de la vitesse du courant cytoplasmique induite à distance par l'irradiation chez *Nitella flexilis* AG. *Experientia*, **19**, 198–199.

GILLET, C. (1964a). Relations entre la dose des rayons X et la vitesse du courant cytoplasmique chez *Nitella flexilis* AG. *Protoplasma*, **58**, 286–293.

GILLET, C. (1964b). Perméabilité des membranes cellulaires et modification du courant cytoplasmique de *Nitella flexilis* après irradiation. *Int. J. Radiat. Biol.*, **8**, 533–539.

GILLET, C. (1965). Restauration du courant cytoplasmique aussitôt après l'irradiation. *Protoplasma*, **60**, 24–30.

GILLET, C. and KLERCKX, L. (1965). Calcium flux from X-irradiated *Nitella* cells. *Life Sci.*, **4**, 1561–1565.

GODWARD, M. B. E. (1954). Irradiation of *Spirogyra* chromosomes. *Heredity, Lond.*, **8**, 293.

GODWARD, M. B. E. (1960). Resistance of algae to radiation. *Nature, Lond.*, **185**, 706.

GODWARD, M. B. E. (1962). Invisible radiations. In R. A. Lewin (Ed.), *Physiology and Biochemistry of Algae*. Academic Press, New York. pp. 551–566.

GOLDBLITH, S. A., PROCTOR, B. S., DAVISON, S. and KAN, B. (1953). Relative bactericidal efficiencies of three types of high-energy ionizing radiations. *Fd Res.*, **18**, 659–677.

GORBMAN, A. M. and JAMES, M. S. (1963). An exploratory study of radiation damage in the thyroids of coral reef fishes from the Enewetok atoll. In V. Schultz and A. W. Klement, Jr. (Eds), *Radioecology*, Reinhold, New York. pp. 385–399.

GOULD, G. W. and ORDAL, Z. J. (1968). Activation of spores of *Bacillus cereus* by γ radiation. *J. gen. Microbiol.*, **50**, 77–84.

GREEN, J. W. and ROTH, J. S. (1955). The effect of radiation from small amounts of P^{32}, S^{35} and K^{42} on the development of *Arbacia* eggs. *Biol. Bull. mar. biol. Lab., Woods Hole*, **108**, 21–28.

GROS, C. M., KEILING, R., BLOCH, J. and VILAIN, J. P. (1958). Influence des rayons X sur la survie d'un poisson (*Carassius carassius*) soumis aux basses températures. *C.r. Séanc. Soc. Biol.*, **152**, 1187–1190.

GROSCH, D. S. (1962). The survival of *Artemia* populations in radioactive sea water. *Biol. Bull. mar. biol. Lab., Woods Hole*, **123**, 302–316.

GROSCH, D. S. and SMITH, Z. H. (1957). X-ray experiments with *Molgula manhattensis*: adult sensitivity and induced zygotic lethality. *Biol. Bull. mar. biol. Lab., Woods Hole*, **112**, 171–179.

GUTKNECHT, J. (1965). Uptake and retention of cesium-137 and zinc-65 by seaweeds. *Limnol. Oceanogr.*, **10**, 58–66.

HALBERSTAEDTER, L. and BACK, A. (1942). The effect of X-rays on single colonies of *Pandorina*. *Br. J. Radiol.*, **15**, 124–128.

HALBERSTAEDTER, L. and LUNTZ, A. (1929). Die Wirkung der Radiumstrahlen auf *Eudorina elegans*. *Arch. Protistenk.*, **68**, 177–186.

HARGIS, W. J., ARRIGHI, M. F., RAMSEY, R. W. and WILLIAMS, R. (1957). Some effects of high frequency X-rays on the oyster drill, *Urosalpinx cinera*. *Proc. natn. Shellfish. Ass.*, **47**, 68–72.

HARLEY, J. (1956). Operation troll. *Rep. atom. Energy Commn. U.S.*, NYO-4656.

HARVEY, R. S. (1964). Uptake of radionuclides by freshwater algae and fish. *Hth Phys.*, **10**, 243–247.

HARVEY, R. S. (1969). Effects of temperature on the sorption of radionuclides by a blue-green alga. In *Symposium on Radioecology*. Proceedings of the Second National Symposium, Ann Arbor, Michigan, May 15–17, 1967. CONF-670503, U.S. Atomic Energy Commission Doc. TID-4500. pp. 266–269.

HAYNES, R. H. (1966). The interpretation of microbial inactivation and recovery phenomena. *Radiat. Res. (Suppl.)*, **6**, 1–29.

HELLWIG, H. (1963). Einwirkung von Röntgenstrahlen auf die Kieselalgae (Diatomee) *Nitzschia linearis*. *Radiat. Bot.*, **3**, 249–257.

HENLEY, C. and COSTELLO, D. P. (1957). The effects of X-irradiation on the fertilized eggs of the annelid, *Chaetopterus*. *Biol. Bull. mar. biol. Lab.*, *Woods Hole*, **112**, 184–195.

HENSHAW, P. S. (1932). Studies on the effect of Röntgen rays on the time of the first cleavage in some marine invertebrate eggs. I. Recovery from Röntgen ray effects in *Arbacia* eggs. *Am. J. Roentg.*, **27**, 890–898.

HILL, R. F. and SIMSON, E. (1961). A study of radiosensitive and radioresistant mutants of *Escherichia coli* strain B. *J. gen. Microbiol.*, **24**, 1–14.

HILLOVÁ, J. (1967). Recovery of X-ray induced damage in *Chlamydomonas reinhardi*. *Bull. Boris Kidric Inst. Nucl. Sci.*, **17**, No. 4/A, Suppl. 1966 (1967). (U.S. Atomic Energy Commission translation, No. AEC-tr-6646/4A, pp. 11–15.)

HILLOVÁ, J. and DRÁŠIL, V. (1967). Inhibitory effect of iodoacetamide on recovery from sublethal damage in *Chlamydomonas reinhardti*. *Int. J. Radiat. Biol.*, **12**, 201–208.

HOLLAENDER, A., STAPLETON, G. E. and MARTIN, F. L. (1951). X-ray sensitivity of *E. coli* as modified by oxygen tension. *Nature, Lond.*, **167**, 103–104.

HOLLAENDER, A. and ZIMMER, E. M. (1945). The effect of ultraviolet radiations and X-rays on mutation production in *Penicillium notatum*. *Genetics, Princeton*, **30**, 8.

HOLWECK, F. and LACASSAGNE, A. (1931a). Action des rayons α sur *Polytoma uvella*. Détermination des 'cibles' correspondant aux principales lésions observées. *C.r. Séanc. Soc. Biol.*, **107**, 812–814.

HOLWECK, F. and LACASSAGNE, A. (1931b). Essai d'interprétation quantique des diverses lésions produites dans les cellules par les radiations. *C. r. Séanc. Soc. Biol.*, **107**, 814–817.

HOPPENHEIT, M. (1969). Strahlenbiologische Untersuchungen an Gammariden (Crustacea, Amphipoda). *Helgoländer wiss. Meeresunters.*, **19**, 163–204.

HORSLEY, R. J., BANERJEE, S. N. and BANERJEE, M. (1967). Analysis of lethal responses in *Oedogonium cardiacum* irradiated at different cell stages. *Radiat. Bot.*, **7**, 465–476.

HORSLEY, R. J. and FUČIKOVSKY, L. A. (1961). Further growth and radiation studies with filamentous green algae. *Int. J. Radiat. Biol.*, **4**, 409–428.

HORSLEY, R. J. and FUČIKOVSKY, L. A. (1963). Variation in radiosensitivity during the cell-cycle of *Oedogonium cardiacum*. *Int. J. Radiat. Biol.*, **6**, 417–429.

HORTOBAGYI, T. and VIGASSY, J. (1967). Micro-organisms in the reactor circuits exposed to radiation of the nuclear reactor Budapest-Csilleberc. *Acta biol. hung.*, **18**, 151–160. (*Nucl. Sci. Abstr.*, **22**, 48748.)

HOWARD, A. (1968). The oxygen requirement for recovery in split-dose experiments with *Oedogonium*. *Int. J. Radiat. Biol.*, **14**, 341–350.

HOWARD, A. and HORSLEY, R. J. (1960). Filamentous green algae for radio-biological study. *Int. J. Radiat. Biol.*, **2**, 319–330.

HOWARD-FLANDERS, P. and ALPER, T. (1957). The sensitivity of microorganisms to irradiation under controlled gas conditions. *Radiat. Res.*, **7**, 518–540.

HUG, O. (1958). Die Auslösung von Fühlerreflexen bei Schnecken durch Röntgen- und Alphastrahlen. *Strahlentherapie*, **106**, 155–160.

HUG, O. (1960). Reflex-like responses of lower animals and mammalian organs to ionizing radiation. In *Immediate and Low Level Effects of Ionizing Radiations. Int. J. Radiat. Biol.* (Spec. Suppl.), **1960**, 217–224.

HUG, O., ESCH, H. and MILTENBURGER, H. (1964). Electrophysiologische Begleiterscheinungen strahleninduzierter Bewegungen bei Mimosen. *Biophysik.*, **1**, 374–379.

HUG, O. and MILTENBURGER, H. (1962). Strahleninduzierte Turgorbewegungen (Radionastium) bei Mimosen und anderen sensitiven Pflanzen. *Naturwissenschaften*, **49**, 499.

HYODO, Y. (1964). Influence of temperature on the survival time and the development of histopathological changes in the intestine of the irradiated goldfish, *Carassius auratus*. In *Annual Report 1963, National Institute of Radiological Sciences*, Science & Technology Agency, Japan, NIRS-**3**, 33–35.

HYODO, Y. (1965). Development of intestinal damage after irradiation and rate of ³H-thymidine incorporation into intestinal epithelial cells of irradiated goldfish, *Carassius auratus*, at different temperatures. In *Annual Report 1964, National Institute of Radiological Sciences*, Science & Technology Agency, Japan, NIRS-**4**, 37–39.

HYODO, Y. (1966). Effects of X-irradiation on phosphatase activity in the intestine of the goldfish, *Carassius auratus*, at different temperatures. In *Annual Report 1965, National Institute of Radiological Sciences*, Science & Technology Agency, Japan, NIRS-**5**, 33 only.

HYODO, Y. and AOKI, K. (1963). Histopathological observations of the intestinal epithelium of goldfish after total-body X-irradiation. In *Annual Report 1962, National Institute of Radiological Sciences*, Science & Technology Agency, Japan, NIRS-**2**, 35–36.

HYODO-TAGUCHI, Y. (1968). Effect of X-irradiation on DNA synthesis and cell proliferation in the intestinal epithelial cells of the goldfish at different temperatures, with special reference to recovery process. In *Annual Report 1967, National Institute of Radiological Sciences*, Science and Technology Agency, Japan, NIRS-**7**, 39–40.

IVANOV, V. N. (1967). Effect of radioactive substances on the embryonic development of fish. (Russ.). Translated from pp. 185–190 of *Voprosy Biookeanografii*. Materialy II. Mezhdunarodnogo Okeanografiiheskogo Kongressa. 30 Maya-9 Iyunya 1966 g, Moscow, Naukova Dumka, Kiev, 1967. (U.S. Atomic Energy Commission Doc. No. AEC-tr-6940, pp. 47–51.)

IWASAKI, T. (1963). Effects of fractionated irradiation on encysted dry eggs of *Artemia salina*. In *Annual Report 1962, National Institute of Radiological Sciences*, Science & Technology Agency, Japan, NIRS-**2**, 33.

JACOBSON, A. P. (1966). Measurement of bioluminescence and cellular respiration during X-ray exposure. Thesis, University of Michigan, Ann Arbor, Michigan. (*Nucl. Sci. Abstr.*, **21**, 34839.)

JACOBSON, B. S. (1957). Evidence for recovery from X-ray damage to *Chlamydomonas*. *Radiat. Res.*, **7**, 394–406.

JACOBSON, B. S. (1962). Relationships between cell division and death in X-irradiated *Chlamydomonas* cultures. *Radiat. Res.*, **17**, 82–91.

KAN, B., GOLDBLITH, S. A. and PROCTOR, B. E. (1958). Effect of γ radiation on bacterial spores that have undergone germination induced by treatment with L-amino acids. *Fd Res.*, **23**, 41–50.

KASINOVA, G. V. (1964). X-ray sensitivity of green and colorless cells of *Euglena gracilis* KLEBE. (Russ.) *Radiobiologiya*, **4**, 4. (Engl. transl.: Israel Program for Scientific Translations, Jerusalem, 1965, 156–157.)

KIMBALL, R. F. (1957). Nongenetic effects of radiation on microorganisms. *A. Rev. Microbiol.*, **11**, 199–220.

KINNE, O. (1966). Physiological aspects of animal life in estuaries with special reference to salinity. *Neth. J. Sea Res.*, **3**, 222–244.

KINNE, O. (1967). Physiology of estuarine organisms with special reference to salinity and temperature: General aspects. In G. H. Lauff (Ed), *Estuaries*, A.A.A.S., Washington, D.C. pp. 525–540. (*Publs Am. Ass. Advmt Sci.*, **83**.)

KÖSSLER, F. (1964). Atmungsdepression nach chronischer Gamma-Bestrahlung und Photoreaktivierung bei *Chlorella*. *Naturwissenschaften*, **51**, 289.

KÖSSLER, F. (1965). Atmung und Photosynthese von *Chlorella* unter chronischer Gamma-Bestrahlung. *Radiat. Bot.*, **5**, 115–128.

KOH, W. Y., MOREHOUSE, C. T. and CHANDLER, V. L. (1956). Relative resistance of micro-organisms to cathodes rays. I. Nonsporeforming bacteria. *Appl. Microbiol.*, **4**, 143–146.

KOROTKOVA, G. P. and TOKIN, B. P. (1965). On the reaction of sponges and coelenterates to β-irradiation. (Russ.). *Radiobiologiya*, **5**, 190–197. (Engl. transl.: U.S.A.E.C. Transl. Ser., AEC-tr-6599).

KULIKOV, I. V., TIMOFEEVA, N. A. and SHISHENKOVA, L. K. (1968). Radiosensitivity of developing tench embryos (*Tinca tinca* L.). (Russ.) *Radiobiologiya*, **8**, 391–395.

KUMAR, H. D. (1964). Effects of radiation on the blue-green algae. II. Effects on growth. *Ann. Bot.*, **28**, 555–564.

LANGENDORFF, H., LANGENDORFF, M. and REUSS, A. (1933). Über die Wirkung von Röntgenstrahlen verschiedener Wellenlänge auf biologische Objekte. *Strahlentherapie*, **46**, 655–662.

LAWRENCE, C. W. (1965). Influence of non-lethal doses of radiation on recombination in *Chlamydomonas reinhardii*. *Nature, Lond.*, **206**, 789–799.

LEA, D. E. (1955). *Action of Radiation on Living Cells*, Cambridge University Press, London.

LEA, D. E., HAINES, R. B. and BRETSCHER, E. (1941). The bactericidal action of X-rays, neutrons, and radioactive radiations. *J. Hyg., Camb.*, **41**, 1–16.

LEA, D. E., HAINES, R. B. and COULSON, C. A. (1936). The mechanism of the bactericidal action of radioactive radiations. *Proc. R. Soc. (B)*, **120**, 40–47.

LEA, D. E., HAINES, R. B. and COULSON, C. A. (1937). The action of radiations on bacteria. III. γ rays on growing and on non-proliferating bacteria. *Proc. R. Soc. (B)*, **123**, 1–21.

LEY, F. J. (1963). The influence of dose rate in the inactivation of microorganisms. *Int. J. appl. Radiat. Isotopes*, **14**, 38–41.

LIU, C. K. (1948). X-radiation effects on the restitution of dissociated *Microciona*. *Biol. Bull. mar. biol. Lab., Woods Hole*, **95**, 259.

LURIA, S. E. (1939). Action des radiations sur le *Bactérium coli*. *C.r. hebd. Séanc. Acad. Sci., Paris*, **209**, 604–606.

LYNCH, W. F. (1958). The effect of X-rays, irradiated sea water, and oxydizing agents on the rate of attachment of *Bugula* larvae. *Biol. Bull. mar. biol. Lab., Woods Hole*, **114**, 215–225.

MacLEOD, R. A. (1965). The question of the existance of specific marine bacteria. *Bact. Rev.*, **29**, 9–23.

MARČENKO, E. (1965). Restoration of irradiated algae after a period of darkness. *Nature, Lond.*, **207**, 542–543.

MARTIN, S. D. and DERSE, P. H. (1968). Use of gamma radiation to control algae. *Environ. Sci. Technol.*, **2**, 1041–1043.

MATSUYAMA, A., NAMIKI, M. and OKOZAWA, T. (1967). Alkali halides as agents enhancing the lethal effect of ionizing radiations on microorganisms. *Radiat. Res.*, **30**, 687–701.

MATSUYAMA, A., THORNLEY, M. J. and INGRAM, M. (1964). The effect of freezing on the radiation sensitivity of vegetative bacteria. *J. appl. Bact.*, **27**, 110–124.

MAUCHLINE, J. and TEMPLETON, W. L. (1964). Artificial and natural radio-isotopes in the marine environment. *Oceanogr. mar. Biol., A. Rev.*, **2**, 229–279.

MORGAN, F. (1960). Fisheries radiobiology and the discharge of radioactive wastes. In *Disposal of Radioactive Wastes*. Proceedings of the Scientific Conference. . . . held at . . . Monaco, 16–21 Nov., 1959. I.A.E.A., Vienna, **2**, 17–24.

MORGAN, F. (1962). The design and development of marine monitoring programmes. In *Seminar on Agricultural and Public Health Aspects of Radioactive Contamination in Normal and Emergency Situations*, Scheveningen, Holland, Dec., 1961. F.A.O./W.H.O./I.A.E.A. 233–237.

MOUTSCHEN, J. (1957). Action des rayonnement X sur la croissance des cellules internodales de l'algue *Chara vulgaris* L. *Experientia*, **13**, 240–244.

MOUTSCHEN, J. and DAHMEN, M. (1956). Sur les modifications de la spermiogenèse de *Chara vulgaris* L. induites par les rayons X. *Revue Cytol. Biol. vég.*, **17**, 3–4.

NADSON, G.-A. and PHILIPPOV, G.-S. (1925). Influence des rayons X sur la sexualité et la formation des mutantes chez les champignons inférieurs (Mucorintes). *C.r. Séanc. Soc. Biol.*, **93**, 473–475.

NADSON, G.-A. and PHILIPPOV, G.-S. (1928). De la formation de nouvelles races stables chez les champignons inférieurs sous l'influence des rayons X. *C.r. hebd. Séanc. Acad. Sci., Paris*, **186**, 1566–1568.

NAKANISHI, Y. H., IWASAKI, T., NAKAZAWA, T. and KATO, H. (1964). Effects of gamma irradiation on the development of *Artemia* eggs. In *Annual Report 1963, National Institute of Radiological Sciences*, Science & Technology Agency, Japan, NIRS-**3**, 31–32.

NAKAZAWA, T. and YASUMASU, I. (1964). Effect of ^{32}P transmutation on the development of *Artemia* egg. In *Annual Report 1963, National Institute of Radiological Sciences*, Science and Technology Agency, Japan, NIRS-**3**, 42–43.

NYBOM, N. (1953). Some experiences from mutation experiments in *Chlamydomonas*. *Hereditas*, **39**, 317–324.

O'BRIEN, J. P. (1946). Studies on the cellular basis of regeneration in *Nais paraguayensis* and the effects of X-rays thereon. *Growth*, **10**, 25–44.

PARK, H. D. (1958). Sensitivity of hydra tissues to X-rays. *Physiol. Zool.*, **31**, 188–193.

PEPPER, R. E., BUFFA, N. T. and CHANDLER, V. L. (1956). Relative resistances of microorganisms to cathode rays. III. Bacterial spores. *Appl. Microbiol.*, **4**, 149–152.

PERSHINA, Z. G., KOZNOVA, L. B., SOBOLEV, S. M. and KRUSHCHOV, V. G. (1966). Influence of dose rate and time factor on bactericidal effect of radiation. (Russ.). In M. P. Domshlak (Ed.), *Voprosy Obshchei Radiobiologii*, Atomizdat, Moscow. (*Nucl. Sci. Abstr.*, **22**, 23685.)

PETROVÁ, J. (1942). Über den Vergleich der α-Strahlenempfindlichkeit von Kern und Plasma. *Ber. dt. bot. Ges*, **60**, 148–151.

PETROVÁ, J. (1963) The 'direct' and 'indirect' effects of ionizing radiation on the alga *Zygnema*. *Folia biol. Prāha*, **9**, 51–59.

POLIKARPOV, G. G. (1961). Ability of some Black Sea organisms to accumulate the fission products. *Science, N.Y.*, **133**, 1127–1128.

POLIKARPOV, G. G. (1966). *Radioecology of Aquatic Organisms*. (Translated, with revisions by the author, from the Russian edition.) North-Holland Publishing Co., Amsterdam.

POLIKARPOV, G. G. and IVANOV, V. N (1961). Action of ^{90}Sr–^{90}Y on developing Khamsa spawn (Russ.). *Voprosi Iktheologii*, **1**, 583–589. (Engl. transl.: U.K.A.E.A., Transl. TRG Inf. Series 166 (W), 1963.)

POLIKARPOV, G. G. and IVANOV, V. N. (1962). Harmful effects of strontium–90–yttrium–90 on the early development of red mullet, wrasse, and anchovy. *Dokl. Proc. Acad. Sci. U.S.S.R. (Biol. Sci.)*, **144**, 219–222.

POLLARD, E. C. (1966). Phenomenology of radiation effects on microorganisms. In A. Zuppinger (Ed.), *Handbuch der medizinischen Radiologie u. Strahlenbiologie*, Vol. II, Pt 2. Springer, Berlin. pp. 1–34.

POLLARD, E. C. and GRADY, L. J. (1967). CsCl density gradient centrifugation studies of intact bacterial cells. *Biophys. J.*, **7**, 205–213.

POLLARD, E. and VOGLER, C. (1961). Radiation action on some metabolic processes in *Escherichia coli*. *Radiat. Res.*, **15**, 109–119.

POLLARD, E. C. and WELLER, P. K. (1968). Postirradiation permeability of *Escherichia coli* cells. *Radiat. Res.*, **35**, 722–739.

POMPER, S. (1965). The physical environment for fungal growth. 4. The effects of radiation. In G. C. Ainsworth and A. S. Sussman (Eds), *The Fungi*, Vol. I. Academic Press, New York. pp. 575–597.

POMPER, S. and ATWOOD, K. C. (1955). Radiation studies on fungi. In A. Hollaender (Ed.). *Radiation Biology*, Vol. II. McGraw Hill, New York. pp. 431–453.

PORTER, J. W. and KNAUSS, H. J. (1954). Inhibition of growth of *Chlorella pyrenoidosa* by β-emitting radioisotopes. *Pl. Physiol., Lancaster*, **29**, 60–63.

PORTER, J. W. and WATSON, M. S. (1954). Gross effects of growth inhibiting levels of tritium oxide on *Chlorella pyrenoidosa*. *Am. J. Bot.*, **41**, 550–555.

POSNER, H. B. (1965). Effects of gamma irradiation on growth, colony forming ability and some cellular constituents of *Chlorella pyrenoidosa*. *Radiat. Bot.*, **5**, 129–141.

POSNER, H. B. and SPARROW, A. H. (1964). Survival of *Chlorella* and *Chlamydomonas* after acute and chronic gamma irradiation. *Radiat. Bot.*, **4**, 253–257.

POWERS, E. L., LYMAN, J. T. and TOBIAS, C. A. (1968). Some effects of accelerated charged particles on bacterial spores. *Int. J. Radiat. Biol.*, **14**, 313–330.

POWERS, E. L., WEBB, R. B. and EHRET, C. F. (1958). Modification of sensitivity to radiation in single cells by physical means. *Proceedings Second International Conference on the Peaceful Uses of Atomic Energy*, **22**, 404. United Nations, New York.

PRAVDINA, G. M. (1966). Fish behavior in the vicinity of a radiation source. (Russ.). In *Voprosy Obshchei Radiobiologii*. Atomizdat, Moscow. 97–103. (Translation REF. ZH. BIOL. 1967 No. 21163.)

PRICE, T. J. (1965). Accumulation and retention of radionuclides and the effects of external radiation on molluscs; LD-50 determinations of oysters and clams. In *Annual Report, Radiobiological Laboratory, U.S. Fish & Wildlife Service* (=Circular 204), **1963**, 13–14.

RALSTON, H. J. (1939). Immediate and delayed action of X-rays on *Dunaliella salina*. *Am. J. Cancer*, **37**, 288–297.

RASTOGI, R. K., HYODO-TAGUCHI, Y. and EGAMI, N. (1968). Autoradiographic analysis of the effects of fractionated whole-body X-irradiation on the cell proliferation in the intestinal epithelium of the goldfish, *Carassius auratus*. In *Annual Report 1967, National Institute of Radiological Sciences*, Science & Technology Agency, Japan, NIRS-7, 38–39.

REES, G. H. (1962). Effects of gamma radiation on two decapod crustaceans, *Palaemonetes pugio* and *Uca pugnax*. *Chesapeake Sci.*, **3**, 29–34.

RICE, T. R. (1965). Long-term effects of cesium-137 on a copepod. In *Annual Report, Radiobiological Laboratory, U.S. Fish & Wildlife Service* (=Circular 204), **1963**, 10.

RUGH, R. (1950). The negation of squid embryo tropism by X-radiation. In *U.S.A.E.C. Report* NYOO-**1502**.

RUGH, R. (1953). The X-irradiation of marine gametes. A study of the effects of X-irradiation at different levels on the germ cells of the clam, *Spisula* (formerly *Mactra*). *Biol. Bull. mar. biol. Lab.*, *Woods Hole*, **104**, 197–209.

RUGH, R. and CLUGSTON, H. (1955). Hydration and radiosensitivity. *Proc. Soc. exp. Biol. Med.*, **88**, 467–472.

RYZNAR, L. and DRÁŠIL, V. (1967). Photoreactivation in *Chlamydomonas reinhardi* DANGEARD. *Folia microbiol.*, *Praha*, **12**, 524–528.

SANSOME, E. R., DEMEREE, M. and HOLLAENDER, A. (1945). Quantative irradiation experiments with *Neurospora crassa*. I. Experiments with X-rays. *Am. J. Bot.*, **32**, 218–226.

SAVANNAH RIVER LABORATORY (1966). *Effect of the Savannah River Plant on Environmental Radioactivity*. Radiological and Environmental Sciences Div., Savannah River Laboratory, E. I. du Pont de Nemours and Co., Aiken, South Carolina. Semiannual Report, July through December 1965 (Feb. 1966). (U.S. Atomic Energy Commission Doc. DPST-66-30-1.)

SAVANT, P., BONÉT-MAURY, P. and DEYSINE, A. (1964). Efficacité biologique des protons de 152 Mev. Determination à partir de la courbe dose-survie de *Salmonella typhi*. *Annls Inst. Pasteur*, *Paris*, **106**, 662–669.

SCHELSKE, C. L., SMITH, W. D. C. and LEWIS, J.-A. (1965). Radioactivity in the estuarine environment. In *Annual Report, Radiobiological Laboratory, U.S. Fish & Wildlife Service* (=Circular 217), **1964**, 8–13.

SCHWINGHAMER, E. A. (1958). Relation of survival to radiation dose in rust fungi. *Radiat. Res.*, **8**, 329–343.

SCOTT, R. (1954). A study of caesium accumulation by marine algae. In J. E. Johnston (Ed.), *Proceedings of Second Radioisotope Conference, Oxford*, Vol. I. Medical and Physiological Applications. Butterworths, London. pp. 373–380.

SENEGER, C. M. (1964). Effect of X-ray irradiation on oxygen uptake by planarians. *Worm Runner's Digest*, **6**, 46–51. (Quoted in: *Nucl. Sci. Abstr.*, **18**, Abstr. 31248).

SERIANNI, R. W. and BRUCE, A. K. (1968). Radioresistance of *Micrococcus radiodurans* during the growth cycle. *Radiat. Res.*, **36**, 193–207.

SERRES, F. J. DE (1964). Genetic analysis of the structure of the *ad-3* region of *Neurospora carssa* by means of irreparable recessive lethal mutations. *Genetics*, *Princeton*, **50**, 21–30.

SERRES, F. J. DE and OSTERBIND, R. S. (1962). Estimation of the relative frequencies of X-ray induced viable and recessive lethal mutations in the *ad-3* region of *Neurospora crassa*. *Genetics, Princeton*, **47**, 793–796.

SERRES, F. J. DE, WEBBER, B. B. and LYMAN, J. T. (1967). Mutation-induction and nuclear inactivation in *Neurospora crassa* using radiations with different rates of energy loss. *Radiat. Res.* (Suppl.), **7**, 160–171.

SEYMOUR, A., HELD, E., LOWMAN, F., DONALDSON, L. and SOUTH, D. (1957). Survey of radioactivity in the sea and in pelagic marine life west of the Marshall Islands, September 1–20, 1956. *Rep. atom. Energy Commn. U.S.*, **UWFL-47**.

SHECHMEISTER, I. L., WATSON, L. J., COLE, V. W. and JACKSON, L. L. (1962). The effect of X-irradiation on goldfish. I. The effect of X-irradiation on survival and susceptibility of the goldfish, *Carassius auratus*, to infection by *Aeromonas salmonicida* and *Gyrodactylus* spp. *Radiat. Res.*, **16**, 89–97.

SHEKHTMAN, YA. L. and ZANIN, V. A. (1966). Ionizing radiation action on metabolism in bacteria. (Russ.). *Radiobiologiya*, **6**, 666–670. (*Nucl. Sci. Abstr.*, **21**, 13086.)

SHEPARD, D. C., GIESE, A. C. and BRANDT, C. L. (1956). Action of X-rays on *Didinium nasutum*. *Radiat. Res.*, **4**, 154–157.

SHEVCHENKO, V. A. (1965). Effect of X-rays on the survival rate and mutational process of *Chlorella*. (Russ.). *Radiobiologiya*, **5**, 253–259. (Engl. transl.: *Rep. Congr. atom. Energy Commn U.S.* AEC-tr-6599. 134–145.)

SIX, E. (1958). Die Wirkung von Strahlen auf *Acetabularia*. III. Die Wirkung von Röntgen-strahlen und ultravioletten Strahlen auf kernhaltige Teile von *Acetabularia mediterranea*. *Z. Naturforsch.*, **13b**, 6–14.

SOLBERG, A. N. (1938). The susceptibility of *Fundulus heteroclitus* embryos to X-radiation. *J. exp. Zool.*, **78**, 441–465.

SPARROW, A. H., UNDERBRINK, A. G. and SPARROW, R. C. (1967). Chromosomes and cellular radiosensitivity. I. The relationship of D_0 to chromosome volume and complexity in seventy-nine different organisms. *Radiat. Res.*, **32**, 915–945.

SPEAR, F. G. (1944). The action of neutrons on bacteria. *Br. J. Radiol.*, **17**, 348–351.

SPIKES, J. D., MAYNE, B. C. and NORMAN, R. W. VAN (1958). Gamma radiation effects on *Chlorella* metabolism. *Pl. Physiol., Lancaster*, **33** (Suppl.), 27.

STAPLETON, G. E. (1952). Variations in the radiosensitivity of *Escherichia coli* during the growth cycle. Thesis, University of Tennessee.

STAPLETON, G. E. and HOLLAENDER, A. (1952). Mechanisms of lethal and mutagenic action of ionizing radiations on *Aspergillus terreus*. II. Use of modifying agents and conditions. *J. cell. comp. Physiol.*, **39** (Suppl. 1), 101–112.

STAPLETON, G. E. and MARTIN, F. L. (1949). Comparative lethal and mutagenic effects of ionizing radiations in *Aspergillus terreus*. *Am. J. Bot.*, **36**, 816.

STÉPHEN-DUBOIS, F. (1956). Evolution du régénerât caudal de *Nereis diversicolor* après irradiation régionale aux rayons X. *Bull. Soc. zool. Fr.*, **81**, 199–207.

STOLETOV, V. N., ZHEVNER, V. D., GARIBYAN, D. V. and SHESTIKOV, S. V. (1965). Pigment mutations in *Anacystis nidulans* induced by nitrosomethylurea. (Russ.). *Genetika*, **6**, 61–66. (*Nucl. Sci. Abstr.*, **20**, 28958.)

TAI, W. H. (1962). Effect of X rays on the accumulation of catalase in conidia of *Aspergillus niger*. *Bull. Inst. Chem. Acad. Sin.*, **6**, 71–78.

TARPLEY, W., ILAVSKY, J., MANOWITZ, B. and HORRIGAN, R. V. (1953). Radiation steriliza-tion. I. The effect of high energy gamma radiation from kilocurie radioactive sources on bacteria. *J. Bact.*, **65**, 305–309.

TEMPLETON, W. L. (1965). Ecological aspects of the disposal of radioactive wastes to the sea. In *Ecology and the Industrial Society*. Fifth Symposium of the British Ecological Society. Blackwell Scientific Publications, Oxford. pp. 65–97.

TETTEH, G. K. and CORMACK, D. V. (1968). Inositol, oxygen, and water effects on the X-ray sensitivity of *Serratia marcescens*. *Radiat. Res.*, **34**, 532–543.

THATCHER, F. S. (1963). Some public health aspects of the microbiology of irradiated foods. *Int. J. appl. Radiat. Isotopes*, **14**, 51–58.

THORNLEY, M. J. (1963a). Radiation resistance among bacteria. *J. appl. Bacteriol.*, **26**, 334–345.

THORNLEY, M. J. (1963b). Metabolic conditions before and after irradiation. In M. Ebert and A. Howard (Eds), *Radiation Effects in Physics, Chemistry, and Biology*. North Holland Publishing Co., Amsterdam. pp. 170–188.

TITANI, T., KONDO, M., SUGURO, H. and TAKAHASHI, T. (1958). Ovicidal effect of ⁶⁰Co gamma rays upon parasitic eggs and bacteria. In *Proceedings of the Second International Conference on the Peaceful Uses of Atomic Energy*, **27**, 430. United Nations, New York.

TRGOVČEVIĆ, Ž. and KUĆAN, Ž. (1969). Correlation between the breakdown of deoxyribonucleic acid and radiosensitivity of *Escherichia coli*. *Radiat. Res.*, **37**, 478–492.

TSURUGA, H. (1962). Occurrence of radioruthenium in the laver, *Porphyra tenera*, and other seaweeds. *Bull. Jap. Soc. Scient. Fish.*, **28**, 372–378.

TSYPIN, A. B. and KHOLODOV, Y. A. (1964). Development in a conditioned reflex to ionizing radiation in fish and rabbits. (Russ.). *Radiobiologiya*, **4**, 402–408. (Engl. transl.: U.S.A.E.C. Transl. Ser., AEC-tr-6599).

UNIVERSITY OF ROCHESTER (1964). *Bases for Genetic Studies in Blue-green Algae*. University of Rochester (N.Y.), U.S. Atomic Energy Commission Report NYO-3046-1.

VIJVER, G. VAN DE (1967). Studies on the metabolism of *Tetrahymena pyriformis* GL. III. Effects of γ irradiation on the respiratory rate. *Enzymologia*, **33**, 331–344.

WALLAUSCHEK, E. and LÜTZEN, J. (1964). Studies of problems relating to radioactive waste disposal into the North Sea. II. General survey on radioactivity in sea water and marine organisms. Health and Safety Office, European Nuclear Energy Agency, Organization for Economic Co-operation and Development, Paris.

WAMBERSIE, A. (1967). Contribution à l'étude de l'efficacité biologique relative des faisceaux de photons et d'electrons de 20 MeV du betatron. *J. belge Radiol. Monogr.*, **1**, 104–124.

WATSON, D. G., GEORGE, L. A. and HACKETT, P. L. (1959). Effects of chronic feeding of phosphorus-32 on rainbow trout. In *United States Atomic Energy Commission Report*, HW-**59500**, 73–77.

WEBBER, B. B. and SERRES, F. J. DE (1965). Induction kinetics and genetic analysis of X-ray induced mutations in the *ad-3* region of *Neurospora crassa*. *Proc. Natn. Acad. Sci. U.S.A.*, **53**, 430–437.

WELANDER, A. D. (1954). Some effects of X-irradiation of different embryonic stages of the trout (*Salmo gairdnerii*). *Growth*, **18**, 227–255.

WELANDER, A. D. (1968). Effects of irradiation, temperature, and other environmental factors on salmonoid embryos. Progress Report, University of Washington, Seattle, Laboratory of Radiation Ecology, June 8, 1968. (U.S. Atomic Energy Commission Doc. No. RLO-2049-1),

WELANDER, A. D., DONALDSON, L. R., FOSTER, R. F., BONHAM, K. and SEYMOUR, A. H. (1948). The effects of Röntgen rays on the embryos and larvae of the chinook salmon. *Growth*, **12**, 203–242.

WELANDER, A. D., DONALDSON, L. R., FOSTER, R. F., BONHAM, K., SEYMOUR, A. H. and LOWMAN, F. G. (1949). The effect of Röntgen rays on adult rainbow trout. In *United States Atomic Energy Commission Report* UWFL-**17**, 20 pp.

WHITE, J. C., JR. (1964). Fractional doses of X-radiation: a preliminary study of effects on teleost embryos. *Int. J. Radiat. Biol.*, **8**, 85–91.

WHITE, J. C., JR. (1965a). Acute and fractionated irradiation of early embryos of mummichog. In *Annual Report, Radiobiological Laboratory, U.S. Fish & Wildlife Service* (= Circular 217) **1964**, 28–30.

WHITE, J. C., JR. (1965b). Continuous low level irradiation of post-larval flounders. In *Annual Report, Radiobiological Laboratory, U.S. Fish & Wildlife Service* (= Circular 217) **1964**, 30–31.

WHITE, J. C., JR. (1969). Interaction of chronic gamma radiation, salinity, and temperature on the morphology of young pinfish. In *Annual Report, Radiobiological Laboratory, U.S. Fish & Wildlife Service* (= Circular 309) **1968**, 41–47.

WHITE, J. C., JR. and ANGELOVIC, J. W. (1966a). Tolerance of several marine species to Co⁻⁶⁰ irradiation. *Chesapeake Sci.*, **7**, 36–39.

WHITE, J. C., JR. and ANGELOVIC, J. W. (1966b). Acute radiation LD-50 values at different times after irradiation for several marine organisms. In *Annual Report, Radiobiological Laboratory, U.S. Fish & Wildlife Service* (= Circular 244) **1965**, 40–42.

WHITE, J. C., JR. and ANGELOVIC, J. W. (1966c). Effects of chronic irradiation on developing teleost embryos. In *Annual Report, Radiobiological Laboratory, U.S. Fish & Wildlife Service* (=Circular 244) **1965**, 43–45.

WHITE, J. C., JR., ANGELOVIC, J. W., ENGEL, D. W. and DAVIS, E. M. (1966). Interactions of radiation, salinity, and temperature on estuarine organisms. In *Annual Report to the Atomic Energy Commission, August 1, 1966.* Radiobiology Laboratory, U.S. Fish & Wildlife Service, Beaufort, N.C., 76–93.

WICHTERMAN, R. (1955). Survival and other effects following X-irradiation of the flagellate, *Euglena gracilis. Biol. Bull. mar. biol. Lab., Woods Hole,* **109,** 371.

WICHTERMAN, R. (1957). Biological effects of radiations on Protozoa. *Bios, Mount Vernon, Iowa,* **28,** 3–20.

WITKIN, E. M. (1947). Genetics of resistance to radiation in *Escherichia coli. Genetics, Princeton,* **32,** 221–248.

WOESE, C. R. (1958). Comparison of the X-ray sensitivity of bacterial spores. *J. Bact.,* **75,** 5–8.

WOOD, E. J. F. (1965). *Marine Microbial Ecology,* Reinhold Publishing Corp., New York.

WOOD, E. J. F. (1967). *Microbiology of Oceans and Estuaries,* Elsevier Publishing Company, New York.

YAMAGUCHI, T. (1962). Changes in the intercellular Na, K and Ca contents of *Paramecium* after γ-irradiation. In *Annual Report 1961, National Institute of Radiological Sciences,* Science & Technology Agency, Japan, NIRS-1, 45–46.

ZAKHAROV, I. A. and TUGARINOV, V. V. (1964). Radiosensitivity of the unicellular alga *Chlorella vulgaris.* (Russ.) *Radiobiologiya,* **4,** 92–95. (Engl. transl. *Rep. Congr. atom. Energy Commn. U.S.,* AEC-tr-**6404,** 126–130.)

ZELLE, M. R. and HOLLAENDER, A. (1955). Effects of radiation on bacteria. In A. Hollaender (Ed.), *Radiation Biology,* Vol. II, McGraw-Hill, New York. pp. 365–430.

ZHDANOVA, N. N. and P'YANKOV, G. N. (1967). Effect of the dose rate of ^{60}Co γ rays on the survival of some Dematiaceae species. (Russ.) *Izv. Akad. Nauk SSSR. (Ser. Biol.),* **3,** 438–442. (*Nucl. Sci. Abstr.,* **21,** 28689.)

ZHESTYANIKOV, V. D. (1966). The times of repair and development of radiation injury of *Escherichia coli* B. In E. Ya. Graevskii, V. I. Ivanov and V. I. Korogodin (Eds), *Zashchita i Vosstauovlenie pri Luchevykh Povrezhdeniyakh.* (Russ.) Moscow Izdatel'stvo Nauka. (*Nucl. Sci. Abstr.,* **22,** 26009.)

ZHESTYANIKOV, V. D. (1967). Factors of radioresistance of the bacterial cell. (Russ.) *Tsitologiya,* **9,** 1460–1480. (*Nucl. Sci. Abstr.,* **21,** 36412).

ZILL, L. P. and TOLBERT, N. E. (1958). The effect of ionizing and ultraviolet radiations on photosynthesis. *Arch. Biochem. Biophys.,* **76,** 196–203.

ZIRKLE, R. E. (1940). The radiobiological importance of the energy distribution along ionization tracks. *J. cell. comp. Physiol.,* **16,** 221–235.

ZOBELL, C. E. (1946). *Marine Microbiology,* Chronica Botanica, Waltham, Mass.

ZOTIKOV, A. A. (1960). Effect of ionizing radiation on the luminous bacteria *B. issatchenkoi;* radiobiological problem of the time factor and of recovery. (Russ.). *Biofizika,* **5,** 170–175. (*Nucl. Sci. Abstr.,* **15,** 1257.)

12. FACTOR COMBINATIONS

RESPONSES OF MARINE POIKILOTHERMS TO ENVIRONMENTAL FACTORS ACTING IN CONCERT

D. F. ALDERDICE

(1) Introduction

It is presumed that most ecologists will recognize the term 'factor combinations' as implying a description of the interrelationships between biological response and multiple physical, chemical or biological environmental stimuli acting in concert. Factor combinations will be discussed by making use of concepts employed in response surface methodology, a technique introduced by Box and WILSON (1951) for examining multivariable phenomena. Discussion of the statistical and mathematical aspects of response surface analysis will be minimized in favour of general concepts. Readers interested in relevant aspects of experimental design and numerical analysis should consult, among others, Box and WILSON (1951), Box (1956), COCHRAN and COX (1957), HILL and HUNTER (1966) and PENG (1967). As the topic is developed, it is hoped that the reader may perceive in the examination of biological response surfaces a means of describing some of the basic biological-environmental interrelations which may underlie many ecological phenomena.

Traditionally, ecologists have been confronted with the problem of defining the relations between attributes of individuals or populations and those of their environment. The analysis of such multidimensional problems most often has been approached in two general ways: (i) by dismemberment of a complex phenomenon and examination of recognized components by univariable techniques; (ii) by qualitative or semi-quantitative description of the multidimensional aspects of a problem, as far as they may be ascertained by observation and deduction. The differences between these approaches are analogous to those existing between laboratory and field investigations of biological problems.

Laboratory studies tend to be concerned with relationships within controlled but simplified environments. Such experimental evidence is likely to document with relative specificity the potential of living processes. However, when ecological implications are generated from such evidence, there is room for considerable speculation on the manner in which a biological potential may or may not be realized in the more complex and unconstrained natural environment. Thus, the simplicity of the laboratory approach leads to a lack of confidence in the ability to interpret ecological phenomena through inductive inference.

On the other hand, field enquiry searches for relationships in the unconstrained environment. The ability of the investigator to predict events from samples obtained within quite limited dimensions of space and time will depend on his success in recognizing the number of biological, chemical or physical entities which are implicit in the association examined. Thus, the complexities inherent in the field approach lead to a suspicion that interpretation of ecological phenomena may be constrained by the natural limits of human deductive capabilities. The generation of ecological inferences from either field or laboratory data implies that increasing attention must be paid to the multidimensionality of ecological processes.

Response surface methodology should help extend inferential power in the laboratory analysis of ecological problems. Other multivariate techniques applied in the field should provide a means of determining not only the component variables which are important in nature, but also the degree to which relationships derived in the laboratory may be expressed or realized in the natural environment.

(a) Univariable Relationships

It is common practice in laboratory studies to attempt to quantify the relationships between an environmental entity, such as salinity or temperature, and a selected response of an organism exposed to defined levels of the entity. The underlying functional relationship between variable and response is generally unknown. As an alternative, an empirical means of describing the relationship is usually employed. The simplest of such empirical relations is expressed in the linear regression of response score (Y) on the levels of one variable (x_1). Thus,

$$Y = b_0 x_0 + b_1 x_1 \tag{1}$$

where $b_0 x_0$ is the value of Y for $x_1 = 0$, and b_1 measures the slope of the linear relation between Y and x_1*.

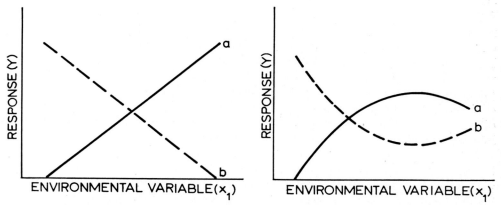

Fig. 12-1: Rectilinear and second order relations between response, Y, and one independent variable, x_1. (Original.)

In some cases, the experimental relationship obtained may resemble that of an exponential form. Transformation of the Y's and x's to logarithms may reduce the expression to rectilinear form, thus providing some convenience in numerical computation or comparison of results. Hence, if $Y = ax^b$, then $Y' = a' + bx'$, where $Y' = \log Y$, $x' = \log x$ and $a' = \log a$.

* In simple linear regression the mean value of Y at $x = 0$ is often expressed as a in the form $Y = a + bx$, where $x = x_1$ of Equation (1). In Equation (1) a 'dummy' variable x_0 has been introduced which always takes the value unity; hence $b_0 x_0 = b_0 = a$.

Where a polynomial expression is utilized, the addition to Equation (1) of a further term for simple curvature yields a second order form

$$Y = b_0 x_0 + b_1 x_1 + b_{11} x_1{}^2 \tag{2}$$

The rectilinear and second order expressions are illustrated in Fig. 12-1. In both cases, the relationship may take two forms (a, b), depending on the signs involved in the expressions. The relationships may be extended further by including higher degree terms (e.g. degree 3, $b_{111} x_1{}^3$). In practice, it is usually unnecessary to attempt to fit to biological data polynomials of degree greater than three.

(b) Multivariable Relationships

When the response of an organism is considered over levels of two variables (x_1, x_2), the straight line relation in the previous section becomes a surface or plane in x_1, x_2. The values of Y obtained will cluster in the level of the plane in x_1, x_2. Furthermore, an interaction term $(b_{12} x_1 x_2)$ may enter the expression: the magnitude of the response (Y) may change over levels of x_1 in a manner dependent on the level of x_2 at which Y is obtained. The expression then becomes

$$Y = b_0 x_0 + b_1 x_1 + b_2 x_2 + b_{12} x_1 x_2 \tag{3}$$

and the straight line relation in Fig. 12-1 may be considered as a section or transect of the plane in the first panel of Fig. 12-2 (actually a curved surface, but linear along sections parallel to the x_1 - or x_2 - axes).

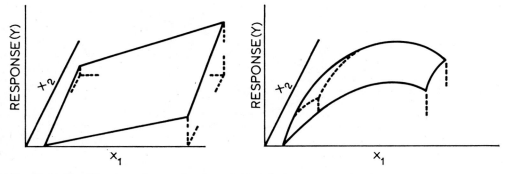

Fig. 12-2: Rectilinear and second order relations between response, Y, and two independent variables, x_1 and x_2. (Original.)

More often, however, biological data, measured with respect to levels of two environmental variables, describe a curved plane in which there is some form of curvature in the relation between Y and both x_1 and x_2. Simple curvature may be described by a second order expression such as

$$Y = b_0 x_0 + b_1 x_1 + b_2 x_2 + b_{11} x_1{}^2 + b_{22} x_2{}^2 + b_{12} x_1 x_2 \tag{4}$$

In general, expansion of the polynomial generates sets of terms (b_0), $(b_1, b_2, \ldots)$, $(b_{11}, b_{22}, \ldots, b_{12}, \ldots)$, $(b_{111}, \ldots, b_{112}, \ldots)$, $\ldots$ which are, respectively, of degree $0, 1, 2, 3, \ldots$ A second order surface, for example, employs a polynomial expanded to include terms up to degree 2 (e.g. Equation (4) for 2 factors). In addition,

b_0, b_1, etc. can be called regression coefficients. With the x_1, x_2, x_1^2, x_2^2, x_1x_2 terms (Equation 4) as independent variables, b_0x_0 measures the mean effect, b_1x_1 and b_2x_2 the linear effects, $b_{11}x_1^2$ and $b_{22}x_2^2$ the quadratic effects related to simple curvature, while $b_{12}x_1x_2$ is a second degree term measuring the linear × linear interaction between x_1 and x_2. The curved line in Fig. 12-1 may be considered as a transect of the curved plane shown in the second panel of Fig. 12-2 (a curved surface, but with curved sections parallel to the x_1- or x_2-axes). Values of Y would tend to cluster about the surface of the curved plane at various levels of x_1 and x_2.

If the curved plane in the second panel of Fig. 12-2 is extended until the response is minimized at high and low levels of both variables x_1 and x_2, the nature of the surface generated is that generalized in Fig. 12-3. Where $Y = 0$, the curved plane meets the x_1, x_2 plane and describes an ellipse with axes ac and db. The point of maximum response e on the surface, when projected onto the x_1, x_2 plane, meets the plane at s, the centre of the ellipse.

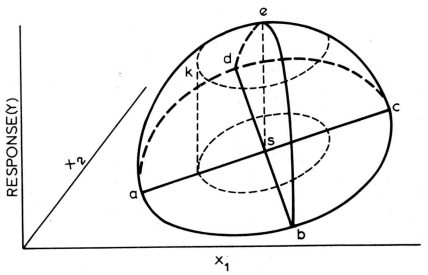

Fig. 12-3: Second order response surface. (Original.)

For purposes of visualization, the second order surface in Fig. 12-3 may be considered as enclosing a solid. If the solid is lowered into a container of water, the water will rise and wet the surface in a series of elliptical boundaries such as that upon which point k rests. If this ellipse is projected onto the x_1, x_2 plane, it will form an isopleth upon which response Y will take a specific value. Any number of isopleths could be described on the x_1, x_2 plane. Each would describe an ellipse on the x_1, x_2 plane and associate a number of combinations of levels of the environmental variables with a specific level of response.

Response surfaces

The configuration illustrated in Fig. 12-3 is an example of what is generally described as a response surface (BOX and WILSON, 1951; HILL and HUNTER, 1966),

relating levels of response to those of independent quantitative variables at which
the response is elicited. The analysis of response surfaces may be considered in
general terms as the expansion as its Taylor series of the response function

$$\eta_u = \phi(x_{1u}, x_{2u}, \ldots, x_{ku}) \tag{5}$$

where η is estimated by measure of response, Y, at $u = 1, \ldots, N$ combinations of
k quantitative variables $x_1, x_2, \ldots, x_k$ (Box and Wilson, 1951). Those seeking
further detail on numerical analysis are urged to consult the review of response
surface methodology by Hill and Hunter (1966), and in particular the excellent
treatment of the subject by Box (1956).

It would appear that second order surfaces of the nature described in Equation
(4) and Fig. 12-3 may often be sufficient for approximating the relationship
between a response and levels of two or more variables within a restricted region
of the factor space spanned by the x variables. Because biological response
phenomena (e.g. survival, growth rate, oxygen consumption) often reach local
maxima within a range of levels of an environmental variable, the second order
expression provides a means of approximating the relationship between levels of
environmental variables and response maxima (or minima).

The fitting of a response surface to biological data may be undertaken using
standard regression analysis. Analysis of variance of such data usually suggests
that a highly significant portion of the total variance is explained by treatment
effects (exposure to various combinations of levels of the environmental x variables). However, a significant portion of the treatment variance often remains
unexplained after fitting a second order regression surface (Forrester and
Alderdice, 1966; Alderdice and Forrester, 1968). It is suspected, in many
cases, that the unexplained variance is associated with departures of the true
response surface from that generated with the emperical expression. For
example, the effects of temperatures near the lower incipient lethal level (Fry,
1947) tend to introduce what appears to be a cubic term (e.g. x_1^3), thereby disturbing the variance ratio used to examine the degree to which the second order
expression fits all of the data (Alderdice, 1963a). Experimental designs accom-
modating cubic effects are often outside the range of experimental economy. On
the other hand, the fitting of second order regression surfaces to experimental
data remains a very useful method for approximating multivariable response
relations, as long as the factor space is limited to examining those levels of environmental variables found in the vicinity of response maxima. A possible resolution to
the problem of fitting a surface to data which are more complex than the constraints of a simple model will allow, is outlined on p. 1710.

Before the exploration of a multivariable biological response surface is begun,
the experimenter first must choose the ranges of the variables which will be
investigated. Often he must rely on prior but incomplete information and hope
that the ranges provisionally set will encompass the region of maximum values
of the response selected for measurement. In some cases, the initial trials may not
find a maximum; responses may be found to increase in some particular direction
near the border, or beyond the limits of the range of the variables employed. The
response surface in such cases is usually described as a ridge system (Fig. 12-4). If
the axis of the ridge system can be calculated, following the axis experimentally

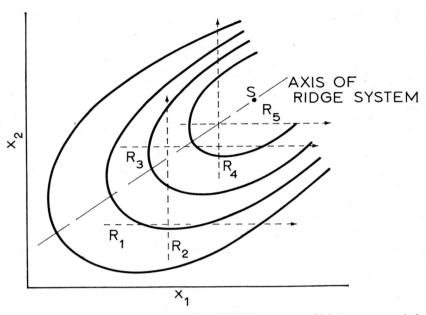

Fig. 12-4: Representative response surface ridge system. If factors are varied
one at a time, test runs R_1 to R_5 will move toward higher response values
but may not locate the centre of the surface, S. (Original.)

toward higher response values will usually locate a maximum. If the maximum is
remote from the original experimental space, a new experimental space would
likely be required, centred around the provisional estimate of the maximum
obtained. Single variable analysis of ridge systems, illustrated by the R_1, R_2, R_3,
R_4 and R_5 test runs in Fig. 12-4 are difficult to interpret and are experimentally
uneconomic in time and effort. Such runs may be repeated on the rising ridge
until the experimenter is unable to distinguish a real change in response from
experimental error. Yet, he still may not find the centre of the response surface.
It is more efficient to use factorial designs employing several environmental
variables simultaneously, followed by calculation of the configuration of the
response surface.

An example of a two-variable second order regression surface is illustrated in
Fig. 12-5 (ALDERDICE, 1963a). In this case, isopleths of median resistance time in
minutes are shown for the effect of sodium pentachlorophenate on young coho
salmon *Oncorhynchus kisutch* at various combinations of levels of salinity and
temperature. Such expressions may be extended to include further environmental
variables.

In the case where response is determined as a function of three environmental
variables, the expression is

$$Y = b_0 x_0 + b_1 x_1 + b_2 x_2 + b_3 x_3 + b_{11} x_1{}^2 + b_{22} x_2{}^2 + b_{33} x_3{}^2 + b_{12} x_1 x_2 + b_{13} x_1 x_3$$
$$+ b_{23} x_2 x_3 \tag{6}$$

In the region of the response domain where response is maximized, the surface may
be approximated by an ellipsoid with principal axes X_1, X_2 and X_3 (Fig. 12-6).

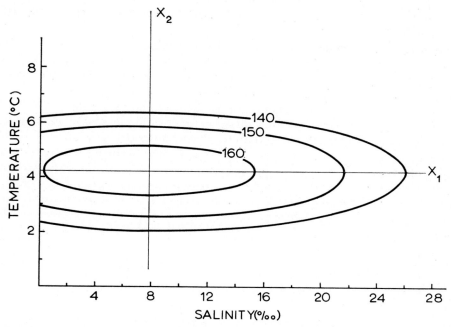

Fig. 12-5: Isopleths of median resistance time in minutes, for juvenile coho salmon *Oncorhynchus kisutch*, exposed to 3 mg/l sodium pentachlorophenate at various salinities and temperatures. (After ALDERDICE, 1963a; modified.)

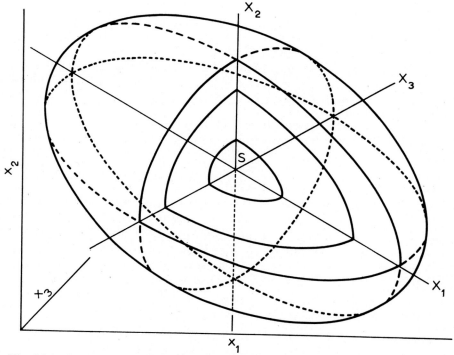

Fig. 12-6: Three-factor second order response surface. A section of the ellipsoid is cut away to show isopleths of response decreasing in three planes from a maximum at the centre. S. X_1, X_2 and X_3 are the axes of the surface. (Original.)

In this case, response isopleths may be considered as concentric 'skins', like those of an onion, within the response solid.

Beyond three independent or environmental variables, such a relationship cannot be visualised geometrically. However, by appropriate numerical treatment (Box, 1956), such multidimensional systems may be handled effectively with the use of appropriate factorial designs (Box, 1956; Box and Hunter, 1961), standard multiple regression analysis, analysis of variance and matrix algebra. When using more than two environmental variables, much of the unwieldy computation involved is eliminated by computer analysis of the numerical data.

The geometric description of such surfaces may be obtained directly from the second order expression, e.g., Equation (4) for two environmental variables. However, these polynomial equations can be expressed in simplest or canonical form by a process of canonical reduction (Box, 1956), which involves the translation of the centre of the experimental space to the centre of the response surface, and rotation of the axes of measurement to coincide with those of the surface. For example:

$$Y = b_0 x_0 + b_1 x_1 + b_2 x_2 + b_{11} x_1^2 + b_{22} x_2^2 + b_{12} x_1 x_2 \qquad (4)$$

reduces to $\quad Y - Y_s = \lambda_{11} X_1^2 + \lambda_{22} X_2^2 \qquad (7)$

where $\qquad\qquad Y =$ response

$Y_s =$ response at the centre of the surface, S

$\lambda_{11}, \lambda_{22} =$ eigenvalues, denoting the rate at which the response changes in moving away from S

$X_1, X_2 =$ axes of the response surface.

Furthermore, the X- and the x-axes are related as follows:

$$\begin{bmatrix} X_1 \\ X_2 \end{bmatrix} = \begin{bmatrix} m_{11} & m_{12} \\ m_{21} & m_{22} \end{bmatrix} \begin{bmatrix} x_1 - x_{1s} \\ x_2 - x_{2s} \end{bmatrix} \qquad (8)$$

where $\qquad [m] =$ eigenvectors, providing for rotation from the x- to the X-axes

$x_{1s}, x_{2s} =$ locus of the centre of the response surface, providing for translation of the centre $x_1, x_2 = 0$ to $X_1, X_2 = 0$.

This short survey of some aspects of the computational analysis is meant only to serve as an introduction to the terminology employed in the following discussions of biological response surfaces.

Basic properties of biological response surfaces

Examination of biological response surfaces in the manner outlined indicates that they are dynamic. Surface configuration and surface location may change with respect to the axes of measurement. These variations may be associated both with genetic and non-genetic differences (i.e., in state of growth, size, physiological condition, previous history, or season). The manner in which such changes can occur is illustrated in Fig. 12-7. The elliptical response surface in the figure is analogous to the surface projected on the x_1-, x_2-axes of Fig. 12-3. The types of surface variation are analogous to four classes of biological phenomena:

(i) *Euryplasticity and stenoplasticity.* These terms, as usually employed, relate the ability of an organism to live successfully either within a broad or a narrow

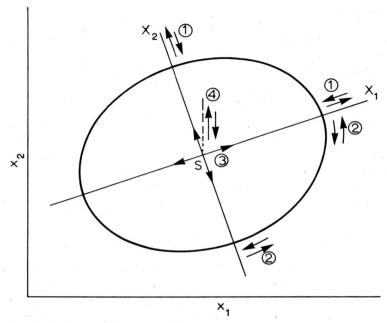

Fig. 12-7: Dynamic properties of response surfaces. (1) changes in the lengths of the X-axes (plasticity), (2) rotation of the X-axes about the centre, S (interactions), (3) translation of the centre of the surface (changes in tolerance or resistance), (4) changes in magnitude of response at the centre, S (changes in capacity). (Original.)

range of levels of an environmental variable. A quantitative connotation is implied in their use, yet not without considerable problems of interpretation. The major problem appears in the effort to relate a quantitative measure of viability to a measure of the range of the environmental variable normally available to the organism. The quantitative relationship would then be expressed, somewhat arbitrarily, in a measure of the extent to which the available range is or is not successfully utilized. The relation may be extended to include a comparison of stages of development (egg, larva, juvenile, adult) for a given species. The concept would appear also to force a distinction between limitations on viability within an available range imposed by the organism, as contrasted with restrictions on an organism which might be imposed by the available range itself (e.g. antarctic species).

The quantitative aspects of response surface analysis invite an attempt to 'measure' plasticity. For example, the relative lengths of the X_1- and X_2-axes of a response surface can vary within and between species for a given set of environmental variables x_1 and x_2 (Fig. 12-7). For a euryplastic organism, for instance, the X_1-axis of the surface may stretch over a considerable range of the environmental variable x_1 normally available to the organism. High response values, centred around S, could then extend in the direction of the X_1-axis over a considerable portion of the natural range of x_1. For a stenoplastic organism, the reverse would be true; the length of the X_2-axis could be short with reference to the range of the x_2 variable normally available to the organism. Thus, the region of high response values around the centre S and in the direction of the X_2-axis could be limited with respect to the natural range of x_2.

The argument will not be forced. Nevertheless, as an example, an organism might be considered euryplastic in terms of a given response if 90% or more of maximum response (e.g. survival) were possible over 50% of the normally available range of the environmental variable considered.

(ii) *Interaction effects.* The principal axes of the surface may rotate about the centre of the surface, S (Fig. 12-7). That is, the maximum response obtainable at a given level of one variable will depend on the level of another variable simultaneously applied. Rotation of response surface axes, therefore, indicates the presence of a coupled action, or interaction of environmental variables. The coupled relationship may be calculated for a surface equation, such as that of Equation (4), by entering a selected value of one variable (e.g. x_1) into the equation, taking the derivative of the fitted equation with respect to the second variable (x_2), setting the equation to zero and solving for x_2.

(iii) *Changes in tolerance or resistance.* The location of the centre of a response surface, S, may change (e.g. with time) in regard to the x_1, x_2 scales of measurement. With growth and development of the organism, maximum response, therefore, may shift to another locus of x_1 and x_2. Hence, as the surface changes location, response scores obtained on a specific pathway in the experimental space may change (e.g. over several levels of x_2 at a constant level of x_1). Therefore, changes in resistance or tolerance would be noted as changes in response scores with respect to the axes of measurement over selected transects of compared surfaces.

(iv) *Changes in capacity.* With growth and development of an organism, the absolute value of the response at S may increase or decrease. Conversely, a change in tolerance or resistance could occur, resulting from a change in location of a response surface, without a change necessarily occurring in the capacity of the surface. Therefore, changes in capacity would be noted by changes in absolute magnitude of response at the centres of compared surfaces.

(2) Application to Biological Phenomena

Having defined the relationship of response to levels of environmental variables in the vicinity of response maxima, the manner may be explored in which such multivariable response surfaces reflect quantitative aspects of biological phenomena.

(a) *Dynamic Properties of Response Surfaces*

As described in the previous section, the configuration of response surfaces has been found to vary in four general ways. These types of variation may be demonstrated by referring again to the salinity-temperature effects on median resistance time for juvenile coho salmon *Oncorhynchus kisutch* exposed to a concentration of 3 mg/l of sodium pentachlorophenate (Fig. 12-5; ALDERDICE, 1963a). In this example the response surfaces were examined in identical fashion at a number of different ages of the young salmon. A series of these surfaces is reproduced in Fig. 12-8. Data pertinent to this comparison are summarized in Table 12-1. From Fig. 12-7 and Table 12-1 the following observations can be made.

Plasticity

In terms of salinity, coho salmon fry (Fig. 12-8A) show high resistance to the pentachlorophenate over a broad range of salinities. The nature of the surface changes, until, in the post-smolt stage in salt water (Fig. 12-8E), the relationship is almost reversed to that of the fry. Where the fry showed resistance to pentachlorophenate over a narrow range of temperatures and a broad range of salinities, the post-smolt in salt water is resistant over a broader range of temperatures but a narrower range of salinities.

Interactions

In Fig. 12-8A, there is virtually no interaction between salinity and temperature: the axes of the response surface are coplanar with those of measurement (salinity and temperature). In panels B and C (Fig. 12-8), there is a rotation of the X-axes of the surfaces, implying the presence of an interaction between the two environmental variables acting on the response. In panel C, the centre of the surface is outside the range of possible salinity experience (below 0‰ S). Such occurrences are not unreasonable and they have been discussed elsewhere (FRY, 1947). In the remaining panels (D, E) a small interaction exists, indicating that, as salinity is increased, maximum resistance times can be maintained only with coupled small decreases in temperature.

Changes in tolerance or resistance

The centre of the surface, hence maximum resistance, moves with respect to the axes of measurement. Transects of the surface at a fixed salinity or temperature level, therefore, show changes in resistance (i.e. with age), as the transects cut across different sections of the response surface.

Changes in capacity

The absolute value of maximum resistance time decreases with increasing size (or age) from 164 mins for coho salmon fry in late summer of their first year to 50 mins in the late summer of their second year.

The changes which are possible in terms of a selected response, when response is measured over a field of levels of several environmental variables, leads one to conclude that organisms may exhibit greater dynamic changes in response to their environment than might be expected from univariable experimental findings. Single factor analysis in the example given would be equivalent to describing resistance time as modified by salinity at a specified temperature.

Fig. 12-9 is an example of the information which would be obtained if resistance time was measured in panels A, B and E of Fig. 12-8 over a range of salinities at a fixed level of temperature (e.g. by taking transects of the response surfaces at 4°C). From the results illustrated in Fig. 12-9, one would conclude little more than the fact that resistance time to the pentachlorophenate was diminishing with growth of the juvenile coho salmon. The point to stress is that

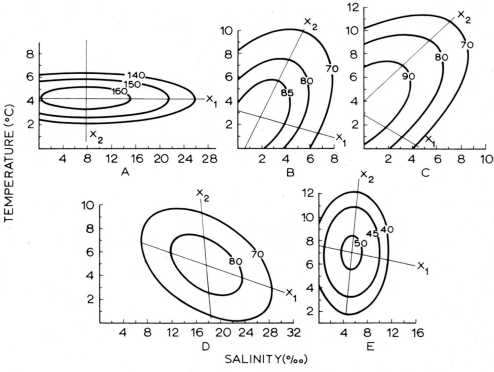

Fig. 12-8: Response of juvenile coho salmon *Oncorhynchus kisutch* to 3 mg/l sodium pentachlorophenate at various salinities and temperatures. Isopleths of median resistance time
in minutes are shown for five different ages of fish, A to E (see also Table 12-1). The surfaces
are sections of ellipsoids with axes X_1, X_2 and X_3 (the latter, involving dissolved oxygen,
not shown), sectioned in the plane of X_1, X_2 with $X_3 = 0$. Hence, the X_1-, X_2- axes are,
but appear not to be, orthogonal. (After ALDERDICE, 1963a; modified.)

Table 12-1

Data relating to tests conducted with five different age groups of juvenile coho
salmon *Oncorhynchus kisutch* exposed to 3 mg/l sodium pentachlorophenate at
various combinations of salinity and temperature (Original)

| Panel | Time of tests | Juvenile stage | Previous history at $12° \pm 1°C$ | Centre of surface, S | | | Mean sample | |
				$S‰$	$°C$	Median resistance time (mins)	Length (cm)	Weight (g)
A	Aug./Sept.	fry	fresh water	7·9	4·2	164	5·26	1·56
B	April/May	smolt	fresh water	1·9	2·7	88	11·06	14·05
C	June/July	post-smolt	fresh water	−0·8	3·2	96	11·43	13·46
D	July/Sept.	post-smolt	fresh water	17·7	4·9	84	11·72	17·17
E	Sept./Oct.	post-smolt	salt water*	5·5	7·0	50	13·14	26·35

* for 5 months, from time of normal migration to salt water, until tested

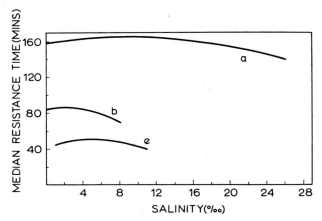

Fig. 12-9: Simulated single factor analysis for surfaces in
Figs 12-8A, B and E. Transects of the surfaces were
taken parallel to the salinity axis at 4°C. Apparent
change in capacity with age is the only effect which
would be demonstrated (curves a, b, e). (Original.)

response surfaces may change dramatically in the test space considered (range of
salinities and temperatures employed). A univariable experiment can cut across
such dynamic surfaces in only one of a very large number of ways. Furthermore,
univariable procedures usually constrain the choice of test conditions to follow
the axes of measurement (e.g. salinity or temperature or dissolved oxygen, etc.),
whereas a description of response in terms of the axes of the dynamic response
surface would be more instructive and meaningful.

In summary, aspects of changes in tolerance and capacity, plasticity of response,
and interaction of environmental variables are best appreciated by employing a
multivariable approach; the nature of these phenomena can be described by
examining response surfaces.

(b) Non-genetic Adaptation (Acclimation)

Non-genetic adaptation or acclimation may be considered as a term embracing
those slower adaptation processes, usually of intact individuals, resulting from
exposure to environmental changes (FISHER, 1958; Chapter 3.31). Studies of
DOUDOROFF (1942, 1945), FRY and co-authors (1942), BRETT (1944) and FRY (1947)
put the concept of temperature acclimation in poikilotherms on a quantitative
basis. By identifying the rates at which fishes acclimated to temperature, it
became possible to quantify measures of temperature tolerance or resistance for
individuals adapted to various acclimation temperatures (HART, 1947, 1952;
BRETT, 1952). The temperature tolerance polygon was introduced as a means of
describing the temperature experience an organism can tolerate (e.g. Fig. 12-18).
The polygon is bounded by those incipient lethal temperatures which, for every
acclimation temperature, allow 50% of a sample of organisms to survive indefinitely (FRY, 1947). Outside of the polygon in the zone of resistance, survival

time is limited, and is a function of the temperature level employed and the period of exposure.

Repeated construction of parts of the temperature polygon for a given species, in some cases, has yielded differences in the boundary conditions relating acclimation and incipient lethal temperature levels (FRY and co-authors, 1946; HART, 1952). Such differences could reflect the influence of minor variations in conditions ancillary to those of testing, such as differences in the chemical properties of test waters. However, the dynamic nature of those biological response surfaces investigated (COSTLOW and co-authors, 1960, 1962, 1966; ALDERDICE, 1963a) leads to the suggestion that temporal changes in response to temperature (as well as other abiotic entities) could be real. Such differences would be associated with changes in physiological state and could be expressed as variations with season or size (TYLER, 1966; HEATH, 1967). In fact, the temperature polygon could be considered as a special section of a response surface. Indeed, BRETT (1958) has suggested that other polygons may occur within the boundaries of the tolerance polygon, relating other functional capacities such as activity, growth, or spawning of fishes to more restricted zones of temperature experience.

PRECHT (1958) has examined differences in the nature of temperature adaptation processes. He considers two characteristics of these processes, namely: **capacity adaptation** (adjustments within a normal range of temperatures) and **resistance adaptation** (adjustments to temperature extremes)*. The insight into the description of acclimation processes afforded by PRECHT's studies makes it desirable to discuss these characteristics in the light of other related work. However, a difficulty of interpretation arises immediately, when such processes are considered from the point of view of response surface terminology so far employed. In fact, the present author would prefer to reserve the term capacity adaptation for a particular type of capacity change which can be extracted from the above classification. Because of this initial problem regarding terminology, it has been decided to describe these adaptation processes in terms of response surface terminology and to review pertinent literature in the light of these descriptions.

Initially it is assumed that there exists, within reasonable biological ranges of environmental variables, a region (or regions) in which the test organism has the greatest capacity to perform certain biological functions. It is assumed further that the environmental conditions associated with biological maxima may be approximated in the area around the region of maximum response by a second order response surface. The following description of adaptation processes is then proposed.

(i) 'Capacity adaptation'** involves a change in magnitude of a biological response resulting from a change in the shape (or capacity) of a response surface such that the absolute magnitude of maximum response either increases or decreases. Such variations in response could result from a change in dimensions of the surface, a change in both its dimensions and position, but not from a change in position alone.

(ii) 'Resistance adaptation' involves a change in magnitude of a biological response resulting from the transfer of an organism from one locus of levels of

---

* The terms ('Leistungsadaptation' and 'Resistenzadaptation') appear to have been defined first in CHRISTOPHERSEN and PRECHT (1953). These authors have frequently collaborated, particularly regarding metabolic processes of micro-organisms. Recent surveys of their work may be found in PRECHT (1964, 1967, 1968), PRECHT and co-authors (1966) and CHRISTOPHERSEN (1967).

** Terminology of the author is indicated by single quotes.

environmental factors to another within a dimensionally stable response domain. Such a change could result either from following a new transect of a stationary response surface (such as an acclimation surface), or through change in position of a response surface with respect to the axes of measurement.

An example of 'capacity adaptation' was noted in the changes in maximum response at the centres of the resistance time surfaces in Fig. 12-8 (see also Table 12-1). Examples of 'resistance adaptation' will be described, but may be previewed by referring to acclimation-, test-temperature surfaces (e.g. Fig. 12-13), or by noting, for example, the change in position of the resistance time surfaces in panels B, C and D of Fig. 12-8. These modified definitions will be considered at length in the following discussion.

BULLOCK (1955) has reviewed the effects of temperature on various biological reactions. His generalized rate-temperature (R-T) relationship (Fig. 12-10A) may be expanded by adding an axis for acclimation state (Fig. 12-10B). The surface then obtained may be considered as a modified second order surface expressing biological response rate as a function of any one of a number of acclimation temperatures when the organism is tested acutely* over a range of temperatures on the test-temperature axis. The surface generated is bounded in this hypothetical example by BULLOCK's cold- and warm-adapted curves, and the surface again is crossed by a rate curve (the acclimation diagonal) upon which the organism is acclimated to all temperatures at which the rate is measured.

PRECHT (1958) has summarized the relations between temperature and various biological rates in terms of acclimation effects, as reproduced in Fig. 12-11. If the two curves of PRECHT's figure relating oxygen consumption to test temperature are expanded by the addition of an axis of temperature acclimation, a relationship is obtained (Fig. 12-12) similar to that pointed out by BULLOCK (Fig. 12-10). On this basis, PRECHT's generalization of cold- and warm-adapted rate relationships again describes two sections of a modified rate surface. A diagonal on this surface (joining the low rate end of the cold-adapted section a to the high rate end of the warm-adapted section b) would again (Fig. 12-10) describe rates under conditions in which acclimation and test temperatures are equal.

In both examples described, the surface of the curved response plane is a rising ridge with highest responses occurring in the region of the high test-temperature end of the cold-adapted section. The two sections shown (a, b) in Fig. 12-12 and the other explanatory marks follow PRECHT's (1958) notations (Fig. 12-11). Section a is equivalent to PRECHT's A_1A_2 curve and section b to his B_1B_2 curve. Thus, if cold-adapted organisms acclimated to temperature t_1 (Fig. 12-11) are held at temperature t_2 on section a until acclimated, their rate response would drop to that found on the equal acclimation-, test-temperature diagonal (see example 3a, Fig. 12-12). If, on the other hand, warm-adapted organisms acclimated to temperature t_2 are held at temperature t_1 on section b until acclimated, their rate response would rise to that found on the equal acclimation-, test-temperature diagonal (see example 3b, Fig. 12-12). The surface illustrated, with minor changes, would accommodate PRECHT's adaptation types 1, 2 and 3.

* Acutely determined responses are those measures of response found at one or more test levels (e.g. temperature) to which the organism is not acclimated, and determined before the organism can acclimate to those levels and thereby modify its response.

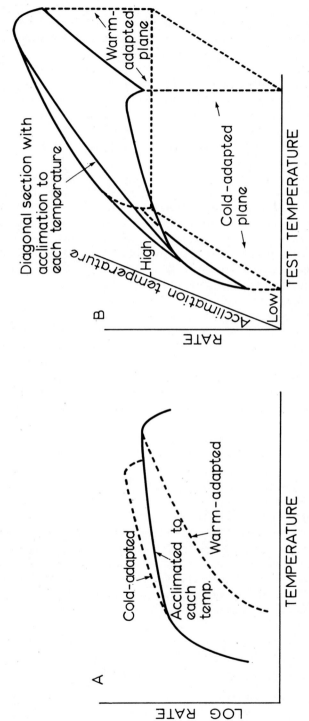

Fig. 12-10: (A) Generalized effects of temperature on rate functions of poikilotherms. Upper and lower curves: acutely determined rates for cold-adapted and warm-adapted individuals. Acclimated curve: individuals acclimated to each temperature at which the rate is determined. (After BULLOCK, 1955; redrawn.) (B) Expansion of the curves from A to an acclimation- and test-temperature surface. (Original.)

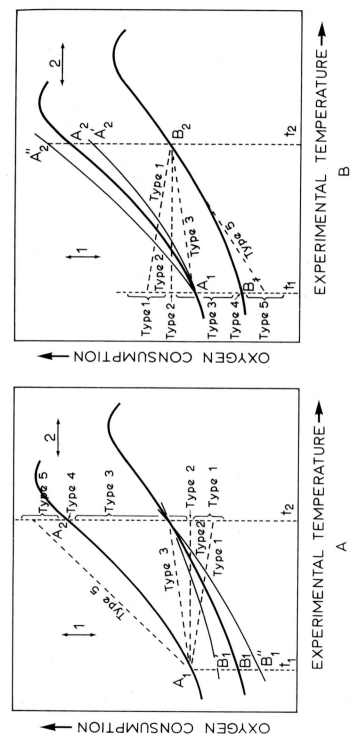

Fig. 12-11: PRECHT's adaptation types. Upper curves (panels A and B): adaptation curves for cold-adapted animals acclimated to t_1; lower curves (both panels): adaptation curves for animals acclimated to t_2. In both cases the rate curves result from acute measurement of rates at other temperatures. Adaptation types 1 to 5: rate changes encountered with (A)—cold-adapted animals moved from t_1 to t_2 and held until acclimated; (B)—warm-adapted animals moved from t_2 to t_1 and held until acclimated. 1: direction of capacity adaptation; 2: direction of resistance adaptation. (After PRECHT, 1958; modified.)

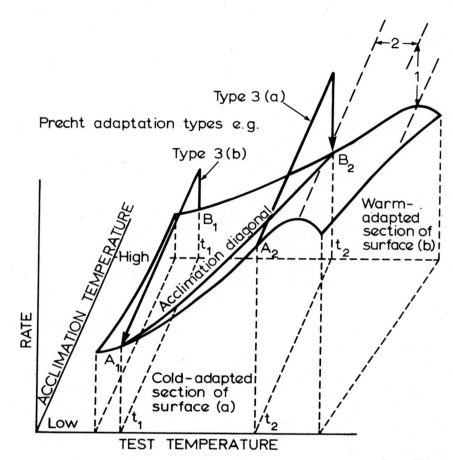

Fig. 12-12: Expansion of PRECHT's adaptation types to a surface. Using PRECHT's notations, two sections of the surface are shown: a cold-adapted section (*a*) with rate curve A_1A_2, and a warm-adapted section (*b*) with rate curve B_1B_2. Rates on the acclimation diagonal are those obtained when temperature acclimation is complete. Two examples of Type 3 adaptation are shown: 3(*a*), obtained on the acclimation diagonal (arrow) for an individual raised from t_1 to t_2 and held until acclimation is complete; 3(*b*), obtained for an individual lowered from t_2 to t_1 and similarly held (arrow). Minor changes in the rate surface would produce adaptation types 1, 2 and 3. In the upper right corner of the figure, 1 designates the difference in magnitude of maximum rates on the cold-adapted and warm-adapted sections of the surface (assumed analogous to capacity adaptation according to PRECHT), 2 the difference in temperatures at which rate maxima occur on the same surface sections (assumed analogous to resistance adaptation according to PRECHT). As redefined, both 1 and 2 are aspects of 'resistance adaptation'. (Original.)

A further description of adaptation types may be obtained from the rising ridge surface illustrated in Fig. 12-12. Several characteristics of the surface should be emphasized. First, response rates rise in magnitude toward the high-temperature end of the cold-adaptation section (a). Second, test temperatures, at which maximum response rates are obtained on the warm-adaptation section, are higher than those for which maximum response is obtained on the cold-adaptation section. This latter displacement on the ridge surface indicates that the axis of the ridge is not parallel to that of acclimation temperature, and that an interaction exists between acclimation and test temperature, as would be expected. Thus, as acclimation temperature increases (in other intermediate sections of the surface between a and b), maximum rates will be found at progressively higher test temperatures, and the magnitude of the maximum will become progressively lower as section b is approached.

The different rate responses, obtained by taking an organism acclimated to one temperature and moving it to another and holding it until acclimated, are analogous to moving the organism over any one of a number of pathways on the response surface. The changes in response are therefore considered as 'resistance adaptations'. They may have two characteristics, (i) a change in magnitude of rate response, which is dependent on the choice of acclimation temperature (Fig. 12-12 arrow 1; Fig. 12-10), (ii) a displacement of these maxima if there is an interaction between acclimation and test temperatures (Fig. 12-12 arrow 2; cf. PRECHT's capacity adaptation).

'Capacity adaptation' then is considered here as a change in the nature of the surface whereby an increase or decrease occurs in the **ultimate response maximum** of the surface. The surfaces shown in Fig. 12-8, with their decreasing resistance maxima at the surface centres, provide an example of what here is called 'capacity adaptation'. It would seem that capacity adaptation, so defined, may be coupled with long-term physiological differences occurring in association with growth, metamorphosis (e.g. smolt transformation), or variations related to season which may be under hormonal control. It follows that, unless the response surface is sufficiently defined, it may be difficult to distinguish between the two processes empirically described as resistance and capacity adaptation.

PRECHT's (1958) adaptation types 4 and 5 (Fig. 12-11), as far as can be judged, would not occur on a response surface of the type illustrated in Fig. 12-12 (based on BULLOCK's and PRECHT's generalized rate relationships with temperature). However, there is no reason to believe that response surfaces are restricted to the one type discussed. For example, isopleths may be constructed on surfaces such as those of Fig. 12-13 (see also Fig. 12-3). With slight rotation of the surface, PRECHT's adaptation types 1, 2 or 3 may be obtained (panel A). If, however, a region of maximum response is located between the equal-temperature diagonal and section a of the surface (panel B), an adaptation type equivalent to PRECHT's type 4 would result. Similarly, if such a region of the surface were located on or near the equal-temperature diagonal (panel C), a type equivalent to PRECHT's adaptation type 5 would be obtained. Moving from acclimation temperature t_1 to t_2 on section a (panel B) would produce no change in rate when the animal became acclimated to temperature t_2 (PRECHT's type 4). Under the same conditions, movement from t_1 to t_2 in the type 5 case (panel C) would result in an increase in

P

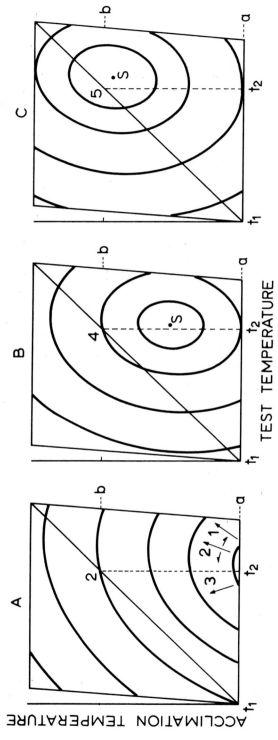

Fig. 12-13: Stylized surfaces and associated PRECHT adaptation types. Elliptical contours are isopleths of rate with maxima at S; rates on the surface diagonal are those obtained when acclimation to each temperature is complete. As in Fig. 12-12, a and b are cold- and warm-adapted surface transects; t_1 and t_2 are test temperatures. (A) surface similar to that of Fig. 12-12; 2: axis of the surface shown (arrow 2); moving an individual from t_1 to t_2 on section a would result in Type 2 adaptation. 1, 3: rotation of surface, as indicated (arrows 1, 3), produces Type 1 or Type 3 adaptation. (B) surface yielding Type 4 adaptation. (C) Type 5 adaptation. (Original.)

rate over that obtained acutely at t_2, when the organism became adapted to temperature t_2.

PRECHT's (1958) models are based on the implication that the R-T surface is curved, and is either a rising ridge or has a local maximum. However, since surface characteristics are not taken into account, the models cannot, by themselves, distinguish between 'resistance adaptation' and 'capacity adaptation'. A moment's consideration of Fig. 12-13 and sketching of two elliptical surfaces, one with and one without interaction effects, will convince the reader that this is true.

It is also important to recognize that PRECHT's (1958) models are based largely on studies of metabolic rates. As pointed out by FRY (1964, p. 722), PRECHT's models consider the capacity of an organism to provide metabolism as a basis for performance; they are not a classification of performance. Thus, metabolism available to an organism under specific sets of external conditions may not be directed in a constant manner toward a particular measured activity. Therefore, metabolic-rate response surfaces may differ basically from performance surfaces with which they might be compared (e.g. Table 12-3).

On the basis of PRECHT's (1958) models, the adaptation types discussed would appear largely to be reflections of differences in the general nature of response surfaces. The five adaptation types outlined are analogous to describing the result of following test-temperature pathways across recognizably different response surfaces, where the differences between acutely determined (unacclimated) and final (acclimated) rates of response are a function of the shape of the surface involved. There could conceivably be a vast number of such acclimation types—intermediate between or equivalent to PRECHT's five types—all dependent for their expression on the species examined, its physiological state, and the particular response investigated. When surfaces are compared in this manner, it would appear that they should be capable of general characterization in terms of the four surface properties listed earlier: plasticity, interaction effects, and changes or differences in tolerance and capacity. It should be evident also that the most meaningful of such comparisons would be obtained in the region of a local response maximum within the biokinetic range.

PROSSER (1958b) has considered patterns of acclimation of biological rate phenomena in another way, according to the presence or absence of rotation or translation in rate-temperature curves. Rotation and translation are two basic properties of surface relations (interaction effects and changes in tolerance) implicit in their description. The five basic types of rate-temperature curves listed by PROSSER (1958b; Fig. 12-14) may be considered as pathways traced across curved response surfaces. Although they too, by themselves, cannot distinguish between resistance and capacity adaptation, they do allow for the interpretation of interaction effects.

It is not obvious, without further definitive data, how rotation of a rate surface (interaction) could be portrayed where both fully acclimated and acutely determined rates are compared on the same surface. Rotation in any section of a surface would be dependent on the acclimation level under which testing is initiated (e.g. warm or cold), and could change as the acclimation level is progressively changed. A clue to this problem appears to lie in the non-linear nature of the axis of maxi-

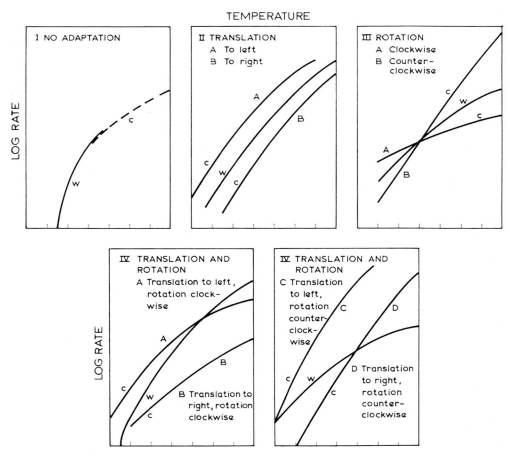

Fig. 12-14: Summary of patterns of rate functions with respect to temperature. For each of the 9 patterns shown there is a set of 'warm-' (w) and 'cold-acclimated' (c) rate curves. Each pair of curves is taken here to represent limiting sections of an acutely determined rate surface where an individual is acclimated either to a 'warm' or a 'cold' temperature (e.g. Fig. 12-13, t_1, t_2) and tested acutely at other temperatures. The processes of translation and rotation refer to the relations implicit between the members of each pair. (After PROSSER, 1958b; modified.)

mum walking rate of the lobster, to be discussed in a later section*. In this respect, consideration of PROSSER's (1958b) acclimation curves as limiting sections of an R-T surface (as in Fig. 12-13) leads to an interesting observation. The rate relationships in Fig. 12-14 may be reconstructed as sections of response surfaces if the following assumption is made: a process occurs, suggestive of progressive rotation or translation of isopleths, as they proceed outward from their centre; that is, the extent of apparent rotation or translation changes progressively for isopleths of diminishing rate. Rate relations similar to PROSSER's acclimation types

* See Fig. 12-19. In this case, the points of contact of tangents to the isopleths drawn parallel to the acclimation- or test-temperature scales do not form linear series, indicating a changing interaction between acclimation temperature and test temperature over the whole surface.

indeed may be generated as transects of such surfaces. A representative set of these surfaces is illustrated in Fig. 12-15.

Two descriptive attributes of response surfaces have now been presented, and both appear to be associated with properties of rotation and translation. In effect, however, they are quite distinct: one has to do with the basic structure of response surfaces, the other with their potential mobility. In distinguishing between them, reference will be made to the 'inner structure' of response surfaces, as suggested by the surfaces generated from PROSSER's (1958b) acclimation curves. These could be developed from rate surfaces subject to 'inner rotation or translation'. On the other hand, the apparent total movement of the surfaces for the various ages of young salmon illustrated in Fig. 12-8 will be termed 'integral mobility'. Properties involved in describing movement of the latter surfaces will be designated as 'integral rotation or translation'.

Consideration of the attributes of integral mobility and inner structure brings a more general perspective to the examination of response surfaces. A simple, non-detailed description of response surfaces may be obtained on the basis of second order polynomials (e.g. Equation 4) or their canonical equivalents (e.g. Equation 7) used as a model. Under this simplified scheme, mobility of surfaces may be examined in terms of rotation or translation of whole surfaces around their centres. Integral mobility of surfaces is amenable to algebraic description in terms of variations of Equation 8 (Table 12-2). However, the specific nature of the surfaces themselves, indicated by consideration of their inner structure, is not demonstrated under the constraints of the simple model employed. Detailed consideration of the specific structure of surfaces requires more powerful models with which to examine essentially what appear to be non-linear scales of biological response to environmental variables. A further consideration of non-linear inner structure in the analysis of response surfaces is presented under *Methodology*.

More complete knowledge of the structure and mobility of response surfaces must await their further detailed description. On the basis of available information (Fig. 12-8, see also Fig. 12-17), surface mobility appears to be associated with developmental stanzas (e.g. embryonic, larval or juvenile stages) or stages of physiological readjustment (e.g. metamorphosis, or sexual maturation). It appears to involve movement of the total surface structure within the test space associated with the levels of environmental variables under examination.

Table 12-2

Algebraic representation of four types of surface variations associated with integral surface mobility (Original)

No translation no rotation	$\begin{bmatrix} X_1 \\ X_2 \end{bmatrix} = \begin{bmatrix} 1 & 0 \\ 0 & 1 \end{bmatrix} \begin{bmatrix} x_1 \\ x_2 \end{bmatrix}$
Rotation	$\begin{bmatrix} X_1 \\ X_2 \end{bmatrix} = \begin{bmatrix} m_{11} & m_{12} \\ m_{21} & m_{22} \end{bmatrix} \begin{bmatrix} x_1 \\ x_2 \end{bmatrix}$
Translation	$\begin{bmatrix} X_1 \\ X_2 \end{bmatrix} = \begin{bmatrix} 1 & 0 \\ 0 & 1 \end{bmatrix} \begin{bmatrix} x_1 - x_{1s} \\ x_2 - x_{2s} \end{bmatrix}$
Translation and rotation	$\begin{bmatrix} X_1 \\ X_2 \end{bmatrix} = \begin{bmatrix} m_{11} & m_{12} \\ m_{21} & m_{22} \end{bmatrix} \begin{bmatrix} x_1 - x_{1s} \\ x_2 - x_{2s} \end{bmatrix}$

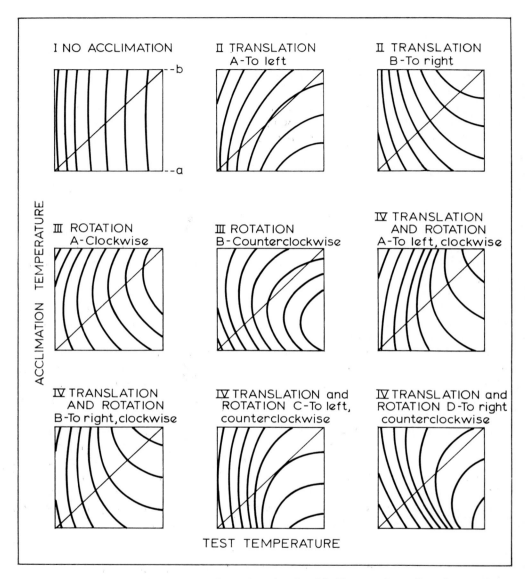

Fig. 12-15: Representative rate surfaces associated with PROSSER's acclimation patterns (Fig. 12-14). Surfaces were generated as follows: (1) Solid models were constructed using pairs of warm- and cold-acclimated rate curves from Fig. 12-14 as limiting sections of an acclimation surface; (2) isopleths of rate were added to the surface of each solid and the surfaces were sketched in plan view; (3) a family of drafting ellipses was used separately to obtain the surfaces developed through progressive inner translation or rotation of the isopleths (ellipses) of the family of ellipses; (4) a portion of each surface (3) was selected, as illustrated, which conformed in general to the characteristics of the acclimation surfaces from (1) and (2). Acclimation types II to IV can be generated from the same family of ellipses. The type I surface is illustrated as a section of a broad elliptical surface with centre remote from and to the right of the rate surface shown (for explanation of surface diagonal see Fig. 12-13). (Original.)

In summary, rate-temperature relationships as outlined may be interpreted as the information obtained on transects of response surfaces. Resistance and capacity adaptations and acclimation types may be considered as interpretive characteristics of acclimation surfaces. The true nature of the inner structure of those surfaces which have been considered in detail appears to be non-linear. It is suggested that response surfaces may also demonstrate integral mobility when various developmental stages of an organism are compared. A more detailed description of rate phenomena awaits further elucidation of the nature of rate surfaces. Their investigation within species should allow a description of changes in response associated with stages of development or physiological state. Conversely, seasonal changes in response (e.g. USHAKOV, 1964) should be considered as a potential source of variation which, if not recognized, could desensitize measures of response made either within or between species.

(c) *Tolerance and Resistance*

Mention was made earlier of FRY's (1947) tolerance polygon which relates acclimation and test conditions, usually in terms of temperature, to the magnitude or rate of response evoked by temperature experience. Considerable use has been made of this device for describing and comparing quantitatively the temperature tolerance and resistance of animals (FRY, 1947; HART, 1947; BRETT, 1952, 1956; McLEESE, 1956; HOFF and WESTMAN, 1966; see also Fig. 12-18). It is inferred here that the FRY polygon can be considered as a section of a response ellipse or rising ridge without disturbing any of the general arguments related to surface analysis. The polygon is usually truncated, since there is an upper limit to those acclimation temperatures which can produce an increase in tolerance to progressively higher temperatures.

Considering the tolerance polygon as a truncated tolerance ellipse implies that part of the ellipse lies outside the biokinetic range. This condition is not as illogical as it might seem. Even those conditions which would lead to maximum biological response may lie outside the biokinetic range, a fact noted earlier by FRY (1947, p. 38 to 39) in the case of the effect of temperature on metabolic scope of the bullhead *Ameiurus nebulosus*. In the example noted earlier (Fig. 12-8C) the effects of salinity and temperature on the resistance of juvenile coho salmon *Oncorhynchus kisutch* to sodium pentachlorophenate were such that maximum resistance to the poison would be found at a salinity below 0‰.

In some cases associated with temperature, there is a departure from the elliptical configuration inferred as underlying the response polygon in the lower incipient lethal temperature range. Such cases often involve compound curvature of the lower incipient lethal boundary of the polygon (e.g. BRETT, 1952; his Figs 22, 23) and suggest that third degree terms may enter into the description of temperature experience in those regions of the polygon. Thus if x_1 and x_2 were defined as acclimation and test temperature, respectively, then more of the total variance associated with temperature effects could be explained by adding further terms to the polynomial used as a regression model. That is,

$$Y = b_0 x_0 + b_1 x_1 + b_2 x_2 + b_{11} x_1^2 + b_{22} x_2^2 + b_{12} x_1 x_2 + b_{111} x_1^3$$
$$+ b_{222} x_2^3 + b_{112} x_1^2 x_2 + b_{122} x_1 x_2^2 \tag{9}$$

Very limited studies have been reported on the multivariable consideration of tolerance or resistance. Many investigations reported in the literature have considered the effects of environmental variables on organisms primarily from a univariable point of view. Some of these provide tantalizing glimpses of the operation of several factors acting in concert on tolerance or resistance, particularly with respect to temperature, salinity, and ionic composition (DOUDOROFF, 1942, 1945; BRETT, 1952; KINNE, 1952, 1954, 1957; GIBSON and HIRST, 1955; RANADA, 1957; DEHNEL, 1960; MORRIS, 1960; TODD and DEHNEL, 1960; LINDSEY, 1962a, b; CRAIGIE, 1963; CRISP and COSTLOW, 1963; LANCE, 1963; SPRAGUE, 1963; BHATNAGAR and CRISP, 1965; PODOLIAK, 1965).

The study by MCLEESE (1956) on the effects of temperature, salinity and dissolved oxygen on tolerance and resistance in the American lobster *Homarus americanus* remains as one of the few examples in which the tolerance polygon of FRY (1947) has been extended to a consideration of more than two environmental factors. MCLEESE was concerned primarily with description of the boundary conditions delimiting the zone of tolerance from the lethal zone (or zone of resistance) as defined by acclimation and test levels of the three factors considered. Reference to his data, however, provides some appreciation of the geometry of the zone of tolerance in three-space (Fig. 12-16). The tolerance solid so constructed resembles a section of an ellipsoid. There is some suggestion that third degree terms would enter into the definition of boundary conditions at high oxygen–low temperature levels and at high temperature–high salinity levels. MCLEESE's

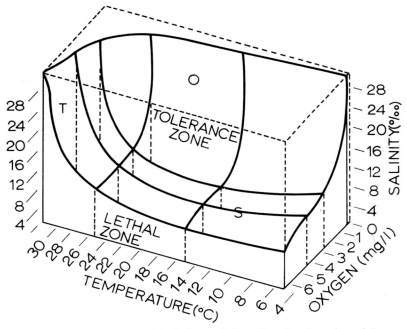

Fig. 12-16: Boundaries of lethal conditions for the American lobster *Homarus americanus*. Regions are shown where temperature (T), salinity (S) and dissolved oxygen (O), respectively, act alone as lethal factors. (After MCLEESE, 1956; redrawn.)

analysis of variance of his data did not include a separation of quadratic (and cubic) effects associated with curvature, although from the nature of the tolerance solid quadratic effects would probably account for a significant portion of the total variance. The three linear × linear interactions (S × T, S × O, T × O) between salinity (S), temperature (T) and dissolved oxygen (O) are of the same order as his error term: thus, the three axes of the ellipsoid could lie close to coplanarity with the axes of measurement. It would appear that the tolerance solid for the lobster, being primarily ellipsoidal in character, could be generated geometrically by employing standard multiple regression methods based on a second order polynomial (Equation 6).

Surface analysis has been explored as a means of documenting effects of water quality alteration on viability of living organisms (ALDERDICE, 1963a, b, 1966). Problems in pollution ecology have resulted very often in the setting of provisional limits to the amount of pollutional change which may be tolerated by aquatic organisms. These changes may involve alteration in natural properties such as increased temperature or reduced dissolved oxygen concentration, or chemical alteration of water by addition of biocides or industrial wastes. Resistance time of juvenile coho salmon *Oncorhynchus kisutch* was found to be dependent not only on concentration of a biocide employed, but also on concomitant levels of salinity, temperature and dissolved oxygen in the test media. With respect to the three latter ancillary variables, resistance time was also found to vary with age or size of the fish tested (Fig. 12-8). These studies affirm an important consideration. In attempting to set maximum permissible concentrations of water-borne pollutants, consideration must be given to sensitivity of the species comprising the biota, the variations in tolerance of organisms related to size, season or state of physiological development, and the range or field of combinations of levels of natural variables under which the biota will experience such changes in water quality (see also Volume V).

In studying biological problems arising from water pollution, it has been common practice to define maximum permissible concentration limits of a given pollutant as a function of thresholds of acute (lethal) response. In such cases, ancillary variables (e.g. temperature, dissolved oxygen) are usually constrained to specific levels, largely as a matter of operational convenience. Developments in this field of enquiry would suggest, however, that for a given species a specific set of levels of those ancillary variables known or suspected to be of importance should be adopted as background requirements for biological assays of tolerance to a selected pollutional substance. Furthermore, it is suggested that this set of levels might be predetermined as that providing the greatest potential for development and survival, for the species and stage considered, in terms of some objective measure or measures of biological capacity. Conversely, it is suggested that the operation of ancillary environmental variables (e.g. temperature, dissolved oxygen) on the response to a pollutional substance should be investigated by multivariable techniques to ensure that threshold levels of these substances would apply to the organism over all environmental loading conditions. The range of loading conditions would be determined by the ranges of the ancillary variables and combinations of their levels normally found in the environment under investigation (ALDERDICE, 1966).

(d) Capacity

Within the range of levels of one or more environmental variables there exists a region bounded by limits beyond which an organism may not survive indefinitely. The zone of temperature tolerance for an organism is one such region. It is inferred that, within the region bounded by survival limits, there may also exist other, more restricted regions which would delimit other measures of biological capacity. These could include, for example, measures of growth, conversion efficiency, locomotion, or metabolic scope. Survival will be considered here as a limiting measure of living capacity. A distinction is made arbitrarily between those studies designed to define boundary conditions, as in the previous case of the American lobster *Homarus americanus* and those in which attention is directed more toward evaluating quantitative variations in survival within an environmental factor space. Furthermore, it is suggested that measures of capacity may often be maximized with respect to rather specific levels of several environmental variables acting in concert on the response measured. Those sets of levels of environmental variables which maximize biological capacity in terms of magnitude of response, or efficiency with respect to metabolic cost, will be defined as optima. Environmental optima are considered in the terminology of FRY (1947, p. 30 to 31) as those conditions under which an animal can perform a certain activity best as judged by objective measurements.

There is a rapidly expanding literature on the subject of response surface analysis (HILL and HUNTER, 1966). Much of this literature, however, is oriented toward the subject of optimization of industrial chemical processes, the topic which served as a means of introducing the subject of response surface methodology (BOX and WILSON, 1951). Nevertheless, by analogy, much of this literature is pertinent to biological enquiry. Several studies in agronomy have employed response surface techniques and are of particular interest, namely those of HADER and co-authors (1957) and MOORE and co-authors (1957) on nutrition and yield of lettuce, WELCH and co-authors (1963) on yield of Bermuda grass, and HERMANSON (1965) on potato production. The techniques, which together are now usually defined as response surface methodology, as yet have found use only to a limited extent in general biological enquiry.

In surveying some of the literature in biology, it is not surprising to find that most studies pertinent to this enquiry are designed primarily to evaluate the effects of single environmental variables on biological response. In a number of cases, such studies provide partial information on the quantitative effects of several variables acting in concert on functional capacities. Among these are studies on survival (KINNE, 1954; GIBSON and HIRST, 1955; DAVISON and co-authors, 1959; TODD and DEHNEL, 1960; BISHAI, 1961; RIVARD, 1961a, b; SILVER and co-authors, 1963; ZEIN-ELDIN and ALDRICH, 1965), on rate of development (BROEKHUYSEN, 1936; SANDOZ and ROGERS, 1944; KINNE, 1953, 1961; GARSIDE, 1959, 1966; RIVARD, 1961a, b; KINNE and KINNE, 1962; HENDERSON, 1963), on reproductive rate and egg production (BIRCH, 1945; KINNE, 1953, 1961; EDMONDSON, 1965), on growth and size (EMERSON, 1947; GIBSON and HIRST, 1955; KINNE, 1956, 1958a, 1960; BLAXTER and HEMPEL, 1961; KINNE and KINNE, 1962; ARAI and co-authors, 1963; SILVER and co-authors, 1963; JITTS and co-authors, 1964;

MADDUX and JONES, 1964; SHUMWAY and co-authors, 1964; SWEET and KINNE, 1964; EINSELE, 1965; ZEIN-ELDIN and ALDRICH, 1965), on metabolism and respiration (FRY and HART, 1948a; KINNE, 1952; GIBSON and FRY, 1954; KANUNGO and PROSSER, 1959; DEHNEL, 1960; HALCROW, 1963; BEAMISH, 1964a, b, c; BRETT, 1964; DEAN and GOODNIGHT, 1964; MCFARLAND and PICKENS, 1965; MCINTIRE, 1966), on nerve-muscle excitation (BENTHE, 1954), on food intake and conversion efficiency (KINNE, 1960, 1962; KINNE and PAFFENHÖFER, 1965; PAFFENHÖFER, 1968; BRETT and co-authors, 1969), on osmoregulation (KINNE, 1952; FLÜGEL, 1960; KINNE and co-authors, 1963), on locomotory activity (FRY and HART, 1948b; GIBSON and FRY, 1954; MCLEESE and WILDER, 1958; ROOTS and PROSSER, 1962; DAVIS and co-authors, 1963; BRETT, 1964), on meristic characteristics and form (BRIAN and OWEN, 1952; KINNE, 1956, 1958b; LINDSEY, 1962a, b; SWEET and KINNE, 1964), and on dynamic aspects of biological responses (BAGGERMAN, 1960; BEAMISH, 1964a, b; HOLLIDAY, 1965; TYLER, 1966).

In addition, pertinent information may be found in the reviews by BULLOCK (1955), PRECHT and co-authors (1955), PROSSER (1955, 1958a) (in particular the contributions in the latter by FISHER, KINNE, PRECHT, PROSSER, and WENT), KINNE (1963a, b, 1964a, b, 1966), PALOHEIMO and DICKIE (1965, 1966a, b) and in Chapters 2, 3 and 4.

In describing biological responses of organisms to multivariable conditions, reference will be made to a few specific examples from the literature. These have been selected either because measures of biological capacity have been approached through response surface methodology, or because the studies have sufficient data to allow some meaningful interpretation of multivariable effects.

COSTLOW and co-authors (1960, 1962, 1966) have examined the effects of salinity and temperature on the development of larval stages of several marine decapods in the laboratory. They employed a second order polynomial (Equation 4) as a regression model and used certain primary aspects of response surface analysis to describe their data. A modified version of their illustrations of the effects of salinity and temperature on mortality in four zoeal stages and the megalops of the crab *Rhithropanopeus harrisii* is presented in Fig. 12-17. Isopleths of response are a function of the regression relationship employed (the model), as well as the data to which they are fitted (by analogy, a straight line can be fitted to curvilinear data). If a maximum (or minimum) region of response is found, the isopleths drawn through equal levels of response will be elliptical. Thus, for the fourth stage larva of *R. harrisii* a region of minimum mortality is indicated in panel D with a centre in the region 15‰ to 20‰ S and 24° to 25°C. However, if a region of maximum survival happens to lie close to the boundary or outside the range of the variables employed (the factor space) then surface configurations can be obtained similar to those in panels A, B and E (a minimax) or to that in panel C (a response minimum, e.g. maximum mortality). If the canonical equations of these surfaces had been calculated (Equation 7), it is presumed that both eigenvalues (λ_{11}, λ_{22}) would have had negative values in the case of the fourth zoeal stage, indicating that a region of maximum response had been located. For the first and second zoeal stages and for the megalops, however, canonical reduction would likely yield a positive eigenvalue associated with temperature, but a negative eigenvalue associated with salinity (panels A, B, E). These latter results

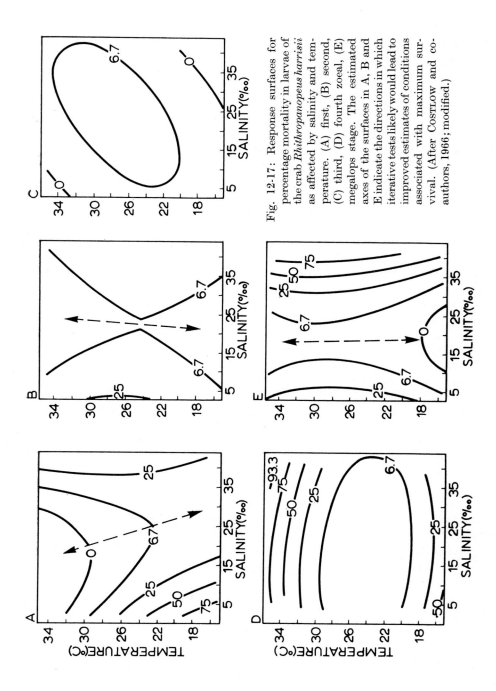

Fig. 12-17: Response surfaces for percentage mortality in larvae of the crab *Rhithropanopeus harrisii* as affected by salinity and temperature. (A) first, (B) second, (C) third, (D) fourth zoeal, (E) megalops stage. The estimated axes of the surfaces in A, B and E indicate the directions in which iterative tests likely would lead to improved estimates of conditions associated with maximum survival. (After COSTLOW and co-authors, 1966; modified.)

would indicate that if a response maximum exists for the stages in question, it would be found in the direction of the estimated surface axis indicated in panels A, B and E.

Judgement in these cases would probably yield a decision to explore other salinity–temperature combinations along the axes (computed, in canonical reduction of the data) in the direction of lower temperatures. Having conducted these further tests, a region of maximum response (minimum mortality) would probably be found at a salinity–temperature combination near or below 15°C. Recalculation of all of the data (or execution of a new series of experiments centred around the new provisional maximum) would then provide an estimate of the salinity–temperature locus associated with minimum mortality. In the third zoeal stage (panel C), however, both eigenvalues in the associated canonical equation would be positive, indicating that if a region of maximum response exists it would be remote from the factor space employed. In this case, judgement would suggest redefining the factor space to include lower temperatures.

It should be emphasized that the most meaningful use of second order surface fitting lies in finding response optima, or combinations of environmental factors which maximize a measure of biological capacity. When a maximum is not found from the initial set of trials, it may often be located by a further linear sequence of trials leading in the direction suggested by axes of the surface computed from the initial trials (Box, 1956, p. 529). Such iterative procedures often may be difficult to accomplish with biological material. Particular stages of development of test organisms may be available only for a limited period each year. Nevertheless, response surfaces are most meaningful in the region of a maximum, and iterative procedures provide a means of finding that region. In the case of *Rhithropanopeus harrisii*, a further series of studies involving lower temperatures would likely find temperature–salinity maxima for all stages considered.

The elliptical type of surface generated by a second order polynomial in the region of response maxima remains as a simplified but useful approximation of the true nature of the response surface. The true surface, in fact, may have a number of irregularities in it or the axes of the surface may be neither linear, nor orthogonal (at right angles to each other). However, the general utility of the fitting procedure is not necessarily destroyed by minor departures of the true surface from that dictated by the regression model employed. As an alternative, increased confidence in the prediction of multivariable relationships would of necessity be based on an exhaustive description of all parts of the surface investigated (e.g. EMERSON, 1947; his Fig. 3). Nevertheless, one is often hard-pressed to appreciate the complexity of inherent interrelationships and interactions when working with multivariable relationships. Response surface analysis provides insight into these complexities; it often reveals information on associations between response and environmental factors which probably would remain unrecognized in other methods of analysis.

An interesting surface to examine in this regard is that which can be constructed from the data of McLEESE and WILDER (1958). In this instance, the effect of temperature is acutely determined on walking rate of lobsters *Homarus americanus*, previously held at a number of acclimation temperatures. The authors demarked a zone of activity for the lobster within the boundaries of its zone of temperature tolerance (Fig. 12-18). The response surface within the zone of activity has been

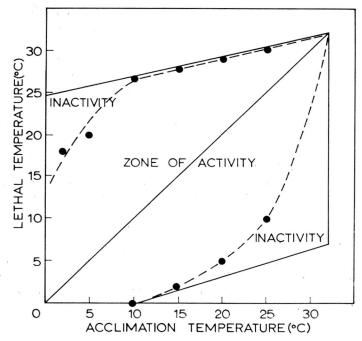

Fig. 12-18: Locomotory activity of *Homarus americanus* as a
function of temperature. Walking rate (ft/min; 1 ft = 30·48 cm)
was determined under stimulus of bright illumination. Zones of
activity and inactivity are demarked by points at which temperature experience resulted in zero activity; they are enclosed
by the temperature tolerance polygon. (After McLEESE and
WILDER, 1958; redrawn.)

approximated by plotting a number of transects of the surface from data given in
the original paper and drawing provisional isopleths in terms of magnitude of
walking rate (Fig. 12-19). Below acclimation temperatures of about 20°C, the data
generally conform to an elliptical pattern. However, the axes of the surface are
neither linear nor orthogonal. Within the total temperature range there are two
regions of maximum walking rate. The small area of temperature experience
encompassed by acclimation levels of about 13° to 17°C, and test temperatures of
about 12° to 22°C, probably could be reasonably well approximated by a second
order surface. However, it would be misleading to attempt fitting a second order
polynomial to the whole surface portrayed as terms higher than second degree
would be required.

PRECHT's (1958) adaptation types may be considered again, on the basis of the
walking-rate response surface illustrated. The summary in Table 12-3 indicates
that adaptation type, as measured, may well depend on the characteristics of the
section of the response surface examined or, in other words, on the levels of
temperature employed.

Some further interpretation of the walking-rate response surface may be
attempted. The two regions of maximum walking rate are analogous to earlier

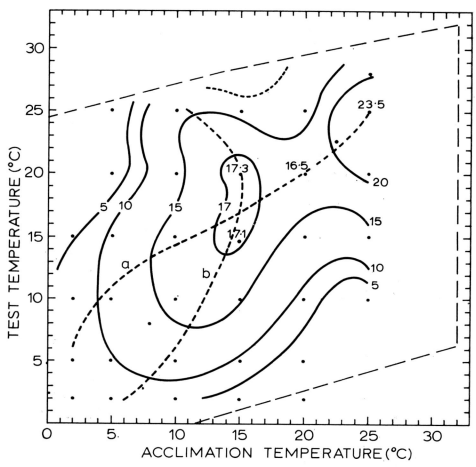

Fig. 12-19: Response surface for walking rate of *Homarus americanus* within the zone of activity (Fig. 12-18). Walking rate isopleths are in ft/min (1 ft = 30.48 cm). Axes associated with maximum walking rates: *a* at each acclimation, *b* at each test temperature. The grid of points locates the original data; several maximum values are indicated. (Based on data by McLeese and Wilder 1958; original.)

examples of bimodal temperature effects on locomotory response, as cited by Fisher (1958; see also Fisher and Sullivan, 1958). Following Fisher's interpretation, the lower of the two maxima would appear to be a function of the integrated individual, associated with temperature selection, and under control of the central nervous system. Fisher also suggests that the temperature associated with the peak probably shifts with acclimation temperature; in the present example this is indeed the case. A second, higher temperature maximum was found by Fisher and Sullivan (1958) at near-lethal temperatures. By destroying part of the cerebellum of speckled trout *Salvelinus fontinalis*, they found that the first locomotory maximum disappears, and locomotory performance increases progressively to a single peak at the second, higher temperature maximum. The second

Table 12-3

Estimated walking rates of lobsters *Homarus americanus* which would be obtained on various transects of the surface in Fig. 12-19. The data illustrate that PRECHT's (1958) adaptation types are dependent, in part, on the choice of temperature levels or, in other words, the sections of the response surface examined (Based on data of McLEESE and WILDER, 1958)

Initial acclimation temperature (°C) t_1	Final temperature (°C) t_2	Walking rate at initial acclimation temperature (m/min; ft/min)	Walking rate at final temperature		PRECHT's adaptation type
			Acute rate (m/min; ft/min) at temperature t_1	After acclimation (m/min; ft/min)	
t_1	t_2	(cold-adapted initially at temperature t_1)			
5	10	2·90; 9·5	3·78; 12·4	5·00; 16·4	5
10	15	5·00; 16·4	5·00; 16·4	5·21; 17·1	?
15	20	5·21; 17·1	5·27; 17·3	5·03; 16·5	1
20	25	5·03; 16·5	4·36; 14·3	7·16; 23·5	?
5	15	2·90; 9·5	3·54; 11·6	5·21; 17·1	5
10	20	5·00; 16·4	4·42; 14·5	5·03; 16·5	?
t_2	t_1	(warm-adapted initially at temperature t_2)			
10	5	5·00; 16·4	3·69; 12·1	2·90; 9·5	5
15	10	5·21; 17·1	4·60; 15·1	5·00; 16·4	3
20	15	5·03; 16·5	4·39; 14·4	5·21; 17·1	1
25	20	7·16; 23·5	6·28; 20·6	5·03; 16·5	5
15	5	5·21; 17·1	2·47; 8·1	2·90; 9·5	3
20	10	5·03; 16·5	2·77; 9·1	5·00; 16·4	2

maximum might, therefore, be associated with general irritability of nervous tissue which, through increased spontaneous locomotory activity, might remove an organism from near-lethal circumstances. Lack of linearity of the surface axes suggests that interactions between acclimation and test temperatures are not constant over the range of temperature tolerance of the lobster *Homarus americanus*. At an acclimation temperature of 2°C (Axis a, Fig. 12–19), the lobster's walking rate is maximal at a test temperature of about 6°C. However, as acclimation temperature increases, this difference decreases. At acclimation temperatures of 20°C and higher, the difference disappears, and maximum locomotory ability appears to occur at the acclimation temperature. It is possible that temperature selection in *Homarus americanus* may follow the axis (a, Fig. 12-19) to a level associated with acclimations between 15° and 20°C. Above 20°C, temperature selection might reach a plateau or 'final preferendum' (FRY, 1947) dictated by the upper area of high temperature performance and an assumed behavioural tendency for an animal to avoid temperatures in that region. Finally, the region of maximum locomotory performance at the lower peak provides a range of high locomotory capability over a range of temperatures (about 8° to 24°C) greater than those to which it is acclimated (about 8° to 19°C). Within this range, therefore, the poikilothermous animal can perform in a manner relatively independent of the temperature level. Maximum locomotory ability is found between approximately 13° and 22°C for an animal acclimated to temperatures between 13° and 17°C. Whether or not this region of temperature experience may be called an optimum would be judged on the results of other objective measures of functional capacity and their possible association with this region.

Similar arguments to those discussed for the walking rate of *Homarus americanus* apply to data of FRY and HART (1948b) for cruising speeds of the goldfish *Carassius auratus**. Their results are reproduced in Fig. 12-20. In Fig. 12-21, these and other data for the goldfish have been replotted on a grid of acclimation and test temperatures. The cruising speed relationships at each acclimation temperature (Fig. 12-20) are seen to be transects of a response surface (Fig. 12-21).

The surface illustrated is based on limited data and is, at best, an approximation of the true nature of the surface. Nevertheless, general characteristics of the surface are deserving of comment. The major (long) axis of the surface is non-linear. It appears to parallel the axis of equal acclimation and test temperatures for the region of maximum and near-maximum cruising speeds (between acclimation temperatures of about 20° and 35°C). At acclimation temperatures below 20°C, the major axis of the surface is rotated clockwise and meets the 0°C acclimation-temperature transect at a test temperature between 15° and 20°C. Cruising speeds, therefore, are partly dependent for their expression on an interaction between acclimation and test temperatures. A recent, more comprehensive analysis of these data (LINDSEY and co-authors, 1970) is in general agreement with the foregoing observations.

In assessing the relation between cruising speed and temperature experience, previous studies have focused on the description of surface transects, rather than

---

* A similar case appears in the data of BENTHE (1954) on the excitability of a foot preparation of the snail *Limnaea stagnalis*. FRY (1967, p. 392) draws attention to two other similar cases: the cruising speeds of the minnow *Notropis cornutus* and the Atlantic salmon *Salmo salar*.

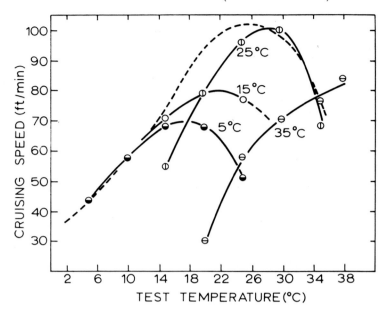

Fig. 12-20: Cruising speeds in ft/min (1 ft = 30·48 cm) of goldfish *Carassius auratus* in relation to temperature. Acutely determined relations are shown for individuals acclimated to 5°, 15°, 25° or 35°C. The broken line indicates cruising speeds for fish acclimated to each test temperature. (After FRY and HART, 1948b; redrawn.)

on those circumstances which define maximum performance over a field of temperature levels. Hence, a further consideration of performance, in relation to acute-temperature and acclimation-temperature experience, may assist in the interpretation of cruising speed maxima.

For example, in Fig. 12-21, there are two 'ridges' of maximum performance on the response surface, both rising to the **ultimate maximum cruising speed** at the centre of the surface. The first of these ridges may be visualized by locating the acclimation temperature of maximum swimming capacity, at a specific test temperature, over a series of acclimation temperatures (e.g. horizontal transects of Fig. 12-21). For example, a tangent to the 12·19 m/min (40 ft/min) isopleth, drawn parallel to the acclimation temperature axis at a test temperature of about 3°C, estimates that maximum cruising speed occurs at an acclimation temperature of about 10°C. Similarly, at a test temperature of 10°C, maximum cruising speed on the 18·29 m/min (60 ft/min) isopleth is found at an acclimation temperature of about 10°C. The acclimation temperatures so estimated, continuing toward the centre of the surface, define a ridge of maximum performance with respect to acclimation. As estimated from the provisional surface (Fig. 12-21), this ridge lies close to the diagonal of equal acclimation- and test-temperature experience, and drops below the diagonal at test temperatures under about 10°C. By analogy, it is this ridge which FRY (1967, p. 393) discusses when he concludes that at a given temperature the best performance is exhibited by the animal acclimated to that temperature.

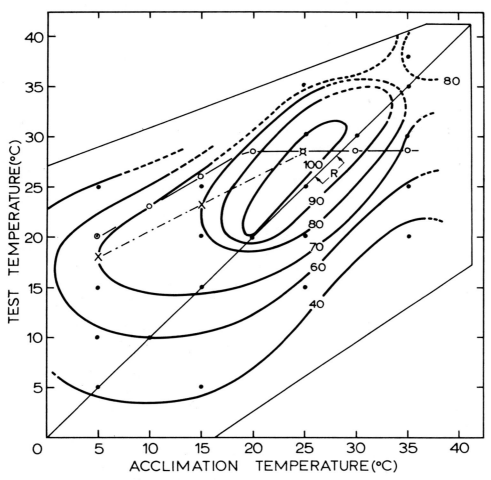

Fig. 12-21: Response surface for cruising speed of *Carassius auratus* (data of Fig. 12-20). Cruising speed isopleths are in ft/min (1 ft = 30·48 cm). The surface is bounded by the temperature tolerance polygon. The grid of points locates the original test conditions. Also indicated: —x— original estimates of maximum cruising speed at 5°, 15° and 25°C; —o— preferred temperatures; R: estimated location of maximum metabolic scope based on measurements for individuals acclimated to each test temperature. (Original, based on various sources.)

The second ridge of maximum performance is found by locating the test temperature of maximum swimming capacity, at a specific acclimation temperature, over a series of test temperatures (e.g. vertical transects of Fig. 12-21). For example, a tangent to the 12·19 m/min (40 ft/min) isopleth, drawn parallel to the test temperature axis at an acclimation temperature of about 1°C, estimates that maximum cruising speed occurs at a test temperature of about 17°C. Similarly, tangents to the other isopleths given, drawn parallel to the test temperature axis, define the location of a ridge which culminates at the centre of the surface. This second ridge estimates maximum swimming ability of *Carassius auratus* under acute temperature conditions.

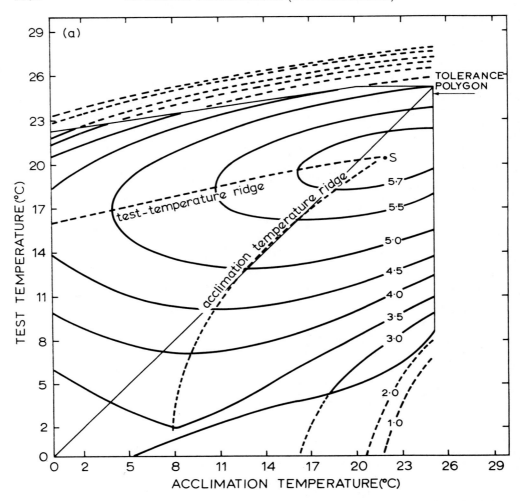

Fig. 12-22 (a): Critical swimming speeds of juvenile coho salmon *Oncorhynchus kisutch* (7·5 to 9·5 cm total length), within the temperature tolerance polygon defined by BRETT (1952). Position of ridges for fish tested in fall and early winter. Isopleths of swimming speed are shown in fish lengths swum per sec. Broken lines converging in S indicate test-temperature and acclimation-temperature ridges of maximum performance. (After GRIFFITHS and ALDERDICE, MS.)

The two ridge systems described have been termed the **acclimation temperature** and **test temperature ridges** of maximum performance, respectively, in a study of swimming performance of the juvenile coho salmon *Oncorhynchus kisutch* (GRIFFITHS and ALDERDICE, MS; Fig. 12-22a, b). On the basis of these data, and those on *Carassius auratus* (FRY and HART, 1948b; LINDSEY and co-authors, 1970), the following tentative statements are made. Ultimate maximum cruising speed occurs at the centre of the performance surface, located on or near the equal acclimation- and test-temperature diagonal. Divergence of the ridge from the diagonal appears to occur at lower temperatures. It is also possible that the ridge may shift position

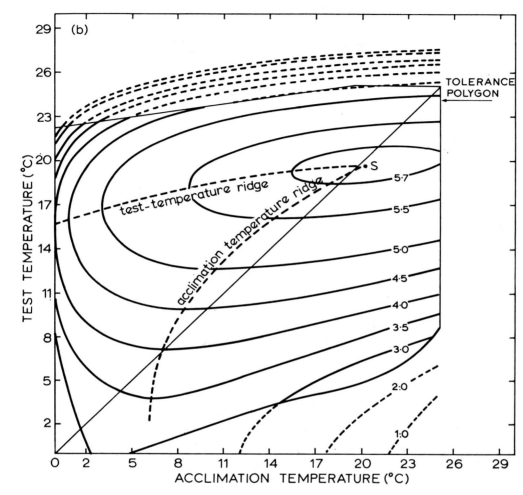

Fig. 12-22 (b): Critical swimming speeds of juvenile coho salmon *Oncorhynchus kisutch* (7·5 to 9·5 cm total length) within the temperature tolerance polygon defined by BRETT (1952). Position of ridges for fish tested in late winter and early spring. For further details consult legend to Fig. 12-22 (a). (After GRIFFITHS and ALDERDICE, MS.)

seasonally, providing a form of **performance compensation** at low winter water temperatures (Fig. 12-22b). Finally, maximum performance on the test temperature ridge, at acclimation temperatures below the surface centre, is found at test temperatures consistently higher than those of acclimation.

A second local region of maximum cruising speeds for *Carassius auratus* appears in the upper right corner of the temperature polygon (Fig. 12-21) between acclimation and test temperatures of 35° and 40°C. Preliminary construction of a three-dimensional model of the surface indicated this local maximum could not be part of the region of maxima between temperatures of 20° and 30°C. The entire surface, in fact, appears basically similar to that for walking rate of *Homarus americanus* (Fig. 12-19). It is presumed that this upper region of maximum cruising speeds again could provide the animal with a capacity to move rapidly under the stimulus of near-lethal temperatures, thereby increasing its chances of encountering lower, more favourable temperature levels.

The temperature region associated with ultimate maximum cruising speeds, centred between acclimation levels of 20° and 30°C, might be considered as a 'physiological optimum' if other measures of response to temperature were to find maximum expression in that region. Of such measures, temperature preferenda (selected temperatures) for the goldfish *Carassius auratus* have been illustrated by FRY (1947; Fig. 12-21). As noted by FRY and HART (1948b), the final preferendum for temperature and ultimate maximum cruising speed coincide at a temperature of 28° to 29°C for fish acclimated to 25°C. In Fig. 12-21 these points lie in the centre of the cruising speed surface. At acclimation temperatures below that at the centre of the surface, preferred temperatures lie reasonably close to the axis of the test temperature ridge of maximum performance. However, below the 20°C acclimation level the position of the preferred temperature 'ridge', relative to acclimation temperature, is located at test temperatures consistently 2 to 3 °C higher than that for cruising speed. The close relationship, yet non-correspondence of the surface transects for cruising speed maxima and preferred temperatures, is puzzling. Nevertheless, both measures apparently coincide in the centre of the cruising speed surface.

In addition, FRY and HART (1948a) investigated metabolic scope of *Carassius auratus* (difference between active and 'standard' levels of oxygen uptake for a series of acclimation temperatures). In this case all metabolic scope measurements, relating to the surface in Fig. 12-21 would lie on the diagonal of equal acclimation and test temperatures. Maximum metabolic scope on that transect was found at a temperature of about 26° to 28°C (Fig. 12-21, R). This maximum lies very near the maxima for ultimate maximum cruising speed and final preferendum. It is suggested that all three measures of response to temperature might be found to coincide if metabolic scope could be determined acutely, i.e., by obtaining estimates of active metabolism at a series of temperatures for each acclimation level, and subtracting routine levels of metabolism obtained under the same temperature conditions. Routine metabolism (mean levels of oxygen consumption for an organism not subject to excitation from external stimuli; see FRY, 1967, p. 394) probably would have to replace estimates of standard metabolism in acute determinations of metabolic scope. Meaningful determinations of standard metabolism measured acutely could be extremely difficult, if not impossible, to obtain (FRY, 1967, p. 398)*.

In the light of such evidence, the centre of the cruising speed surface may be considered a physiological optimum with respect to temperature. It would appear that *Carassius auratus* may prefer those (or slightly higher) temperatures at which it can be most active, while enjoying the greatest latitude for respiratory capacity above those levels required for routine metabolism.

Comparable associations are suggested by data of GRAHAM (1949) and FERGUSON

---

* Difficulties with acute measurement of routine levels might also be rather formidable. Such measurements would have to consider initial excitation of an individual when moved from one temperature to another, the time course of metabolic acclimation (Chapters 3.31, 3.32), and possible bias resulting from different levels of spontaneous activity at different temperatures (FRY, 1964, p. 725).

(1958) for the speckled trout *Salvelinus fontinalis*, and a similar relationship appears to hold also for the sockeye salmon *Oncorhynchus nerka*. At about 15°C, temperature-acclimated (5° to 25°C) juvenile sockeye show maximum locomotory capacity (BRETT, 1967), maximum metabolic scope (BRETT, 1964), an apparent final preferendum (BRETT, 1952), as well as maximum growth when fed to satiation at intervals daily (BRETT and co-authors, 1969).

Finally, it is of interest to note the dissimilar curves for the relation between acclimation temperatures and preferred temperatures in various species of fishes (FERGUSON, 1958; ZAHN, 1962; FRY, 1964). It is speculated that the low temperature end of these preference curves may be related to responses which optimize biological capacity, such as cruising speed; however, the possible relation to metabolic scope, as suggested in the previous example, remains undetermined with respect to the axes of maximum capacity of the response surface. The upper portion of those curves which reach a plateau may reflect the entry of a behavioural component which, through increased activity, tends to remove the organism from near-lethal conditions. It seems certain that detailed development of response surfaces for acutely determined measures of capacity, and their association with temperature preference, could shed further light on these important yet not fully determined relationships.

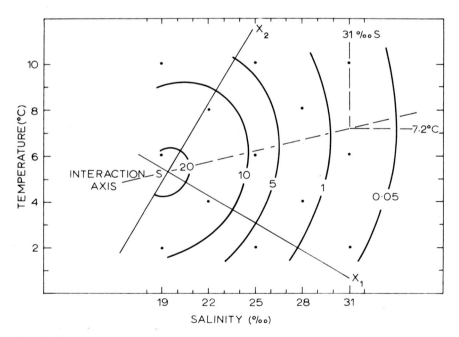

Fig. 12-23: Response surface calculated for percentage survival to hatching, in eggs of Pacific cod *Gadus macrocephalus*, as a function of incubation salinity and temperature. Interaction axis indicates that maximum survival at higher salinities is maintained by coupled increases in temperature (e.g. at 31‰S, maximum survival is calculated to be at a temperature of 7·2°C). The grid of points locates original test conditions. (After FORRESTER and ALDERDICE, 1966; modified.)

A number of recent studies deal with development of marine fish eggs (BLAXTER and HEMPEL, 1961, 1966; HOLLIDAY and co-authors, 1964; BLAXTER, 1965; HEMPEL, 1965; HOLLIDAY, 1965; HOLLIDAY and JONES, 1965; LASKER, 1965); in two such studies, response surface analysis has been employed (FORRESTER and ALDERDICE, 1966; ALDERDICE and FORRESTER, 1968), determining the effects of salinity and temperature on survival and development respectively for eggs of Pacific cod *Gadus macrocephalus* and English sole *Parophrys vetulus*. In the latter case, iterative testing was employed, the initial description of the response surface being expanded around conditions leading to both maximum larval size and maximum survival to hatching.

Maximum survival of *Gadus macrocephalus* eggs (Fig. 12-23) was found to be associated with salinity and temperature levels near 19‰S and 5°C, conditions also closely approximating those at which maximum size of newly hatched larvae was obtained. Interaction between salinity and temperature was found to produce an appreciable effect on both survival and larval size. To maintain maximum survival for salinities of 19‰ up to 31‰, a corresponding increase in temperature was required ranging from about 5° to 7°C. FORRESTER and ALDERDICE (1966) experienced difficulty with mass transfer of oxygen to the developing eggs, leading to low maximum survival at the centre of the salinity–temperature surface (about 25% survival). Further studies (ALDERDICE and FORRESTER, 1971) incorporating dissolved oxygen as a third variable overcame these difficulties; although maxi-

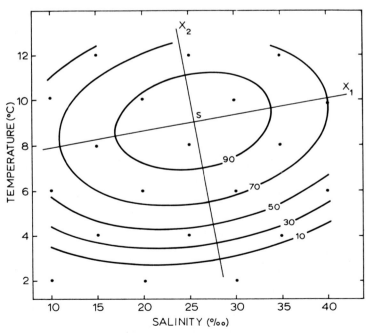

Fig. 12-24: Response surface calculated for percentage survival to hatching, for eggs of the English sole *Parophrys vetulus*, as a function of incubation salinity and temperature. The grid of points locates original test conditions. (After ALDERDICE and FORRESTER, 1968.)

mum survival rose to over 80%, no appreciable change in the location of the centre of the salinity–temperature surface (19‰S, 5°C) could be detected*.

Effects of salinity and temperature on development of *Parophrys vetulus* eggs (ALDERDICE and FORRESTER, 1968) resulted in the definition of a salinity–temperature surface having a computed centre for maximum survival (about 97%) at 25·6‰S and 9·0°C (Fig. 12-24; see also under *Methodology*). A semi-quantitative analysis of all of the data produced the viability relationships illustrated in Fig. 12-25. Multiple regression analysis was conducted on total

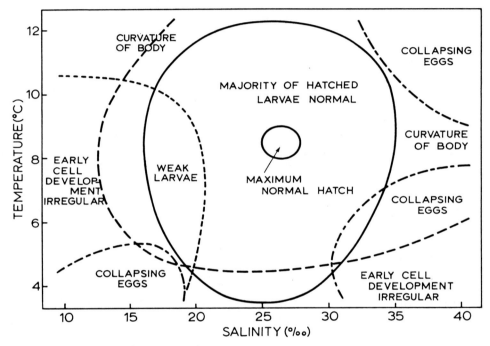

Fig. 12-25: Effects of incubation salinity and temperature on viability of developing eggs of *Parophrys vetulus*. Various gross developmental anomalies occurred more frequently in certain areas of the salinity–temperature test space, as indicated. (After ALDERDICE and FORRESTER, 1968.)

hatch, viable hatch and size of newly hatched larvae, followed by canonical reduction and construction of response surfaces. All three response measures were maximized within the same salinity–temperature region. In addition, incubation times were shorter (for given temperatures) at salinities within the same range as those associated with the above maxima (Table 12-4). It is possible that the salinity and temperature levels at the computed centres for total hatch, viable hatch and larval size are identical, and that the differences between them are the result of errors of estimation. Nevertheless, multiple response maxima would not

---

* Recent analysis of these results estimates the conditions for maximum survival to hatching near: (a) 18·1‰S, 5·5°C, 9·7 ppm O_2 using the model in Equation 6; (b) 15·6‰S, 3·7°C, 9·0 ppm O_2 using a non-linear model (LINDSEY and co-authors, 1970; see also *Methodology*).

Table 12-4

Salinity–temperature levels considered optimal with respect to four measures of response associated with egg development of the English sole *Parophrys vetulus* (After ALDERDICE and FORRESTER, 1968; modified)

End point measured	Salinity (‰)		Temperature (°C)	
	Mean	Range*	Mean	Range*
Minimum incubation period (6° to 12°C)**		23·0–27·0		
Maximum larval length	28·1		7·9	
Maximum total hatch	25·6	25·2–26·2	9·0	8·9–9·4
Maximum viable hatch	25·9	24·9–26·9	8·4	7·9–8·5

* based on variation in estimates of surface centre calculated from individual replicates.
** at given temperatures within the range indicated.

necessarily coincide; the factors and their levels which may govern one developmental criterion may differ from those governing or controlling another.

The fact that incubation periods were shortened in the range of salinities associated with maximum hatch of eggs of the English sole *Parophrys vetulus* suggests that such salinities might minimize the energy requirements for osmoregulation while diverting more energy into growth and development. KINNE (1963b, p. 311) states that

'in general, the osmoregulatory capacity of a given species appears to be greatest at near-optimum temperatures or somewhat below that, and to decrease in supra-normal temperatures.'

In a related case, maximum per cent hatch of eggs of the Atlantic herring *Clupea harengus* appears to be at a salinity near or somewhat above 20‰ (HOLLIDAY and BLAXTER, 1960; their Fig. 3), while the species appears to require least osmotic regulation at or somewhat above 17·5‰S (HOLLIDAY and JONES, 1965; their Fig. 3). It is possible that the osmoregulatory capacity of developing *Parophrys vetulus* eggs may be most efficient at temperatures of 8° to 9°C at salinities between approximately 26‰ and 28‰. ALDERDICE and FORRESTER (1968) are led to speculate that 'maximum viability potential' of *Parophrys vetulus* eggs may be found at those levels of salinity and temperature at which metabolic cost of osmoregulation is minimized, resulting in maximum production of viable larvae of maximum size in a minimum period of incubation.

Further points of interest emerge from inspection of the foregoing salinity-temperature hatching surfaces (Figs 12-23, 12-24). On the basis of such surfaces, what can be said regarding plasticity of the eggs? It may be noted for *Parophrys vetulus* eggs that salinities and temperatures associated with 90% or greater survival on the hatching surface are those between about 17‰ and 34‰S and 7° and 11°C. Field evidence shows that the adult English sole is found within the ranges of 20‰ to 34‰S, and 2·3° to 18°C (ALDERDICE and FORRESTER, 1968). In terms of this measure of viability, development of the *Parophrys vetulus* egg is highly successful over all of the salinity range but over less than one-third of the temperature range occupied by the adult. In comparison with the adult, therefore, *Parophrys vetulus* eggs could be termed euryhaline and stenothermal.

It is obvious that unique response surfaces could exist for each species, and for various stages of development (e.g. Fig. 12-17). Furthermore, salinity–temperature response surfaces could also be influenced by other variables, and their levels, operating as ancillary variables on the determination of salinity or temperature tolerance. For example, if the survival of *Parophrys vetulus* eggs was studied over various salinities but at a fixed temperature, the eggs might be considered euryhaline (Fig. 12-24) at 9°C, but stenohaline at 7° or 11°C. If a third axis were added to the hatching surface (e.g. dissolved oxygen), descriptions of plasticity might again be altered. The point to be stressed is that tolerance with respect to one environmental variable cannot be measured realistically without acknowledging the operation of other important variables which play a role in its expression. In this respect the use of fixed test conditions for comparison of thermo- or halinoplasticity between species or developmental stages may be a convenience to description, while at the same time invalidating the comparison sought. This is true particularly where interactions occur. For example, eggs of both *Parophrys vetulus* and *Gadus macrocephalus* exhibit an interaction such that, with increases in salinity, maximum survival is maintained only with coupled increases in temperature (i.e. beneficial effects of low/low–high/high combinations; KINNE, 1964a, p. 324). Therefore, it is argued that the ranges of salinity tolerance (plasticity) of these eggs would be compared more properly at temperatures found at the centres of the response surfaces, and in a direction relative to the axes of response rather than to those of measurement.

In his extensive review on the effects of salinity–temperature combinations, KINNE (1964a) has discussed (p. 321 *et seq.*) numerous observations similar to those presented here. Many of his conclusions may be translated directly into surface relationships. Thus, where subnormal salinities tend to decrease and supranormal salinities tend to increase tolerance to high temperatures, a low/low–high/high interaction is indicated between salinity and temperature. Further, KINNE concludes for the amphipod *Gammarus duebeni* that the range of salinities tolerated is widest at optimal temperatures, and that the range of temperatures tolerated is widest at optimal salinities. Fig. 12-23 or 12-24 could represent such a tolerance surface.

In summary, 'plasticity' remains, at best, a relative term, although its dimensions could be described more effectively in terms of the relative areas of compared response surfaces and the ranges associated with the lengths of the surface axes. These attributes are intimately associated with the likely dependence of response on multiple variables, and the effect of interactions in rotating surfaces away from the axes of measurement.

As an example of the possible application of response surface techniques to the investigation of biological phenomena, reference is made to two papers by KINNE (1960, 1962) on growth and development of the desert pupfish *Cyprinodon macularius*. KINNE's data are not extensive enough to allow for complete examination of surface features. Nevertheless, the data are of considerable interest and are sufficient to allow some speculation on associated interrelationships. KINNE's data recently have been used further by PALOHEIMO and DICKIE (1966a, b) in a comprehensive survey of the relations between food and growth in fishes.

KINNE (1960) examined the effects of salinity and temperature on growth or size attained, growth rate, appetite and food conversion efficiency, generally at 2-week intervals within the period of 6 to 40 weeks of age of the euryplastic *Cyprinodon macularius*. In Fig. 12-26 these four biological responses are re-examined at two age levels, and associated salinity–temperature surfaces have been freely estimated for each response. In some cases, relevant data from KINNE (1962) have been added to aid interpretation of the surfaces. In each of the four panels (Fig. 12-26) transects are shown for temperature (15° to 35°C at 35‰S) and for salinity (0‰ to 35‰ or 55‰S at 30°C) at ages approximating 12 and 22 weeks following hatching. Each transect suggests that the four implied response surfaces could be represented by a two-factor salinity, temperature second-order polynomia (Equation 4). Each surface has been freely estimated in that fashion.

It must be pointed out that estimation of the four surfaces presented will not necessarily provide an interpretation of functional relationships which may underlie the associations between salinity and temperature and the four dependent responses. Such interpretation requires knowledge of the interrelations existing between the four dependent responses listed, as well as others such as metabolic scope and locomotory activity. Functionally speaking, it is these biological activities which probably, in large measure, would determine total energy flow through the system.

The data for growth (size achieved) are sufficient to provide a reasonable estimate of salinity–temperature effects (Fig. 12-26A). A centre of maximum growth is estimated in the region of 35‰S and 30°C, and is maintained at both 12 and 22 weeks of age. The region of maximum growth has a broad, flat peak at 12 weeks, ranging from approximately 25° to 35°C and 15‰ to more than 35‰S for a fish of a total length of 20·0 mm. At 22 weeks the region of maximum growth is more restricted around the same centre conditions of 35‰S and 30°C. A low/low–high/high interaction exists; maximum growth at low temperatures is attained when there are coupled decreases in salinity.

Growth-rate surface isopleths at 12 weeks (Fig. 12-26B) appear to correspond generally with the surface for growth (Fig. 12-26A). At 12 weeks, growth rate is maximal near 35‰S or higher, and 30°C; at 22 weeks, the region of the maximum is broader, with a flatter peak, and encompasses salinities of about 15‰ to 35‰ and temperatures of 20° to 30°C. Again there is a salinity–temperature interaction similar to that for growth; higher growth rates at lower temperatures are achieved in coupled, lower salinities.

Appetite (Fig. 12-26C) is a measure of daily voluntary food (*Enchytraeus albidus*) consumption (KINNE, 1960, p. 292). Again, surface isopleths approximate those for growth and growth rate. At both 12 and 22 weeks, maximum food intake is found at or near 35‰S and 30°C. A similar low/low–high/high interaction exists between salinity and temperature. Because of the similarity in configuration of the three surfaces outlined, it is reasonable to conclude that growth, growth rate and appetite are directly correlated.

Food conversion efficiency (Fig. 12-26D) surface isopleths differ from those considered above. Maximum conversion efficiency appears to occur from near

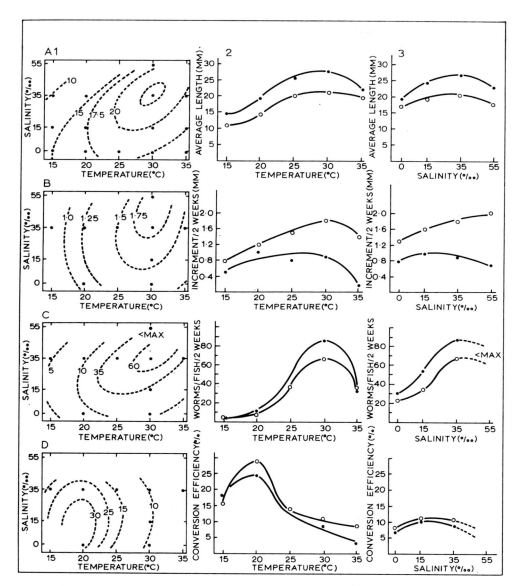

Fig. 12-26: Effects of salinity and temperature on the euryplastic desert pupfish *Cyprinodon macularius*. Four responses are shown: (A) growth (size attained), (B) growth rate, (C) appetite (voluntary food intake), (D) food conversion efficiency. Columns: (1) approximate salinity–temperature surfaces for each response, (2) transects of surface in (1) at 35‰S, (3) transects of surface in (1) at 30°C. Age of fish (post-hatching) for surface transects: A: 12 weeks (—○—) and 22 weeks (—●—); B to D: 12 to 14 weeks (—○—) and 20 to 22 weeks (—●—). Response surfaces in each case are for the younger fish. Data compiled from tables and estimated from graphs for fish hatched from eggs spawned in 35‰S and fed on a restricted food supply. (After KINNE, 1960, 1962; modified.)

fresh water to 30‰S and at 20°C. An interaction is indicated similar to that for the other three surfaces considered; maximum conversion efficiency at higher temperatures is attained with coupled increases in salinity. At both 12 and 22 weeks the region of maximum conversion efficiency is a broad, flat peak. Although a maximum appears to occur at or near 15‰S and 20°C, the plateau of near-maximum conversion efficiencies includes salinities ranging from 0‰ to 35‰.

The complementary nature of salinity–temperature interaction effects for conversion efficiency on the one hand, and growth, growth rate and appetite on the other, provides the following speculative conclusion. Under conditions promoting maximum conversion efficiency (0‰ to 30‰S, 20°C), low growth, growth rate and appetite occur, yet these responses are maximal for salinities at that temperature (20°C). Under conditions promoting maximum growth, growth rate and appetite (35‰S, 30°C), food conversion efficiency is low, yet the latter is maximal or near-maximal for salinities at that temperature (30°C). Furthermore, maximum growth appears to be maintained at both 12 and 22 weeks at 35‰S and 30°C. Thus, even though conversion efficiency is low at 30°C (35‰S) the level of gross food turnover associated with maximum food intake at 30°C is sufficient to produce a maximum advantage in terms of growth and growth rate. By analogy, an automobile (fish) will travel furthest (attain greatest length) at high speed (growth rate) as long as fuel (food) requirements can be satisfied, even though its fuel (food) consumption is high and its power/performance ratio (conversion efficiency) is low.

Finally, attention is directed to the regions of the salinity–temperature surfaces at which growth rate and conversion efficiency are maximized or near-maximal. At both 12 and 22 weeks these regions (e.g. Fig. 12-26B1, 1·5 mm isopleth; Fig. 12-26D1, 25% isopleth) form broad plateaus with maximum or near-maximum responses spread out in the direction of the interaction axis for each surface. The net result appears to be a broadening of the ranges of salinity (from near zero to 35‰S) and temperature (20° to 30°C) over which growth rate and conversion efficiency jointly retain maximum values; in other words, salinity–temperature combinations of maximum benefit to the fish, in terms of growth rate and conversion efficiency, are those found on the axis of maximum responses common to both surfaces. This axis is approximated by a line joining the surface centres and is approximately the axis of low/low–high/high interaction between salinity and temperature for either surface.

In a related paper, KINNE (1962) reported on the effects of the salinity of the spawning medium on eggs of *Cyprinodon macularius*. Growth, growth rate, appetite and conversion efficiency were followed for eggs spawned, incubated and reared in various salinity–temperature combinations. His values for growth rate and conversion efficiency are set out in Fig. 12-27 for fish of ages comparable to those in the preceding section. As noted (KINNE, 1960), growth rate is maximal at incubation conditions at or near 35‰S and 30°C (i.e. 9·2 and 14·5 mg increase in dry weight, upper right, rear corners of the cubes for both ages, Fig. 12-27). It is interesting to note that, even though the relationships between incubation temperature and salinity are retained at different spawning salinities (compare rear and front faces of each cube, Fig. 12-27), ultimate maximum growth rate is attained by fish from eggs spawned in fresh water (compare 14·5 and 9·2 mg increase in dry weight for 13 to 16-week fish, Fig. 12-27A). Conversely, as before, high

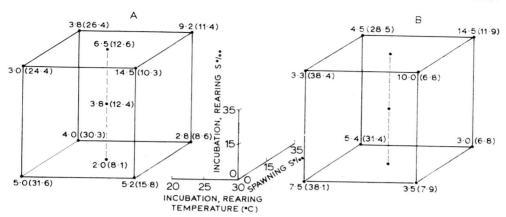

Fig. 12-27: Effects of salinity of the spawning medium, and salinity and temperature of incubation and rearing, on the development of *Cyprinodon macularius*. Two post-hatching age groups are shown: (A) 13 to 16 weeks, (B) 19 to 20 weeks. Each doublet of figures shows growth (initial figure; as increase in dry weight, mg) and conversion efficiency (in brackets; in per cent). The age groups shown are approximate and represent a grouping of available data into categories of younger and older fish of the age ranges given. The trends indicated in the figure are consistent with those for average values given in the original paper. (Based on data from KINNE, 1962.)

growth rates are associated with low food conversion efficiencies and the relationship of conversion efficiency to salinity and temperature appears to be the same as that indicated earlier (Fig. 12-26D). Furthermore, conversion efficiency reaches an apparent ultimate maximum for eggs spawned in fresh water (compare 31·6 and 30·3% conversion efficiency for 13 to 16-week fish, Fig. 12-27A; 38·1 and 31·4% for 19 to 20-week fish, Fig. 12-27B). For the older fish (19 to 20 weeks) a broad plateau for maximum conversion efficiency is evident in the fact that conversion efficiencies at 20°C are similar at rearing salinities of 0‰ and 35‰ (38·1 and 38·4%, respectively). The similarity between these data (KINNE, 1962) and the previous example (KINNE, 1960) is such that the front and rear faces of the factor spaces in Fig. 12-27 may be considered to have surface isopleths for growth and conversion efficiency similar to those illustrated in Fig. 12-26A and D.

KINNE (1962) held newly laid eggs of *Cyprinodon macularius* in several spawning salinities for a period of 3 to 6 hrs, and observed the effect of such exposure on subsequent egg development. His review of the evidence indicates the likelihood of an 'irreversible imprinting' by electrolyte concentration of effects beneath the plasma membrane as a result of early salinity experience (see also HOLLIDAY and BLAXTER, 1960). These changes may be associated with a period of high permeability of the plasma membrane at the time of and for a short period following oviposition.

Experimental points on the vertices of each cube in Fig. 12-27 represent a 2^3 factorial design. Hence the main effects of spawning salinity, incubation salinity and incubation temperature may be calculated for growth, in terms of increase in dry weight, from the data in the figure. For example, in panel A, the main effect for spawning salinity is one-eighth of the difference between the dry weights on the upper and lower spawning salinity planes (35‰ and 0‰). The main effect of spawning salinity ($b_1 \simeq -1.0$), although about one-half the magnitude of those for incubation salinity ($b_2 \simeq 1.7$) and temperature ($b_3 \simeq 2.0$), probably is biologically significant.

The data are taken to indicate that for a given level of spawning salinity there exists a region of incubation and rearing salinity–temperature conditions which is more favourable for growth and development. Furthermore, a spawning salinity at or near 0‰ appears to be the most favourable, for it is upon this spawning salinity plane that growth maxima are themselves maximized—at incubation and rearing conditions at or near 35‰S and 30°C (the plane with vertices 3·0, 14·5, 5·2 and 5·0 mg increase in dry weight, Fig. 12-27A).

Thus, it is presumed that the salinity of the spawning medium may act on the immediate environment of the developing embryo and thereby influence its subsequent development. Salinity (and temperature) experience throughout incubation also has significant effects on embryonal growth and size at hatching

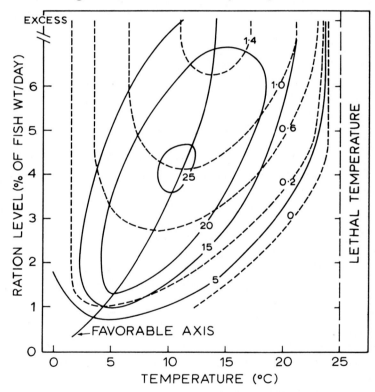

Fig. 12-28: Relation between growth rate (%/day, dotted ellipses) and gross conversion efficiency (%, solid ellipses) for young sockeye salmon *Oncorhynchus nerka*, relative to temperature and level of ration fed. Growth rate is optimized on excess rations near 15°C. Conversion efficiency is optimized at a ration of 4 to 4½% of fish weight/day at 11° to 11·5°C. The 'favorable axis' between these response centres describes a common ridge, in terms of temperature and ration level, upon which growth rate and conversion efficiency jointly reach maximum and near-optimum values. (After BRETT and co-authors, 1969; modified.)

(e.g. see KINNE and SWEET, 1965). Both of these environmental entities could influence embryonal development and the apportioning of energy flow among growth, activity of the embryo and osmoregulation, effects which could be expressed later in post-hatching development.

Some of the growth relations indicated in the foregoing example on *Cyprinodon macularius* have been documented for the young sockeye salmon *Oncorhynchus nerka* (BRETT and co-authors, 1969). In this study, the relation between growth rate and conversion efficiency is considered with respect to temperature and level of ration fed. While maximum growth on excess rations occurs at 15°C (see p. 1699), maximum gross conversion efficiency occurs at lower ration levels at a temperature near 11·5°C (Fig. 12-28). In BRETT's terminology, the two dependent variables are maximized on a 'favorable axis'. In other words, growth rate and conversion efficiency are maximized on a common ridge with respect to the two independent variables, ration level and temperature. Hence, as in the previous example (p. 1706), interaction with respect to the environmental variables provides a range of conditions over which both growth rate and conversion efficiency maintain near-optimal values. Thus, a change in the level of food available to the young sockeye could be compensated for by an adjustment in environmental temperature.

On the basis of the foregoing laboratory findings, it is of interest to examine other evidence concerning the daily vertical migrations performed by young sockeye salmon (NARVER, 1967, MS) in the thermally stratified waters of temperate lakes which they inhabit as juveniles. The young salmon feeds in the upper, warmer layers, and then moves to colder, deeper waters (to which it is presumably acclimated) where digestion would occur. At higher temperatures (e.g. 15°C), the greater indicated swimming capacity (p. 1699) of the young sockeye could enhance its ability to gather food. On the other hand, gross conversion efficiency, and growth on more limited rations, would be maximized at lower temperatures (e.g. 11·5°C). It is speculated that circumstances provided by diel migrations may maximize food gathering ability, and its utilization, and may be associated, respectively, with the test (acute) temperature and acclimation temperature ridges of maximum locomotory performance defined earlier (pp. 1694–6). Furthermore, the extent and intensity of diel migration could be governed, at least in part, by the interactive relationship which would be expected between existing environmental temperatures and food availability.

(3) Extensions and Implications

(a) *Methodology*

The theme of the arguments presented has been that of indicating how greater insight may be gained in the description of relationships between biological responses and environmental factors if explored on a multiple factor basis. Investigation of configuration and dynamic properties of a response surface is more instructive than the description of a transect of the surface made with reference to arbitrary scales of measurement. Responses are seen to exist in dynamic, multidimensional domains and the problem arises as to how these domains may be described quantitatively. Response surface analysis provides a means of approaching such a description. Functional relationships have been

Q

explored with the aid of these techniques (Box and YOULE, 1955; Box and COUTIE, 1956) although lack of information on basic mechanisms usually precludes such attempts. Nevertheless, response domains, including those in biological enquiry, may be reasonably approximated empirically using second order polynomials as a starting point for regression analysis. Where more than three environmental factors are involved in the determination of response, the dimensions of the domain cannot be portrayed graphically. However, reduction of the polynomial to its canonical equivalent allows interpretation of the domain through inspection of the numerical properties of the canonical equation.

In general, response surfaces generated on the basis of second degree polynomials (e.g. Equation 4) are found to have properties which reflect biological phenomena relating to plasticity, interaction effects, tolerance, and capacity. These simple models also provide an indication of potential surface mobility, and therefore demonstrate the dynamic properties of response surfaces.

However, detailed examination of response surfaces indicates that true surface configuration, or inner structure, may be much more complex than the approximation provided by the simple polynomial, used as a model (e.g. walking rate surface, Fig. 12-19). PROSSER's (1958b) acclimation patterns (p. 1679) indicate that the inner structure of surfaces itself is associated at least with the properties of (inner) rotation and translation. Furthermore, the regression model employed in the fitting procedure may require higher degree terms for an adequate fit than those provided by the second order model. For example, there are two response maxima on each of the surfaces illustrated in Figs 12-19 and 12-21, while portions of each surface show types of compound curvature associated with cubic effects.

Difficulties associated with the approximation of true inner structure of response surfaces may be overcome, at least in part, by employing transformations of the dependent (response) or independent (environmental) variables (Box and COX, 1964; DRAPER and HUNTER, 1969). Box and TIDWELL (1962) describe a procedure of this nature which undoubtedly will find considerable use in the further description of both inner structure and mobility of biological response surfaces. They present a method by which a second order polynomial in the transformed environmental variables is employed to obtain a more adequate representation of the surface (than that provided by the untransformed second order polynomial). This is accomplished without the need of trying to fit higher order polynomials (e.g. Equation 9) in the untransformed variables. Furthermore, the transformed polynomial may be considered as a non-linear model describing the observed data. LINDSEY and co-authors (1970) recently employed this approach, using as the non-linear model (e.g. for two independent variables):

$$Y^\gamma = b_0 x_0 + b_1 x_1^{\alpha_1} + b_2 x_2^{\alpha_2} + b_{11} x_1^{2\alpha_1} + b_{22} x_2^{2\alpha_2} + b_{12} x_1^{\alpha_1} x_2^{\alpha_2} \qquad (10)$$

where the vector of coefficients b is analogous to that in Equation (4) and the γ and α are power parameters calculated by maximum likelihood procedure. Among several examples of biological surfaces fitted by LINDSEY and co-authors in this manner* are the data of ALDERDICE and FORRESTER (1968) for total hatch of *Parophrys vetulus* eggs as a function of salinity (x_1) and temperature (x_2) (p. 1701).

---

* LINDSEY and SANDNES (1970) have produced a programme for IBM computer which provides analysis and graphical output for response surfaces involving two, or three, independent variables.

The fitted linear untransformed and non-linear transformed models were compared in terms of the original uncoded data. For the linear model

$$Y = -186\cdot6 + 7\cdot1x_1 + 43\cdot4x_2 - 0\cdot2x_1{}^2 - 2\cdot7x_2{}^2 + 0\cdot2x_1x_2,$$

while for the non-linear model

$$Y^{0\cdot85} = -2448\cdot5 + 1527\cdot7x_1{}^{0\cdot19} + 3735\cdot6x_2{}^{-0\cdot24} - 366\cdot5x_1{}^{0\cdot38} - 2540\cdot2x_2{}^{-0\cdot47}$$
$$- 362\cdot7x_1{}^{0\cdot19}x_2{}^{-0\cdot24}.$$

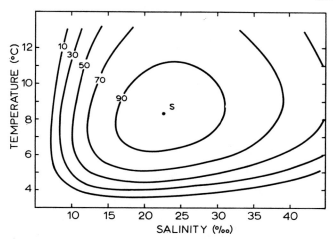

Fig. 12-29: Percentage survival to hatching of *Parophrys vetulus* eggs as a function of salinity and temperature. The response surface is based on the same data illustrated in Fig. 12-24, but on the basis of the non-linear model of Equation 10. (After LINDSEY and co-authors, 1970; modified.)

The estimated centre of the surface in terms of the two models lies at 25·9‰ S, 9·1°C, and at 22·5‰ S, 8·3°C, respectively. Even though the terms in the linear equation (x_1, x_2, $x_1{}^2$, $x_2{}^2$, x_1x_2) explain a highly significant portion of the variance associated with treatment effects (salinity and temperature), a significant component remains (deviations from regression) which is not explained by regression on the above terms. However, the variance associated with 'lack of fit' with the linear model is reduced considerably by employing the non-linear model. Moreover, the relative likelihood or plausibility of the linear model as opposed to the non-linear model is very low (LINDSEY and co-authors, 1970). The salinity–temperature surface generated for the data obtained on eggs of *Parophrys vetulus*, assuming the non-linear model, is illustrated in Fig. 12-29; it may be compared for the same data, assuming a linear model (Equation 4), with the surface presented in Fig. 12-24*.

* Note that by analogy the properties of rotation and translation set out for linear equations in Table 12-2 may be extended to non-linear surfaces. The axes of the non-linear surface ($X_1{}'$, $X_2{}'$) may be described by replacing each x in the equations of Table 12-2 by its x^x equivalent. For example, in the first case shown in the table, the non-linear equivalent becomes

$$\begin{bmatrix} X_1{}' \\ X_2{}' \end{bmatrix} = \begin{bmatrix} 1 & 0 \\ 0 & 1 \end{bmatrix} \begin{bmatrix} x_1{}^{x_1} \\ x_2{}^{x_2} \end{bmatrix}$$

Another attribute of the approach adopted in Equation (10) is the provision of insight into alternative units of measurement which might be employed to simplify the fitted relationship. By considering other possible estimates of the parameters, most of the commonly employed transformations may be tried in place of the maximum likelihood estimates which they may approximate. Some of these alternative estimates may be adopted in the fitted model, provided their relative likelihoods (e.g. ratio of the likelihood of a particular estimate to the likelihood of the maximum likelihood estimate of the parameter) retain high enough values. Thus, if the maximum likelihood estimate of γ could be replaced by 0·5, $\sqrt{Y}$ is an approximation of Y^γ; similarly if the maximum likelihood estimate of α could be replaced by zero, $\log x$ is an approximation of x^α, and so forth. For data obtained on eggs of *Parophrys vetulus*, the relative likelihoods for the parameters taking other selected values (e.g. $\gamma = 1$, $\alpha_1 = 0$, $\alpha_2 = -0·5$) lead to the suggestion that the relation of total hatch of eggs to salinity and temperature could be represented nearly as well, given the data and the non-linear model, by

$$Y = b_0 x_0 + (b_1 + b_{11}) \log x_1 + \frac{b_2}{\sqrt{x_2}} + \frac{b_{22}}{x_2} + \frac{b_{12}}{\sqrt{x_2}} \log x_1$$

The application of response surface techniques to biological problems requires a broader utilization among biologists of the values inherent in good experimental design (BOX and WILSON, 1951; BOX, 1956, and BOX and HUNTER, 1961). For example, the addition of four more trials to the 11 experimental combinations of KINNE (1962; Fig. 12-27), and minor adjustments to the salinity and temperature levels of some of these trials, would permit a full three-factor response surface analysis of the data. Applying an orthogonal composite factorial design (BOX and WILSON, 1951; BOX, 1956) to the factor space implicit in KINNE's study would generate the experimental combinations (trials) shown in Table 12-5. In this case, the actual salinity and temperature levels would be designated by the coding relationship

$$x = \text{Factor} \ (-) \ \text{Base Level/Unit}$$

where

Factor = salinity or temperature level

Base Level = level at the centre of the design (coded level 0)

Unit = difference in units of the factor between the design centre and those at (equal) coded intervals of $+1$ or -1.

For example, the coded value of 20°C (Table 12-5) is $x_1 = (20–25)/5 = -1$. In Table 12-5, the factor levels were generated under the provision that the lowest salinity level employed would be 0‰. The calculated salinity and temperature levels shown are rounded off to the nearest tenth.

(b) Ecological Implications

In considering relationships between organisms and their environment, both field and laboratory approaches are essential parts of the total enquiry. It is only reasonable to imply that interpretation of ecological events in the field would be aided by knowledge gained through laboratory study of the living organisms involved. Laboratory studies tend to define fundamental attributes of organisms;

Table 12-5

Summary of salinity and temperature levels for a three-factor orthogonal composite factorial design utilizing the general factor space employed by KINNE (1962). Coding relationships are shown at the bottom of the table (Original)

Trial	Experimental design coded levels			Corresponding temperature and salinity levels		
	x_1	x_2	x_3	Incubation, rearing temp., °C (x_1)	Incubation, rearing salinity,‰ (x_2)	Spawning salinity, ‰ (x_3)
1	−1	−1	−1	20	3·2	3·2
2	−1	−1	1	20	3·2	33·2
3	−1	1	−1	20	33·2	3·2
4	−1	1	1	20	33·2	33·2
5	1	−1	−1	30	3·2	3·2
6	1	−1	1	30	3·2	33·2
7	1	1	−1	30	33·2	3·2
8	1	1	1	30	33·2	33·2
9	0	0	0	25	18·2	18·2
10	−1·215	0	0	18·9	18·2	18·2
11	1·215	0	0	31·1	18·2	18·2
12	0	−1·215	0	25	0	18·2
13	0	1·215	0	25	36·5	18·2
14	0	0	−1·215	25	18·2	0
15	0	0	1·215	25	18·2	36·5

Factor	Coded factor level					Base level	Unit
	−1·215	−1	0	1	1·215		
x_1	18·9	20	25	30	31·1	25	5
x_2	0	3·2	18·2	33·2	36·5	18·2	15
x_3	0	3·2	18·2	33·2	36·5	18·2	15

field studies tend to discover whether or not, or to what extent, those attributes may be utilized by the organism under habitat conditions. Yet, there continues to exist some gaps between classical field ecology and the disciplines (e.g. physiology, biochemistry, genetics, systematics) which could contribute to the understanding of ecological events. In both areas of endeavour, research techniques are becoming more powerful. The need for a wider scope of interdisciplinary research in ecology is well recognized (BULLOCK, 1958; BUZZATI-TRAVERSO, 1958; RAE, 1958; REDFIELD, 1958). Even so, the greater potential for enrichment of ecological interpretation remains as a hopeful future fulfilment of laboratory studies.

Singular measures of physiological capacity obtained in simplified laboratory environments may serve to differentiate species or groups of organisms in a general way (BRETT, 1956), yet leave many ecological implications unresolved. Furthermore, biological capacity may rarely find total expression in the natural environment; unique capabilities may be suppressed or overshadowed by other factors, either unknown or unaccounted for in the laboratory evaluation. Interrelationships may depend on behavioural as well as physiological phenomena. Biotic relationships may overwhelm physiological potential; predation or competition may lead to an apparent under-exploitation of an environmental range which a species has the physiological capacity to occupy.

Environments are multidimensional, and investigational techniques which attempt to emulate this condition should provide ecologically more meaningful information on relationships between biological responses and environmental factors. Non-correspondence between laboratory and field evidence should not discount the value of laboratory study of biological attributes. It should indicate merely that total ecological relationships are more complex than laboratory studies of limited scope will estimate. Thus, non-correspondence can be a useful consequence of laboratory–field comparisons, leading naturally to a further search for other important elements in the total relationship investigated. Such comparisons are complementary, and provide a means of iterating toward improved knowledge.

The dimensions of a response domain for any organism should be investigated with respect to the number of environmental factors, and their ranges, which influence the ability of the organism to survive and reproduce successfully. Of these factors a certain restricted set may be found to be of commanding importance. Their initial selection would depend on judgement and previous knowledge of the organism investigated.

The most fundamental biological response domain is that of tolerance, defining the limits within which survival is possible. Inside the tolerance domain exist other domains within which other biological capacities find their maximum expression. Conditions which influence locomotory activity and metabolic scope may affect the ability of an organism to escape from predators, its ability to obtain food, its appetite, food conversion efficiency, and rate of growth. A reproductive domain characterizes the conditions under which an organism may reproduce, through influencing viability, size, or number of its reproductive products or offspring. Egg, embryo, larva, or sub-adult stages may have different survival and developmental domains than the adult (expressed in terms of viability, rate of development, final size or meristic characteristics).

Several multivariable response domains taken together may largely determine the distribution of an organism—or its likely equivalent—the ability to survive, grow and reproduce in a particular environment. The approach outlined in this enquiry, therefore, traverses multivariable techniques and concludes on the threshold of multivariate techniques. Thus a number of distinct measures of biological responses (multivariate), measured over a limited but important set of environmental variables (multivariable), may associate in a manner permitting estimation of optimum conditions for the physiological potential of a given organism or developmental stage. These optimal relationships should be relatively invariant and could, therefore, serve as a reasonably stable basis for assessing complex ecological relationships.

(4) Conclusions

This chapter considers responses of marine poikilotherms to environmental factors acting in concert. On the basis of response surface methodology, quantitative interrelationships are evaluated between organismic responses and concomitant intensities of environmental factors. Generated on the basis of second order regression models, response surfaces may reflect biological phenomena associated with properties of plasticity, interaction, tolerance and capacity. Biological response surfaces are dynamic and appear to vary not only with the species tested, but also with size or age, physiological state and previous history of the individual.

Expanding BULLOCK's (1955) R–T relationship to a surface, PRECHT's (1958) adaptation types are found to be equivalent to specific transects of a continuum of possible transects across various rate surfaces. Examined in terms of rate surfaces, CHRISTOPHERSEN and PRECHT's (1953) concepts of resistance adaptation and capacity adaptation are modified and extended. Resistance adaptation is associated with changes in location of a response domain relative to the axes of measurement, capacity adaptation with changes in domain dimensions. On the same surfaces and in terms of presence $(+)$ or absence $(-)$ of two surface properties—rotation and translation—four basic acclimation types are resolved: $(-, -)$, $(+, -)$, $(-, +)$ and $(+, +)$; these lead to PROSSER's (1958b) acclimation patterns.

Biological response surfaces are discussed further in terms of two attributes suggested through examination of pertinent literature. These have been termed **integral mobility** and **inner structure**. The former involves movement of a response surface as a dimensional unit within a factor space and appears to be associated with differences in age or physiological state of an organism. The latter draws attention to the specific structure of a response surface and is indicated in the basic configuration of rate surfaces involving temperature acclimation. Both response surface attributes have in common at least the descriptive properties of rotation and translation.

Interaction between acclimation and test temperatures may be found on an R–T surface ('inner rotation'); hence, the axis of maximum rate of response is a function of surface configuration. Maximum rates are not necessarily found on the equal (acclimation and test) temperature diagonal of such surfaces. In this sense, from studies of FRY and co-workers, an association is reviewed between the axes of maximum swimming speed, metabolic scope, and temperature preference. Physiological temperature optima thereby are suggested.

Within the limits of a tolerance domain, other, more restricted domains are presumed to exist in which other biological capacities may attain maximum expression (locomotory activity, metabolic scope, growth, appetite, food conversion efficiency, reproductive capacity, embryological development).

The general arguments regarding plasticity, interaction, tolerance, and capacity are considered by employing a second order polynomial as a linear model. In this manner, response surface analysis applied to biological data provides a simplified means of describing multivariable relationships. However, inadequacy of the model can reduce the precision of these empirical descriptions; in some cases, terms higher than second degree may be necessary to represent limited sections of biological surfaces adequately. Furthermore, in several cases where biological response surfaces could be examined in detail, the closer approximation showed true surface configuration to be asymmetric. In such cases the general arguments may be brought into the analytical scheme by employing non-linear models for empirical estimation of surface structure.

The methods employed lead to a proposition that optimum conditions for physiological potential could be estimated for given species or developmental stages. These relationships, assumed relatively invariant, could serve as a basis for exploring more complex ecological phenomena.

Acknowledgement. Much of the work reviewed in this chapter is that of Dr. F. E. J. FRY, University of Toronto, or of his students. Dr. FRY's tutorage lead the writer to application of response surface techniques in environmental physiology. His inspiration and continued support are warmly acknowledged.

Literature Cited (Chapter 12)

ALDERDICE, D. F. (1963a). Some effects of simultaneous variation in salinity, temperature and dissolved oxygen on the resistance of juvenile coho salmon (*Oncorhynchus kisutch*) to a toxic substance. PhD thesis, University of Toronto.

ALDERDICE, D. F. (1963b). Some effects of simultaneous variation in salinity, temperature and dissolved oxygen on the resistance of young coho salmon to a toxic substance. *J. Fish. Res. Bd Can.*, **20**, 525–550.

ALDERDICE, D. F. (1966). The detection and measurement of water pollution. Biological assays. In *Pollution and Our Environment.* A conference held in Montreal from Oct. 31 to Nov. 4, 1966. Canadian Council of Resource Ministers, Ottawa. Vol. 3 (Paper D25–1–3) (also in: *Can. Fish. Rep.*, **9**, 33–39, 1967.)

ALDERDICE, D. F. and FORRESTER, C. R. (1968). Some effects of salinity and temperature on early development and survival of the English sole (*Parophrys vetulus*). *J. Fish. Res. Bd Can.*, **25**, 495–521.

ALDERDICE, D. F. and FORRESTER, C. R. (1971). Effects of salinity, temperature and dissolved oxygen on early development of the Pacific cod (*Gadus macrocephalus*). *J. Fish. Res. Bd. Can.* **28**, (in press).

ARAI, M. N., COX, E. T. and FRY, F. E. J. (1963). An effect of dilutions of seawater on the lethal temperature of the guppy. *Can. J. Zool.*, **41**, 1011–1015.

BAGGERMAN, B. (1960). Salinity preference, thyroid activity and the seaward migration of four species of Pacific salmon (*Oncorhynchus*). *J. Fish. Res. Bd Can.*, **17**, 295–322.

BEAMISH, F. W. H. (1964a). Respiration of fishes with special emphasis on standard oxygen consumption. II. Influence of weight and temperature on respiration of several species. *Can. J. Zool.*, **42**, 177–188.

BEAMISH, F. W. H. (1964b). Seasonal changes in the standard rate of oxygen consumption of fishes. *Can. J. Zool.*, **42**, 189–194.

BEAMISH, F. W. H. (1964c). Respiration of fishes with special emphasis on standard oxygen consumption. IV. Influence of carbon dioxide and oxygen. *Can. J. Zool.*, **42**, 847–856.

BENTHE, H. F. (1954). Über die Temperaturabhängigkeit neuromuskulärer Vorgänge. *Z. vergl. Physiol.*, **36**, 327–351.

BHATNAGAR, K. M. and CRISP, D. J. (1965). The salinity tolerance of nauplius larvae of cirripedes. *J. Anim. Ecol.*, **34**, 419–428.

BIRCH, L. C. (1945). A contribution to the ecology of *Calandra orycae* L. and *Rhizopertha dominica* FAB. (Coleoptera) in stored wheat. *Trans. R. Soc. S. Aust.*, **69**, 140–149.

BISHAI, H. M. (1961). The effect of salinity on the survival and distribution of larval and young fish. *J. Cons. perm. int. Explor. Mer*, **26**, 166–179.

BLAXTER, J. H. S. (1965). The feeding of herring larvae and their ecology in relation to feeding. *Rep. Calif. coop. oceanic Fish. Invest.*, **10**, 79–88.

BLAXTER, J. H. S. and HEMPEL, G. (1961). Biologische Beobachtungen bei der Aufzucht von Heringsbrut. *Helgoländer wiss. Meersunters.*, **7**, 260–283.

BLAXTER, J. H. S. and HEMPEL, G. (1966). Utilization of yolk by herring larvae. *J. mar. biol. Ass. U.K.*, **46**, 219–234.

BOX, G. E. P. (1956). The determination of optimum conditions. In O. L. Davies (Ed.), *Design and Analysis of Industrial Experiments*. Oliver and Boyd, London. pp. 495–578.

BOX, G. E. P. and COUTIE, G. A. (1956). Application of digital computers in the exploration of functional relationships. *Inst. electl Engr.*, **103** (Part B, Suppl. 1), 100–107.

BOX, G. E. P. and COX, D. R. (1964). An analysis of transformations. *Jl R. statist. Soc. (B)*, **26**, 211–252.

BOX, G. E. P. and HUNTER, J. S. (1961). The 2^{k-p} fractional factorial designs. Part I. *Technometrics*, **3**, 311–351.

BOX, G. E. P. and TIDWELL, P. W. (1962). Transformation of the independent variables. *Technometrics*, **4**, 531–550.

BOX, G. E. P. and WILSON, K. B. (1951). On the experimental attainment of optimum conditions. *Jl R. statist. Soc. (B)*, **13**, 1–45.

BOX, G. E. P. and YOULE, P. V. (1955). The exploration and exploitation of response surfaces: an example of the link between the fitted surface and the basic mechanism of the system. *Biometrics*, **11**, 287–323.

BRETT, J. R. (1944). Some lethal temperature relations of Algonquin Park fishes. *Univ. Toronto Stud. biol. Ser.*, **52**, 1–49. (*Publs Ont. Fish. Res. Lab.*, **63**.)

BRETT, J. R. (1952). Temperature tolerance in young Pacific salmon, genus *Oncorhynchus*. *J. Fish. Res. Bd Can.*, **9**, 265–323.

BRETT, J. R. (1956). Some principles in the thermal requirements of fishes. *Q. Rev. Biol.*, **31**, 75–87.

BRETT, J. R. (1958). Implications and assessments of environmental stress. In P. A. Larkin (Ed.), *The Investigation of Fish-power Problems*. Institute of Fisheries, University of British Columbia. pp. 69–97.

BRETT, J. R. (1964). The respiratory metabolism and swimming performance of young sockeye salmon. *J. Fish. Res. Bd Can.*, **21**, 1183–1226.

BRETT, J. R. (1967). Swimming performance of sockeye salmon (*Oncorhynchus nerka*) in relation to fatigue time and temperature. *J. Fish. Res. Bd Can.*, **24**, 1731–1741.

BRETT, J. R., SHELBOURN, J. E. and SHOOP, C. T. (1969). Growth rate and body composition of fingerling sockeye salmon *Oncorhynchus nerka* in relation to temperature and ration size. *J. Fish. Res. Bd Can.*, **26**, 2363–2394.

BRIAN, M. V. and OWEN, G. (1952). The relation of the radula fraction to the environment in *Patella*. *J. Anim. Ecol.*, **20**, 241–249.

BROEKHUYSEN, G. J. (1936). On development, growth and distribution of *Carcinides maenas* (L.). *Archs néerl. Zool.*, **2**, 257–399.

BULLOCK, T. H. (1955). Compensation for temperature in the metabolism and activity of poikilotherms. *Biol. Rev.*, **30**, 311–342.

BULLOCK, T. H. (1958). Homeostatic mechanisms in marine organisms. In A. A. Buzzati-Traverso (Ed.), *Perspectives in Marine Biology*. University of California Press, Berkeley. pp. 199–210.

BUZZATI-TRAVERSO, A. A. (1958). Perspectives in marine biology. In A. A. Buzzati-Traverso (Ed.), *Perspectives in Marine Biology*. University of California Press, Berkeley. pp. 613–621.

CHRISTOPHERSEN, J. (1967). Adaptive temperature responses in microorganisms. In C. L. Prosser (Ed.), *Molecular Mechanisms of Temperature Adaptation. A Symposium . . .* A.A.A.S., Washington, D.C. pp. 327–348. (*Publs Am. Ass. Advmt Sci.*, **84**.)

CHRISTOPHERSEN, J. and PRECHT, H. (1953). Die Bedeutung des Wassergehaltes der Zelle für Temperaturanpassungen. *Biol. Zbl.*, **71**, 104–119.

COCHRAN, W. G. and COX, G. M. (1957). *Experimental Designs*. (Chapter 8A). Wiley, New York. pp. 335–375.

COSTLOW, J. D., JR., BOOKHOUT, C. G. and MONROE, R. (1960). The effect of salinity and temperature on larval development of *Sesarma cinereum* (BOSC) reared in the laboratory. *Biol. Bull. mar. biol. Lab., Woods Hole*, **118**, 183–202.

COSTLOW, J. D., JR., BOOKHOUT, C. G. and MONROE, R. (1962). Salinity-temperature effects on the larval development of the crab, *Panopeus herbstii* MILNE-EDWARDS, reared in the laboratory. *Physiol. Zool.*, **35**, 79–93.

COSTLOW, J. D., JR., BOOKHOUT, C. G. and MONROE, R. J. (1966). Studies on the larval development of the crab *Rhithropanopeus harrisii* (GOULD). I. The effect of salinity and temperature on larval development. *Physiol. Zool.*, **39**, 81–100.

CRAIGIE, D. E. (1963). An effect of water hardness in the thermal resistance of the rainbow trout, *Salmo gairdnerii* RICHARDSON. *Can. J. Zool.*, **41**, 825–830.

CRISP, D. J. and COSTLOW, J. D., JR. (1963). The tolerance of developing cirripede embryos to salinity and temperature. *Oikos*, **14**, 22–34.

DAVIS, G. E., FOSTER, J., WARREN, C. E. and DOUDOROFF, P. (1963). The influence of oxygen concentration on the swimming performance of juvenile Pacific salmon at various temperatures. *Trans. Am. Fish. Soc.*, **92**, 111–124.

DAVISON, R. C., BREESE, W. P., WARREN, C. E. and DOUDOROFF, P. (1959). Experiments on the dissolved oxygen requirements of cold-water fishes. *Sewage ind. Wastes*, **31**, 950–966.

DEAN, J. M. and GOODNIGHT, C. J. (1964). A comparative study of carbohydrate metabolism in fish as affected by temperature and exercise. *Physiol. Zool.*, **37**, 280–299.

DEHNEL, P. A. (1960). Effect of temperature and salinity on the oxygen consumption of two intertidal crabs. *Biol. Bull. mar. biol. Lab., Woods Hole*, **118**, 215–249.

DOUDOROFF, P. (1942). The resistance and acclimatization of marine fishes to temperature changes. I. Experiments with *Girella nigricans* (AYRES). *Biol. Bull. mar. biol. Lab., Woods Hole*, **83**, 219–244.

DOUDOROFF, P. (1945). The resistance and acclimatization of marine fishes to temperature changes. II. Experiments with *Fundulus* and *Atherinops*. *Biol. Bull. mar. biol. Lab., Woods Hole*, **88**, 194–206.

DRAPER, N. R. and HUNTER, W. G. (1969). Transformations: some examples revisited. *Technometrics*, **11**, 23–40.

EDMONDSON, W. T. (1965). Reproductive rate of planktonic rotifers as related to food and temperature in nature. *Ecol. Monogr.*, **35**, 61–111.

EINSELE, W., (1965). Problems of fish-larvae survival in nature and the rearing of economically important middle European freshwater fishes. *Rep. Calif. coop. oceanic Fish. Invest.*, **10**, 24–30.

EMERSON, S. (1947). Growth responses of a sulfonamide-requiring mutant strain of *Neurospora*. *J. Bact.*, **54**, 195–207.

FERGUSON, R. G. (1958). The preferred temperature of fish and their midsummer distribution in temperate lakes and streams. *J. Fish. Res. Bd Can.*, **15**, 607–624.

FISHER, K. C. (1958). An approach to the organ and cellular physiology of adaptation to temperature in fish and small mammals. In C. L. Prosser (Ed.), *Physiological Adaptation*. Ronald Press, New York. pp. 3–49.

FISHER, K. C. and SULLIVAN, C. M. (1958). The effect of temperature on the spontaneous activity of speckled trout before and after various lesions of the brain. *Can. J. Zool.*, **36**, 49–63.

FLÜGEL, H. (1960). Über den Einfluß der Temperatur auf die osmotische Resistenz und die Osmoregulation der decapoden Garnele *Crangon crangon* L. *Kieler Meeresforsch.*, **16**, 186–200.

FORRESTER, C. R. and ALDERDICE, D. F. (1966). Effects of salinity and temperature on embryonic development of the Pacific cod (*Gadus macrocephalus*). *J. Fish. Res. Bd Can.*, **23**, 319–340.

FRY, F. E. J. (1947). Effects of the environment on animal activity. *Univ. Toronto Stud. biol. Ser.*, **55**, 1–62. (*Publs Ont. Fish. Res. Lab.*, **68**).

FRY, F. E. J. (1964). Animals in aquatic environments: fishes. In D. B. Dill, E. F. Adolph and C. G. Wilber (Eds), *Handbook of Physiology*. Sect. 4. Adaptation to the environment. American Physiological Society, Washington, D.C. pp. 715–728.

FRY, F. E. J. (1967). Responses of vertebrate poikilotherms to temperature. In A. H. Rose (Ed.), *Thermobiology*. Academic Press, London. pp. 375–409.

FRY, F. E. J., BRETT, J. R. and CLAWSON, G. H. (1942). Lethal limits of temperature for young goldfish. *Revue can. Biol.*, **1**, 50–56.

FRY, F. E. J. and HART, J. S. (1948a). The relation of temperature to oxygen consumption in the goldfish. *Biol. Bull. mar. biol. Lab., Woods Hole*, **94**, 66–77.

FRY, F. E. J. and HART, J. S. (1948b). Cruising speed of goldfish in relation to water temperature. *J. Fish. Res. Bd Can.*, **7**, 169–175.

FRY, F. E. J., HART, J. S. and WALKER, K. F. (1946). Lethal temperature relations for a sample of young speckled trout, *Salvelinus fontinalis*. *Univ. Toronto Stud. biol. Ser.*, **54**, 1–35. (*Publs Ont. Fish. Res. Lab.*, **66**.)

GARSIDE, E. T. (1959). Some effects of oxygen in relation to temperature on the development of lake trout embryos. *Can. J. Zool.*, **37**, 689–698.

GARSIDE, E. T. (1966). Effects of oxygen in relation to temperature on the development of embryos of brook trout and rainbow trout. *J. Fish. Res. Bd Can.*, **23**, 1121–1134.

GIBSON, E. S. and FRY, F. E. J. (1954). The performance of the lake trout, *Salvelinus namaycush*, at various levels of temperature and oxygen pressure. *Can. J. Zool.*, **32**, 252–260.

GIBSON, M. B. and HIRST, B. (1955). The effect of salinity and temperature on the pre-adult growth of guppies. *Copeia*, **1955**, 241–243.

GRAHAM, J. M. (1949). Some effects of temperature and oxygen pressure on the metabolism and activity of the speckled trout, *Salvelinus fontinalis*. *Can. J. Res.* (D), **27**, 270–288.

GRIFFITHS, J. S. and ALDERDICE, D. F. (MS). Effects of acclimation and acute temperature experience on the swimming speed of juvenile coho salmon. Biological Station, Nanaimo.

HADER, R. J., HARWARD, M. E., MASON, D. D. and MOORE, D. P. (1957). An investigation of some of the relationships between copper, iron and molybdenum in the growth and nutrition of lettuce: I. Experimental design and statistical methods for characterizing the response surface. *Proc. Soil Sci. Soc. Am.*, **21**, 59–64.

HALCROW, K. (1963). Acclimation to temperature in the marine copepod, *Calanus finmarchicus* (GUNNER.). *Limnol. Oceanogr.*, **8**, 1–8.

HART, J. S. (1947). Lethal temperature relations of certain fish of the Toronto region. *Trans. R. Soc. Can.*, **41** (Sect. V), 57–71.

HART, J. S. (1952). Geographic variations of some physiological and morphological characters in certain freshwater fish. *Univ. Toronto Stud. biol. Ser.*, **60**, 1–79. (*Publs Ont. Fish. Res. Lab.*, **72**.)

HEATH, W. G. (1967). Ecological significance of temperature tolerance in Gulf of California shore fishes. *J. Ariz. Acad. Sci.*, **4**, 172–178.

HEMPEL, G. (1965). On the importance of larval survival for the population dynamics of marine food fish. *Rep. Calif. coop. oceanic Fish. Invest.*, **10**, 13–23.

HENDERSON, N. E. (1963). Influence of light and temperature on the reproductive cycle of the eastern brook trout, *Salvelinus fontinalis* (MITCHILL). *J. Fish. Res. Bd Can.*, **20**, 859–897.

HERMANSON, H. P. (1965). Maximization of potato yield under constraint. *Agron. J.*, **57**, 210–213.

HILL, W. J. and HUNTER, W. G. (1966). A review of response surface methodology: a literature survey. *Technometrics*, **8**, 571–590.

HOFF, J. G. and WESTMAN, J. R. (1966). The temperature tolerances of three species of marine fishes. *J. mar. Res.*, **24**, 131–140.

HOLLIDAY, F. G. T. (1965). Osmoregulation in marine teleost eggs and larvae. *Rep. Calif. coop. oceanic Fish. Invest.*, **10**, 89–95.

HOLLIDAY, F. G. T. and BLAXTER, J. H. S. (1960). The effects of salinity on the developing eggs and larvae of the herring. *J. mar. biol. Ass. U.K.*, **39**, 591–603.

HOLLIDAY, F. G. T., BLAXTER, J. H. S. and LASKER, R. (1964). Oxygen uptake of developing eggs and larvae of the herring (*Clupea harengus*). *J. mar. biol. Ass. U.K.*, **44**, 711–723.

HOLLIDAY, F. G. T. and JONES, M. P. (1965). Osmotic regulation in the embryo of the herring (*Clupea harengus*). *J. mar. biol. Ass. U.K.*, **45**, 305–311.

JITTS, H. R., McALLISTER, C. D., STEPHENS, K. and STRICKLAND, J. D. H. (1964). The cell division rates of some marine phytoplankters as a function of light and temperature. *J. Fish. Res. Bd Can.*, **21**, 139–157.

KANUNGO, M. S. and PROSSER, C. L. (1959). Physiological and biochemical adaptation of goldfish to cold and warm temperatures. I. Standard and active oxygen consumption of cold- and warm-acclimated goldfish at various temperatures. *J. cell. comp. Physiol.*, **54**, 259–263.

KINNE, O. (1952). Zur Biologie und Physiologie von *Gammarus duebeni* LILLJ., V: Untersuchungen über Blutkonzentration, Herzfrequenz und Atmung. *Kieler Meeresforsch.*, **9**, 134–150.

KINNE, O. (1953). Zur Biologie und Physiologie von *Gammarus duebeni* LILLJ. I. *Z. wiss. Zool.*, **157**, 427–491.

KINNE, O. (1954). Experimentelle Untersuchungen über den Einfluß des Salzgehaltes auf die Hitzeresistenz von Brackwassertieren. *Zool. Anz.*, **152**, 10–16.

KINNE, O. (1956). Über den Einfluß des Salzgehaltes und der Temperatur auf Wachstum, Form und Vermehrung bei dem Hydroidpolypen *Cordylophora caspia* (PALLAS), Athecata, Clavidae. *Zool. Jb. (Abt. allg. Zool. Physiol. Tiere)*, **66**, 565–638.

KINNE, O. (1957). A programmatic study of the comparative biology of marine and brackish water animals. *Année biol.*, **33**, 87–92.

KINNE, O. (1958a). Über die Reaktion erbgleichen Coelenteraten-Gewebes auf verschiedene Salzgehaltes- und Temperaturbedingungen. *Zool. Jb. (Abt. allg. Zool. Physiol. Tiere)*, **67** 407–486.

KINNE, O. (1958b). Adaptation to salinity variations—some facts and problems. In C. L. Prosser (Ed.), *Physiological Adaptation*. Ronald Press, New York. pp. 92–106.

KINNE, O. (1960). Growth, food intake, and food conversion in a euryplastic fish exposed to different temperatures and salinities. *Physiol. Zool.*, **33**, 288–317.

KINNE, O. (1961). Growth, molting frequency, heart beat, number of eggs, and incubation time in *Gammarus zaddachi* exposed to different environments. *Crustaceana*, **2**, 26–36.

KINNE, O. (1962). Irreversible nongenetic adaptation. *Comp. Biochem. Physiol.*, **5**, 265–282.

KINNE, O. (1963a). Adaptation, a primary mechanism of evolution. In H. B. Whittington and W. D. I. Rolfe (Eds), *Phylogeny and Evolution of Crustacea*. Museum of Comparative Zoology, Cambridge, Mass., 27–50. (Spec. Publ.)

KINNE, O. (1963b). The effects of temperature and salinity on marine and brackish water animals. I. Temperature. *Oceanogr. mar. Biol. A. Rev.*, **1**, 301–340.

KINNE, O. (1964a). The effects of temperature and salinity on marine and brackish water animals. II. Salinity and temperature salinity combinations. *Oceanogr. mar. Biol. A. Rev.*, **2**, 281–339.

KINNE, O. (1964b). Non-genetic adaptation to temperature and salinity. *Helgoländer wiss. Meeresunters.*, **9**, 433–458.

KINNE, O. (1966). Physiological aspects of animal life in estuaries with special reference to salinity. *Neth. J. Sea Res.*, **3**, 222–244.

KINNE, O. and KINNE, E. M. (1962). Rates of development in embryos of a cyprinodont fish exposed to different temperature-salinity-oxygen combinations. *Can. J. Zool.*, **40**, 231–253.

KINNE, O. and PAFFENHÖFER, G.-A. (1965). Hydranth structure and digestion rate as a function of temperature and salinity in *Clava multicornis* (Cnidaria, Hydrozoa). *Helgoländer wiss. Meeresunters.*, **12**, 329–341.

KINNE, O., SHIRLEY, E. K. and MEEN, H. E. (1963). Osmotic responses of hermit crabs (*Pagurus longicarpus* SAY) exposed to various constant temperatures and salinities. *Crustaceana*, **5**, 317.

KINNE, O. and SWEET, J. G. (1965). Die Umweltabhängigkeit der Körperform frischgeschlüpfter *Cyprinodon macularius* (Teleostei). *Naturwissenschaften*, **3**, 69–70.

LANCE, J. (1963). The salinity tolerance of some estuarine planktonic copepods. *Limnol. Oceanogr.*, **8**, 440–449.

LASKER, R. (1965). The physiology of Pacific sardine embryos and larvae. *Rep. Calif. coop. oceanic Fish. Invest.*, **10**, 96–101.

LINDSEY, C. C. (1962a). Experimental study of meristic variation in a population of threespine sticklebacks (*Gasterosteus aculeatus*). *Can. J. Zool.*, **40**, 271–312.

LINDSEY, C. C. (1962b). Observations on meristic variation in ninespine sticklebacks, *Pungitius pungitius*, reared at different temperatures. *Can. J. Zool.*, **40**, 1237–1247.

LINDSEY, J. K. and SANDNES, A. M. (1970). Program for the analysis of non-linear response surfaces (extended version). *Tech. Rep. Fish. Res. Bd Can.*, **173**, 1–94. Biological Station, Nanaimo.

LINDSEY, J. K., ALDERDICE, D. F., and PIENAAR, L. V. (1970). Analysis of nonlinear models— The nonlinear response surface. *J. Fish. Res. Bd Can.*, **27**, 765–791.

McFARLAND, W. N. and PICKENS, P. E. (1965). The effects of season, temperature, and salinity on standard and active oxygen consumption of the grass shrimp, *Palaemonetes vulgaris* (SAY). *Can. J. Zool.*, **43**, 571–585.

McINTIRE, C. D. (1966). Some factors affecting respiration of periphyton communities in lotic environments. *Ecology*, **47**, 918–930.

McLEESE, D. W. (1956). Effects of temperature, salinity and oxygen on the survival of the American lobster. *J. Fish. Res. Bd Can.*, **13**, 247–272.

McLEESE, D. W. and WILDER, D. G. (1958). The activity and catchability of the lobster (*Homarus americanus*) in relation to temperature. *J. Fish. Res. Bd Can.*, **15**, 1345–1354.

MADDUX, W. S. and JONES, R. F. (1964). Some interactions of temperature, light intensity and nutrient concentration during the continuous culture of *Nitzschia closterium* and *Tetraselmis* sp. *Limnol. Oceanogr.*, **9**, 79–86.

MOORE, D. P., HARWARD, M. E., MASON, D. D., HADER, R. J., LOTT, W. L. and JACKSON, W. A. (1957). An investigation of some of the relationships between copper, iron and molybdenum in the growth and nutrition of lettuce: II. Response surfaces of growth and accumulation of Cu and Fe. *Proc. Soil Sci. Soc. Am.*, **21**, 65–74.

MORRIS, R. W. (1960). Temperature, salinity and southern limits of three species of Pacific cottid fishes. *Limnol. Oceanogr.*, **5**, 175–179.

NARVER, D. W. (MS 1967). Diel vertical movements of pelagial sockeye salmon juveniles. *MS Rep. Ser. Fish. Res. Bd Can.*, **949**, 24–28.

PAFFENHÖFER, G.-A. (1968). Nahrungsaufnahme, Stoffumsatz und Energiehaushalt des marinen Hydroidpolypen *Clava multicornis*. *Helgoländer wiss. Meeresunters.*, **18**, 1–44.

PALOHEIMO, J. E. and DICKIE, L. M. (1965). Food and growth of fishes. I. A growth curve derived from experimental data. *J. Fish. Res. Bd Can.*, **22**, 521–542.

PALOHEIMO, J. E. and DICKIE, L. M. (1966a). Food and growth of fishes. II. Effects of food and temperature on the relation between metabolism and body weight. *J. Fish. Res. Bd Can.*, **23**, 869–908.

PALOHEIMO, J. E. and DICKIE, L. M. (1966b). Food and growth of fishes. III. Relations among food, body size and growth efficiency. *J. Fish. Res. Bd Can.*, **23**, 1209–1248.

PENG, K. C. (1967). *The Design and Analysis of Scientific Experiments*. Ch. 8: Response surface designs. Addison-Wesley, Reading, U. S.A. pp. 150–172.

PODOLIAK, H. A. (1965). Some effects of stress on the osmotic tolerance of fingerling brown trout. *Fish. Res. Bull. N.Y.*, **28**, 71–82.

PRECHT, H. (1958). Concepts of the temperature adaptation of unchanging reaction systems of cold-blooded animals. In C. L. Prosser (Ed.), *Physiological Adaptation*. Ronald Press, New York. pp. 50–78.

PRECHT, H. (1964). Über die Resistenzadaptation wechselwarmer Tiere an extreme Temperaturen und ihre Ursachen. *Heloglönder wiss. Meeresunters.*, **9**, 392–411.

PRECHT, H. (1967). A survey of experiments on resistance-adaptation. In A. S. Troshin (Ed.), *The Cell and Environmental Temperature*. Pergamon Press, Oxford. pp. 307–321.

PRECHT, H. (1968). Der Einfluß 'normaler' Temperaturen auf Lebensprozesse bei wechselwarmen Tieren unter Ausschluß der Wachstums und Entwicklungs-prozesse. *Helgoländer wiss. Meeresunters.*, **18**, 487–548

PRECHT, H., BASEDOW, T., BERECK, R., LANGE, F., THIEDE, W. and WILKE, L. (1966). Reaktionen und Adaptationen wechselwarmer Tiere nach einer Änderung der Anpassungstemperatur und der zeitliche Verlauf. *Helgoländer wiss. Meeresunters.*, **13**, 369–401.

PRECHT, H., CHRISTOPHERSEN, J. and HENSEL, H. (1955). *Temperatur und Leben*, Springer, Berlin.

PROSSER, C. L. (1955). Physiological variation in animals. *Biol. Rev.*, **30**, 229–262.

PROSSER, C. L. (Ed.) (1958a). *Physiological Adaptation*, Ronald Press, New York.

PROSSER, C. L. (1958b). General summary: The nature of physiological adaptation. In C. L. Prosser (Ed.), *Physiological Adaptation*. Ronald Press, New York. pp. 167–180.

RAE, K. M. (1958). Parameters of the marine environment. In A. A. Buzzati-Traverso (Ed.), *Perspectives in Marine Biology*. University of California Press, Berkeley. pp. 3–16.

RANADA, M. R. (1957). Observations on the resistance of *Tigriopus fulvus* (FISCHER) to changes in temperature and salinity. *J. mar. biol. Ass. U.K.*, **36**, 115–119.

REDFIELD, A. C. (1958). The inadequacy of experiment in marine biology. In A. A. Buzzati-Traverso (Ed.), *Perspectives in Marine Biology*. University of California Press, Berkeley. pp. 17–26.

RIVARD, I. (1961a). Influence of temperature and humidity on mortality and rate of development of immature stages of the mite *Tyrophagus putrescentiae* (SCHRANK) (Acarina: Acaridae) reared on mold cultures. *Can. J. Zool.*, **39**, 419–426.

RIVARD, I. (1961b). Influence of temperature and humidity on longevity, fecundity and rate of increase of the mite *Tyrophagus putrescentiae* (SCHRANK) (Acarina: Acaridae) reared on mold cultures. *Can. J. Zool.*, **39**, 869–876.

ROOTS, B. I. and PROSSER, C. L. (1962). Temperature acclimation and the nervous system in fish. *J. exp. Biol.*, **39**, 617–629.

SANDOZ, M. and ROGERS, R. (1944). The effect of environmental factors on hatching, moulting, and survival of zoea larvae of the blue crab *Callinectes sapidus* RATHBUN. *Ecology*, **25**, 216–228.

SHUMWAY, D. L., WARREN, C. E., and DOUDOROFF, P. (1964). Influence of oxygen concentration and water movement on the growth of steelhead trout and coho salmon embryos. *Trans. Am. Fish. Soc.*, **93**, 342–356.

SILVER, S. J., WARREN, C. E. and DOUDOROFF, P. (1963). Dissolved oxygen requirements of developing steelhead trout and chinook salmon embryos at different water velocities. *Trans. Am. Fish. Soc.*, **92**, 327–343.

SPRAGUE, J. B. (1963). Resistance of four freshwater crustaceans to lethal high temperature and low oxygen. *J. Fish. Res. Bd Can.*, **20**, 387–415.

SWEET, J. G. and KINNE, O. (1964). The effects of various temperature-salinity combinations on the body form of newly hatched *Cyprinodon macularius* (Teleostei). *Helgoländer wiss. Meeresunters.*, **11**, 49–69.

TODD, M.-E. and DEHNEL, P. A. (1960). Effect of temperature and salinity on heat tolerance in two grapsoid crabs, *Hemigrapsus nudus* and *Hemigrapsus oregonensis*. *Biol. Bull. mar. biol. Lab., Woods Hole*, **118**, 150–172.

TYLER, A. V. (1966). Some lethal temperature relations of two minnows of the genus *Chrosomus*. *Can. J. Zool.*, **44**, 349–364.

USHAKOV, B. (1964). Thermostability of cells and proteins of poikilotherms and its significance in speciation. *Physiol. Rev.*, **44**, 518–560.

WELCH, L. F., ADAMS, W. E. and CARMON, J. L. (1963). Yield response surfaces, isoquants and economic fertilizer optima for coastal Bermudagrass. *Agron. J.*, **55**, 63–67.

ZAHN, M. (1962). Die Vorzugstemperaturen zweier Cypriniden und eines Cyprinodonten und die Adaptationstypen der Vorzugstemperatur bei Fischen. *Zool. Beitr.* (N.F.), **7**, 15–25.

ZEIN-ELDIN, Z. P. and ALDRICH, D. V. (1965). Growth and survival of postlarval *Penaeus aztecus* under controlled conditions of temperature and salinity. *Biol. Bull. mar. biol. Lab., Woods Hole*, **129**, 199–216.

AUTHOR INDEX

Numbers in italics refer to those pages on which the Author's work is stated in full.

TAXONOMIC INDEX

R

SUBJECT INDEX